Special Functions and Analysis of Differential Equations

Special Functions and Analysis of Differential Equations

Edited by
Praveen Agarwal
Ravi P. Agarwal
Michael Ruzhansky

CRC Press
Taylor & Francis Group
Boca Raton London New York

CRC Press is an imprint of the
Taylor & Francis Group, an **informa** business
A CHAPMAN & HALL BOOK

First edition published 2021
by CRC Press
6000 Broken Sound Parkway NW, Suite 300, Boca Raton, FL 33487-2742

and by CRC Press
2 Park Square, Milton Park, Abingdon, Oxon, OX14 4RN

CRC Press is an imprint of Taylor & Francis Group, LLC

ISBN: 978-0-367-33472-7 (hbk)
ISBN: 978-0-429-32002-6 (ebk)

Typeset in Palatino
by codeMantra

Contents

Preface

Differential Equations are very important tools in Mathematical Analysis. These are widely found in mathematics itself and in its applications to statistics, computing, electrical circuit analysis, dynamical systems, economics, biology, and so on. A non-linear differential equation is simply a differential equation where some non-linearity is applied to either the inputs or the outputs of the equation. Many times during the solution of complex problems of differential equations in engineering and sciences, we treat with important functions defined by improper integrals and series (or infinite products). Those functions are generally called special functions. Special functions contain a very old branch of mathematics. For example, trigonometric functions have been studied for more than a thousand years, due mainly to their numerous applications in astronomy. Yet the origins of their unified and rather complete theory date back to the 19th century. From the application point of view, special functions such as important mathematical tools rest on usefulness for the applied scientists and engineers due to their remarkable properties, as Paul Turan once remarked that special functions would be more appropriately labeled useful functions. Various special functions—such as Bessel and all cylindrical functions; the Gauss, Kummer, confluent, and generalized hypergeometric functions; the classical orthogonal polynomials, the incomplete Gamma and Beta functions and error functions; and the Airy and Whittaker functions—will provide solutions to integer order differential equations and systems used as mathematical models. However, recently there has been an increasing interest in and widely extended use of differential equations and systems of fractional order (that is, of arbitrary order) as better models of phenomena of various physics, engineering, automatization, biology and biomedicine, chemistry, earth science, economics, nature, and so on. Now, new unified presentation and extensive development of special functions associated with fractional calculus (a branch of Mathematical Analysis) are necessary tools, being related to the theory of differentiation and integration of arbitrary order (i.e., fractional calculus) and to the fractional order (or multi-order) differential and integral equations.

This edited volume is mainly based on talks delivered during an International Workshop on Computational Science and Applications at Sharda University, Greater Noida, Knowledge Park-3, India, during, 5–6 February 2018, organized under the auspices of the Centre for Advanced Research in Applied Mathematics and Physics. The Chancellor of Sharda University, Mr. P. K. Gupta, and Vice Chancellor of Sharda University, Prof. B. S. Panwar, graced the inaugural function of the workshop besides Prof. K. L. Chopra, former director of IIT Kharagpur (Padma Shri and recipient of Bhatnagar award), Prof. N. K. Gupta (Padma Shri and former vice president INSA), and several senior professors of IIT Delhi and other national institutions.

The topics covered in the edited volume are the *ingredients for advances in special functions and analysis of differential equations* and their applications.

There are 17 chapters in this edited volume. Chapter 1, contributed by professors Arman Dabiri, Behrouz Parsa Moghadam, and J. A. Tenreiro Machado, has presented a Chebyshev spectral collocation method for the numerical solution of the fractional Fokker–Planck equation (FFPE) with Riesz derivatives. In Chapter 2, professors Giovanni Mingari

Scarpello and Daniele Ritelli have presented their study on elastica planar macro-problems and their link to special functions. They will concern the deflections of a supported rod bent by axial compression.

Chapter 3, by professors M. H. Heydari, A. Shakiba, Z. Avazzadeh, and Carlo Cattani, is devoted to how the second-kind Chebyshev wavelets (SKCWs) are applied to compute an approximate solution for the variable-order space-time fractional telegraph equation (V-S-FTE). The SKCW's operational matrix of variable-order fractional derivative (OMV-FD) is derived and used in the proposed method. The main advantage of this approach is that it reduces these kinds of problems to those of solving systems of algebraic equations and simplifies.

Professors J. E. Restrepo, R. A. Higuita, and Shilpi Jain have established some types of stabilities for some classes of Riemann–Liouville and Caputo fractional Volterra type integro-differential equations and established some sufficient conditions to guarantee Hyers–Ulam–Rassias and Hyers–Ulam stabilities for the considered fractional equations. The sufficient conditions are founded by means of the use of the Banach Fixed Point Theorem in terms of a generalized metric in Chapter 4. Chapter 5 deals with the work of professors Muhammad Saqib, Ilyas Khan, and Sharidan Shafie on the application of the Atangana–Baleanu derivative fractional into the convective flow of Carboxy-Methyl-Cellulose (CMC) based carbon nanotubes (CNT's) nanofluid in a vertical microchannel. The magnetohydrodynamic (MHD) flow through a porous medium together with heat transfer is considered.

In Chapter 6, professors Rúben Sousa, Manuel Guerra, and Semyon Yakubovich have discussed their work on a unified framework for the construction of convolutions and product formulas associated with a general class of regular and singular Sturm–Liouville boundary-value problems. Professor Yasuhiro Abe has worked on integral representations of the generalized hypergeometric functions in terms of twisted homology and cohomology in Chapter 7.

In Chapter 8, professors F.-H. Li and Shigeru Kanemitsu worked on assembled basic facts about robust control theory with a treatment of the behavior of the boundary functions of the plant. They give some enlightening examples from mathematics, in general, by which we show that an analogy can be a leading navigator for the expedition to a new field.

In Chapter 9, professor Arsen Pskhu provided overviews on two adjoint integral transforms involving the Wright function as the kernel; discussed their basic properties, especially emphasizing those related to operators of fractional integration and differentiation; and described their application to solve evolutionary equations of fractional order and to study special functions arising in fractional calculus. In Chapter 10, professors Yusuke Kamata, Tingli Ma, Yong Sun, and Shigeru Kanemitsu have worked on electric current as a continuous flow. They show the flow-part by reducing the Maxwell equations to two partial DEs and one wave equation. The continuous-part is shown by the equation of continuity as with fluid flows.

Chapter 11 deals with the work of professors M. Emin Özdemir, Ahmet Ocak Akdemir, Erhan Set, and Alper Ekinci on some fractional derivatives and related integral inequalities. In the context of this study, definitions and properties of Riemann–Liouville fractional integral, conformable fractional integral, Katugampola fractional integral, and generalized fractional integral, as well as how integral inequalities in the literature related to investigated fractional integrals were investigated and brought together. Some new integral inequalities have been performed via generalized fractional integrals. In Chapter 12, professor Dmitrii B. Karp has given a simpler and shorter proof of the expansion around the positive singular point and further identifies more precisely the abscissa of convergence of the underlying inverse factorial series. Our new proof hinges on a slight generalization of a particular case of Braaksma's theorem about analytic continuation of Fox's H function. Chapter 13 is based on a relation between non-integer integro-differentiation operators and Riemann's zeta function. The properties of these differential operators give off the concepts used in an "engineering" approach of the Riemann Hypothesis. This link is based on the self-similarity of the bi-categories at play, given by professors Philippe Riot, Alain Le Méhauté, and Dmitrii Tayurskii. In Chapter 14, professors Haci Mehmet Baskonus, Carlo Cattani, and Armando Ciancio have defined the modified exponential function method and various classifications of the traveling wave solutions to the (2 + 1)-dimensional Boussinesq dynamical equation. The complete result shows that there exist many surfaces of traveling wave solutions to the (2 + 1)-dimensional Boussinesq dynamical model, such as complex solutions, hyperbolic solutions, and rational and double periodic solutions. These solutions differ from any of those studied in advance.

In Chapter 15, professors Anum Shafiq, Zakia Hammouch, Tabassum Naz Sindhu, and Dumitru Baleanu worked on the impact of mixed convective flow of an incompressible viscoelastic third-grade fluid along an exponentially stretched surface. Heat transfer analysis is carried out in the presence of convective boundary condition and Joule heating effects. Series solutions of the governing boundary-value problems are obtained via Homotopy Analysis Method (HAM). Convergence of the problem is explicitly discussed. Statistical methodology (via correlation coefficient and probable error) is used to check the significance of the pertinent parameters on skin friction and Nusselt number. Chapter 16 is based on the study of the boundary-value problems for a third-order mixed-type loaded equation with type-changing line and Caputo fractional differential operator. Using the method of integral equations, a unique solvability of the formulated problem has been proved by Praveen Agarwal, Umida Baltaeva, and Jessada Tariboon. In Chapter 17, professors J. E. Solís-Pérez, J. F. Gómez-Aguilar, R. F. Escobar-Jiménez, and J. Reyes-Reyes have worked on the state-observer–based approach to synchronize fractional conformable chaotic systems. A new conformable operator based on the Riemann–Liouville derivative is proposed to design the observer. A unidirectional topology (master/slave) is employed to facilitate complete synchronization. The fractional conformable attractor is represented by a state-space model, while a fractional conformable state observer is designed to represent the slave. A numerical method based on the Grünwald–Letnikov fractional derivative and the Euler method are used to solve the numerical simulations of the fractional conformable derivatives. Numerical simulations are presented to show the effectiveness of the methodology.

We take this opportunity to thank Ms Aastha Sharma of CRC Press/Taylor & Francis Group for providing continuous guidance and help. We will be failing in our duty if we don't express our gratitude to Mr Manoj Mittal, Chairman, Anand ICE, Jaipur, for providing valuable financial help for organizing an academic activity yielding a valuable volume. I also pay my sincere regards to my parents, Dr N. K. Agarwal and Smt Rukmani Devi Agarwal. This work was supported by the Praveen Agarwal research grants under the SERB (project TAR/2018/000001), DST(project DST/INT/DAAD/P-21/2019, and INT/RUS/RFBR/308).

Prof. Praveen Agarwal
Anand ICE, India

Prof. Ravi P. Agarwal
Texas A&M University, USA

Prof. Michael Ruzhansky
Ghent University, Belgium

Editors

Praveen Agarwal is a professor at the Department of Mathematics, Anand International College of Engineering, Jaipur, India. Being on the editorial boards of several journals of repute, Agarwal has conducted and attended a number of conferences. Recently, he has received the Most Outstanding Researcher 2018 Award for his contribution to mathematics from the Union Minister of Human Resource Development of India. He has received numerous international research grants. He has published over 250 articles related to special functions, fractional calculus, and mathematical physics in several leading mathematics journals. His latest research has focused on partial differential equations, fixed point theory, and fractional differential equations.

Ravi P. Agarwal received his PhD in mathematics from the Indian Institute of Technology, Madras, India. He is a professor of mathematics at the Florida Institute of Technology. His research interests include numerical analysis, inequalities, fixed point theorems, and differential and difference equations. He is the author/co-author of over 1000 journal articles and more than 25 books, and actively contributes to over 500 journals and book series in various capacities.

Michael Ruzhansky is a professor at the Department of Mathematics, Imperial College London. He has published over 100 research articles in several leading international journals. He has also published five books and memoirs and nine edited volumes. He has researched topics related to pseudo-differential operators, harmonic analysis, and partial differential equations. More recently, he has worked on boundary-value problems and their applications. He has been on the editorial board of many respected international journals and served as the president of the International Society of Analysis, Applications, and Computations (ISAAC) in the period 2009–2013.

Contributors

Yasuhiro Abe
Cereja Technology Co., Ltd.
Tokyo, Japan

Praveen Agarwal
International Center for Basic and Applied Sciences
Jaipur, India
and
Department of Mathematics
Anand International College of Engineering
Jaipur, India
and
Harish-Chandra Research Institute (HRI)
Allahabad India
and
Netaji Subhas University of Technology
Delhi, India

Ahmet Ocak Akdemir
Department of Mathematics
Ağrı İbrahim Çeçen University
Ağrı, Turkey

Zakieh Avazzadeh
School of Mathematical Science
Xi'an Jiaotong-Liverpool University
Suzhou, China

Dumitru Baleanu
Department of Mathematics and Computer Sciences, Faculty of Arts and Sciences
Cankara University
Ankara, Turkey

Haci Mehmet Baskonus
Department of Mathematics and Science Education
Harran University
Sanliurfa, Turkey

Umida Baltaeva
Department of Exact Sciences
Khorezm Mamun Academy
Khiva, Uzbekistan
and
Department of Applied Mathematics
Urgench State University
Urgench, Uzbekistan

Carlo Cattani
Engineering School (DEIM)
University of Tuscia
Viterbo, Italy

Armando Ciancio
Department of Biomedical and Dental Sciences and Morphofunctional Imaging
University of Messina
Messina, Italy

Arman Dabiri
School of Engineering
Southern Illinois University
Edwardsville

Alper Ekinci
Department of Mathematics
Ağrı İbrahim Çeçen University
Ağrı, Turkey
and
Bandırma Vocational High School
Bandırma 17 Eylül University
Bandırma, Turkey

R. F. Escobar-Jiménez
Tecnológico Nacional de México/CENIDET
Morelos, México

J. F. Gómez-Aguilar
Tecnológico Nacional de México/CENIDET
Morelos, México

Manuel Guerra
CEMAPRE and ISEG (School of Economics and Management)
Universidade de Lisboa
Lisbon, Portugal

Zakia Hammouch
Département de Mathématique
Faculte des Sciences et Techniques Errachidia
Université Moulay Ismaïl
Meknes, Morocco

Mohammad Hossein Heydari
Department of Mathematics
Shiraz University of Technology
Shiraz, Iran

R. A. Higuita
Institute of Mathematics
University of Antioquia
Antioquia, Colombia

Abd-Allah Hyder
Faculty of Engineering
Al-Azhar University
Cairo, Egypt

Shilpi Jain
Department of Mathematics
Poornima College of Engineering
Jaipur, Rajasthan

Yusuke Kamata
Graduate School of Life Science and Systems Engineering
Kyushu Institute of Technology
Kitakyushu, Japan

Shigeru Kanemitsu
Faculty of Engineering, Department of Applied Science for Integrated System Engineering
Kyushu Institute of Technology
Kitakyushu, Japan

Dmitrii B. Karp
Holon Institute of Technology
Holon, Israel

Ilyas Khan
Ton DucThang University
Ho Chi Minh City, Vietnam
and
Department of Mathematics, College of Science Al-Zulfi
Majmaah University
Al-Majmaah, Saudi Arabia

F.-H. Li
Henan Suda Electric Vehicle Technology Co., Ltd.
Sanmenxia, P.R. China

Tingli Ma
Graduate School of Life Science and Systems Engineering
Kyushu Institute of Technology
Kitakyushu, Japan

J. A. Tenreiro Machado
Department of Electrical Engineering
Institute of Engineering of Porto
Porto, Portugal

Alain Le Méhauté
Materials Design Inc.
Montrouge, France
and
Department of Physics
Kazan Federal University
Kazan, Russia

Behrouz Parsa Moghadam
Department of Mathematics, Lahijan Branch
Islamic Azad University
Lahijan, Iran

M. Emin Özdemir
Department of Mathematics Education
Bursa Uludağ University
Bursa, Turkey

Arsen Pskhu
Fractional Calculus Department
Institute of Applied Mathematics and Automation of Kabardino-Balkar Scientific Center Russian Academy of Science (KBSC RAS)
Nalchik, Russia

J. E. Restrepo
Regional Center of Southern Federal University
Rostov Oblast, Russia
and
Nazarbayev University
Astana, Kazakhstan

J. Reyes-Reyes
Tecnológico Nacional de México/CENIDET
Morelos, México

Philippe Riot
Zeta Innovation
Rue Gutemberg, Paris

Daniele Ritelli
Dipartimento di Scienze Statistiche P. Fortunati
Università di Bologna
Bologna, Italy

Muhammad Saqib
Department of Mathematical Sciences, Faculty of Science
Universiti Teknologi Malaysia
Kuala Lumpur, Malaysia

Giovanni Mingari Scarpello
Dipartimento di Scienze Statistiche P. Fortunati
Università di Bologna
Milano, Italy

Erhan Set
Department of Mathematics
Ordu University
Ordu, Turkey

Sharidan Shafie
Department of Mathematical Sciences, Faculty of Science
Universiti Teknologi Malaysia
Kuala Lumpur, Malaysia

Anum Shafiq
Department of Mathematics
Preston University
Karachi, Pakistan

Ali Shakiba
Department of Computer Science
Vali-e-Asr University of Rafsanjan
Rafsanjan, Iran

Tabassum Naz Sindhu
Department of Statistics
Quaid I Azam University
Islamabad, Pakistan

J. E. Solís-Pérez
Tecnológico Nacional de México/CENIDET
Morelos, México

Rúben Sousa
Departamento de Matemática, Faculdade de Ciências
Universidade do Porto
Porto, Portugal

Yong Sun
Department of Applied Science for Integrated System Engineering
Kyushu Institute of Technology
Kitakyushu, Japan

Jessada Tariboon
Department of Mathematics, Faculty of Applied Science
King Mongkut's University of Technology North Bangkok
Bangkok, Thailnad

Dmitrii Tayurskii
Physics Department
Kazan Federal University
Kazan, Russia

Semyon Yakubovich
Departamento de Matemática, Faculdade de Ciências
Universidade do Porto
Porto, Portugal

M. Zakarya
Faculty of Engineering
Al-Azhar University
Cairo, Egypt

1

A Chebyshev Spatial Discretization Method for Solving Fractional Fokker–Planck Equation with Riesz Derivatives

Arman Dabiri
Southern Illinois University

Behrouz Parsa Moghadam
Islamic Azad University

J. A. Tenreiro Machado
Institute of Engineering of Porto

CONTENTS

1.1 Introduction

Einstein [1] proposed that the Brownian motion, representing continuous small random fluctuations in the position of particles, can be described by the linear evolution of the mean square displacement using a diffusion constant, i.e., $\langle(\Delta x)^2\rangle \propto t$. The theory yields a partial differential equation (PDE) initially introduced by Fokker and Planck [2], and presently called the Fokker–Planck equation (FPE). This PDE is constructed by examining the time evolution of the velocity distribution function of particles in the presence of random external forces. The FPE has found applications in a variety of scientific areas, such as biology [3], physics [4,5], finance [6], and others [7,8].

In anomalous (non-Fickian) diffusion, the mean square displacement in the Einstein's equation is described by the power law evolution such that $\langle(\Delta x)^2\rangle \propto t^\alpha$, wherein $\alpha \in (0,1)$ and $\alpha \in (1,2)$ describe the subdiffusion and superdiffusion processes, respectively. This results in obtaining FPEs with fractional derivatives (FDs), so-called fractional FPEs (FFPEs). It has been shown that FFPEs are elegant formulations for modeling natural dynamical processes with hereditary properties, such as the particle transport dynamics of anomalous diffusion and relaxation patterns [9], Lévy processes with heavy-tailed probability distributions [10,11], continuous random walks [12], the motion of ribosomes as well as mRNA, and pattern formation [13–15]. Moreover, related research is also growing rapidly in different areas, e.g., [16–20].

Many different types of FFPEs have been introduced in terms of the probability density function of the corresponding process [9,21]. FFPEs can be classified into three main types, namely (i) time FFPEs (TFFPEs) with a temporal FD (see Ref. [9]), (ii) space FFPEs (SFFPEs) with spatial FDs (see Refs. [10,11]), and (iii) time-space FFPEs (TSFFPEs) with spatial and temporal FDs (see Ref. [14]). In all the previous studies, the fractional time-derivative operator is chosen as the left-sided Riemann–Liouville (RL) or Caputo derivatives of order $\alpha \in (0,1]$. However, the SFFPEs or TSFFPEs are similar to the second-order advection-dispersion equation having the fractional spatial derivatives with orders $\beta \in (1,2]$ defined such that the left- and right-sided FDs need to satisfy the Dirichlet boundary conditions. The exact solution of FFPEs cannot be generally granted, and hence, effective numerical methods need to be considered for approximating the solution. However, tackling fractional PDEs (FPDEs) is more laborious than handling PDEs because of the convolution form of the derivative operator and the weak singularity kernel in FDs. The most popular numerical methods that have been proposed to solve FFPEs can be classified as the finite-difference algorithms [15,22–24], method of lines [24,25], and spectral techniques [26].

In finite difference schemes, the solution domain is initially divided into many smaller subdomains. Afterwards, a few numbers of locally trial well-defined functions are used to approximate the solution in each subdomain. On the other hand, spectral methods employ a large number of basis (or trial) functions to approximate the solution globally in a domain of interest. The result of using these local and global approximations is to obtain exponential convergence rates for spectral methods in comparison to linear convergence rates for the finite difference schemes [27]. More importantly, it has been shown that in special problems, such as linear differential equations with the smooth solution, spectral methods require significantly smaller computation time in contrast to the finite difference schemes. In spectral methods, substitution of a truncated series of basis functions into the original differential equation leads to a residual, which can be minimized using several different techniques. Consequently, spectral methods are classified into three categories, namely Galerkin, Tau, and Collocation techniques, that are based on the choice of basis functions and the employed minimization technique [27]. Spectral collocation methods are convenient due to their straightforward form of constructions and have been used in different problems. Spectral collocation methods were first applied to study the electronic energy bands in metals [28]. The effect of the proper choice of trial functions and the distribution of collocation points for obtaining an accurate solution were initially discussed in [29]. Moreover, the foundation of using orthogonality in collocation methods was established in the same study. The application of collocation methods in solving PDEs was initially suggested in [30,31] using the Fourier method.

In this study, we find the approximated solution of SFFPEs by its spatial discretization at shifted Chebyshev–Gauss–Lobatto (CGL) collocation points. This yields a system of linear differential equations that can be solved by standard ordinary differential equation (ODE) algorithms. For this purpose, two operational matrices of left- and right-sided RL fractional differentiation are derived at the collocation points to obtain the left- and right-sided approximations, respectively. The proposed method is used to solve SFFPEs with constant or varying coefficients. It is shown that the method is computationally efficient in achieving the solution with spectral convergence.

The outline of this chapter is as follows. In Section 1.2, the fundamental definitions in fractional calculus are reviewed. Then, the fractional Fokker–Planck for advection-reaction-dispersion is introduced to simulate the Lévy motion with α-stable densities. In Section 1.3, the framework of the method of lines using Chebyshev spectral collocation method is proposed for the FFPE spatial discretization. The error analysis and the stability of the proposed numerical method are discussed in Section 1.4. The efficiency and accuracy of the proposed method are illustrated with two numerical examples in Section 1.5. Finally, Section 1.6 outlines the conclusions.

1.2 Preliminaries and Problem Statement

In this section, we introduce the necessary notations, and basic definitions from polynomial interpolation and fractional calculus that are used throughout this chapter. We also recall relevant properties of the Chebyshev polynomials.

Definition 1.1. Let Ω be an open subset of the Euclidean space $\mathbb{R}^n$ and $\mathbf{x} = (x_1, \ldots, x_n)$ represent an element of this subset. In addition, the positive integrable function $w(\mathbf{x}) : \Omega \mapsto \mathbb{R}^+$ denotes a weight function and $1 \le p \le +\infty$. A $\mathcal{L}^p_w(\Omega)$ space represents a space of square-integrable functions $\phi(\mathbf{x}) : \Omega \mapsto \mathbb{R}$ whose absolute value raised to the p-power is well-defined and bounded:

$$\mathcal{L}^2_w(\Omega) := \left\{\phi(\mathbf{x}) : \Omega \mapsto \mathbb{R} \mid \phi(\mathbf{x}) \text{ is measurable and } \|\phi(\mathbf{x})\|_{\mathcal{L}^2_w(\Omega)} < \infty\right\}, \tag{1.1}$$

where the corresponding norm $\|\cdot\|_{\mathcal{L}^p_w(\Omega)}$ is

$$\|\cdot\|_{\mathcal{L}^p_w(\Omega)} := \sqrt[p]{\langle\cdot,\cdot\rangle_w}, \tag{1.2}$$

and the inner product $\langle\cdot,\cdot\rangle_w$ is given by the Lebesgue integral

$$\langle\phi(\mathbf{x}), \psi(\mathbf{x})\rangle_w := \int_\Omega \phi(\mathbf{x})\psi(\mathbf{x})w(\mathbf{x})d\mathbf{x}. \tag{1.3}$$

Note 1.1. It is clear that $\|\cdot\|_{\mathcal{L}^\infty_w(\Omega)} \equiv \|\cdot\|_{\mathcal{L}^\infty(\Omega)} \equiv \sup_{\mathbf{x}\in\Omega}|\cdot|$ for any weighted function $w(\mathbf{x}) \in \mathbb{R}^+$.

Note 1.2. If $w(\mathbf{x}) = 1$, then we represent the space of measurable functions $\mathcal{L}^p_1(\Omega)$ simply by $\mathcal{L}^p(\Omega)$.

Definition 1.2. Let $\mathcal{AC}^n(\Omega)$ denote a space of measurable functions on $\Omega = [a, b]$ for which they have continuous derivatives up to $n-1$ order on Ω.

Definition 1.3. Let $\Omega \subset \mathbb{R}^n$ be a bounded interval with elements represented by $\mathbf{x} = (x_1, \ldots, x_n)$, and let the set of nonnegative numbers $\boldsymbol{s} = \{s_i \in \mathbb{Z}^+ \mid i = 0, \ldots, n\}$ represent a nonnegative multi-index set such that $s = \sum_{i=1}^n s_i \in \mathbb{N}$. The weighted Sobolev space $\mathcal{W}^r_{w,p}(\Omega)$, $r \ge 0$, is a space of all functions $\phi(\mathbf{x}) : \Omega \mapsto \mathbb{R}$ such that the function itself and all its weak derivatives up to the order r are in the Lebesgue space $\mathcal{L}^p_w(\Omega)$. That is

$$\mathcal{W}^r_{w,p}(\Omega) := \{\phi(\mathbf{x}) \in \mathcal{L}^p_w(\Omega) : \forall s,\ 0 \le s \le r, D^s\phi(\mathbf{x}) \in \mathcal{L}^p_w(\Omega)\}, \tag{1.4}$$

where

$$D^s\phi(\mathbf{x}) := \frac{\partial^s\phi(\mathbf{x})}{\partial^{s_1}x_1\partial^{s_2}x_2\ldots\partial^{s_n}x_n}. \tag{1.5}$$

Note 1.3. For simplicity, we skip the subscript p in the weighted Sobolev space $\mathcal{W}^r_{w,p}(\Omega)$ when $p=2$.

Definition 1.4. The Hilbert space $\mathcal{H}^r_w(\Omega)$ is the Sobolev space $\mathcal{W}^r_w(\Omega)$, i.e., $\mathcal{W}^r_{w,p}(\Omega)$ when $p=2$, and it is associated with the following norm and seminorm

$$\|\phi(\mathbf{x})\|^2_{\mathcal{H}^r_w(\Omega)} := \sum_{k=0}^{r} \left\|D^k\phi(\mathbf{x})\right\|^2_{\mathcal{L}^2_w(\Omega)}, \tag{1.6a}$$

$$|\phi(\mathbf{x})|^2_{\mathcal{H}^{r;m}_w(\Omega)} := \sum_{k=\min\{r,m+1\}}^{r} \left\|D^k\phi(\mathbf{x})\right\|^2_{\mathcal{L}^2_w(\Omega)}. \tag{1.6b}$$

Proposition 1.1. *The following statement is true for the weighted Hilbert spaces:*

$$\mathcal{H}^r_w(\Omega) \subset \mathcal{H}^{r-1}_w(\Omega) \subset \dots \mathcal{H}^2_w(\Omega) \subset \mathcal{H}^1_w(\Omega) \subset \mathcal{H}^0_w(\Omega) \equiv \mathcal{L}^2_w(\Omega). \tag{1.7}$$

Proposition 1.2. *For* $m+1 \geq r$, *we have*

$$|\phi(\mathbf{x})|_{\mathcal{H}^{r;m}_w(\Omega)} \equiv \|D^r\phi(\mathbf{x})\|_{\mathcal{L}^2_w(\Omega)}\,. \tag{1.8}$$

1.2.1 Fractional Calculus

Definition 1.5 (Section 1.2.1 in [32]). The left- and right-sided RL integrals of order $\alpha \in \mathbb{C}$, $\mathrm{R}(\alpha) > 0$, for the function $\phi(x) : \Omega \mapsto \mathbb{R}$, $\Omega = [a,b]$, are defined by

$$\left({}_a\mathcal{J}^\alpha_x\phi\right)(x) := \frac{1}{\Gamma(\alpha)}\int_a^x \phi(\zeta)\,(x-\zeta)^{\alpha-1}\,d\zeta, \tag{1.9a}$$

$$\left({}_x\mathcal{J}^\alpha_b\phi\right)(x) := \frac{1}{\Gamma(\alpha)}\int_x^b \phi(\zeta)\,(\zeta-x)^{\alpha-1}\,d\zeta, \tag{1.9b}$$

respectively, where $\Gamma(z) : \mathbb{C} \mapsto \mathbb{C}$ is the Gamma function defined by

$$\Gamma(z) := \int_0^\infty x^{z-1}e^{-x}\,dx, \tag{1.10}$$

such that $\Gamma(n) = (n-1)!$, $n \in \mathbb{N}$.

Property 1.1 (Lemma 2.3 and Property 2.2 in [32]). *Let* $\Omega = [a,b]$, $\phi(x) \in \mathcal{L}^p(\Omega)$, $\alpha \in \mathbb{C}$ $(\Re(\alpha) > 0)$, *and* $\beta \in \mathbb{C}$ $(\Re(\beta) > 0)$. *Then, the following equation holds at almost any point* $x \in \Omega$, *and for all points in* Ω *if* $\alpha + \beta > 1$*:*

$$\left({}_a\mathcal{J}^\alpha_x\left({}_a\mathcal{J}^\beta_x\phi\right)\right)(x) = \left({}_a\mathcal{J}^{\alpha+\beta}_x\phi\right)(x), \tag{1.11a}$$

$$\left({}_x\mathcal{J}^\alpha_b\left({}_x\mathcal{J}^\beta_b\phi\right)\right)(x) = \left({}_x\mathcal{J}^{-\alpha+\beta}_b\phi\right)(x). \tag{1.11b}$$

Definition 1.6 ([32]). The left- and right-sided RL fractional-order derivatives of order $\alpha > 0$ for $\phi(x) : \Omega \mapsto \mathbb{R}$, $\Omega = [a,b]$, are defined by applying the $\lceil\alpha\rceil$th derivative operator, denoted by $\partial^{\lceil\alpha\rceil}_x(\cdot)$, on the α-order left- and right-sided RL fractional integrals of $\phi(x)$, respectively, namely

$$\left({}^R_a\mathcal{D}^\alpha_x\phi\right)(x) := \left(\partial^{\lceil\alpha\rceil}_x\left({}_a\mathcal{J}^{\lceil\alpha\rceil-\alpha}_x\phi\right)\right)(x), \quad x > a, \tag{1.12a}$$

$$\left({}^R_x\mathcal{D}^\alpha_b\phi\right)(x) := (-1)^{\lceil\alpha\rceil}\left(\partial^{\lceil\alpha\rceil}_x\left({}_x\mathcal{J}^{\lceil\alpha\rceil-\alpha}_b\phi\right)\right)(x), \quad x < b, \tag{1.12b}$$

where $\lceil\alpha\rceil$ the well-known ceiling function returns the smallest integer larger than α.

Property 1.2 (Property 2.1 in [32]). *Let $\alpha, \beta \in \mathbb{C}$ such that $\Re(\alpha) > 0$ and $\Re(\beta) > 0$. Then, the left-sided RL integral and derivative of the power function $(x-a)^\beta$, $x > a$, are*

$$ {}_a\mathcal{J}_x^\alpha\left((x-a)^\beta\right) = \frac{\Gamma(\beta+1)}{\Gamma(\beta+1-\alpha)}(x-a)^{\beta+\alpha}, \tag{1.13a}$$

$$ {}_a\mathcal{D}_x^\alpha\left((x-a)^\beta\right) = \frac{\Gamma(\beta+1)}{\Gamma(\beta+1-\alpha)}(x-a)^{\beta-\alpha}, \tag{1.13b}$$

and the right-sided RL integral and derivative of the power function $(b-x)^\beta$, $x < b$, are

$$ {}_x\mathcal{J}_b^\alpha\left((b-x)^\beta\right) = \frac{\Gamma(\beta+1)}{\Gamma(\beta+1-\alpha)}(b-x)^{\beta+\alpha}, \tag{1.14a}$$

$$ {}_x\mathcal{D}_b^\alpha\left((b-x)^\beta\right) = \frac{\Gamma(\beta+1)}{\Gamma(\beta+1-\alpha)}(b-x)^{\beta-\alpha}. \tag{1.14b}$$

Remark 1.1 ([32]). *For any $n \in \mathbb{N}$ and $x \in \Omega$, $\Omega = [a,b]$, we have*

$$ \left({}_a\mathcal{D}_x^n\phi\right)(x) \equiv \left({}_x\mathcal{D}_b^n\phi\right)(x) \equiv \left(\partial_x^n\phi\right)(x). \tag{1.15}$$

Definition 1.7. The finite Riesz integral and derivative of order $\alpha > 0$ for $x \in \Omega$, $\Omega = [a,b]$, are defined by

$$\begin{aligned}
\left(|\mathcal{J}|_{(a,b)}^\alpha\,\phi\right)(x) &:= c_\alpha \int_b^a \phi(\zeta)\,|x-\zeta|^{\alpha-1}\,d\zeta \\
&= c_\alpha\left({}_a\mathcal{J}_x^\alpha + {}_x\mathcal{J}_b^\alpha\right)\phi(x), && (1.16a)\\
\frac{\partial^\alpha}{\partial|x|^\alpha}\phi(x) \equiv \left(|\mathcal{D}|_{(a,b)}^\alpha\,\phi\right)(x) &:= \partial_x^{\lceil\alpha\rceil}|\mathcal{J}|_{(a,b)}^{\lceil\alpha\rceil-\alpha}\,\phi(x) && (1.16b)\\
&= -c_{\lceil\alpha\rceil-\alpha}\left({}_a\mathcal{D}_x^\alpha + {}_x\mathcal{D}_b^\alpha\right)\phi(x) && (1.16c)\\
&= -c_{\lceil\alpha\rceil-\alpha}\partial_x^{\lceil\alpha\rceil}\int_a^b \phi(\zeta)\,|\zeta-x|^{\lceil\alpha\rceil-\alpha-1}\,d\zeta, && (1.16d)
\end{aligned}$$

where $c_\alpha = \dfrac{1}{2\Gamma(\alpha)\cos\left(\frac{\alpha\pi}{2}\right)}$.

Definition 1.8. The Riesz potential operators represented by $|\mathcal{D}'|_{(a,b)}^\alpha$ and $|\mathcal{J}'|_{(a,b)}^\alpha$ are equivalent to the Riesz operators when $c_\alpha = 1/2$.

1.2.2 Polynomial Interpolation

Definition 1.9. Let $\mathbb{P}_N(\Omega)$ represent the space spanned by the orthogonal polynomials $\{p_i(t)\}_{i=0}^{N-1}$ with the inner product $\langle\cdot,\cdot\rangle_w$, where $w(t) \in \mathcal{L}_w^2(\Omega)$ is the weight function. Then, we denote by $\mathcal{P}_N : \mathcal{L}_w^2(\Omega) \mapsto \mathbb{P}_N(\Omega)$ the orthogonal projections for any square-integrable function $\phi(x) \in \mathcal{L}_w^2(\Omega)$, namely

$$ \left(\mathcal{P}_N\phi\right)(x) := \sum_{i=0}^{N-1} \tilde{\phi}_i\,P_i(x), \tag{1.17}$$

where $\tilde{\phi}_i \in \mathbb{R}$, and the truncated (or finite) series (1.17) represents an approximation for $\phi(x)$ in $\mathbb{P}_N(\Omega)$. Moreover, for any $\phi(x) \in \mathcal{L}_w^2(\Omega)$, we have

$$ \left\langle\left(\mathcal{P}_N\phi\right)(x) - \phi(x), P_n(x)\right\rangle_w = 0, \quad n = 0, \ldots, N-1. \tag{1.18}$$

Proposition 1.3. *The completeness of the space $\mathbb{P}_N(\Omega)$ holds*

$$\lim_{N\to\infty} \|(\mathcal{P}_N\phi)(x) - \phi(x)\|_{\mathcal{L}^2_w(\Omega)} = 0. \tag{1.19}$$

Definition 1.10. Consider the orthogonal polynomials $\{p_i(x)\}_{i=0}^{N-1}$ defined over $\mathcal{L}^2_w(\Omega)$ with the weight function $w(x)$. The quadrature weight $\{w_i\}_{i=0}^{N-1}$ and nodes $\{x_{d_i}\}_{i=0}^{N-1}$ for the orthogonal polynomials $\{p_i(x)\}_{i=0}^{N-1}$ are defined such that

$$\langle P_n(x), P_m(x)\rangle_w = \langle P_n(x), P_m(x)\rangle_N := \sum_{i=0}^{N-1} P_n(x_{d_i})P_m(x_{d_i})w_i. \tag{1.20}$$

Definition 1.11. The interpolation operator $\mathcal{I}_N : \mathcal{L}^2_w(\Omega) \mapsto \mathbb{P}_N(\Omega)$, for any square-integrable function $\phi(x) \in \mathcal{L}^2_w(\Omega)$, is defined by

$$((\mathcal{I}_N\phi))(x) := \sum_{i=0}^{N-1} \hat{\phi}_i\, P_i(x), \tag{1.21}$$

such that

$$((\mathcal{I}_N\phi))(x_{d_i}) = \phi(x_{d_i}), \quad i = 0, 1, \ldots, N-1, \tag{1.22}$$

where x_{d_i} are the quadrature points, and $\hat{\phi}_i$ are given by

$$\hat{\phi}_i = \frac{\langle \phi(x), P_i(x)\rangle_N}{\|P_i(x)\|_N^2}, \tag{1.23}$$

and $\|P_i(x)\|_N^2 = \langle P_i(x), P_i(x)\rangle_N$.

Lemma 1.1. *For the discrete interpolation operator and truncated interpolation operator, we have*

$$(\mathcal{I}_N\phi)(x) = (\mathcal{P}_N\phi)(x) + (\mathcal{R}_N\phi)(x), \tag{1.24}$$

where the aliasing error $(\mathcal{R}_N\phi)(x)$ is

$$(\mathcal{R}_N\phi)(x) = \sum_{i=0}^{N-1} \frac{P_i(x)}{\|P_i(x)\|_N^2} \left\langle \sum_{j=N}^{\infty} \tilde{\phi}_j P_j(x), P_i(x) \right\rangle_N. \tag{1.25}$$

Proof. Using the infinite series of $\phi(x)$ and Equation (1.23) along with the discrete orthogonality of the polynomials yields

$$\hat{\phi}_i = \frac{\left\langle \sum_{j=0}^{\infty} \tilde{\phi}_j\, P_j(x), P_i(x)\right\rangle_N}{\|P_i(x)\|_N^2} = \tilde{\phi}_i + \frac{\left\langle \sum_{j=N}^{\infty} \tilde{\phi}_j\, P_j(x), P_i(x)\right\rangle_N}{\|P_i(x)\|_N^2}. \tag{1.26}$$

Substituting the obtained $\hat{\phi}_i$ in the discrete interpolation yields

$$(\mathcal{I}_N\phi)(x) := \sum_{i=0}^{N-1} \left(\tilde{\phi}_i + \frac{\left\langle \sum_{j=N}^{\infty} \tilde{\phi}_j\, P_j(x), P_i(x)\right\rangle_N}{\|P_i(x)\|_N^2} \right) P_i(x), \tag{1.27}$$

which completes the proof by using Definition 1.9. □

Proposition 1.4. *The interpolation error is always larger than the truncated error.*

Proof. Using the fact that the aliasing error is orthogonal to the truncation error $\phi(x) - (\mathcal{P}_N\phi)(x)$ along Lemma 1.1, one can write

$$\|u - (\mathcal{I}_N\phi)(x)\|_{\mathcal{L}^2_w(\Omega)} = \|u - (\mathcal{P}_N\phi)(x)\|_{\mathcal{L}^2_w(\Omega)} + \|(\mathcal{R}_N\phi)(x)\|_{\mathcal{L}^2_w(\Omega)}, \tag{1.28}$$

which completes the proof. □

Theorem 1.1 (Section 1.2.2.2 in [33]). *The orthogonal projection of $\phi(x)$ in the space $\mathbb{T}_N(\Omega)$, i.e., $\mathcal{P}_N\phi(x)$, converges to $\phi(x)$ by increasing N. That is*

$$\lim_{N\to\infty} \|(\mathcal{P}_N\phi)(x) - \phi(x)\|_{\mathcal{L}^2_w(\Omega)} \to 0. \tag{1.29}$$

Proof. The proof is given simply by the fact that the space $\mathbb{P}_N(\Omega)$ is complete when $N \to \infty$. □

1.2.3 Problem Statement

The 1-D mass transport of a single chemical with a source term is described by the second-order advection-reaction-dispersion equation (ADE)

$$(\partial_t u)(x,t) = \partial_x \left(-\gamma u + (\partial_x\, \eta(x)\, u)\right)(x,t), \tag{1.30}$$

where $u(x,t)$ is the solute concentration at the point x, and γ and $\eta(x)$ represent the diffusion and drift coefficients, respectively. The ADE (1.30) can be represented by the FPE since it describes the change of the probability density function of the random movements of particles (i.e., the so-called Brownian motion) in both space and time. If the movement of the particles is dependent on past movements, such as the travel distance along the fractal paths requires power law times, then the ADE (1.30) can be derived as [11]

$$(\partial_t u)(x,t) = -\partial_x (\gamma u)(x,t) + \left(\frac{1}{2} + \frac{\beta}{2}\right)_a \mathcal{D}^{\alpha}_x (du)(x,t) + \left(\frac{1}{2} - \frac{\beta}{2}\right)_x \mathcal{D}^{\alpha}_b (\eta u)(x,t), \tag{1.31}$$

where $1 < \alpha \leq 2$ is the order of the fractional RL derivatives and $\beta \in [-1,1]$ denotes the relative weight of forward versus backward transition probability. In the case that $\beta = 0$, the fractional ADE (1.31) describes a spatially symmetric transition, and the FD terms can be represented by the Riesz potential a spatially derivative operator

$$(\partial_t u)(x,t) = -\gamma\partial_x (u)(x,t) + \eta(|\mathcal{D}|^{\alpha}_{(a,b)}\, u)(x,t). \tag{1.32}$$

If $\beta = -1$, then the fractional ADE (1.31) determines the positively skewed transition.

In general, let $\mathcal{F}^{\alpha}_x(\cdot)$ denote a linear operator of spatial differentiation operators in which α is the maximum order of differentiation operators and $s(x,t)$ represents the source terms. Then, the general form for the motion of a concentration field $u(x,t)$ is given by the linear SFFPE

$$(\partial_t u)(x,t) = (\mathcal{F}^{\alpha}_x u)(x,t) + s(x,t), \tag{1.33}$$

subject to the initial and boundary conditions

$$u(x,0) = f(x), \quad x \in \Omega, \tag{1.34a}$$

$$u(a,t) = p(t), \;\; u(b,t) = q(t), \quad t \in \Theta, \tag{1.34b}$$

where $\Omega^* = [a,b]$ and $\Theta = [0,\tau]$.

Hereafter, we will assume that $u(x,t)$ and $s(x,t)$ belong to $\mathcal{H}^r_{w^*}(\Omega^*) \times \mathcal{C}(\Theta)$, $r \geq \lceil \alpha \rceil$, $p(t) \in \mathcal{C}(\Theta)$, and $q(t) \in \mathcal{C}(\Theta)$. Since the domain of $\mathcal{F}^{\alpha}_x$, denoted by $D(\mathcal{F}^{\alpha}_x)$, is assumed to be a dense subspace of $\mathcal{H}^r_{w^*}(\Omega^*)$, this operator is a linear map from $D(\mathcal{F}^{\alpha}_x)$ to $\mathcal{H}^r_{w^*}(\Omega^*)$.

1.3 Chebyshev Spatial Discretization Method for Solving Linear SFFPEs

In the follow-up, we shall adopt the method of lines by discretizing the function in the spatial domain by using the shifted CGL points to solve the linear SFFPE (1.33). For this purpose, the spatial differentiation operator is discretized using the shifted Chebyshev polynomials and their orthogonality property. As a result, the FPDE (1.33) is converted into a system of linear ODEs, where the temporal error can be bounded by using well-developed ODE solvers [34].

1.3.1 Chebyshev Operational Matrix of Fractional Operators

The Chebyshev polynomials of the first kind are a set of eigenfunctions for a Sturm–Liouville equation, and hence they are orthogonal on $\Omega = [-1, 1]$. These orthogonal polynomials are denoted by $T_n(x)$, where n represents the degree of the Chebyshev polynomial, and they satisfy the following three-term recurrence relation according to Favard's theorem (Theorem 1.2 [35]):

$$T_n(x) = 2xT_{n-1}(x) - T_{n-2}(x), \quad n = 2, 3, \dots, \tag{1.35}$$

with $T_0(x) = 1$ and $T_1(x) = x$.

Definition 1.12. The nth-degree shifted Chebyshev polynomial of the first kind $T_n^*(y) : \Omega^* \mapsto [-1, 1]$, $\Omega^* = [a, b]$, is defined by the affine mapping $x = \mu(y - a) + 1$, $\mu = \frac{-2}{b-a}$. Consequently, we have the following recurrence relation

$$T_n^*(y) = 2\left(\mu(y - a) + 1\right) T_{n-1}^*(y) - T_{n-2}^*(y), \tag{1.36}$$

for $n = 2, 3, \dots$, so that, $T_0^*(y) = 1$ and $T_1^*(y) = y$. In addition, the shifted Chebyshev polynomials of the first kind can be also given by

$$T_n^*(y) = 2\left(-\mu(b - x) - 1\right) T_{n-1}^*(y) - T_{n-2}^*(y) \tag{1.37}$$

such that $T_0^*(y) = 1$ and $T_1^*(y) = -\mu(b - y) - 1$.

Proposition 1.5. *Let $\mathbb{T}_N(\Omega)$ define the space of all Chebyshev polynomials $\{T_i(x)\}_{i=0}^{i=N-1}$ that are complete in $\mathcal{L}_w^2(\Omega)$ with respect to the weight function $w(x) = 1/\sqrt{1 - x^2}$. Then, we have*

$$\langle T_n, T_m \rangle_{\mathcal{L}_w^2(\Omega)} = c_n \frac{\pi}{2} \delta_{nm}, \tag{1.38}$$

where δ_{nm} is the well-known Kronocker delta function, $c_n = 2$ for $n = 0$, and $c_n = 1$ for $n \neq 0$.

Proposition 1.6. *Let $\mathbb{T}_N^*(\Omega^*)$ define the space of all shifted Chebyshev polynomials $\{T_i^*(x)\}_{i=0}^{i=N-1}$ that are complete in $\mathcal{L}_{w^*}(\Omega^*)$ with respect to the weight function $w^*(x) = w\left(\mu(x - a) + 1\right)$. Then, we have*

$$\langle T_n^*, T_m^* \rangle_{\mathcal{L}_{w^*}(\Omega^*)} = \frac{1}{\mu} \langle T_n, T_m \rangle_{\mathcal{L}_{w^*}^2(\Omega^*)}. \tag{1.39}$$

Definition 1.13. The regular and shifted CGL points are defined by $\mathbf{x}_{d_k} = \left\{\cos\left(\frac{k\pi}{N-1}\right)\right\}_{k=0}^{N-1} \subseteq \Omega$ and $\mathbf{x}_d^* = \left\{\frac{1}{\mu}\left(\cos\left(\frac{k\pi}{N-1}\right) - 1\right) + a\right\}_{k=0}^{N-1} \subseteq \Omega^*$, respectively.

Proposition 1.7. *The interpolation operator $\mathcal{T}_N^* : \mathcal{L}_{w^*}(\Omega^*) \mapsto \mathbb{T}_N^*(\Omega^*)$ for any square-integrable function $\phi(x) \in \mathcal{L}_{w^*}(\Omega^*)$ can be written in the matrix form*

$$\left(\mathcal{T}_N^* \phi\right)(x) = {\mathbf{T}_N^*}^T(x) H_N^* \boldsymbol{\phi}_d, \tag{1.40}$$

where $\mathbf{T}_N^*(x) = [T_0^*(x), \ldots, T_N^*(x)]^T$, the column-vector $\boldsymbol{\phi}_d = [\phi(x_{d_0}^*), \ldots, \phi(x_{d_{N-1}}^*)]^T$ is the discretized values of $\phi(x)$ at all the shifted CGL points $\mathbf{x}_d^* = \left\{x_{d_i}^*\right\}_{k=0}^{N-1}$ declared in Definition 1.13, and H_N^* is a constant $N \times N$ matrix defined by

$$H_N^* = \frac{1}{N}\begin{bmatrix} \frac{1}{2} & 1 & \cdots & 1 & \frac{1}{2} \\ (-1)^1 & 2T_1^*\left(x_{d_1}^*\right) & \cdots & 2T_1^*\left(x_{d_{N-1}}^*\right) & 1 \\ \vdots & \vdots & \cdots & \vdots & \vdots \\ (-1)^{N-2} & 2T_{N-1}^*\left(x_{d_1}^*\right) & \cdots & 2T_{N-1}^*\left(x_{d_{N-1}}^*\right) & 1 \\ \frac{1}{2}(-1)^{N-1} & T_N^*\left(x_{d_1}^*\right) & \cdots & T_N^*\left(x_{d_{N-1}}^*\right) & \frac{1}{2} \end{bmatrix}. \tag{1.41}$$

Proposition 1.8 (Equation (5.3.2) in [33]). *The following inequality exists between the discretized norms and regular norm in $\mathbb{T}_N(\Omega)$:*

$$\|\cdot\|_{\mathcal{L}_w^2(\Omega)} \leq \|\cdot\|_N \leq \sqrt{2}\,\|\cdot\|_{\mathcal{L}_w^2(\Omega)}\,. \tag{1.42}$$

Definition 1.14. A family of elementary $\alpha-$power functions is defined by

$$\mathbf{E}_N^\alpha(x) = \left[x^\alpha, x^{\alpha+1}, \ldots, x^{\alpha+N-1}\right]^T. \tag{1.43}$$

In addition, the left- and right-sided shifted elementary $\alpha-$degree functions in $\Omega^* = [a, b]$ are defined by $\mathbf{E}_N^\alpha(x-a)$ for $x \geq a$, and $\mathbf{E}_N^\alpha(b-x)$ for $x \leq b$, respectively. In the case that $\alpha = 0$, this family constructs a basis of elementary polynomials, i.e., $\mathbf{E}_N(x) = \{x^n\}_{n=0}^{N-1}$.

Proposition 1.9. *The vector Chebyshev polynomial $\mathbf{T}_N(x) = [T_0(x), \ldots, T_{N-1}(x)]^T$ is a linear map of the elementary polynomials $\mathbf{E}_N(x) = \left[1, x, \ldots, x^{N-1}\right]^T$ using the $N \times N$ map C defined by*

$$\mathbf{T}_N(x) = C_N\, \mathbf{E}_N(x), \tag{1.44}$$

whose first row is $[1, 0, \ldots, 0]$ and the other elements $[C_N]_{nk}$, $k \in \mathbb{N}$ and $n \geq 2$, are defined by

$$[C_N]_{nk} = \begin{cases} \frac{n-1}{2}\frac{(-1)^k 2^{n-2k-1}(n-k-2)!}{k!(n-2k-1)!}, & k \leq \lfloor \frac{n-1}{2} \rfloor, \\ 0, & k > \lfloor \frac{n-1}{2} \rfloor. \end{cases} \tag{1.45}$$

Proposition 1.10. *A square-integrable function $\phi(x) \in \mathcal{L}_{w^*}(\Omega^*)$ can be approximated by the interpolation operator $\mathcal{T}_N^*(\cdot)$ as*

$$\left(\mathcal{T}_N^* \phi\right)(x) = {\mathbf{E}_N^0}^T(x-a)\, {C_{N+}^*}^T\, H_N^*\, \boldsymbol{\phi}_d, \tag{1.46a}$$

$$\left(\mathcal{T}_N^* \phi\right)(x) = {\mathbf{E}_N^0}^T(b-x)\, {C_{N-}^*}^T\, H_N^*\, \boldsymbol{\phi}_d, \tag{1.46b}$$

where $C_{N\pm}^* = C_N\; B_{N\pm}$ and the $N \times N$ matrices $B_{N\pm}$ are determined by means of the binomial theorem so that

$$[B_{N\pm}]_{nk} = \begin{cases} (\pm\mu)^k \begin{pmatrix} n \\ n-k \end{pmatrix}, & k \leq n, \\ 0, & k > n. \end{cases} \tag{1.47}$$

Definition 1.15. The fractional Chebyshev differentiation matrices of the left- and right-sided RL, denoted by D^{α}_{N+} and D^{α}_{N-}, respectively, are $N \times N$ matrices that map the discretized values of any function $\phi(x) \in \mathcal{H}^{r}_{w^*}(\Omega^*)$, $\Omega^* = [a, b]$, to the discretized values of the left- and right-sided α-derivatives of $\phi(x)$ at the shifted CGL collocation points, respectively. Thus,

$$\mathrm{D}^{\alpha}_{N+}\, \boldsymbol{\phi}_d = [({}_a\mathcal{D}^{\alpha}_x\, \phi)(x^*_{d_0}), \ldots, ({}_a\mathcal{D}^{\alpha}_x\, \phi)(x^*_{d_{N-1}})]^T, \tag{1.48a}$$

$$\mathrm{D}^{\alpha}_{N-}\, \boldsymbol{\phi}_d = [({}_x\mathcal{D}^{\alpha}_b\, \phi)(x^*_{d_0}), \ldots, ({}_x\mathcal{D}^{\alpha}_b\, \phi)(x^*_{d_{N-1}})]^T, \tag{1.48b}$$

where $x^*_{d_i}$, $i = 0, \ldots, N-1$, are the shifted Gauss–Lobatto Chebyshev (GLC) collocation points formulated in Definition 1.13.

Lemma 1.2. *The fractional Chebyshev differentiation matrices of the left- and right-sided RL are constructed as*

$$\mathrm{D}^{\alpha}_{N\pm} = {E^{\alpha}_{N\pm}}^T\, \Lambda^{\alpha}_N\, {C^*_{N\pm}}^T\, H^*_N, \tag{1.49}$$

where the binomial matrix Λ^{α}_N is a diagonal matrix with elements $[\Lambda]_{kk} = {}^{\Gamma(k+1)}/_{\Gamma(k+1-\alpha)}$, $k = 0, \ldots, N-1$, the matrices $C^*_{N\pm}$ are defined in Equation (1.46),

$$E_+ = \begin{bmatrix} (x^*_{d_0} - a)^{-\alpha} & (x^*_{d_0} - a)^{1-\alpha} & \cdots & (x^*_{d_0} - a)^{N-1-\alpha} \\ (x^*_{d_1} - a)^{-\alpha} & (x^*_{d_1} - a)^{1-\alpha} & \cdots & (x^*_{d_1} - a)^{N-1-\alpha} \\ \vdots & \vdots & \cdots & \vdots \\ (x^*_{d_{N-2}} - a)^{-\alpha} & (x^*_{d_{N-2}} - a)^{1-\alpha} & \cdots & (x^*_{d_{N-2}} - a)^{N-1-\alpha} \\ (x^*_{d_{N-1}} - a)^{-\alpha} & (x^*_{d_{N-1}} - a)^{1-\alpha} & \cdots & (x^*_{d_{N-1}} - a)^{N-1-\alpha} \end{bmatrix}, \tag{1.50}$$

and E_- is an $N \times N$ matrix whose rows are flipped rows of E_+ in the up-down direction.

Proof. Let $\phi(x) \in \mathcal{L}_{w^*}(\Omega^*)$, $\Omega^* = [a, b]$. According to Proposition 1.10, the function can be approximated by using the left- and right-sided shifted elementary zero-degree functions given in Equation (1.46). In addition, by using Property 1.2, it can be shown that the left- and right-sided FDs of $\mathbf{E}^0_N(x-a)$ and $\mathbf{E}^0_N(b-x)$ in the sense of RL are

$${}_a\mathcal{D}^{\alpha}_x\left(\mathbf{E}^0_N(x-a)\right) = \Lambda^{\alpha}_N\, \mathbf{E}^{-\alpha}_N(x-a), \tag{1.51}$$

$${}_x\mathcal{D}^{\alpha}_b\left(\mathbf{E}^0_N(b-x)\right) = \Lambda^{\alpha}_N\, \mathbf{E}^{-\alpha}_N(b-x). \tag{1.52}$$

Consequently, we have

$$\left({}_a\mathcal{D}^{\alpha}_x\left(\mathcal{T}_N\phi\right)\right)(x) = {}_a\mathcal{D}^{\alpha}_x\left(\mathbf{E}^T_N(x-a)\right)\, {C^*_+}^T\, H^*_N\, \boldsymbol{\phi}_d = {\mathbf{E}^{-\alpha}_N}^T(x-a)\, \Lambda^{\alpha}_N\, {C^*_+}^T\, H^*_N\, \boldsymbol{\phi}_d \tag{1.53}$$

and

$$\left({}_x\mathcal{D}^{\alpha}_b\left(\mathcal{T}_N\phi\right)\right)(x) = {}_x\mathcal{D}^{\alpha}_b\left(\mathbf{E}^T_N(b-x)\right)\, {C^*_-}^T\, H^*_N\, \boldsymbol{\phi}_d = {\mathbf{E}^{-\alpha}_N}^T(b-x)\, \Lambda^{\alpha}_N\, {C^*_-}^T\, H^*_N\, \boldsymbol{\phi}_d. \tag{1.54}$$

Evaluating the Equations (1.53) and (1.54) at each point of the shifted CGL points gives Equation (1.49). Finally, the symmetric behavior of $\mathbf{E}^0_N(x-a)$ and $\mathbf{E}^0_N(b-x)$ at the shifted-CGL collocation points in $\Omega^* = [a, b]$ completes the proof. □

Consequently, the fractional Chebyshev differentiation matrix in the sense of Riesz is given by

$$|\mathrm{D}_N|^{\alpha} = -c_{\lceil\alpha\rceil-\alpha}\left(\mathrm{D}^{\alpha}_{N+} + \mathrm{D}^{\alpha}_{N-}\right). \tag{1.55}$$

1.3.2 Spatial Discretization of SFFPEs

Let $u^N(x,t)$ represent a short notation for the spatial interpolation operator $(\mathcal{I}_{N,x}u)(x,t)$ of $u(x,t) \in \mathcal{H}^r_{w^*}(\Omega^*) \times \mathcal{C}(\Theta)$, where the subscript "$x$" in the interpolation operators indicates the interpolation in the spatial space only. Then, using the interpolation operator, we can approximate the solution $s(x,t)$ and the source term as

$$u(x,t) \approx u^N(x,t) = (\mathcal{I}_{N,x}u)(x,t) = \mathbf{T}^{*T}(x) H^*_N \, \boldsymbol{u}_d(t), \tag{1.56a}$$

$$s(x,t) \approx s^N(x,t) = (\mathcal{I}_{N,x}s)(x,t) = \mathbf{T}^{*T}(x) H^*_N \, \boldsymbol{s}_d(t), \tag{1.56b}$$

where $\boldsymbol{u}_d(t) = [u_{d_0}(t), \ldots, u_{d_{N-1}}(t)]^T$ and $\boldsymbol{s}_d(t) = [s_{d_0}(t), \ldots, s_{d_{N-1}}(t)]^T$, so that $u_{d_i}(t)$ and $s_{d_i}(t)$ stand for the short representations for $u(x^*_{d_i}, t)$ and $s(x^*_{d_i}, t)$, respectively, and $x^*_{d_i}$, $i = 0, \ldots, N-1$, are the shifted CGL points in Definition 1.13.

Substituting the approximate solution in the SFFPE (1.33) gives

$$R^N(x,t) = \left(\partial_t u^N\right)(x,t) - (\mathcal{F}^\alpha_x u^N)(x,t) + s^N(x,t), \tag{1.57}$$

for which $R^N(x^*_j, t) = 0$, $j = 0, \ldots, N-1$. Since $\mathbf{T}^{*T}(x) H^*_N$ is equivalent to the Lagrange interpolation polynomials $\{l_i(x)\}_{i=0}^{N-1}$ satisfying $l_i(x^*_j) = \delta_{ij}$, we have

$$\dot{u}_{d_j}(t) = \mathcal{F}^\alpha_x \left(\mathbf{T}^{*T}(x^*_j)\,\mathrm{col}_j(H^*_N) u_{d_j}\right)(t) + \mathbf{T}^{*T}(x^*_j)\,\mathrm{col}_j(H^*_N)\; s_{d_j}(t), \tag{1.58}$$

for $j = 0, \ldots, N-1$, where $\mathrm{col}_n(A)$ gives the nth column of matrix A. Moreover, the spatial differentiation operator $\mathcal{F}^\alpha_x$ is linear, and we have

$$\mathcal{F}^\alpha_x \left(\mathbf{T}^{*T}(x^*_j) H^*_N\right) \boldsymbol{u}_d(t) = \mathrm{F}_N \, \boldsymbol{u}_d(t), \tag{1.59}$$

where the operator matrix of the spatial differentiation $\mathcal{F}_x(\cdot)$ is defined as

$$\mathrm{F}^\alpha_N = \begin{bmatrix} \mathcal{F}^\alpha_x(T^*_0)(x^*_{d_0}) & \mathcal{F}^\alpha_x(T^*_1)(x^*_{d_0}) & \cdots & \mathcal{F}^\alpha_x(T^*_{N-1})(x^*_{d_0}) \\ \mathcal{F}^\alpha_x(T^*_0)(x^*_{d_1}) & \mathcal{F}^\alpha_x(T^*_1)(x^*_{d_1}) & \cdots & \mathcal{F}^\alpha_x(T^*_{N-1})(x^*_{d_1}) \\ \vdots & \vdots & \cdots & \vdots \\ \mathcal{F}^\alpha_x(T^*_0)(x^*_{d_{N-1}}) & \mathcal{F}^\alpha_x(T^*_1)(x^*_{d_{N-1}}) & \cdots & \mathcal{F}^\alpha_x(T^*_{N-1})(x^*_{d_{N-1}}) \end{bmatrix} H^*_N. \tag{1.60}$$

Thus, the result of discretization will be the system of ODEs

$$\dot{\mathbf{u}}_d(t) = \mathrm{F}^\alpha_N \, \boldsymbol{u}_d(t) + \mathbf{s}_d(t), \tag{1.61}$$

that can be solved by well-known numerical integration methods, such as Runge–Kutta algorithms. To consider the boundary condition (1.34b), let I_N represent a $N \times N$ identity matrix, $\bar{\underline{I}}_N$ stands for a $N \times N$ identity matrix whose first and last elements are replaced by zero, $\bar{K}_N$ be a $N \times 1$ matrix of zeros whose first element is replaced by one, and $\underline{K}_N$ be a $N \times 1$ matrix of zeros whose last element is replaced by one. Thus, the approximate solution of SFFPEs (1.33) is obtained by the system of ODEs (1.62)

$$\dot{\mathbf{u}}_d(t) = \bar{\underline{I}}_N \, \mathrm{F}^\alpha_N \, \boldsymbol{u}_d(t) + \bar{\underline{I}}_N \mathbf{s}_d(t) + \bar{K}_N \dot{p}(t) + \underline{K}_N \dot{q}(t) \tag{1.62}$$

subject to the initial condition $u_{d_i}(0) = f_{d_i}$, for $i = 0, \ldots, N-1$.

1.4 Convergence Analysis

The convergence analysis of FDEs is more complex than the one for ODEs, and different approaches have been proposed to bound their approximation errors [36–41]. In this section, we prove that the proposed spatial discretization scheme converges in the spatial coordinate if all the exact solution and the coefficients belong to $\mathcal{H}^r_{w^*}(\Omega^*)$ such that $r \geq \lceil\alpha\rceil$.

Theorem 1.2 (Section 5.5.3 in [33]). *Suppose that $\mathcal{H}^r_w(\Omega)$ is the Hilbert space associated with the Sobolev norm $\|\cdot\|_{\mathcal{H}^r_w(\Omega)}$ and seminorm $|\cdot|_{\mathcal{H}^{r;m}_w(\Omega)}$ presented in Definition 1.4. Also, let $\tilde{\phi}(x)$ denote the discretization error of $\phi(x)$, i.e., $\tilde{\phi}^N(x) = \phi(x) - \phi^N(x)$. Then, the interpolation error $\tilde{\phi}^N(x)$ at the GLC points for the function $\phi(x) \in \mathcal{H}^r_w(\Omega)$ holds the following inequality*

$$\left\|\tilde{\phi}^N(x)\right\|_{\mathcal{H}^l_w(\Omega)} \leq cN^{2l-r}\left|\phi(x)\right|_{\mathcal{H}^{r;N}_w(\Omega)}, \quad 0 \leq l \leq r, \tag{1.63}$$

where $c \in \mathbb{R}^+$.

Corollary 1.1. *Suppose that $\mathcal{H}^r_{w^*}(\Omega^*)$ is the Hilbert space associated with the Sobolev norm $\|\cdot\|_{\mathcal{H}^r_{w^*}(\Omega^*)}$ and seminorm $|\cdot|_{\mathcal{H}^{r;m}_{w^*}(\Omega^*)}$ formulated in Definition 1.4. Also, let $\tilde{\phi}(x)$ denote the discretization error of $\phi(x)$, i.e., $\tilde{\phi}^N(x) = \phi(x) - \phi^N(x)$. Then, the interpolation error $\tilde{\phi}^N(x)$ at the shifted GLC points for the function $\phi(x) \in \mathcal{H}^r_{w^*}(\Omega^*)$ holds the following inequality*

$$\left\|\tilde{\phi}^N(x)\right\|_{\mathcal{H}^l_{w^*}(\Omega^*)} \leq cN^{2l-r}\mu^{-2l+\min(r,N)}\left|\phi(x)\right|_{\mathcal{H}^{r;N}_{w^*}(\Omega^*)}, \quad 0 \leq l \leq r, \tag{1.64}$$

where $c \in \mathbb{R}^+$ and μ is defined in Equation (1.36).

Proof. The proof is obtained by replacing the integral in the norm definition over $\Omega = [-1, 1]$ by the one over $\Omega^* = [a, b]$. □

Proposition 1.11 ([33]). *Let $2 \leq p \leq \infty$ and $\phi(x) \in \mathbb{T}_N(\Omega)$. For any integer $r \geq 1$, there exists a positive constant c such that*

$$\left\|\partial^r_x\phi(x)\right\|_{\mathcal{L}^p_w(\Omega)} \leq cN^{2r}\left\|\phi(x)\right\|_{\mathcal{L}^p_w(\Omega)}. \tag{1.65}$$

Theorem 1.3. *Let $\Omega^* = [a, b]$. For any $\phi(x) \in \mathcal{H}^l_{w^*}(\Omega^*)$, $l \geq \lceil\alpha\rceil$, with $\phi(a) = 0$, we have*

$$\left\|\left({}^R_a\mathcal{D}^\alpha_t\phi\right)(x)\right\|_{\mathcal{H}^{l-\lceil\alpha\rceil}_{w^*}(\Omega^*)} \leq c\left\|\phi(x)\right\|_{\mathcal{H}^l_{w^*}(\Omega^*)}, \tag{1.66}$$

where $c = \frac{(b-a)^{\lceil\alpha\rceil-\alpha+0.5}}{\Gamma(\lceil\alpha\rceil-\alpha+1)\sqrt{(2\lceil\alpha\rceil-2\alpha+1)}}$.

Proof. Using the Sobolev norm Definition 1.3, we have

$$\begin{aligned}
\left\|\left({}^R_a\mathcal{D}^\alpha_x\phi\right)(x)\right\|^2_{\mathcal{H}^r_{w^*}(\Omega^*)} &= \sum_{k=0}^{r}\left\|\left(\partial^k_x\left({}^R_a\mathcal{D}^\alpha_x\phi\right)\right)(x)\right\|^2_{\mathcal{L}^2_{w^*}(\Omega^*)} \\
&= \sum_{k=0}^{r}\left\|\left({}^R_a\mathcal{D}^{k+\alpha}_x\phi\right)(x)\right\|^2_{\mathcal{L}^2_{w^*}(\Omega^*)} \\
&= \sum_{k=0}^{r}\int_{\Omega^*}\left(\frac{1}{\Gamma(\lceil\alpha\rceil-\alpha)}\partial^{k+\lceil\alpha\rceil}_x\int_a^x\frac{\phi(\tau)}{(x-\tau)^{\alpha+1-\lceil\alpha\rceil}}d\tau\right)^2 w^*(x)dx \\
&= \left|\frac{1}{\Gamma(\lceil\alpha\rceil-\alpha)}\int_a^x\frac{\phi(\tau)}{(x-\tau)^{\alpha+1-\lceil\alpha\rceil}}d\tau\right|^2_{\mathcal{H}^{r+\lceil\alpha\rceil,\lceil\alpha\rceil-1}_{w^*}(\Omega^*)}.
\end{aligned} \tag{1.67}$$

Using the integration-by-part technique and then the Cauchy–Schwarz inequality yields

$$\begin{aligned}
\left\|\left({}_{a}^{R}\mathcal{D}_{x}^{\alpha}\phi\right)(x)\right\|_{\mathcal{H}_{w^*}^{r}(\Omega^*)}^{2} &= \left|\frac{1}{\Gamma(\lceil\alpha\rceil-\alpha+1)}\int_{a}^{x}\partial_{\tau}\phi(\tau)(x-\tau)^{\lceil\alpha\rceil-\alpha}d\tau\right|_{\mathcal{H}_{w^*}^{r+\lceil\alpha\rceil,\lceil\alpha\rceil-1}(\Omega^*)}^{2} \\
&\leq \sum_{k=\lceil\alpha\rceil}^{r+\lceil\alpha\rceil} C^2\int_{\Omega}\left(\partial_x^k\phi(x)\right)^2 w^*(x)dx \\
&= C^2\sum_{k=0}^{r}\int_{\Omega}\left(\partial_x^k\partial_x^{\lceil\alpha\rceil}\phi(x)\right)^2 w^*(x)dx \\
&= C^2\left\|\partial_x^{\lceil\alpha\rceil}\phi(x)\right\|_{\mathcal{H}_{w^*}^{r}(\Omega^*)}^{2} = C^2\left\|\phi(x)\right\|_{\mathcal{H}_{w^*}^{r+\lceil\alpha\rceil}(\Omega^*)}^{2}. \qquad (1.68)
\end{aligned}$$

Replacing r by $l-\lceil\alpha\rceil$ ends the proof of Theorem 1.3. □

Corollary 1.2. *Let $\Omega^* = [a,b]$. For any $\phi(t) \in \mathcal{H}_{w^*}^{r+\lceil\alpha\rceil}(\Omega^*)$ when $\phi(a)=\phi(b)=0$, we have*

$$\left\|\left({}_{x}^{R}\mathcal{D}_{b}^{\alpha}\phi\right)(x)\right\|_{\mathcal{H}_{w^*}^{r}(\Omega^*)} = \left\|\left({}_{a}^{R}\mathcal{D}_{x}^{\alpha}\phi\right)(x)\right\|_{\mathcal{H}_{w^*}^{r}(\Omega^*)}. \qquad (1.69)$$

Proof. The proof is similar to the one for Theorem 1.3. □

Theorem 1.4. *Suppose that $u(x,t) \in \mathcal{H}_{w^*}^{r}(\Omega^*)$, where $r \geq \lceil\alpha\rceil$ and $\tilde{u}^N(x,t)$ denotes the spatial discretization error of $u(x,t)$, i.e., $\tilde{u}^N(x,t) = u(x,t) - u^N(x,t)$. Then, for $\forall t \in \Theta$,*

$$\lim_{N\to\infty}\left\|\tilde{u}^N(x_{d_i},t)\right\|_{\mathcal{L}_{w^*}^{2}(\Omega^*)} = 0, \quad i = 0,1,\ldots,N-1, \qquad (1.70)$$

where the x_{d_i}'s represent the shifted collocation points.

Substituting the approximate solution $u^N(x,t)$ in the SFFPE (1.33) gives the residual function (1.57) which is vanished at the shifted collocation points using the proposed method, i.e., $R^N(x_j^*,t) = 0$, $j = 0,\ldots,N-1$. Subtracting the resultant equation from the SFFPE (1.33) one obtains

$$\left(\partial_t\tilde{u}^N\right)(x,t) = R^N(x,t) + \left(\mathcal{F}_x^{\alpha}\tilde{u}^N\right)(x,t) + \tilde{s}^N(x,t). \qquad (1.71)$$

Taking the norm $\|\cdot\|_{\mathcal{H}_{w^*}^{l}(\Omega^*)}$ and considering the $\mathcal{F}_x(\cdot)$ is a linear operator of spatial differentiation operators along with using Definition 1.4, and Theorems 1.2 and 1.3 yields

$$\begin{aligned}
\partial_t\left\|\tilde{u}^N(x,t)\right\|_{\mathcal{H}_{w^*}^{l}(\Omega^*)} &\leq k(x,t) + c_{F,\alpha}(t)\left\|\tilde{u}^N(x,t)\right\|_{\mathcal{H}_{w^*}^{l-\lceil\alpha\rceil}(\Omega^*)} \\
&\leq k(x,t) + c_{F,\alpha}(t)\left\|\tilde{u}^N(x,t)\right\|_{\mathcal{H}_{w^*}^{l}(\Omega^*)}, \qquad (1.72)
\end{aligned}$$

where $c_s(t)$ and $c_{F,\alpha}$ are positive bounded functions in $\mathbb{R}^+$ and

$$k(x,t) = \left\|R^N(x,t)\right\|_{\mathcal{H}_{w^*}^{l}(\Omega^*)} + c_s(t)N^{2l-r}\mu^{-2l+\min(r,N)}\left|s(x,t)\right|_{\mathcal{H}_{w^*}^{r;N}(\Omega^*)}. \qquad (1.73)$$

Integrating from both sides yields

$$\left\|\tilde{u}^N(x,t)\right\|_{\mathcal{H}_{w^*}^{l}(\Omega^*)} \leq K(x,t) + \int_a^t c_{F,\alpha}(s)\left\|\tilde{u}^N(x,s)\right\|_{\mathcal{H}_{w^*}^{l}(\Omega^*)}ds,$$

where $K(x,t) = \int_a^t k(x,s)ds$. Using the Grönwall's inequality gives

$$\left\|\tilde{u}^N(x,t)\right\|_{\mathcal{H}_{w^*}^{l}(\Omega^*)} \leq K(x,t) + \int_a^t K(x,s)c_{F,\alpha}(s)\exp\left(\int_s^t c_{F,\alpha}(r)\,\mathrm{d}r\right)\mathrm{d}s. \qquad (1.74)$$

Next, we prove that $k(x,t)$ goes to 0 at the shifted collocation points when $N \to \infty$. The residual function $R(x,t)$ vanishes at the shifted collocation points, and hence

$$k(x_{d_i},t) = c_s(t)N^{2l-r}\mu^{-2l+\min(r,N)} \left|s(x_{d_i},t)\right|_{\mathcal{H}_{w^*}^{r;N}(\Omega^*)}. \tag{1.75}$$

Setting $l = 0$ to define the $\|\cdot\|_{\mathcal{L}^2_{w^*}(\Omega^*)}$-norm, and considering that $r \geq \lceil \alpha \rceil$, we verify that $k(x_{d_i},t)$ goes to zero for large enough N. Consequently, $K(x_{d_i},t)$ and $\left\|\tilde{u}^N(x,t)\right\|_{\mathcal{L}^2_{w^*}(\Omega^*)}$ converge to zero, and this completes the proof of Theorem 1.4.

1.5 Numerical Examples

In this section, the convergence and performance of the proposed scheme are illustrated by means of two examples. The results are compared with the same recently proposed methods in [42]. All the computations are carried out by MATLAB® running on a computer with Intel(R) Core(TM) i7-8650U @1.90–2.11 GHz CPU, 16 GB RAM, under the 64-bit version of Windows 10 operating system. In addition, we use the function `ode45` in MATLAB, which is an efficient standard solver for ODE initial value problems using the explicit Runge–Kutta method with variable step. The `ode45`-solver is used with the relative error tolerance of 10^{-10} and the absolute error tolerance of 10^{-10}. We calculate the values of $\max_{t\in\Theta}\{\|\cdot\|_{\mathcal{L}^2_{w^*}(\Omega^*)}\}$-norm error and temporal and spatial convergence orders. We use Proposition 1.8 to compute this norm.

Example 1.1. Consider the linear SFFPE (1.33) when $T = 10$, $a = 0$, $b = 1$, $\mathcal{F}_x^\alpha \equiv d\frac{\partial^\alpha}{\partial|x|^\alpha}$, $d = 1$, and the source term is

$$\begin{aligned} s(x,t) = {} & \frac{D}{2\cos\left(\frac{\pi\beta}{2}\right)}\left(\frac{2\,(t+1)^\beta\,(1-x)^{2-\beta}}{\Gamma(3-\beta)} - \frac{12\,(t+1)^\beta\,(1-x)^{3-\beta}}{\Gamma(4-\beta)}\right. \\ & \left. + \frac{24\,(t+1)^\beta\,(1-x)^{4-\beta}}{\Gamma(5-\beta)} + \frac{2\,x^{2-\beta}\,(t+1)^\beta}{\Gamma(3-\beta)} - \frac{12\,x^{3-\beta}\,(t+1)^\beta}{\Gamma(4-\beta)} + \frac{24\,x^{4-\beta}\,(t+1)^\beta}{\Gamma(5-\beta)}\right) \\ & + \beta(t+1)^{\beta-1}\,x^2\,(x-1)^2 \end{aligned} \tag{1.76}$$

subject to the initial and boundary conditions

$$f(x) = (x(1-x))^2, \quad p(t) = q(t) = 0. \tag{1.77}$$

The analytical solution of this problem is $u(x,t) = (t+1)^\beta(x(1-x))^2$ that is used to compare with the numerical approximation. This problem is solved in [42] using the McCormack numerical method, and the obtained results were, at the best, in the order of 10^{-4}.

Figure 1.1a shows the obtained numerical results for Example 1.1 using the proposed method for $N = 24$ and $\beta = 2$. Moreover, Figure 1.1b shows the approximations for different values of β when $t = 10$. In all the cases of β, the response increases over time symmetrically in the spatial coordinate.

Figure 1.2 shows the variation of $\max_{t\in\Theta}\{\|\cdot\|_{\mathcal{L}^2_{w^*}(\Omega^*)}\}$-norm for the approximation error and the computation time versus different values of N and β. The order of errors is around 10^{-11} for all the cases, and increasing β increases the error in general. It is shown that increasing the number of collocation points up to $N = 6$ reduces the error exponentially. After that the

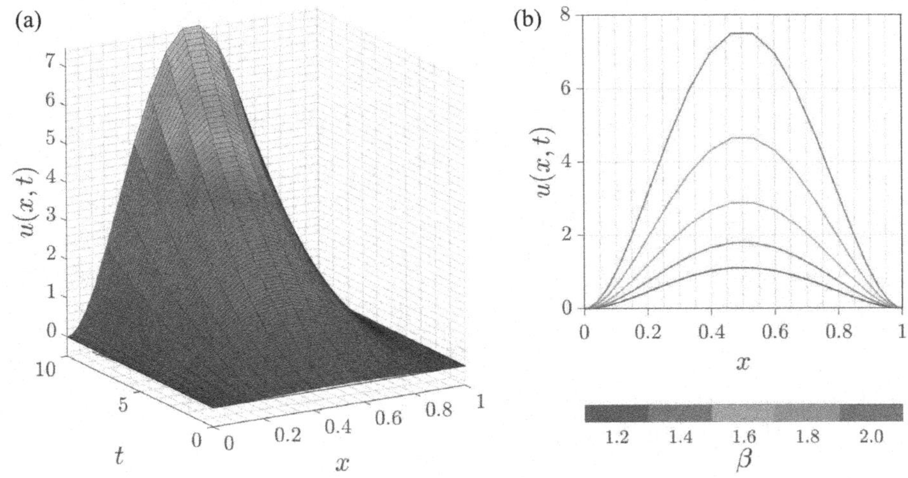

FIGURE 1.1
The approximate solution $u(x,t)$ of Example 1.1 obtained by the proposed method when $N = 24$; (a) $\beta = 2$, (b) $t = 10$ and $\beta = \{1.2, 1.4, 1.6, 1.8, 2.0\}$.

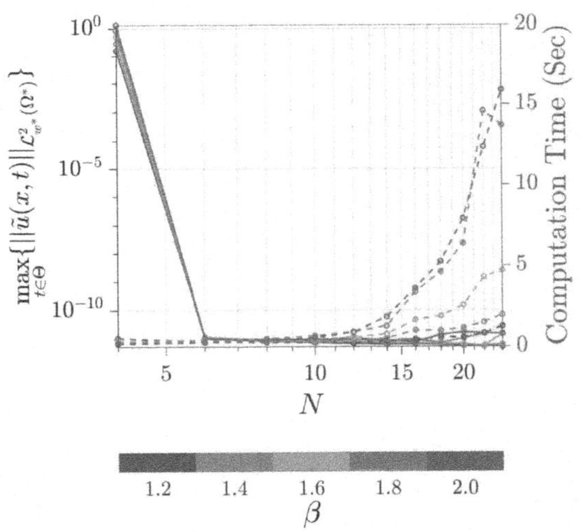

FIGURE 1.2
The maximum error (solid lines) and the computation time (dashed lines) of Example 1.1 obtained by the proposed method for different values of N and $\beta = \{1.2, 1.4, 1.6, 1.8, 2.0\}$.

error stays at the order of the ode-solver absolute tolerance, i.e., 10^{-10}. The main reason for obtaining this steady-state overall error for smaller Ns is because the constant coefficient model in Example 1.1 has the exact solution whose spatial part is a polynomial of degree 4 that is exact in the space $\mathbb{T}^*_N(\Omega*)$ wherein $N \geq 4$. Moreover, the error increment for $N > 20$ is due to the growth of the round-off error for calculating the differentiation matrix as well as the increase in the size of the system of ODEs.

Example 1.2. In this example, we intentionally replace the drift coefficient in Example 1.1 with a varying one as $d(x,t) = (\log(x+2))^\beta$ and set the source term as

$$
\begin{aligned}
s(x,t) = {} & \frac{(\log(x+2))^{\beta}}{\cos\left(\frac{\pi\beta}{2}\right)} \left(\frac{12(t+1)^{\beta}(1-x)^{4-\beta}}{\Gamma(5-\beta)} - \frac{180(t+1)^{\beta}(1-x)^{5-\beta}}{\Gamma(6-\beta)} \right. \\
& + \frac{1080(t+1)^{\beta}(1-x)^{6-\beta}}{\Gamma(7-\beta)} - \frac{2520(t+1)^{\beta}(1-x)^{7-\beta}}{\Gamma(8-\beta)} + \frac{3x^{3-\beta}(t+1)^{\beta}}{\Gamma(4-\beta)} \\
& - \frac{48x^{4-\beta}(t+1)^{\beta}}{\Gamma(5-\beta)} + \frac{360x^{5-\beta}(t+1)^{\beta}}{\Gamma(6-\beta)} - \frac{1440x^{6-\beta}(t+1)^{\beta}}{\Gamma(7-\beta)} \\
& \left. + \frac{2520x^{7-\beta}(t+1)^{\beta}}{\Gamma(8-\beta)} \right) + \beta x^{3}(t+1)^{\beta-1}(x-1)^{4}
\end{aligned} \tag{1.78}
$$

to depict the exponential convergence in the spatial coordinate. In addition, we set $T = 10$, $a = 0$, $b = 1$, and the initial and boundary conditions as

$$f(x) = x^3(1-x)^4, \quad p(t) = q(t) = 0, \tag{1.79}$$

The analytical solution of this problem is $u(x,t) = (t+1)^{\beta}(1-x)^4x^3$, is now of degree 7, and is expected to reach the exact solution in $\mathbb{T}^*_N(\Omega*)$ when $N \geq 7$.

Figure 1.3a shows the obtained results for Example 1.1 using the proposed scheme when $N = 26$ and $\beta = 2$. Figure 1.3b illustrates the final time response for different values of β. Increasing the fractional order increases the response over time.

Figure 1.4 shows that the approximation error exhibits exponential convergence for different values of N and β when $N \leq 7$. Similar to the previous example, when $N \geq 7$, the approximated solution in the spatial space is exact as the spatial part of the exact solution is a polynomial of degree 7.

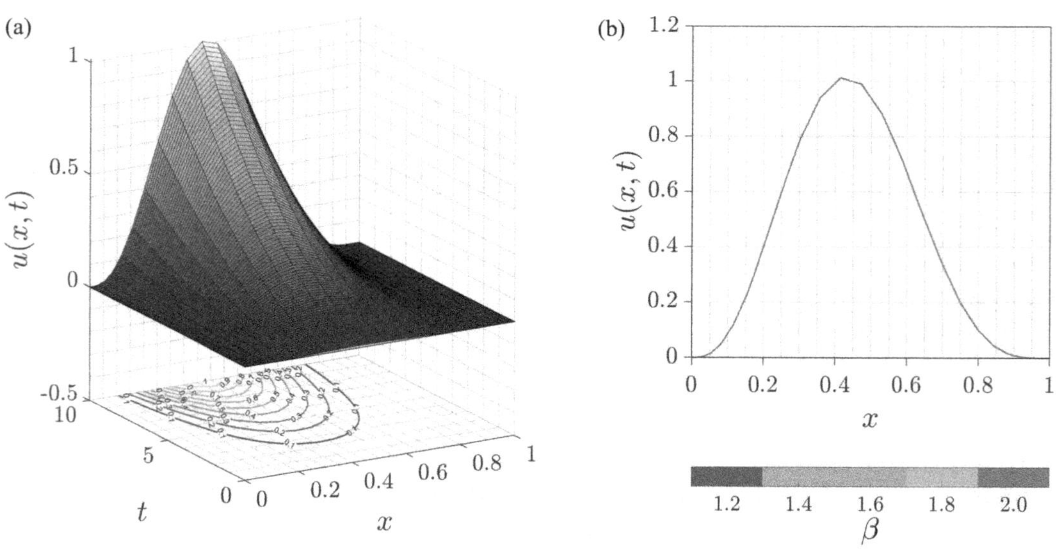

FIGURE 1.3
The approximate solution of Example 1.2 obtained by the proposed method when $N = 26$; (a) $\beta = 2$, (b) $t = 10$ and $\beta = \{1.2, 1.4, 1.6, 1.8, 2.0\}$.

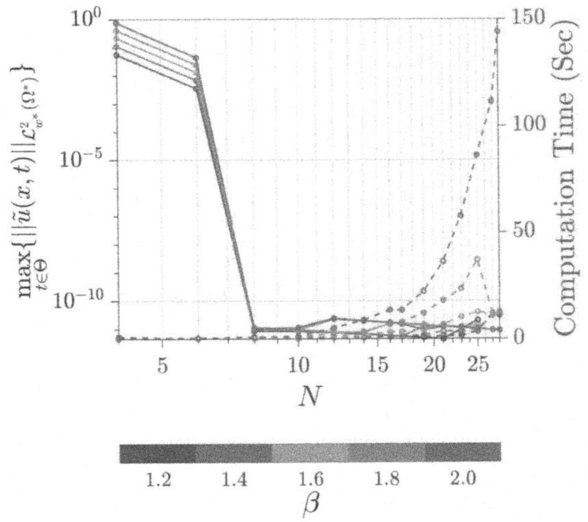

FIGURE 1.4
The maximum error (solid lines) and the computation time (dashed lines) of Example 1.1 obtained by the proposed method for different values of N and β.

1.6 Conclusion

Anomalous diffusion processes can be described by FPDEs known as FFPEs. Solving FPDEs requires more attention than regular PDEs because of the complexity in the definition of FD operators. In this chapter, we considered a semispectral collocation method to solve SFFPEs with fractional operators applied only on the spatial coordinate. This was accomplished by approximating the solution on the spatial coordinate using shifted-Chebyshev polynomials of the first kind. Moreover, the operational matrices of fractional differentiation for the left- and right-sided FDs in the sense of RL were constructed. Consequently, the SFFPEs are converted to a system of ODEs after considering the boundary conditions, that can be solved by any conventional ODE solvers. Moreover, we analytically showed that the discretization scheme exhibits spectral convergence in the spatial space regardless of the choice of the ODE solver. Two examples were presented to illustrate the efficiency of the proposed method for solving SFFPEs with Riesz operators.

References

[1] A. Einstein, "On the motion of small particles suspended in liquids at rest required by the molecular-kinetic theory of heat," *Annalen Der Physik*, vol. 17, pp. 549–560, 1905.

[2] H. Risken, *Fokker-Planck Equation.* Springer, New York, 1996.

[3] Y. Liu and B. Xin, "Numerical solutions of a fractional predator-prey system," *Advances in Difference Equations*, vol. 2011, no. 1, p. 190475, 2011.

[4] H. Rudolf, *Applications of Fractional Calculus in Physics*. World Scientific, Singapore, 2000.

[5] I. M. Sokolov and J. Klafter, "Field-induced dispersion in subdiffusion," *Physical Review Letters*, vol. 97, no. 14, p. 140602, 2006.

[6] J. Janczura and A. Wylomańska, "Subdynamics of financial data from fractional Fokker–Planck equation," 2009.

[7] H. Scher and M. Lax, "Stochastic transport in a disordered solid. I. theory," *Physical Review B*, vol. 7, no. 10, p. 4491, 1973.

[8] T. Solomon, E. R. Weeks, and H. L. Swinney, "Observation of anomalous diffusion and Lévy Flights in a two-dimensional rotating flow," *Physical Review Letters*, vol. 71, no. 24, p. 3975, 1993.

[9] R. Metzler and J. Klafter, "The random walk's guide to anomalous diffusion: A fractional dynamics approach," *Physics Reports*, vol. 339, no. 1, pp. 1–77, 2000.

[10] D. A. Benson, S. W. Wheatcraft, and M. M. Meerschaert, "The fractional-order Governing equation of Lévy Motion," *Water Resources Research*, vol. 36, no. 6, pp. 1413–1423, 2000.

[11] D. A. Benson, S. W. Wheatcraft, and M. M. Meerschaert, "Application of a fractional advection-dispersion equation," *Water Resources Research*, vol. 36, no. 6, pp. 1403–1412, 2000.

[12] P. Straka, "Variable order fractional Fokker–Planck equations derived from continuous time random walks," *Physica A: Statistical Mechanics and Its Applications*, vol. 503, pp. 451–463, 2018.

[13] E. Heinsalu, M. Patriarca, I. Goychuk, G. Schmid, and P. Hänggi, "Fractional Fokker–Planck dynamics: Numerical algorithm and simulations," *Physical Review E*, vol. 73, no. 4, p. 046133, 2006.

[14] Q. Yang, F. Liu, and I. Turner, "Computationally efficient numerical nethods for time- and space-fractional Fokker–Planck equations," *Physica Scripta*, vol. 2009, no. T136, p. 014026, 2009.

[15] P. Zhuang, F. Liu, V. Anh, and I. Turner, "Numerical treatment for the fractional Fokker–Planck equation," *Anziam Journal*, vol. 48, pp. 759–774, 2006.

[16] M. M. Meerschaert and A. Sikorskii, *Stochastic Models for Fractional Calculus*, vol. 43. Walter De Gruyter, Berlin, 2011.

[17] M. M. Meerschaert, E. Nane, P. Vellaisamy, "Fractional Cauchy problems on bounded domains," *The Annals of Probability*, vol. 37, no. 3, pp. 979–1007, 2009.

[18] M. M. Meerschaert, E. Nane, and P. Vellaisamy, "Distributed-order fractional Cauchy problems on bounded domains," ArXiv Preprint ArXiv:0912.2521, 2009.

[19] M. Meerschaert, E. Nane, and P. Vellaisamy, "Transient anomalous sub-diffusion on bounded domains," *Proceedings of the American Mathematical Society*, vol. 141, no. 2, pp. 699–710, 2013.

[20] J. Mijena and E. Nane, "Strong analytic solutions of fractional Cauchy problems," *Proceedings of the American Mathematical Society*, vol. 142, no. 5, pp. 1717–1731, 2014.

[21] R. Metzler, E. Barkai, and J. Klafter, "Anomalous diffusion and relaxation close to thermal equilibrium: a fractional Fokker–Planck equation approach," *Physical Review Letters*, vol. 82, no. 18, p. 3563, 1999.

[22] S. Chen, F. Liu, P. Zhuang, and V. Anh, "Finite difference approximations for the fractional Fokker–Planck equation," *Applied Mathematical Modelling*, vol. 33, no. 1, pp. 256–273, 2009.

[23] Y. Jiang, "A new analysis of stability and convergence for finite difference schemes solving the time fractional Fokker–Planck equation," *Applied Mathematical Modelling*, vol. 39, no. 3-4, pp. 1163–1171, 2015.

[24] F. Liu, V. Anh, and I. Turner, "Numerical solution of the space fractional Fokker–Planck equation," *Journal of Computational and Applied Mathematics*, vol. 166, no. 1, pp. 209–219, 2004.

[25] W. Deng, "Numerical algorithm for the time fractional Fokker–Planck equation," *Journal of Computational Physics*, vol. 227, no. 2, pp. 1510–1522, 2007.

[26] Y. Yang, Y. Huang, and Y. Zhou, "Numerical solutions for solving time fractional Fokker–Planck equations based on spectral collocation methods," *Journal of Computational and Applied Mathematics*, vol. 339, pp. 389–404, 2018.

[27] C. Canuto, M. Y. Hussaini, A. Quarteroni, A. Thomas Jr, et al., *Spectral Methods in Fluid Dynamics*. Springer Science & Business Media, New York, 2012.

[28] J. C. Slater, "Electronic energy bands in metals," *Physical Review*, vol. 45, no. 11, p. 794, 1934.

[29] C. Lanczos, "Trigonometric interpolation of empirical and analytical functions," *Journal of Mathematics and Physics*, vol. 17, no. 1–4, pp. 123–199, 1938.

[30] H.-O. Kreiss and J. Oliger, "Comparison of accurate methods for the integration of hyperbolic equations," *Tellus*, vol. 24, no. 3, pp. 199–215, 1972.

[31] S. A. Orszag, "Comparison of pseudospectral and spectral approximation," *Studies in Applied Mathematics*, vol. 51, no. 3, pp. 253–259, 1972.

[32] A. A. Kilbas, H. M. Srivastava, and J. J. Trujillo, *Theory and Applications of Fractional Differential Equations*, vol. 204. Elsevier Science Inc., New York, 2006.

[33] C. Canuto, M. Hussaini, A. Quarteroni, and T. Zang, *Spectral Methods: Fundamentals in Single Domains*. Springer-Verlag, Berlin Heidelberg, 2007.

[34] L. F. Shampine and M. W. Reichelt, "The Matlab ODE Suite," *SIAM Journal on Scientific Computing*, vol. 18, no. 1, pp. 1–22, 1997.

[35] T. S. Chihara, *An Introduction to Orthogonal Polynomials*. Courier Corporation, North Chelmsford, MA, 2011.

[36] M. A. Zaky and I. G. Ameen, "On the rate of convergence of spectral collocation methods for nonlinear multi-order fractional initial value problems," *Computational and Applied Mathematics*, vol. 38, no. 3, p. 144, 2019.

[37] B. Duan, Z. Zheng, and W. Cao, "Spectral approximation methods and error estimates for Caputo fractional derivative with applications to initial-value problems," *Journal of Computational Physics*, vol. 319, pp. 108–128, 2016.

[38] S. Nemati and S. Sedaghat, "Matrix method based on the second kind Chebyshev polynomials for solving time fractional diffusion-wave equations," *Journal of Applied Mathematics and Computing*, vol. 51, no. 1–2, pp. 189–207, 2016.

[39] Y. Yang, Y. Chen, Y. Huang, and H. Wei, "Spectral collocation method for the time-fractional diffusion-wave equation and convergence analysis," *Computers & Mathematics with Applications*, vol. 73, no. 6, pp. 1218–1232, 2017.

[40] A. Bhrawy, M. Zaky, and D. Baleanu, "New numerical approximations for space-time fractional Burgers' equations via a legendre spectral-collocation method," *Romanian Reports in Physics*, vol. 67, no. 2, pp. 340–349, 2015.

[41] M. Zaky, S. Ezz-Eldien, E. Doha, J. T. Machado, and A. Bhrawy, "An efficient operational matrix technique for multidimensional variable-order time fractional diffusion equations," *Journal of Computational and Nonlinear Dynamics*, vol. 11, no. 6, p. 061002, 2016.

[42] A. R. Haghighi, A. Dadvand, and H. H. Ghejlo, "Solution of the fractional diffusion equation with the Riesz fractional derivative using McCormack method," *Communications on Advanced Computational Science with Applications*, vol. 2014, pp. 1–11, 2014.

2

Special Functions and Their Link with Nonlinear Rod Theory

Giovanni Mingari Scarpello and Daniele Ritelli

Università di Bologna

CONTENTS

2.1 Introduction

There are many physical models or theories in which some functions appear in natural and regular fashion, starting with polynomials and algebraic functions and progressing through the "elementary" transcendental ones, almost all already known to the ancients. But between the end of 17th and the early 18th century, several scientists (Newton, Leibniz, Jakob Bernoulli, Johann Bernoulli) realized that the elementary functions had their limitations, namely, for too many important physical models, like the planetary orbital motion, the oscillation of a suspended chain, or the gravitational potential of a nearly spherical body (e.g., the Earth oblate ellipsoid), the relevant solutions often could not be described in algebraic form. And for the sake of the truth, they rarely could.

Furthermore, even in the realm of pure mathematics, some quantities, such as the arc length of an ellipse or hyperbola, were also impossible to be discussed in closed form; but they were expressed as infinite series, without any theory of convergence, or as integrals or as solutions to differential equations. Such "special" most frequent mathematical objects were given names and notations that have come into common usage: Bessel, Airy, Mathieu functions, the spherical harmonics, the Gamma and Beta functions, and so on. Most of these special functions of Mathematical Physics are generated from the study of those ordinary differential equations (ODEs)in which the separation of variables breaks many categories of linear partial differential equations, mainly that of Laplace. These ODEs depend on the choice of the coordinate system, which in turn is influenced by the problem of physical configuration. They all have some orthogonality relation, and frequently form a basis for functions on whatever domain they're associated with. Among the many special functions that satisfy these second order ODEs we find the polynomials which bear the names of Tchebychev, Hermite, Jacobi, Laguerre, and the Whittaker functions, and the parabolic cylinder functions. As for the Bessel functions, one can study their infinite series, recursion formulae, generating functions, asymptotic series, integral representation theorems, and so on, generating an almost infinite literature for them. Attempts have been made to unify this rich topic, but not one has been completely successful: in spite of the many similarities among these functions, each has peculiar properties that must be studied separately.

Some relationships can be developed by introducing the hypergeometric function $y = y(x; a, b, c)$, which satisfies the Gauss differential equation

$$x(1-x)y'' + [c - (a+b+1)x]\, y' - aby = 0.$$

whose importance relies on the notorious fact that many special functions can be expressed in terms (or as particular cases) of hypergeometric functions. Historically and practically the special functions and their applications arise primarily in Mathematical Physics; nevertheless they have many other uses in both Pure and Applied Mathematics.

For example, Bessel functions are useful in solving certain types of random-walk problems; they also find applications in the theory of numbers. The hypergeometric functions are met in constructing the so-called conformal mappings of polygonal regions whose sides are circular arcs and so on. However, there are further special functions not stemming from second order linear ODEs. We refer to the Jacobi elliptic functions, introduced in 1827, sn, cn, *and*dn whose use in Mechanics is pervasive and whose reference ODEs are nonlinear, of first order:

$$\begin{cases} \mathrm{cn}'u = -\mathrm{sn}u \ \mathrm{dn}u \\ \mathrm{sn}'u = \mathrm{cn}u \ \mathrm{dn}u \\ \mathrm{dn}'u = -k^2\mathrm{sn}u \ \mathrm{cn}u \end{cases}$$

and coupled with each other. Their name is being obtained by inversion of elliptic integrals which have the form

$$\int R\left(t, \sqrt{p(t)}\right) \, \mathrm{d}t$$

where R is a rational function and p is a polynomial of a single variable of degree three or four without multiple factors. They first occurred in the arc length of the ellipse; but also arise elsewhere, e.g., looking for the arc lengths of hyperbola and many other lines, the period of the simple pendulum, the motion about a fixed point, and the *deflection of a thin elastic rod.*

When these problems first arose in the late 17th century, they posed an obstacle again to Leibniz program of integration in *closed form* or by *elementary functions.* But all efforts to express elliptic integrals in finite terms failed, and as early as 1694, Jakob Bernoulli conjectured that the task was impossible, as eventually proved by Liouville in 1833. The key that unlocked many of the secrets of elliptic integrals was the *lemniscate of Bernoulli,* a quartic of cartesian equation $(x^2 + y^2)^2 = x^2 - y^2$. The first to consider it was Jakob Bernoulli, see [7], who showed its arc length is expressed by the elliptic integral

$$\int_0^x \frac{\mathrm{d}t}{\sqrt{1-t^4}}$$

now referred as the lemniscatic integral. The elastica was another of Jakob Bernoulli discoveries: it is nothing but the curve assumed by a thin flexible rod stretched by outside loads. He showed the rectification of his elastica drove just to the above integral and introduced the lemniscate because its arc length is expressed just by that integral.

The problem of "rectification of quadratures" was generally encountered in the setting of ODEs that cannot be solved in closed algebraic form or "by quadratures".[1] The importance of reducing quadratures to rectifications is attested in both words and deeds by all the major figures involved in the early Leibnizian calculus, but the motivation was different. The most common is that "the dimension of the line is simpler than that of an area", as Leibniz repeatedly stressed. In a different way, Jakob Bernoulli solving the ODE of the paracentric isochrone was led (again) to the same elliptic integral met in elastica rectification. For a deep discussion on this subject the reader is addressed to [9]. These investigations led to the seminal discoveries of Fagnano and Gauss on elliptic integrals. In addition, Euler's work on elliptic integrals was also stimulated by the elastica he dealt as a problem of Calculus of variations, so that he gave in 1743 some pictures of elasticæ showing their periodic forms, namely the first images of elliptic function's periodicity.

[1]It would be better named *quadrature through a rectification* that is, when encountering an integral which cannot be evaluated in algebraic form, the pioneers of the Leibnizian calculus thought, instead of leaving it as an area, to express it in terms of the arc length of an auxiliary curve, namely trying to rectify the cubic of equation $y = x^3/3$.

We are convinced that, in the fields of heat transmission, electromagnetism, and elasticity, special functions are always necessary as today as at Maxwell's time. It is true that highly refined models can only be computed numerically, but it is equally true that they too are always an abstraction with respect to the reality that in itself remains unattainable. Therefore a model less adherent to reality but mastered, thanks to the logical and calculation rules, i.e., the special functions, allows a much deeper knowledge of how the system works than the superficial, even if by far more powerful, effects of the computer codes. We will show by several examples how the original Bernoulli–Euler *nonlinear* theory, namely without the approximation of small deformations, produces a set of differential problems whose solution requires massive recourse to elliptic, hypergeometric, and Bessel functions. We will make use of their integral representation theorems (e.g., see (2.87) for the Lauricella functions) in solving our differential problems which are of dual nature, namely of Cauchy and Dirichlet type: the latter being dealt by the shooting method which leads, as subproblems, to some algebraic high-degree (or to transcendental) equations.

2.1.1 A Short Sketch of the Elastica History

In order to show the contribution of the theory of rods to the development of Mathematical Physics, let us point out some main facts. In 1691 Jakob Bernoulli faced the problem of computing the deformed centerline of a thin, homogeneous, straight, and flexible rod under forces and couples applied at its ends: that planar line was named *elastica*. Nevertheless the term beam strain had appeared much earlier in science: Galilei (*Discorsi e dimostrazioni matematiche intorno a due nuove scienze*, 1638) and Mariotte (*Traité du mouvement des eaux et des autres corps fluides*, 1686) had investigated the strength of beams, but Jakob Bernoulli was the one to study the planar deflection by means of leibnizian calculus geometrically. In such a way, in *Curvatura laminae elasticæ*, 1694, he established elastica's differential equation, proving that the centerline is shaped as a cross section of a flexible cylindric surface filled with water, *lintearia*. The subject was also treated by his brother Johann in lectures XLIV and XLV of *Lectiones Mathematicæ de methodo integralium*, see [8], written 50 years before their publication of 1742. Meanwhile, Daniel Bernoulli, Johann's son, in 1738 had obtained an integral expression of potential energy stored in a bent rod. He guessed to Euler in 1742 that a rod would attain such a shape as to minimize the functional of its squared curvature, namely the bending energy stored in it. Accordingly, Euler dealt elastica as an isoperimetric problem of Calculus of Variations, *De curvis elasticis*, 1744, arriving at elastica's ODE, and to geometrical shape of them. The subject attracted Maclaurin too, who pointed out that elastica was connected to elliptic integrals, as by title itself of Section 927 of his*Treatise on fluxions*, 1742, namely: *The construction of the elastic curve, and of other figures, by the rectification of the conic sections.* In 1757, *Sur la force des colonnes*, Euler approached the critical load through a simplified expression to elastica's curvature. In 1770 Lagrange, *Sur la figure des colonnes*, did not limit himself to critical loads already discussed by Euler, but went on to investigate the deflections when the load exceeds its critical value (post-buckling behavior), *using the exact curvature*, and integrating the elastica's ODE in series. Euler went over the problem in 1770 and 1775, always in the field of bending: the reader can see [15], [3], and [6]. Analytic solutions to elastic planar curves through elliptic integrals of the first and second kind can be seen at [42], [48]. Treatments are also centered on Jacobi elliptic functions, [10], [24], while [3] is founded on the Weierstrass $\wp$ function. The close link between elastica and elliptic functions induced later to plot them through suitable curved rubber rods (G. Greenhill: *Graphical representation of the elliptic functions by means of a bent elastic beam*, 1876). For more recent developments, see [12], [18], [35], and [36]: special planar problems are faced in [46], which includes the effects of transverse deformation; in [37] a variety of loads (terminal, transverse, continuous) are tackled; whereas the effect of large loads can be seen in [23]. Unextensibility of a circular rod is numerically computed in [13] where the load is a uniform centrally directed force. A role has been found [39] to elastica in some areas of differential geometry (Willmore surfaces, pseudospherical surfaces) and in several applications, ranging from traditional ones (structural mechanics [17], engineering [28]) to biochemistry [27] and to some new ones to biology [47].

The study of coiling of suboceanic cables, filamentary structures of biomolecules (like DNA) and bacterial fibers, led to a renewed interest in the 3D elastica, whose groundings can be found back in Lagrange's *Mécanique Analytique* 2nd edition (1811), but without integration. That memory was followed by *Mémoire sur l'intégration des équations de la courbe élastique à double courbure* by Binet, 1844; and by *Note sur l'intégration des équations de la courbe élastique à double courbure,* Wantzel, 1844. In 1859 Kirchhoff, *Über das Gleichgewicht und die Bewegung eines unendlich dűnnen elastischen Stabes*, [24], demonstrated that under forces applied only to ends of a rod, the relevant equilibrium equations are identical with the dynamical ones of a rigid body moving around a fixed point (spinning top). Anyway the 3D-elastica problem grew during time, see [2] or [22]: in our previous paper [29] one can find a specific bibliography and a solution under a particular load condition.

2.2 A Supported Rod Inflected by Axial Thrust

2.2.1 The Problem

We compute *the exact curvature* using the planar elastica of a thin, flexible, and supported rod loaded by axial compression. The flexure is induced in the rod by the strain: the moment's arm is just the transverse deflection y. Equilibrium elastica is assumed to be non-isoperimetrical: we have a Dirichlet two-point boundary value problem. It is solved by the "shoooting" method: two auxiliary initial value problems, one for each support A,B, are considered, with ω being the unknown slope of the elastica at support A and, in absolute value, at B. Imposing somewhere the "welding" of the solutions[2], we are led to a transcendental equation for ω. We will obtain the cartesian elastica equation $x = x(y)$ through elliptic integrals of the first and third kind. Even if a lot of literature is available, we did not read any cartesian *non-parametric* solution to the supported compressed rod; our solution shall then be considered original.

2.2.2 The Elastica Nonlinear ODE

We establish our planar model with its auxiliary conditions. We tackle a thin, homogeneous, straight-linear rod in its unstressed configuration, simply supported at its ends A, B. A horizontal thrust of constant size, direction, and sense, say $\vec{\mathrm{F}}$, pushes it in B: then a stationary strain field, by equilibrium of active load $\vec{\mathrm{F}}$ and reactive force $-\vec{\mathrm{F}}$ takes place. The rod is hinged at A, so that the axial thrust cannot induce any translation. And just

[2]Looking for the origin of such a method, at first attempt one finds only articles of the last quarter of the 20th century. On the contrary, its origin goes back to the Italian mathematician C. Severini (1872–1951), who established the method in two articles of 1904, the former [44] concerning differential equations of higher order, the latter [43] specialized to the second order ODEs.

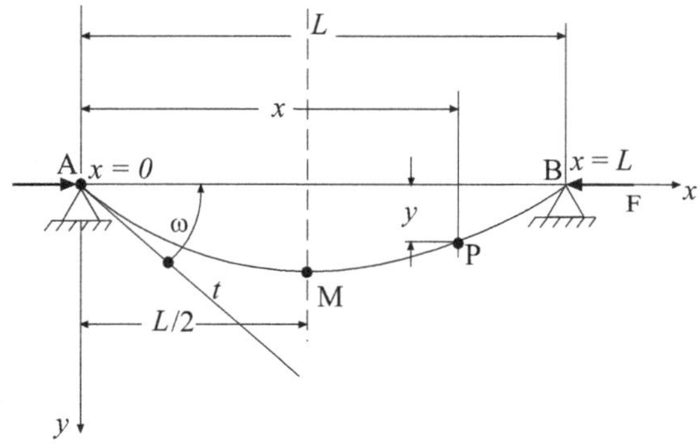

FIGURE 2.1
Straight simply supported rod bent as a consequence of deflection and axial compression.

at A we fix a (x, y) cartesian reference frame with x along the unstrained rod, from A to B; and y downwards, normal at A to x, see Figure 2.1. Our main assumptions are as follows:

- A_1 the rod is thin, initially straight, and homogeneous, with uniform cross section and flexural rigidity EJ, where E is the Young modulus, and J is the cross section moment of inertia about a "neutral" axis normal to the plane of bending and passing through the central line;
- A_2 the rod is unloaded along all its external surface and loaded only by active compression force $\vec{F} =$ const. at its end B;
- A_3 transverse deformation is ignored, and the rod is freely extendible with linear elasticity;
- A_4 due to rest of static equilibrium, no rod element undergoes acceleration.
- A_5 gravity is ignored: it would act as a uniform load along the rod.

The rod shall be thought to stay on two fixed points at distance L, hinged in A, and simply supported in B. Beyond B there is a overhang, and the force is applied at the end of it in such a way that the rod is free of being stretched from the force and its moment so that its length will pass from L to L_{def}.

The flexure formula $\pm EJc = M$ connects flexural rigidity EJ (inertia), the bending moment M (cause), and curvature c (effect) of deformed centerline. In our reference frame (x, y) curvature has always opposite sign of the bending moment. Through exact curvature, if $y = y(x)$ is the unknown deflection of the x-section, we get the elastica fundamental problem:

$$\boxed{EJ\frac{y''}{(1+y'^2)^{3/2}} = -M} \tag{2.1}$$

where prime means derivative with respect to x. Its solution requires the imposition of suitable boundary conditions by the constraints' nature and the knowledge of loads determining M.

Notice that in the case we are going to deal with, M is a function of y, but other possibilities will be met afterwards: e.g., in (2.48), (2.45), and (2.68) the moment will depend *directly* on x according to a linear, quadratic, and cubic law, respectively; furthermore in (2.72) we will meet a special load such that M will be tied not only to x but also to $\frac{dy}{dx}$ simultaneously. We will solve Equation (2.1) assuming $y(x)$ to be invertible and then *monotonic*, say, increasing for the first stage of our problem. Remembering the identity:

$$\frac{-y''}{(1+y'^2)^{3/2}} = \frac{\mathrm{d}}{\mathrm{d}y}\left(\frac{1}{\sqrt{1+y'^2}}\right),$$

obtained assuming y' as a function of y, (2.1) becomes

$$\frac{\mathrm{d}}{\mathrm{d}y}\frac{1}{\sqrt{1+y'^2}} = \frac{M(y)}{EJ}. \tag{2.2}$$

For the axial thrust $\vec{\mathrm{F}}$, the deflection $y(x)$ will be the moment arm, $M(y) = Fy$, and we have

$$\frac{\mathrm{d}}{\mathrm{d}y}\frac{1}{\sqrt{1+y'^2}} = 2\beta^2 y, \tag{2.3}$$

where $2\beta^2 = \left|\vec{\mathrm{F}}\right|/(EJ)$. Observe that this is a consequence of having neglected the rod weight. Taking into account a q (N/m) linear load intensity, the bending moment would become $M(y) = Fy - (q/2)\,x^2$, leading to a loss of autonomy of the problem. For the strength of materials the bending moment is *exogenous,* and then a completely known function depends on loads and on x. In our problem, the bending moment depends on deflection y only, and then (2.2) is autonomous, see (2.3). Let us now enter the auxiliary conditions: with both supports A and B being perfect, i.e., zero yielding, we would have a Dirichlet two-point boundary value problem conditioned by $y(0) = y(L) = 0$. Notice that Love provides a classification of the elasticæ (see [24], art. 263) according to whether they are inflexional (when some bending moment's root falls within $(0, L)$) or not. Our case is the latter, as the bending moment M is never zero outside the support B.

2.2.3 Phase Portrait Analysis

We consider the first kind, two-point, nonlinear Boundary Value Problem:

$$\begin{cases} y''(x) = -2\beta^2 y(x)\left(1+y'^{\,2}(x)\right)^{3/2}, \; 0 < x < L, \\ y(0) = y(L) = 0. \end{cases} \tag{2.4}$$

In our case, even if the right-hand side is continuous, both its partial derivatives hold y', namely the elastica's slope which cannot be *a priori* bounded: then the classic result of [38] does not apply, and (2.4) will not possess a unique solution. A confirmation comes by direct inspection: if (2.4) has a nontrivial solution, one can see that it allows the opposite also. In Theorem 2.2 we will show that nontrivial solutions of (2.4) are two, provided that a proper relationship holds between coefficient β and rod's length L. We begin with the following:

Theorem 2.1. *Any nontrivial solution of the initial value problem*

$$\begin{cases} y''(x) = -2\beta^2 y(x)\left(1+y'^2(x)\right)^{3/2}, \\ y(0) = 0, \; y'(0) = a \in \mathbb{R}, \end{cases} \tag{2.5}$$

is periodic.

Proof. The case $a = 0$ leads to a identically zero solution by which the thesis comes down trivially. Then we will suppose $a \neq 0$. Let ω be the angle between the tangent to elastica in A and the x direction. We get $a = \tan\omega$ so that $a > 0$ leads to $0 < \omega < \pi/2$ and $a < 0$ implies $-\pi/2 < \omega < 0$. We obtain by (2.5) the relevant first order ODE system of phase portrait with $q = y$, $p = y'$:

$$\begin{cases} q'(x) = p(x), \\ p'(x) = -2\beta^2 q(x)\left(1 + p^2(x)\right)^{3/2}, \\ q(0) = 0,\ p(0) = \tan\omega. \end{cases} \tag{2.6}$$

By eliminating x and minding $q(0) = 0$:

$$\begin{cases} \dfrac{\mathrm{d}p}{\mathrm{d}q} = -\dfrac{2\beta^2 q\left(1 + p^2(x)\right)^{3/2}}{p}, \\ p(0) = \tan\omega. \end{cases} \tag{2.7}$$

By separating variables

$$\int_{\tan\omega}^{p} \frac{\xi}{\left(1+\xi^2\right)^{3/2}}\mathrm{d}\xi = \int_0^q -2\beta^2\eta\,\mathrm{d}\eta \iff -\frac{1}{\sqrt{1+p^2}} + \frac{1}{\sqrt{1+\tan^2\omega}} = -\beta^2 q^2,$$

and orbits, the cartesian equation is as follows:

$$p^2(q) = \frac{1}{\left(\beta^2 q^2 + \cos\omega\right)^2} - 1. \tag{2.8}$$

In any occurrence locus (2.8) is a net of infinite sextics, *closed* curves, what closes the proof. □

Relationship (2.8) helps the search of solutions of (2.4): the period of the closed orbits of (2.6) must be consistent with boundary condition in (2.4). We infer from (2.8) that q shall meet the constraint $\beta^2 q^2 + \cos\omega \leq 1$. We will find again this condition in the next theorem 2.2. Notice that in the trivial case $\cos\omega = 1$ (2.8) delivers exactly the zero solution $q \equiv 0$, $p \equiv 0$. Formula (2.8) defines the closed phase portrait of Equation (2.5), namely (2.6). If q^* is the q-value such that $\beta^2 q^{*2} + \cos\omega = 0$, the orbital breadth $Q = 2q^* = 2\left(1 - \cos\omega\right)^{1/2}/\beta$ will equate the period of inverse function obtained by welding both restrictions $x = x(y)$ for two y-ranges, or $Q = 2y_{\max}$.

2.2.4 Integration

Theorem 2.2. *For any couple of values β, $L > 0$ such that the transcendental inequality holds:*

$$2\mathbf{E}\left(\frac{1}{\sqrt{2}}\right) - \mathbf{K}\left(\frac{1}{\sqrt{2}}\right) \leq \frac{\beta L}{\sqrt{2}} \leq \frac{\pi}{2}, \tag{2.9}$$

where $\mathbf{E}$ and $\mathbf{K}$ denote the complete elliptic integrals of the second and first kind, then problem (2.4) has two nontrivial solutions of opposite signs. If $\beta^2 y^2 + \cos\omega \leq 1$, we define $0 \leq y \leq y_{\max}$:

$$\varphi(y) = \arcsin\frac{\beta\sqrt{2}\,y}{\sqrt{(1-\cos\omega)\left(1+\cos\omega+\beta^2 y^2\right)}}, \tag{2.10}$$

$$\boxed{\mathcal{H}(y) = \frac{1}{\beta\sqrt{2}}\left[2E\left(\varphi(y), \sin\frac{\omega}{2}\right) - \frac{\sin^2\frac{\omega}{2}\sin\left[2\varphi(y)\right]}{\sqrt{1-\sin^2\frac{\omega}{2}\sin^2\varphi(y)}} - F\left(\varphi(y), \sin\frac{\omega}{2}\right)\right]} \tag{2.11}$$

then positive solutions can be evaluated:

1. for $0 < x < L/2$ the solution to (2.4) is: $x = \mathcal{H}(y)$,
2. for $L/2 < x < L$ the solution to (2.4) is: $-(x - L) = \mathcal{H}(y)$.

Remark 2.1. *From Theorem (2.2) one can immediately infer that solution of* (2.4) *is symmetrical with respect to the axis* $x = L/2$.

Proof. Proof is founded upon integrating two auxiliary initial value problems:

$$\begin{cases} y''(x) = -2\beta^2 y(x)\left(1 + y'^2(x)\right)^{3/2} \\ y(0) = 0, \\ y'(0) = \tan\omega, \omega \in \left]0, \pi/2\right[, \end{cases} \tag{2.12}$$

$$\begin{cases} y''(x) = -2\beta^2 y(x)\left(1 + y'^2(x)\right)^{3/2}, \\ y(L) = 0, \\ y'(L) = -\tan\omega, \omega \in \left]0, \pi/2\right[. \end{cases} \tag{2.13}$$

They are stemming from the physical reason that each simple support is a constraint which allows rotation to the section. It is then possible to assign in A the elastica's slope, namely a measure of terminal rod rotation. In such a way, solving each of the initial value problems, one gets functions $y_1(x)$ and $y_2(x)$ each valid for a half rod. Based on (2.20), one can weld at $x_{\max} = L/2$ both above solutions of (2.12) and (2.13) so that $y_1'(L/2) = y_2'(L/2) = 0$, by getting the symmetrical solution to (2.4) called in thesis. *In primis* using the change of variable $X := y$, $Y = y'$. Then, being $y'' = Y\,\mathrm{d}Y/\mathrm{d}X$, we arrive at the separable equation:

$$\begin{cases} Y\dfrac{\mathrm{d}Y}{\mathrm{d}X} = -2\beta^2 X\left(1 + Y^2\right)^{3/2} \\ Y(0) = \pm\tan\omega \end{cases}$$

where $+$ comes from (2.12) and $-$ comes from (2.13). Integrating:

$$Y^2 = \frac{1 - \left(\beta^2 X^2 + \cos\omega\right)^2}{\left(\beta^2 X^2 + \cos\omega\right)^2}. \tag{2.14}$$

Now, going back to the old variables, we get two separable initial value problems where (again) $+$ comes from (2.12) and $-$ comes from (2.13):

$$\begin{cases} y'(x) = \dfrac{\sqrt{1 - \left(\beta^2 y^2(x) + \cos\omega\right)^2}}{\beta^2 y^2(x) + \cos\omega}, \\ y(0) = 0 \end{cases} \tag{2.15}$$

$$\begin{cases} y'(x) = -\dfrac{\sqrt{1 - \left(\beta^2 y^2(x) + \cos\omega\right)^2}}{\beta^2 y^2(x) + \cos\omega}, \\ y(L) = 0 \end{cases} \tag{2.16}$$

Integrating again:

$$\int_0^y \frac{\left(\beta^2\eta^2 + \cos\omega\right)\mathrm{d}\eta}{\sqrt{1 - \left(\beta^2\eta^2 + \cos\omega\right)^2}} = x, \quad \text{or} \quad \int_0^y \frac{\left(\beta^2\eta^2 + \cos\omega\right)\mathrm{d}\eta}{\sqrt{1 - \left(\beta^2\eta^2 + \cos\omega\right)^2}} = -(x - L). \tag{2.17}$$

In both cases integration is allowed being $|\beta^2 y^2 + \cos\omega| = \beta^2 y^2 + \cos\omega < 1$. Furthermore $y_{\max} = (1-\cos\omega)^{1/2}/\beta$ is a rest point, where $y' = 0$. Then (2.14) leads to understand how each branch of a nontrivial solution of (2.4), going out either $x = 0$ or $x = L$ respectively, has only one stationary point. In fact the ordinate of such a point is necessary given by $y_{\max}$, because if another one were available, the solution should be a constant. Putting $\tau = \beta^2\eta^2 + \cos\omega$, we find

$$\int_0^y \frac{(\beta^2\eta^2+\cos\omega)\,\mathrm{d}\eta}{\sqrt{1-(\beta^2\eta^2+\cos\omega)^2}} = \frac{1}{2\beta}\int_{\cos\omega}^{\cos\omega+\beta^2y^2} \frac{\tau\mathrm{d}\tau}{\sqrt{(1-\tau)(\tau-\cos\omega)(\tau-(-1))}}. \tag{2.18}$$

For side integral in (2.18): see [19] formula 3.132 number 4 page 51, and remember (2.10); its expression is

$$\sqrt{2}\left[(1+\cos\omega)\,\Pi\left(\varphi(y),\sin^2\frac{\omega}{2},\sin\frac{\omega}{2}\right) - F\left(\varphi(y),\sin\frac{\omega}{2}\right)\right]. \tag{2.19}$$

Afterwards, inserting in (2.19) the identity, [11] nr. 434.01 page 235, the first "special case" which holds for $|k| < 1, 0 \le \phi \le \pi/2$:

$$\Pi\left(\phi,k^2,k\right) = \frac{1}{1-k^2}\left[E(\phi,k) - \frac{k^2\sin(2\phi)}{2\sqrt{1-k^2\sin^2\phi}}\right],$$

then (2.18) becomes, see (2.11):

$$\int_0^y \frac{\beta^2\eta^2+\cos\omega}{\sqrt{1-(\beta^2\eta^2+\cos\omega)^2}}\mathrm{d}\eta = \frac{1}{\beta\sqrt{2}}\mathcal{H}(y).$$

After solving (2.12) and (2.13), if one requires the common rest point $y_{\max}$ to be attained at $x^* = L/2$, both solutions are welded smoothly, providing a complete solution to (2.4). Inserting $y_1'(L/2) = y_2'(L/2) = 0$ in (2.12), (2.13), by comparing their right-hand sides, one gets $y_1(L/2) = y_2(L/2)$. Being $\varphi(y_{\max}) = \pi/2$, we find

$$\int_0^{y_{\max}} \frac{\beta^2\eta^2+\cos\omega}{\sqrt{1-(\beta^2\eta^2+\cos\omega)^2}}\mathrm{d}\eta = \frac{1}{\beta\sqrt{2}}\left[2\mathbf{E}\left(\sin\frac{\omega}{2}\right) - \mathbf{K}\left(\sin\frac{\omega}{2}\right)\right],$$

welding condition becomes

$$\frac{\beta L}{\sqrt{2}} = 2\mathbf{E}\left(\sin\frac{\omega}{2}\right) - \mathbf{K}\left(\sin\frac{\omega}{2}\right), \tag{2.20}$$

which allows to get ω through β and L so that ω will disappear from expressions (2.10) and (2.11), so that $x = x(y;\beta,L)$. For a given rod length L, not all the loads are possible in the sense that the problem cannot admit a solution. If $\omega \in [0,\pi/2]$ then $k := \sin(\omega/2) \in [0, 1/\sqrt{2}]$. We consider the function $f(k) = 2\mathbf{E}(k) - \mathbf{K}(k)$, and applying the derivative rules for complete elliptic integrals with respect to the modulus, [11] 710.00 e 710.02 page 230, we get

$$\frac{\mathrm{d}f}{\mathrm{d}k} = \frac{(1-2k^2)\mathbf{E}(k) - (1-k^2)\mathbf{K}(k)}{k(1-k^2)}.$$

Being $\mathbf{E}(k) < \mathbf{K}(k)\ \forall k \in [0,1]$, and $k^2 < 1/2$, we infer $f(k)$ is decreasing, then the condition (2.9) ensures existence and uniqueness for solution of (2.20), what concludes the proof. □

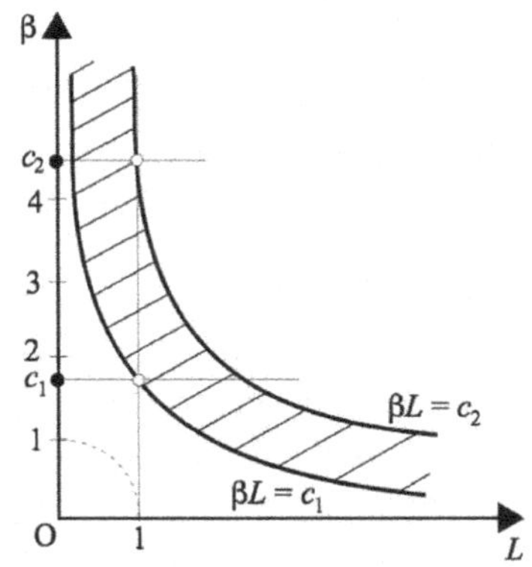

FIGURE 2.2
The shadowed two-hyperbola strip depicts the (β, L) compatibility region for solution existence.

In a cartesian frame (β, L), $\beta > 0, L > 0$, the *compatibility constraint* we found is depicted by the planar shadowed region bounded by the equilateral hyperbolæ, see Figure 2.2:

$$\beta L = \sqrt{2}\left(2\mathbf{E}\left(\frac{1}{\sqrt{2}}\right) - \mathbf{K}\left(\frac{1}{\sqrt{2}}\right)\right) := c_1 \text{ and } \beta \mathrm{L} = \pi\sqrt{2} := \mathrm{c}_2.$$

Remark 2.2. *The elastica arc length correspondent to* (2.4) *solution is* $L_{\text{def}} = \frac{\sqrt{2}}{\beta}\mathbf{K}\left(\sin\frac{\omega}{2}\right)$.

Proof. In fact by (2.14) one gets, for $0 \le y \le y_{\max}$, $x'(y) = (\beta^2 y^2 + \cos\omega)(1 - (\beta^2 y^2 + \cos\omega)^2)^{-1/2}$. Then:

$$L_{\text{def}} = 2\int_0^{y_{\max}} \sqrt{1 + x'^2(y)}\, \mathrm{d}y = 2\int_0^{y_{\max}} \frac{\mathrm{d}y}{\sqrt{1 - (\beta^2 y^2 + \cos\omega)^2}}$$
$$= \frac{1}{\beta}\int_{\cos\omega}^{1} \frac{\mathrm{d}\tau}{\sqrt{(1-\tau)(\tau - \cos\omega)(\tau - (-1))}}.$$

Thesis follows by formula 236.00 page 79 [11]. □

Remark 2.3. *What above implies that bending moment succeeds in stretching the rod, namely* $L_{\text{def}} > L$.

Proof. In fact, see [30], by (2.20) we get

$$L_{\text{def}} - L = \frac{\sqrt{2}}{\beta}\mathbf{K}\left(\sin\frac{\omega}{2}\right) - \frac{\sqrt{2}}{\beta}\left(2\mathbf{E}\left(\sin\frac{\omega}{2}\right) - \mathbf{K}\left(\sin\frac{\omega}{2}\right)\right)$$
$$= \frac{2\sqrt{2}}{\beta}\left(\mathbf{K}\left(\sin\frac{\omega}{2}\right) - \mathbf{E}\left(\sin\frac{\omega}{2}\right)\right) > 0.$$

□

Remark 2.4. *The approximate equation for such a case is* $y'' + \frac{R}{EJ}y = 0$ *and then would lead to an elastica of type*

$$y = k_1 \cos\left(\sqrt{\frac{R}{EJ}}\,x\right) + k_2 \sin\left(\sqrt{\frac{R}{EJ}}\,x\right)$$

with k_1 and k_2 arbitrary constants.

2.3 Cantilever Laded at Its Tip: Elastica Parametrized via Elliptic Functions

2.3.1 The Problem

A thin, flexible, one side-built-in rod under a concentrated terminal force is studied in its elastic equilibrium configuration. A proper set of state variables is chosen, and we solve a second order, nonlinear autonomous boundary value problem, in the rotation $\varphi(s)$ pertaining to each s-section. The search of the free end rotation φ_0, following the isoperimetric assumption, leads to a numerical subproblem. Furthermore, if $x(s)$ and $y(s)$ are the elastica coordinates parametrized on the arc length s, one obtains $x'(s)$ and $y'(s)$ as elliptic functions of s. Finally, some minor changes allow to analyze a more general free end load combination, consisting of an inclined force and a moment.

2.3.2 A 3D Rod Model

Main difficulties about elastica are as follows: the curvature intrinsic nonlinearity under not small displacements; geometric complexity of the unstressed configuration; constraints' effect such as isometry (non-stretching); and finally boundary conditions. The peculiarity of a thin rod under terminal forces and/or terminal couples is that *it buckles rather than breaks*: our common experience tells us that a compressed rod can buckle in any lateral direction. This is the strongest motivation for a spatial modeling of it: let Ωxy be a reference system with origin Ω at rod's free end; the Ωx axis, superimposed to the unstrained rod, is oriented towards the clamping, and Ωy downwards. Let it be $y = y(x)$ the unknown cartesian equation of the *xy projection* of a 3D-elastica. If the rod's free end is loaded by a slender force, its axial component does induce at each transverse s-section a flexure of moment M_a whose arm is depending on how much it is deformed at s. The curvilinear abscissa along the 3D deflection curve Γ is $s \geq 0$ with $s = 0$ at Ω, and $s > 0$ at $P(s)$. We establish the following assumptions:

- A_1 the rod is thin, initially straight, homogeneous, with uniform cross section and flexural rigidity EJ, where E is the Young modulus, and J the cross section second moment of inertia about a "neutral" axis normal to the plane of bending and passing through the central line;
- A_2 the rod is unloaded along all its body, and loaded by a force $\vec{F} = \text{const.}$ at its free end. Then all the strain is due to the loads' bending moment whilst the shearing stress effects are completely ignored.
- A_3 we are faced with linear elasticity, but transverse deformation is ignored. Anyway the rod is not freely extendible: there is a isoperimetric constraint, and L is the value of its invariant length.
- A_4 due to rest of static equilibrium, no rod element undergoes acceleration (Figure 2.3).

Following [22], the rod 3D-strain analysis consists of two equations. The first (forces) is

$$EJ\frac{\mathrm{d}}{\mathrm{d}s}(P-\Omega) \times \frac{\mathrm{d}^3}{\mathrm{d}s^3}(P-\Omega) = \vec{F} \times \frac{\mathrm{d}}{\mathrm{d}s}(P-\Omega),$$

where $\times$ is the cross product of two vectors in $\mathbb{R}^3$. Putting (see Figure 2.4) $(P-\Omega) := \vec{r}$, we have

$$EJ\frac{\mathrm{d}\vec{r}}{\mathrm{d}s} \times \frac{\mathrm{d}^3\vec{r}}{\mathrm{d}s^3} = \vec{F} \times \frac{\mathrm{d}\vec{r}}{\mathrm{d}s} \tag{2.21}$$

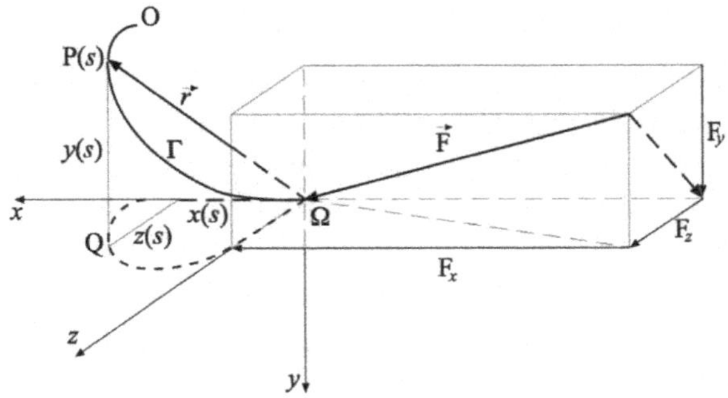

FIGURE 2.3
Sketch of a 3D rod under a load $\vec{F}$: frame of reference Ωxyz, elastica $OP\Omega$, each $s-$section identified by vector $(P - \Omega) := \vec{r}$.

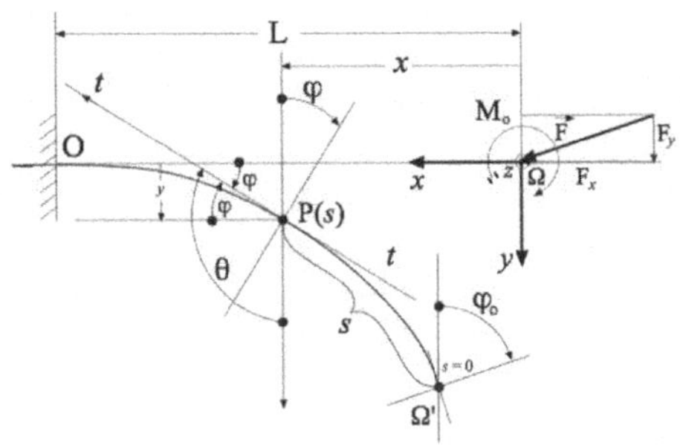

FIGURE 2.4
Sketch of a thin cantilever under a planar load: OΩ unstrained rod whose elastica is $OP\Omega'$, each $s-$section rotates of φ.

The second one (moments) is

$$EJ\frac{\mathrm{d}\vec{r}}{\mathrm{d}s} \times \frac{\mathrm{d}^2\vec{r}}{\mathrm{d}s^2} = \vec{M}, \tag{2.22}$$

where $\vec{M}$ is the external bending moment. Let us specialize both of them to our elastic system depicted at Figure 2.4, where a one side built-in, L -long, thin, straight-linear flexible rod is loaded by a force $\vec{F}$ at its free end, and all the system is planar. Recalling the Ωxy reference system feature, the increasing s is from Ω to the wall, so that the tangent tt to the elastica at whichever point $P(s)$ has its positive sense towards the built-in end O. Finally, Ωz is orthogonal to the elastica plane. Let θ be the angle of tt with the oriented axis Ωy.

Figure 2.4 shows the cantilever simultaneously loaded by a slender thrust and by a bending clockwise couple M_0. The first one $(F_x > 0, F_y > 0)$ induces a global strain in the rod so that its free end, initially parallel to the wall, undergoes a (positive) clockwise rotation. The same happens as a consequence of the couple M_0. Let then be $\varphi_0 > 0$ the unknown total rotation of the free end. A straightforward calculation from (2.22) leads to non-zero projections on Ωz only:

$$M_z = -EJ\frac{\mathrm{d}\theta}{\mathrm{d}s}, \tag{2.23}$$

whilst (2.21) gives

$$-EJ\,\theta''(s) + F_y \sin\theta - F_x \cos\theta = 0. \tag{2.24}$$

2.3.3 A Two-Points Boundary Value Problem

The above vectorial system of (2.21) and (2.22) generally provides six scalar equations, which in a planar system collapses in two scalar differential equations (2.23), (2.24) concerning the flexion angle $\theta(s)$. It is convenient to introduce the rotation $\varphi = \theta - \frac{\pi}{2}$, which the s-section tangent to the elastica makes with its original position. In such a way, Equation (2.24) becomes

$$-EJ\,\varphi''(s) + F_y \cos\varphi + F_x \sin\varphi = 0.$$

At the built-in end $s = L$ no rotation is allowed, and then $\varphi = 0$. At the free end $s = 0$ the bending moment is zero, and therefore $\varphi'(s)$ shall be zero then. So we have the nonlinear two-points boundary value problem:

$$\begin{cases} -EJ\,\varphi''(s) + F_y \cos\varphi + F_x \sin\varphi = 0, \\ \varphi'(0) = 0,\ \varphi(L) = 0. \end{cases} \tag{2.25}$$

Equation (2.25) looks like that of the damped pendulum, but with three differences. First, the independent variable is not time, but the arc length $s \geq 0$ of the strained centerline. Second, we are faced with a trinomial equation due to the simultaneous presence of sinus and cosine with different weights according to which x or y load component is prevailing. Finally, the third is that we don't have a initial value problem, but a Dirichlet or boundary value problem, being the limiting conditions on the ends $s = 0$ and $s = L$. Such a boundary value problem is of mixed type (second kind), due to different variables prescribed at the ends: the rotation at $s = L$ and its derivative at $s = 0$. Looking to the following, (2.25) is better written as

$$\begin{cases} \varphi''(s) = \dfrac{1}{2}\Big(\beta_y \cos\varphi(s) + \beta_x \sin\varphi(s)\Big), \\ \varphi'(0) = 0,\ \varphi(L) = 0, \end{cases} \tag{2.26}$$

where $\beta_x = 2F_x/EJ$ and $\beta_y = 2F_y/EJ \in \mathbb{R}$ should not be both zero simultaneously. In the following we provide the explicit integration of (2.26).

2.3.4 Rod Local Rotations Parametrized through the Arc: $\varphi = \varphi(s)$

We prove that solution $\varphi(s)$ of (2.26) is strictly increasing and its derivative $\varphi'(s)$ is given by

$$\varphi'(s) = \sqrt{\beta_y\,(\sin\varphi - \sin\varphi_0) - \beta_x\,(\cos\varphi - \cos\varphi_0)}. \tag{2.27}$$

Anyway, we shall integrate the equations $x'(s) = \cos\varphi(s),\ \ y'(s) = -\sin\varphi(s)$ from the $s = 0$ (free end rotates of φ_0) to the generic s-section φ, and then

$$\mathrm{d}x = \frac{\mathrm{d}\varphi\,\cos\varphi}{\dfrac{\mathrm{d}\varphi}{\mathrm{d}s}}, \quad \mathrm{d}y = -\frac{\mathrm{d}\varphi\,\sin\varphi}{\dfrac{\mathrm{d}\varphi}{\mathrm{d}s}}$$

so that

$$x(\varphi) = \int_{\varphi_0}^{\varphi} \frac{\cos\phi\,\mathrm{d}\phi}{\sqrt{\beta_y\,(\sin\phi - \sin\varphi_0) - \beta_x\,(\cos\phi - \cos\varphi_0)}}$$

$$y(\varphi) = -\int_{\varphi_0}^{\varphi} \frac{\sin\phi\,\mathrm{d}\phi}{\sqrt{\beta_y\,(\sin\phi - \sin\varphi_0) - \beta_x\,(\cos\phi - \cos\varphi_0)}}$$

The above formulæ provide parametric expressions to elastica's coordinates as functions of the φ-rotation of the rod s-section. The relevant integrals can be carried out through [11] (formulæ 293.06 page 180, and 293.05 page 179). Nevertheless the free end rotation φ_0 requires a numerical approach and the relevant value will heavily affect the integration. It is then better to detect the equation to φ_0 , leaving to the above $x(\varphi)$ and $y(\varphi)$ expressions for possible numerical checks. In order to gain closed form expressions, first we will solve the boundary value problem (2.26): in such a way, by $\varphi(s)$ both $x'(s)$ and $y'(s)$ will come as functions of s; a further integration will provide x and y through s. What above, after having computed the free end rotation φ_0, will allow to complete the shooting, arriving to its analytical expression in terms of a first kind elliptic integral.

2.3.4.1 The Elastica Arc Length

What can be inferred about the strained rod $\mathfrak{R}$? If, for instance , we take into account only $\vec{F}$ shortening and extension effects on the rod (neglecting shear), what is the change $|\Delta L_N|$ in its arc length? It will be

$$|\Delta L_N| = \frac{1}{EA}\int_{\mathfrak{R}} N(s)\mathrm{d}s$$

Figure 2.5 shows that at (s,φ) of elastica, the total normal stress N is

$$N(\varphi) = F_y \sin\varphi - F_x \cos\varphi.$$

By Figure 2.4 we get: $x'(s) = \cos\varphi, x'(s) = -\sin\varphi$, and then

$$\Delta L_N = -\frac{1}{EA}(F_y y_{\Omega'} + F_x (L_0 - x_{\Omega'})).$$

But the coordinates $x_{\Omega'}$ and $y_{\Omega'}$ of the free end position Ω' of $\mathfrak{R}$, are unknown, and ΔL_N, generally speaking, cannot be computed easily. Then the rod *isometric* constraint is assumed: while the rod's shape is changing and its free end is pushed towards the wall, the rod's arc length is kept constant throughout all the elastic flow till equilibrium.

2.3.4.2 The Free End Rotation φ_0

The above equations to $x(\varphi), y(\varphi)$ require to compute the free end equilibrium rotation $\varphi(0) := \varphi_0$ of the (L, E, J) system under the load $\vec{F} \equiv (F_x, F_y, 0)$. From now on, we will assume β_x, $\beta_y > 0$, what is not affecting the generality, the other cases being tractable likewise. On the contrary, it is mandatory that the inequality

$$0 < L < \sqrt[4]{\frac{4}{\beta_x^2 + \beta_y^2}} \tag{2.28}$$

holds, as ensuring that (2.26) has one and only one solution, according to [38] Theorem 6.1 p. 93. Problem (2.25) does not possess one solution if whichever rod (L, E, J) is loaded

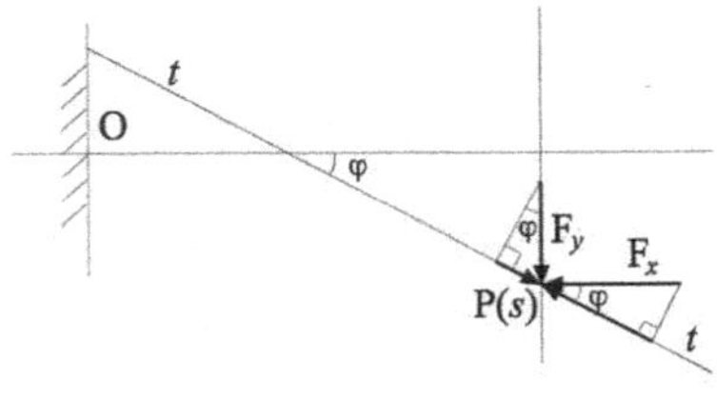

FIGURE 2.5
Geometric elements for evaluating the axial stress at $P(s, \varphi)$.

by whichever $\vec{F}$. In fact, by the previous things we argue that the force's size cannot be willingly large, but shall admit the upper bound $F < EJ/L^2$. In order to face (2.26), one shall previously solve the auxiliary shooting initial value problem:

$$\begin{cases} \varphi''(s) = \dfrac{1}{2}\Big(\beta_y \cos\varphi(s) + \beta_x \sin\varphi(s)\Big), \\ \varphi(0) = \varphi_0 \in [-\pi, \pi]\,,\ \varphi'(0) = 0, \end{cases} \tag{2.29}$$

detecting which conditions shall be met by φ_0 so that the solution to (2.29) is also solution to (2.26) too. Notice that (2.29) is a Weierstraß equation, which, following [1], pages 287–292, is brought to the quadratures putting

$$\begin{aligned} \Phi(\varphi) &= \beta_y\left(\sin\varphi - \sin\varphi_0\right) - \beta_x\left(\cos\varphi - \cos\varphi_0\right), \\ \int_{\varphi_0}^{\varphi} \mathrm{d}s &= \operatorname{sign}\left(\beta_y \cos\varphi_0 + \beta_x \sin\varphi_0\right) \int_{\varphi_0}^{\varphi} \frac{\mathrm{d}\phi}{\sqrt{\Phi(\phi)}}, \end{aligned} \tag{2.30}$$

where $\Phi(\varphi) \geq 0$. Furthermore, due to initial condition rest point of the shooting problem (2.29), it shall be $\Phi(\varphi_0) = 0$, and then φ_0 is a stationary point (maximum or minimum) for the solution to (2.29), solution which will either decrease or increase till to the next (simple) root of $\Phi(\varphi)$, where the monotonicity reverses, and this is due to periodicity of the solutions to (2.29). It is therefore essential to understand how the solutions of (2.29) really behave for different φ_0 values, in order to perform the φ_0 choice capable of solving (2.26). For the purpose, the following Lemma defines the range described by the solutions to (2.29) when the expression $\beta_y \cos\varphi_0 + \beta_x \sin\varphi_0$ changes sign.

Lemma 2.1. *If $\varphi(s)$ solves (2.29), then, defining $r(\beta_x, \beta_y) = \beta_x\left(\beta_x^2 + \beta_y^2\right)^{-1/2}$, we have*

1. if $\beta_x \sin\varphi_0 + \beta_y \cos\varphi_0 > 0$ then for any $s \in \mathbb{R}$:

$$\varphi_0 \leq \varphi(s) \leq 2\arccos\left[-r\left(\beta_x, \beta_y\right)\right] - \varphi_0; \tag{2.31}$$

2. if $\beta_x \sin\varphi_0 + \beta_y \cos\varphi_0 < 0$, then only one of the following holds:
   (a) if $-\pi \leq \varphi_0 < -\arccos\left[r\left(\beta_x, \beta_y\right)\right]$ then:

$$\arccos\left[-r\left(\beta_x, \beta_y\right)\right] - 4\pi - \varphi_0 \leq \varphi(s) \leq \varphi_0;$$

   (b) if $\arccos\left[-r\left(\beta_x, \beta_y\right)\right] < \varphi_0 \leq \pi$ then:

$$2\arccos\left[-r\left(\beta_x, \beta_y\right)\right] - \varphi_0 \leq \varphi(s) \leq \varphi_0.$$

The case $\beta_x \sin\varphi_0 + \beta_y \cos\varphi_0 = 0$ is trivial: solution of (2.29) is stationary: $\varphi(s) = \varphi_0$ for any $s \in \mathbb{R}$.

What above decides about the free end rotation φ_0.

Remark 2.5. *Let $\varphi(s)$ be the solution to the initial value problem (2.29), with $\varphi(0) = \varphi_0$.*

- If $\beta_x \sin\varphi_0 + \beta_y \cos\varphi_0 > 0$ and $\varphi_0 > 0$, then $\varphi(s)$ has no roots.
- If $\beta_x \sin\varphi_0 + \beta_y \cos\varphi_0 > 0$ and $\varphi_0 < 0$, then $\varphi(s)$ has one zero because $0 \in \left[\varphi_0, 2\arccos\left[-r\left(\beta_x, \beta_y\right)\right] - \varphi_0\right]$.
- If $\beta_x \sin\varphi_0 + \beta_y \cos\varphi_0 < 0$, then $\varphi(s)$ has no roots, whatever is φ_0.

We can then state

Theorem 2.3. *If φ_0 is such that*

$$\beta_x \sin\varphi_0 + \beta_y \cos\varphi_0 > 0, \quad -\frac{\pi}{2} < 2\arctan\left(\frac{\beta_x - \sqrt{\beta_x^2+\beta_y^2}}{\beta_y}\right) < \varphi_0 < 0,$$

then, if $L > 0$ meets (2.28), then the solution to (2.29) will solve (2.26) also, if and only if φ_0 is a root of

$$L = \sqrt[4]{\frac{4}{\beta_x^2+\beta_y^2}}\, F\left(\psi(\varphi_0), k(\varphi_0)\right) \tag{2.32}$$

with

$$\psi(\varphi_0) = \arccos\left(\frac{\beta_y + \left(\beta_x + \sqrt{\beta_x^2+\beta_y^2}\right)\tan\frac{\varphi_0}{2}}{\beta_y + \left(\beta_x - \sqrt{\beta_x^2+\beta_y^2}\right)\tan\frac{\varphi_0}{2}}\right), \; k^2(\varphi_0) = \frac{1}{2} + \frac{\beta_x\cos\varphi_0 - \beta_y\sin\varphi_0}{2\sqrt{\beta_x^2+\beta_y^2}}.$$

Furthermore, the common solution to (2.29) and (2.26) is strictly increasing.

Proof. See pp. 12-13 of [31]. □

Corollary 2.1. *If φ_0 is found by (2.32), then, putting*

$$N(s) = s\sqrt[4]{\frac{\beta_x^2+\beta_y^2}{4}}, \quad k^2 = \frac{1}{2} + \frac{\beta_y\cos\varphi_0 - \beta_x\sin\varphi_0}{2\sqrt{\beta_x^2+\beta_y^2}},$$

the solution to (2.26) will be

$$\boxed{\varphi(s) = 2\arctan\frac{p(s)}{q(s)}} \tag{2.33}$$

where

$$\begin{aligned} p(s) &= \left[\sqrt{\beta_x^2+\beta_y^2}\,(1-\cos\varphi_0 \operatorname{cn}(N(s)|k)) + \beta_y\cos\varphi_0\left(\operatorname{cn}(N(s)|k)-1\right)\right]\beta_x \\ &\quad + \beta_y\left(\beta_y - \sqrt{\beta_x^2+\beta_y^2}\right)\operatorname{cn}(N(s)|k)\sin\varphi_0 + \beta_x^2\sin\varphi_0 \\ q(s) &= \left[\cos\varphi_0\operatorname{cn}(N(s)|k)+1\right]\beta_x^2 - \beta_y\left(\sqrt{\beta_x^2+\beta_y^2} - \beta_y\right)(\cos\varphi_0+1) \\ &\quad + \left[(\operatorname{cn}(N(s)|k)-1)\beta_y + \sqrt{\beta_x^2+\beta_y^2}\right]\beta_x\sin\varphi_0 \end{aligned}$$

Proof. It is enough to replace the generic value $\varphi(s)$ to zero at the upper bound of integration. □

2.3.5 Elastica Coordinates $x(s)$, $y(s)$ Parametrized through Its Arc

Formula (2.33) ruling $\varphi(s)$ will be our tool in order to integrate differential equations:

$$x'(s) = \cos\varphi(s), \quad y'(s) = -\sin\varphi(s). \tag{2.34}$$

For instance, inserting (2.33) in (2.34), we get

$$x'(s) = \frac{q^2(s)-p^2(s)}{p^2(s)+q^2(s)}, \qquad y'(s) = -\frac{2p(s)\,q(s)}{p^2(s)+q^2(s)}$$

and then:

$$x'(s) = \frac{a_0 + a_1 \operatorname{cn}(N(s) \mid k) + a_2 \operatorname{cn}^2(N(s) \mid k)}{b_0 + b_2 \operatorname{cn}^2(N(s) \mid k)}, \tag{2.35}$$

$$y'(s) = \frac{c_0 + c_1 \operatorname{cn}(N(s) \mid k) + c_2 \operatorname{cn}^2(N(s) \mid k)}{d_0 + d_2 \operatorname{cn}^2(N(s) \mid k)}, \tag{2.36}$$

where the formulae linking to constants a_0, a_1, a_2, b_0, b_2, c_0, c_1, c_2, d_0, d_2, to the process data, can be found at pp. 14–15 of [31]. After this preliminaries, let us pass to integrate (2.35) and (2.36).

Theorem 2.4. *The elastica parametric equations are*

$$\boxed{x(s) = \frac{a_2}{b_2}s + \frac{\sqrt{2}}{(b_0 + b_2)\sqrt[4]{\beta_x^2 + \beta_y^2}}\left[\frac{a_0 b_2 - a_2 b_0}{b_2}\Pi\left(k^2, k, N(s)\right) + a_1 \operatorname{sd} N(s)\right]} \tag{2.37}$$

$$\boxed{y(s) = \frac{c_2}{d_2}s + \frac{\sqrt{2}}{(d_0 + d_2)\sqrt[4]{\beta_x^2 + \beta_y^2}}\left[\frac{c_0 d_2 - c_2 d_0}{d_2}\Pi\left(k^2, k, N(s)\right) + c_1 \operatorname{sd} N(s)\right]} \tag{2.38}$$

where Π is the third kind of incomplete elliptic integral and, following Glaisher, $\operatorname{sd}x = \operatorname{sn}x/\operatorname{dn}x$.

Proof. Both integrations are of the same nature, and then we will consider the case (2.35), proving (2.37) only. Splitting in partial fractions, and omitting the modulus k, by identities:

$$\frac{a_0 + a_1 z + a_2 z^2}{b_0 + b_2 z^2} = \frac{a_2}{b_2} + \frac{\dfrac{a_0 b_2 - a_2 b_0}{b_2} + a_1 z}{b_0 + b_2 z^2}, \quad b_0 + b_2 \operatorname{cn}^2\sigma = (b_0 + b_2)\left(1 - \frac{b_2}{b_0 + b_2}\operatorname{sn}^2\sigma\right),$$

one finds

$$x(s) = \frac{a_2}{b_2}s + \frac{\sqrt{2}}{(b_0 + b_2)\sqrt[4]{\beta_x^2 + \beta_y^2}}\left[\frac{a_0 b_2 - a_2 b_0}{b_2} I_{x1}(s) + a_1 I_{x2}(s)\right]$$

with

$$I_{x1}(s) = \int_0^{N(s)} \frac{1}{1 - k^2 \operatorname{sn}^2\sigma}\, \mathrm{d}\sigma, \qquad I_{x2}(s) = \int_0^{N(s)} \frac{\operatorname{cn}\sigma}{1 - k^2 \operatorname{sn}^2\sigma}\, \mathrm{d}\sigma.$$

The first integral is given by [11] formula 336.01: $I_{x1}(s) = \Pi\left(k^2, k, N(s)\right)$. In order to compute $I_{x2}(s)$, putting $\operatorname{sn}^2\sigma = w$, we get $2 \operatorname{sn}\sigma \operatorname{cn}\sigma \operatorname{dn}\sigma\, d\sigma = \mathrm{d}w$, where $\operatorname{dn}\sigma$ is the Jacobi delta amplitude elliptic function $\operatorname{dn}^2\sigma = 1 - k^2 \operatorname{sn}^2\sigma$. We find out

$$I_{x2}(s) = \frac{1}{2}\int_0^{\operatorname{sn}^2 N(s)} \frac{1}{(1 - k^2 w)\sqrt{w(1 - k^2 w)}}\, \mathrm{d}w.$$

Putting $z = \frac{1}{k}\sqrt{\frac{1-k^2 w}{w}}$, we get $I_{x2}(s) = \operatorname{sd}N(s)$. □

2.3.6 A Meaningful Generalization about Loads

What if a bending couple $\mathfrak{M}_0$ is added as a further load to the free end? Can our solution be extended to the free end loaded simultaneously by a force and a couple $\mathfrak{M}_0$? Of course the couple is not whichever, but its plane shall be parallel to that of rod and force. In such a case (2.23) shall be replaced by

$$M_z(s) \pm \mathfrak{M}_0 = -EJ\frac{\mathrm{d}\varphi(s)}{\mathrm{d}s},$$

hence (2.25) is affected only at the free end boundary condition, which becomes

$$\frac{\mathrm{d}\varphi}{\mathrm{d}s}(0) = \frac{\mp\mathfrak{M}_0}{EJ}.$$

Then we will find a different $\varphi(s)$. The φ_0 evaluation is not involved by the $\mathfrak{M}_0$ addition. On the contrary, the x and y functional expressions will be conditioned by the new $\varphi(s)$ only.

Kirchhoff demonstrated that under forces applied only to ends of a rod, the relevant equilibrium equations are identical to those of a rigid body moving around a fixed point (spinning top), so that each generalization/progress in such a field of Elasticity will affect the rigid Dynamics too.

2.4 The Cantilever Deflections by Means of the Lauricella Hypergeometric Functions

Five elasticæ are analytically solved of a thin cantilever bent by different coplanar "dead" loadings[3]: namely whose direction is fixed in the space during the rod deformation. They are concentrated couple, tip shear, evenly continuous load, hydrostatic load, and sinusoidal load. Some care has been put (tip sheared rod) in providing analytically, via the isometric assumption, the free end position after the strain, parametrizing its coordinates as a function of the load. Some of our solutions, specially that to the heavy rod by the Lauricella functions, have enriched the literature on the elastica.

2.4.1 The Main Assumptions

We tackle a L-long, thin, weightless rod whose end A is clamped and B free. We put at B the origin O of a (x, y) cartesian reference frame with x along the undeformed rod, from B to A; and y downwards, normal at B to x. Our basic assumptions are as follows:

- A_1 the rod is thin, initially straight, homogeneous, with uniform cross section and flexural stiffness EJ, where E is the Young modulus, and J the cross section moment of inertia about a "neutral" axis normal to the plane of bending and passing through the central line;
- A_2 the slender rod is charged by coplanar dead loads;
- A_3 Linear constitutive load: the induced curvature is proportional via $1/(EJ)$ to the bending moment. The rod undergoes large deflections so that its curvature cannot be linearized.

[3]The opposite case is that of the *follower load* which is being unchanged in its relationship with the rod, e.g., keeping its direction perpendicular, or inclined, to it during and after the strain.

A_4 The shear transverse deformation is ignored, and the rod is assumed to keep its length constant (unextensibility)

A_5 A stationary strain field, by isostatic equilibrium of active loads and reactive forces takes place, and, due to rest of static equilibrium, no rod element undergoes acceleration.

The key Bernoulli–Euler linear constitutive law connects flexural stiffness EJ and bending moment M to the consequent curvature c of the deflected centerline $\pm EJc = M(x)$. With our (traditional) reference frame (x, y), the curvature always has an opposite sign to the bending moment:

$$EJ\frac{y_{xx}}{(1+y_x^2)^{3/2}} = -M(x) \tag{2.39}$$

where $y = y(x)$ is the unknown elastica's cartesian equation and $y_x = \mathrm{d}y/\mathrm{d}x$, $y_{xx} = \mathrm{d}^2y/\mathrm{d}x^2$. Equation (2.39) can be written in a more useful form as

$$\frac{\mathrm{d}}{\mathrm{d}x}\left(\frac{y_x}{\sqrt{1+y_x^2}}\right) = -\frac{1}{EJ}M(x)$$

to be solved together to the perfect clamping boundary conditions $y(L) = 0$, $y'(L) = 0$. For each of our five situations within this section we will solve the Cauchy problem:

$$\begin{cases} \dfrac{\mathrm{d}}{\mathrm{d}x}\left(\dfrac{y_x}{\sqrt{1+y_x^2}}\right) = -\dfrac{1}{EJ}M(x), \quad 0 \le x \le L \\ y(L) = 0 \; y_x(L) = 0 \end{cases} \tag{2.40}$$

Putting

$$H(x) := \int_x^L M(s)\mathrm{d}s$$

the solution of (2.40) is

$$y(x) = -\int_x^L \frac{H(s)}{\sqrt{(EJ)^2 - H^2(s)}}\,\mathrm{d}s. \tag{2.41}$$

In the sequel we are going to use dimensionless variables $\xi = x/L$, $\eta = y/L$, so that the previous relationship becomes

$$\begin{cases} \eta_\xi = \dfrac{H(L\xi)}{\sqrt{(EJ)^2 - H^2(L\xi)}} \\ \eta(1) = 0 \end{cases} \tag{2.42}$$

whose solution is

$$\eta(\xi) = \int_1^\xi \frac{H(L\zeta)}{\sqrt{(EJ)^2 - H^2(L\zeta)}}\,\mathrm{d}\zeta. \tag{2.43}$$

In the following subsections we will integrate (2.43) in five physical different situations concerning the large elastic deformations of a rod cantilever.

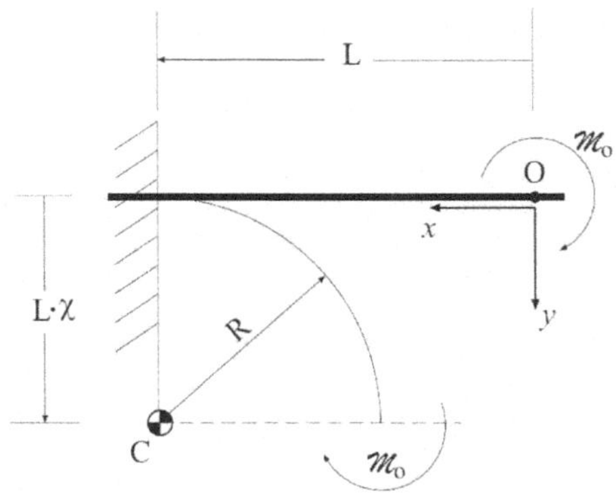

FIGURE 2.6
Horizontal cantilever rod inflected by a constant bending couple at its tip.

2.4.2 Cantilever Inflected by a Constant Bending Couple at Its Tip

The clamped section is at $x = L$, see Figure 2.6. Then equation to be integrated is (2.40), with $M(x) = -\mathcal{M}_0$, so that (2.43) provides, in non-dimensional form:

$$\boxed{\eta(\xi) = \frac{1}{\chi} - \sqrt{\frac{1}{\chi^2} - (\xi - 1)^2}} \tag{2.44}$$

where $\chi = \mathcal{M}_0 L/EJ > 0$. The elastica is better seen reverting at a dimensional form: $x^2 + y^2 - 2Lx - 2\chi Ly + L^2 = 0$, equation of a circle[4] of center $C(L, L\chi)$ and radius $L\chi$. Figure 2.6 shows only its quarter of our concern.

2.4.3 The Heavy Cantilever

The classic elastica theory started by the "lamina", namely a *thin rod without weight but capable to withstand the flexure* and loaded at its free end only. We quote chapter five of [16] on the specific problem of the horizontal cantilever under uniform load gives (page 173)

> This problem... has shown considerable resistance to a closed form solution and all investigations dealing with it have employed approximate methods only.

Accordingly, in [16] are obtained elastica coordinates as parametric functions of the arc length s, stopping their Maclaurin expansions to s^{10}. Let us consider at Figure 2.7a horizontal, flexible, homogeneous, heavy cantilever: it will be inflected by the uniform evenly continuous load q (weight for unity span length). Being

$$M(x) = -\frac{1}{2}qx^2, \tag{2.45}$$

it follows, in non-dimensional coordinates, using (2.43)

$$\eta(\xi) = \frac{\nu}{3}\int_0^{\xi^3-1} \frac{\sigma^{\frac{1}{2}}}{(1+\sigma)^{\frac{2}{3}}(1-\nu\sigma)^{\frac{1}{2}}(1+\nu\sigma)}\, d\sigma$$

[4]This is of course stemming from the Euler–Bernoulli statement: *EJ times the curvature equals moment.* Nevertheless the traditional engineers formulation leads to a parabola as a consequence of linearization.

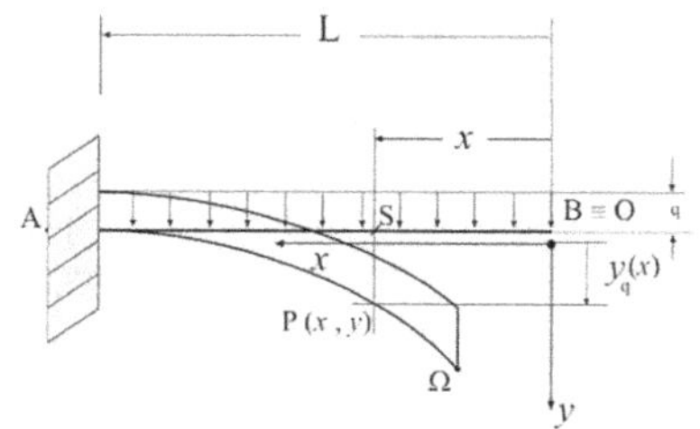

FIGURE 2.7
Horizontal cantilever rod bent by a uniform load.

where $\nu = \frac{qL^3}{6EJ}$. Finally, passing to the variable $\lambda = \sigma/(\xi^3 - 1)$:

$$\eta(\xi) = \frac{\nu\left(\xi^3 - 1\right)^2}{3} \int_0^1 \frac{\lambda}{\left[1 - \lambda\left(1 - \xi^3\right)\right]^{\frac{2}{3}} \sqrt{\left[1 - \lambda\nu\left(\xi^3 - 1\right)\right]\left[1 - \lambda\nu\left(1 - \xi^3\right)\right]}} \, \mathrm{d}\lambda \tag{2.46}$$

By identification, we obtain the solution to our elastica through a Lauricella function of three variables $x_1 = 1 - \xi^3$, $x_2 = \nu\left(\xi^3 - 1\right)$, *and* $x_3 = \nu\left(1 - \xi^3\right)$, so that

$$\boxed{\eta(\xi) = \frac{\nu}{6}\left(\xi^3 - 1\right)^2 F_D^{(3)}\left(\begin{array}{c} 2, \frac{2}{3}, \frac{1}{2}, \frac{1}{2} \\ 3 \end{array}\middle|\, 1 - \xi^3, \nu\left(\xi^3 - 1\right), \nu(1 - \xi^3)\right)} \tag{2.47}$$

For being compliant with the integral representation theorem requirements, it shall be $|\hat{x}_k| < 1$, and then $\nu < 1$. Otherwise, another method will be necessary. We got three ν-ranges. The first is the very narrow one, where the small deflections assumption is true, first order theory. By $|\eta'(\xi)| < h^2, h \in (0, 1)$, we get

$$\nu\left|\xi^3 - 1\right| < h^2\sqrt{1 - \nu^2\left(\xi^3 - 1\right)^2},$$

which at $\xi = 0$, gives: $\nu < h^2/(\sqrt{1 + h^4})$. For instance, if $h = 1/10$, then

if $0 < \nu < 0.0099$ the first order approximation can be accepted;

if $0.0099 < \nu < 1$, by representation theorem the hypergeometric function $F_D^{(3)}$ is capable of describing our solution (2.46)

if $\nu \geq 1$, solution of (2.46) has to be computed numerically.

Formula (2.47), founded on the Lauricella $F_D^{(3)}$ hypergeometric functions, provides a very compact explicit solution to the elastica of a thin clamped heavy rod.

Remark 2.6. *Its elastica, according to the approximate beam theory, would consist of a fourth order rational function of x, i.e.,*

$$y(x) = \frac{1}{EJ}\left(-\frac{q}{24}x^4 + c_1 x + c_2\right)$$

2.4.4 A Tip Sheared Horizontal Cantilever

We refer to Figure 2.8, where the force R is transversal (in [34] is studied the inclined end load) but not a follower one, and we have

$$M(x) = Rx, \tag{2.48}$$

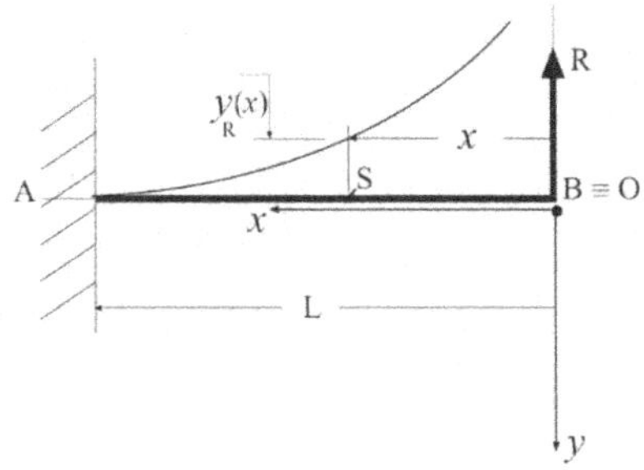

FIGURE 2.8
Horizontal cantilever rod bent under tip-concentrated upwards load.

so that, after defining the non-dimensional $\mu = RL^2/2EJ$, we obtain

$$\eta(\xi) = \mu \int_{\xi}^{1} \frac{1-\zeta^2}{\sqrt{(1-\mu(1-\zeta^2))(1+\mu(1-\zeta^2))}} \mathrm{d}\zeta, \tag{2.49}$$

the bending moment being always greater than zero, so will be $H(x)$, and then it shall be $\eta_\xi > 0$. For this load too, the hypergeometric treatment of the above hyperelliptic integral is possible.

2.4.4.1 A Hypergeometric Treatment

On the purpose, let us pass from ζ to λ , with $\lambda = (1-\zeta^2)/(1-\xi^2)$, we obtain

$$\eta(\xi) = \frac{\mu}{2}\left(1-\xi^2\right)^2 \int_0^1 \frac{\lambda}{\sqrt{[1-\lambda(1-\xi^2)][1-\lambda\mu(1-\xi^2)][1-\lambda\mu(\xi^2-1)]}} \mathrm{d}\lambda \tag{2.50}$$

Recognizing the Lauricella parameters $a = 2$, $b_i = 1/2$, $c = 3$:

$$\boxed{\eta(\xi) = \frac{\mu}{4}\left(1-\xi^2\right)^2 F_D^{(3)}\left(\begin{array}{c} 2; \frac{1}{2}, \frac{1}{2}, \frac{1}{2} \\ 3 \end{array} \middle| \left(1-\xi^2\right), \mu\left(\xi^2-1\right), \mu\left(1-\xi^2\right)\right)} \tag{2.51}$$

is compliant to the representation theorem requirements if $\mu < 1$. For this load we have to Lauricella's the alternative of the elliptic integrals (not applicable to the heavy nor to the submerged rod), so that we will cover this approach with much more details.

2.4.4.2 A Treatment by Elliptic Integrals

Investigating the elastica of a thin cantil ever, the majority of Authors (Landau [22], Love [24], Mathieu [26], Burgatti [10], Panayotounakos and Theocaris [36], [37] and [35], Chucheepsakul and Phungpaigram [12], Basoco, [3], Frisch-Fay, [16]) use the slope-arc length formulation (ϑ, s). We too did so in Section 2.3. The approach consists of integrating the first time for getting ϑ as a function of s. But $\mathrm{d}x = \mathrm{d}s \cos\vartheta(s)$ and $\mathrm{d}y = \mathrm{d}s \sin\vartheta(s)$, so that, through a further integration the elastica's x and y coordinates turn to be parametrized on s. Some other (x, y) parametric equations can be found, e.g., in Saalschütz, [41].

Anyway, a treatment leading to explicit sole cartesian coordinates not seems to be ever established: and it is exactly what we are doing. An old reference, rather fit to Jakob Bernoulli (1694) and Euler (1744) formulae, is that of Kiepert (1912), who arrives at

$$y(x) = \pm \int \frac{C_1 \pm x^2}{\sqrt{a^4 - (C_2 \pm x)^2}} \mathrm{d}x$$

but stops without reducing the above integral to any standard form. It will be just the elliptic integral (2.49) our startup. Let us change its variable passing from ζ to $t = 1 - \zeta^2$, obtaining

$$\eta(\xi) = \frac{1}{2}\int_0^{1-\xi^2} \frac{t}{\sqrt{\left(\frac{1}{\mu} - t\right)(1-t)\left(t + \frac{1}{\mu}\right)}}\,\mathrm{d}t \tag{2.52}$$

For the reader convenience we fix some notation. Let $a > b > x > y > c$. Consider the elliptic integral:

$$\mathcal{I}(a,\, b,\, c, x,\, y) = \int_y^x \frac{t}{\sqrt{(a-t)(b-t)(t-c)}}\,\mathrm{d}t$$

Combining formulae given in [11]: integral 234.16 page 76 and integral 340.01 page 205, introducing

$$k^2 = \frac{b-c}{a-c}, \quad \varphi(x) = \arcsin\sqrt{\frac{(a-c)(b-x)}{(b-c)(a-x)}},$$

we obtain

$$\mathcal{I}(a,\, b,\, c, x,\, y) = \frac{2}{\sqrt{a-c}}\left\{a\left[F(\varphi(y),k) - F(\varphi(x),k)\right] - (a-b)\left[\Pi(\varphi(y),k^2,k)\right.\right.$$
$$\left.\left. - \,\Pi(\varphi(x),k^2,k)\right]\right\}$$

where $F(\varphi, k)$ and $\Pi(\varphi, k^2, k)$ are the first and third kind incomplete elliptic integrals. Now assuming $\mu < 1$ we have the identification

$$\begin{array}{ccccc} a & b & x & y & c \\ \updownarrow & \updownarrow & \updownarrow & \updownarrow & \updownarrow \\ 1/\mu & 1 & 1-\xi^2 & 0 & -1/\mu \end{array}$$

which allows to give (2.49) via (2.52) by means of the integration formula:

$$\boxed{\eta(\xi;\mu) = \sqrt{\frac{\mu}{2}}\left[\frac{1}{\mu}\mathcal{F}\,(\xi,\mu) + \left(1 - \frac{1}{\mu}\right)\mathcal{P}(\xi,\mu)\right]} \tag{2.53}$$

where

$$\mathcal{F}(\xi,\mu) = F\left(\varphi(\mu), \sqrt{\frac{1+\mu}{2}}\right) - F\left(\psi(\mu), \sqrt{\frac{1+\mu}{2}}\right) \tag{2.54}$$

$$\mathcal{P}(\xi,\mu) = \Pi\left(\varphi(\mu), \frac{1+\mu}{2}, \sqrt{\frac{1+\mu}{2}}\right) - \Pi\left(\psi(\mu), \frac{1+\mu}{2}, \sqrt{\frac{1+\mu}{2}}\right) \tag{2.55}$$

being

$$\varphi(\mu) = \arcsin\sqrt{\frac{2\mu}{1+\mu}}, \quad \psi(\mu) = \arcsin\sqrt{\frac{2\mu}{1+\mu}\,\frac{\xi^2}{1-\mu(1-\xi^2)}} \tag{2.56}$$

Formula (2.53) links the dimensionless deflection η to the position ξ and to load μ. From it one could obtain the relevant contour plots cutting by planes of $\mu =$ constant. The thing is shown below, where the clamped end is on the left. The deflection curves are given for some values of $\mu : 0.2; 0.4; 0.6; 0.8$. Each of them is compared with the dot curve coming from the

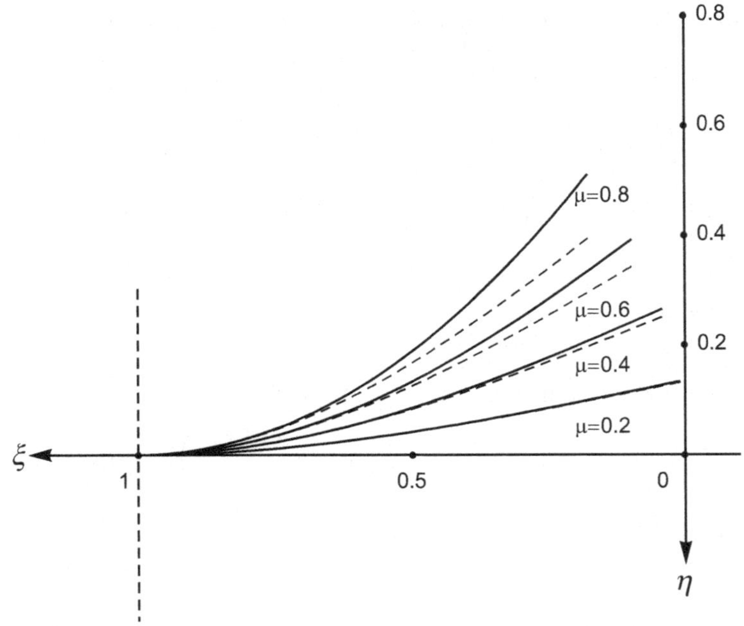

FIGURE 2.9
The sheared cantilever: solution to (2.49) with some μ values compared to the dotted lines of the linearized theory.

approximate theory[5]. Mind that ξ is ranged as $\xi_\Omega < \xi < 1$, and ξ_Ω has to be found from (2.62). Accordingly, each single μ-curve has been plotted between $\xi = 1$ and $\xi_\Omega(\mu)$. From the physical viewpoint, the lines[6] comparison shows at the low loads an agreement which is becoming worst for increasing loads when the approximate theory is progressively failing (Figure 2.9).

2.4.4.3 Consistency with the Literature

It is worth noting that the power series expansion of (2.53) drives back to the approximate expression for $\eta(\xi)$ used in literature. Of course we deeply exploited the computer algebra power for this task. First we computed the Maclaurin third order expansion of (2.53) with respect to ξ obtaining

$$\eta(\xi) = \frac{\mu}{3\,(1-\mu^2)^{3/2}}\,\xi^3 - \frac{\mu}{\sqrt{1-\mu^2}}\,\xi + \sqrt{\frac{\mu}{2}}\,A(\mu)$$

with

$$A(\mu) = \frac{1}{\mu}\,F\left(\varphi(\mu), \sqrt{\frac{\mu+1}{2}}\right) + \left(1 - \frac{1}{\mu}\right)\Pi\left(\frac{\mu+1}{2}; \varphi(\mu), \sqrt{\frac{\mu+1}{2}}\right)$$

[5] Under transversal load at the cantilever tip this problem in Strength of Materials is solved in cartesian coordinates following the linear approximation: the solution is found (e.g., see [5] p. 305) to be $|\eta(\xi)| = (\mu/3)(2 - 3\xi + \xi^3)$ in dimensionless form.

[6] Notice that the plots are coming after Mathematica®'s computation of our explicit solution (2.53), but adding the map $\xi \to 1 - \xi$ in order to provide the geometrical true configuration of the stressed rod.

where ($\varphi(\mu)$ has been introduced in (2.56)). Now we expand the elliptic integrals of the first and third kind with respect to μ, with initial point $\mu_0 = 0$:

$$F\left(\varphi(\mu), \sqrt{\frac{\mu+1}{2}}\right) = \sqrt{2}\,\mu^{1/2} + \frac{31}{60\sqrt{2}}\mu^{5/2} + O\left(\mu^{7/2}\right)$$

$$\Pi\left(\frac{\mu+1}{2}; \varphi(\mu), \sqrt{\frac{\mu+1}{2}}\right) = \sqrt{2}\,\mu^{1/2} + \frac{\sqrt{2}}{3}\mu^{3/2} + O\left(\mu^{5/2}\right)$$

arriving at

$$\sqrt{\frac{\mu}{2}}\,A(\mu) = \frac{2}{3}\mu + O\left(\mu^2\right) \tag{2.57}$$

Moreover recalling the Maclaurin first order expansion with respect to μ

$$\frac{\mu}{3\left(1-\mu^2\right)^{3/2}}\xi^3 - \frac{\mu}{\sqrt{1-\mu^2}}\xi = \mu\left(\frac{\xi^3}{3} - \xi\right) + O\left(\mu^2\right) \tag{2.58}$$

adding (2.57) and (2.58) we obtain eventually

$$\eta(\xi) = \frac{\mu}{3}\left(2 - 3\xi + \xi^3\right) + O\left(\mu^2\right) \tag{2.59}$$

i.e., the usual approximation. Going again to formula (2.53), his peculiarities are: it is non-dimensional; it is new, in closed form, exact, and explicit. Moreover, for being implemented, no previous transcendental equation has to be solved: it requires only to compute the dimensionless parameter $\mu = (RL^2)/(2EJ)$ holding all the rod physical data.

2.4.4.4 How to Compute the Tip Position after the Strain

Considering the small deflections, the deformed centerline's curvature is approximately equated to the deflection's second derivative, assuming the rod deformation field to proceed along purely vertical displacements. If it is so, the free end O ($\xi = 0$, $\eta = 0$) will be mapped to $\xi = 0$, with $\eta(0) \neq 0$, but having the elastica as horizontal projection its full unstrained length, then the rod will undergo extension.

With large displacements all this is not true any more: and a further unknown will appear, namely the abscissa ξ_Ω of the new position Ω where O has been mapped to. Rod's behavior under strain can be modeled in several ways. For instance in [12] after integrating the elastica, the authors write five congruence relationships involving geometric quantities before and after the strain, and then leave the rod free of extension. Following a completely different method, in [40] some possible alternatives like rotational load, load always perpendicular to the tangent at the tip, and so on are taken into account. Anyway, the simplest criterion is the inextensibility assumption A4: the rod keeps its previous length. For the horizontal weightless cantilever inflected by a tip shear (see Figure 2.8), the elastica $\eta\left(\xi\right)$ has been provided by means of elliptic integrals, see formula (2.53). Wishing to compute the tip deflection f, the approximate beam theory, would answer $f = \eta(0)$; but with large deflection we have to first detect ξ_Ω what is done assuming the unextensibility:

$$\int_{\xi_\Omega}^{1} \sqrt{1+\eta_\xi^2}\,\mathrm{d}\xi = 1. \tag{2.60}$$

We see ξ_Ω is the sole unknown, being known η_ξ; so that, after ξ_Ω we will compute $f = \eta\left(\xi_\Omega\right)$. Let us provide the full treatment for this case. Plugging

$$\eta_\xi = \mu\frac{1-\xi^2}{\sqrt{\left(1-\mu\left(1-\xi^2\right)\right)\left(1+\mu\left(1-\xi^2\right)\right)}}$$

in (2.60), after the change $1 - \xi^2 = t$, we get

$$2\mu = \int_0^{1-\xi_\Omega^2} \frac{\mathrm{d}t}{\sqrt{\left(\frac{1}{\mu} - t\right)(1-t)\left(\frac{1}{\mu} + t\right)}}. \tag{2.61}$$

Our unknown appears in the upper integration limit of (2.61): we omit the relevant treatment—which the reader can find at pp. 10–11 of [32]—giving the solution:

$$\xi_\Omega = \sqrt{\frac{1-\mu^2}{2\mu}} \left\{ \frac{\operatorname{sn}\left(\sqrt{2\mu} - F\left(\arcsin\sqrt{\frac{2\mu}{\mu+1}}, \sqrt{\frac{\mu+1}{2}}\right), \sqrt{\frac{\mu+1}{2}}\right)}{\operatorname{dn}\left(\sqrt{2\mu} - F\left(\arcsin\sqrt{\frac{2\mu}{\mu+1}}, \sqrt{\frac{\mu+1}{2}}\right), \sqrt{\frac{\mu+1}{2}}\right)} \right\} \tag{2.62}$$

Eventually η_Ω is obtained plugging (2.62) in (2.53).

2.4.5 The Cantilever Loaded by Sinusoidal Bending Moment

Let us analyze the effect of the sinusoidal load, whose half-wave is only taken, omitting all the possible variants or subcases. Let $q(x)$ be the continuous transverse load (whose highest value is Q) for unity span length of a L-length flexible cantilever: $q(x) = Q\sin\left(\frac{\pi}{L}x\right)$. Integrating twice, one arrives at the bending moment affecting the x-section:

$$M(x) = -\frac{L^2 Q}{\pi^2}\sin\left(\frac{\pi}{L}x\right).$$

In such a way, introducing $\varepsilon = L^3 Q/(\pi^3 EJ) > 0$, with[7] $\varepsilon \ll 1$, following (2.43) we find

$$\eta(\xi) = -\varepsilon\int_1^{\xi} \frac{1+\cos(\pi\zeta)}{\sqrt{1-\varepsilon^2\left[1+\cos(\pi\xi)\right]^2}}\mathrm{d}\zeta. \tag{2.63}$$

If $1 + \cos(\pi\zeta) = u/\varepsilon$, we get the integral:

$$\eta(\xi) = \frac{1}{\pi}\int_0^{\varepsilon[1+\cos(\pi\xi)]} \sqrt{\frac{u}{(1-u)(1+u)(2\varepsilon - u)}}\,\mathrm{d}u. \tag{2.64}$$

Using entries 254.12 page 113 and 337.01 page 201 of [11], we can establish an integration formula, embedding (2.64) as a particular case: let $a > b \geq y > c > d$ and define

$$k^2 = \frac{(a-d)(b-c)}{(a-c)(b-d)}, \quad \alpha^2 = \frac{b-c}{b-d}, \quad \varphi = \arcsin\sqrt{\frac{(b-d)(y-c)}{(b-c)(y-d)}}$$

then:

$$\int_c^y \sqrt{\frac{u-c}{(a-u)(b-u)(u-d)}}\,\mathrm{d}u = \frac{2(b-c)(c-d)}{(b-d)\alpha^2}\sqrt{(a-c)(b-d)}\,\left[\Pi(\varphi,\alpha^2,k) - F(\varphi,k)\right] \tag{2.65}$$

where $\Pi(\varphi, \alpha^2, k)$ is the elliptic integral of the third and $F(\varphi, k)$ of the first kind, respectively.

[7] Inserting in ε some numerical values, e.g., $L = 5$ m, $q = 6$ N/m, $EJ = 100$ N/m^2, $g = 9.81$ m/s^2, one gets $\varepsilon \approx 3 \times 10^{-2}$, so that, unless of assuming unrealistic lengths, ε is essentially much less than unity.

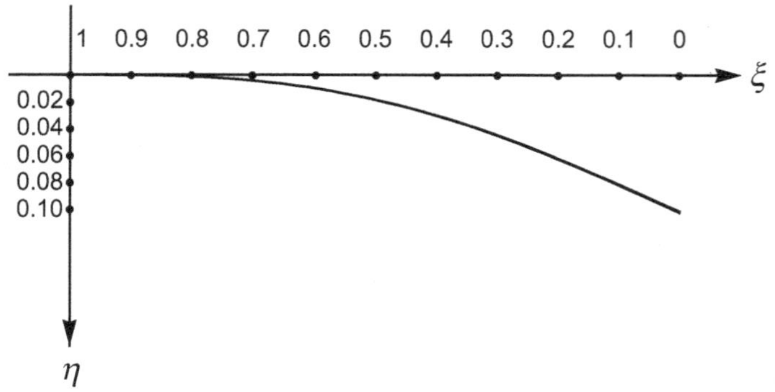

FIGURE 2.10
Overlap of integral (2.63) evaluated numerically and by formula (2.66) with $\varepsilon = 0.1$. The plot shall be conveniently oriented: the rod's clamped end is at the right side, i.e., $\xi = 1$, and the load is thought to act upwards.

Being $2\varepsilon < 1$, in (2.64) we do the identification:

$$\begin{array}{ccccc} a & b & y & c & d \\ \updownarrow & \updownarrow & \updownarrow & \updownarrow & \updownarrow \\ 1 & 2\varepsilon & \varepsilon\left[1+\cos(\pi\xi)\right] & 0 & -1 \end{array}$$

so that our elastica is

$$\boxed{\eta(\xi;\varepsilon) = \frac{2}{\pi\sqrt{2\varepsilon+1}}\left[\Pi\left(\theta(\xi,\varepsilon), \frac{2\varepsilon}{2\varepsilon+1}, 2\sqrt{\frac{\varepsilon}{2\varepsilon+1}}\right) - F\left(\theta(\xi,\varepsilon), 2\sqrt{\frac{\varepsilon}{2\varepsilon+1}}\right)\right]} \tag{2.66}$$

where

$$\theta(\xi,\varepsilon) := \arcsin\left(\frac{\sqrt{2\varepsilon+1}}{\sqrt{\varepsilon\cos(\pi\xi)+\varepsilon+1}}\cos\left(\pi\xi/2\right)\right) \tag{2.67}$$

Figure 2.10, obtained through Mathematica®, provides the overlapping of the integral (2.63) computed numerically and the plot consequent to our compact formula (2.66).

2.4.6 The Cantilever Inflected by Hydrostatic Pressure

We look for the shape of an (L, E, J) thin cantilever clamped at the bottom of a pool of calm water. The rod weight has been deliberately ignored, being partially balanced by Archimedes buoyancy. The rod is submerged at full length, with the adjustments for a rod leaning out of the water being obvious or whose tip is under the sheet of water. We assume, as in Figure 2.11, a reference with the y axis downwards and x axis at the water surface. The hydrostatic head, with ρ being the liquid mass density and g the gravity, will have a linearly deep-growing profile (Stevin law), so that, integrating twice such a head, one gets the bending moment:

$$M(x) = \frac{1}{6}\rho g L x^3 \tag{2.68}$$

By the usual identification, we obtain the solution to our elastica through the Lauricella function of three variables $x_1 = \sigma\left(\xi^4 - 1\right), x_2 = \sigma\left(1 - \xi^4\right), x_3 = 1 - \xi^4$, where $\sigma = \rho g L^5/24EJ$, so that

$$\boxed{\eta(\xi) = \frac{\sigma}{8}\left(1-\xi^4\right)^2 F_D^{(3)}\left(\begin{array}{c} 2, \frac{1}{2}, \frac{1}{2}, \frac{3}{4} \\ 3 \end{array}\middle|\, \sigma\left(\xi^4-1\right), \sigma\left(1-\xi^4\right), 1-\xi^4\right)} \tag{2.69}$$

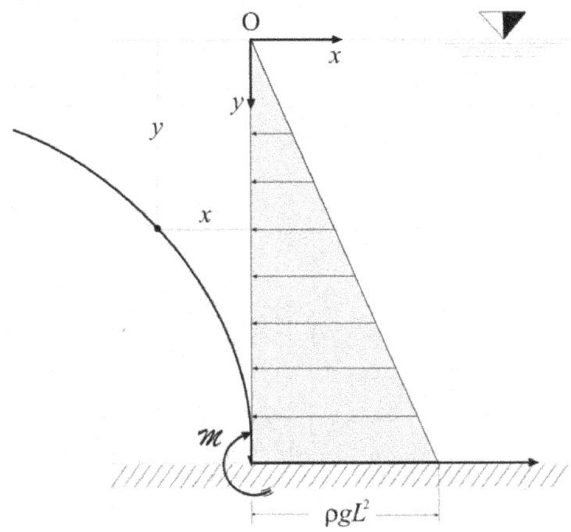

FIGURE 2.11
Submerged vertical rod under hydrostatic pressure.

For being compliant with the representation theorem requirements, it shall be $|x_k| < 1$, and then $\sigma < 1$, which is nothing but a compatibility condition between the rod (L, E, J) and the liquid environment (ρ).

Remark 2.7. *For such a vertical cantilever the approximate solution, in dimensional form, is*

$$y(x) = \frac{1}{EJ}\left(-\frac{\rho g L}{120}x^5 + c_1 x + c_2\right).$$

2.5 A Thin Heavy Flagpole Bent under a Transverse Wind: Its Elastica through the Bessel Functions

2.5.1 The Problem

We compute the planar shape assumed by a slim, vertical flagpole, curved under the simultaneous actions of its own weight, and a transverse wind striking it horizontally. If large deflections occur, its elastica can only be found numerically. But considering the small deflections, we get a third order, variable coefficients, linear ODE. A change of variables leads to a non-homogeneous Airy equation whose integration provides the elastica's slope η_ξ at each ξ through the first kind Bessel functions of rational order. A third integration will lead to the elastica, namely the deflection $\eta(\xi)$, finally obtained through a ${}_1F_2$ hypergeometric function, *which is the sum of a power series defined and convergent for each value of its argument.* This problem was studied for the first time by Greenhill with some simplifications [20]. Anyway, we think our solution is new in the literature.

2.5.2 The Heavy Flagpole under a Transverse Wind

Figure 2.12 shows a L-long pole clamped at G. The average wind speed will increase with height[8] and will attain its maximum value (*gradient speed* w), and it is named so as it is ruled solely by the air pressure gradient.

[8]This is due to the net decreasing obstacles by air stream lifting progressively.

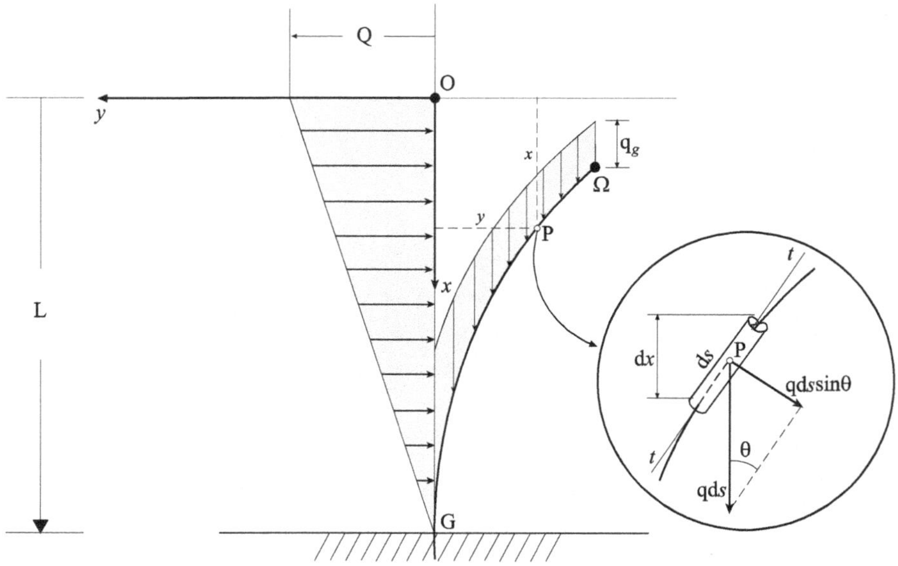

FIGURE 2.12
The heavy flagpole bent by a transverse wind.

In evaluating the wind pressure against the pole, a linear law (viscous drag) will connect the specific load q_w and w, so that q_w will grow linearly with the elevation from the ground to the top[9]:

$$\frac{q_w(x)}{Q} = \frac{L - x}{L}$$

In order to get its contribution to flexure, mind that the $x-$ shear stress due to wind is

$$T_w(x) = Q \int_0^x \left(1 - \frac{u}{L}\right) \mathrm{d}u$$

so that

$$\mathrm{d}M_w = Qx\left(1 - \frac{x}{2L}\right) \mathrm{d}x, \tag{2.70}$$

Let q_g be the weight unity span load: then a bending moment due to such a weight will act. The specific gravity force q_g is uniformly *distributed along the arc length* $\mathrm{d}s$, so that the relevant shear stress on it will be: $\mathrm{d}T_g = q_g \mathrm{d}s \sin\theta$, namely $q_g \mathrm{d}x$, and then, $T_g(x) = q_g x$. But the rotation equilibrium of a elementary rod slice provides

$$\mathrm{d}M_g = q_g x \mathrm{d}y, \tag{2.71}$$

and the total rate of the bending moment by (2.71) and (2.70) results to be

$$\frac{\mathrm{d}M}{\mathrm{d}x} = Qx\left(1 - \frac{x}{2L}\right) + q_g x \frac{\mathrm{d}y}{\mathrm{d}x}. \tag{2.72}$$

Then, even knowing the loads Q, q_g, the bending moment is not integrable as depending on $\mathrm{d}y/\mathrm{d}x$. Taking the derivative with respect to x of (2.39) and injecting there (2.72), we get

$$\frac{\mathrm{d}^2}{\mathrm{d}x^2}\left(\frac{y_x}{\sqrt{1 + y_x^2}}\right) = -Qx\left(1 - \frac{x}{2L}\right) - q_g x y_x$$

[9]Knowing the speed wind w, one can know its kinetic pressure by $p = 1/2\,\varrho w^2$ with ϱ being the air density. For instance if $w = 25$ m/s and $\varrho = 1.225$ kg/m^3, then $p = 383$N/m^2, and, against a breadth of 0.03 m, we will get $Q \approx 11$ N/m.

Let $\xi = x/L$, $\eta = y/L$, $k^2 = q_g L^3/(EJ)$, $\beta = QL^3/(EJ)$; then:

$$\frac{\mathrm{d}^2}{\mathrm{d}\xi^2}\left(\frac{\eta_\xi}{\sqrt{1+\eta_\xi^2}}\right) = -k^2\,\xi\eta_\xi - \beta\xi\left(1-\frac{\xi}{2}\right), \tag{2.73}$$

which is a third order, nonlinear, non-dimensional, non-autonomous ODE to the unknown function η. The boundary conditions to (2.73), assuming, for instance, a *perfect clamping*, are

1. $\eta(1) = 0$, i.e., no clamping displacement;
2. $\eta_\xi(1) = 0$, i.e., no clamping rotation;
3. $\eta_{\xi\xi}(0) = 0$, i.e., zero bending moment at the top.

Putting $\dfrac{\mathrm{d}\eta}{\mathrm{d}\xi} = p(\xi)$, we reduce (2.73) to a second order nonlinear Dirichlet boundary value problem

$$\begin{cases} \dfrac{\mathrm{d}^2}{\mathrm{d}\xi^2}\left(\dfrac{p}{\sqrt{1+p^2}}\right) = -k^2\xi p - \beta\xi\left(1-\dfrac{\xi}{2}\right), \\ p(1) = 0, \quad \text{i.e., no clamping rotation;} \\ p_\xi(0) = 0, \quad \text{i.e., zero bending moment at the top.} \end{cases} \tag{2.74}$$

Not having found any closed form solutions to (2.74), we forget the large rod deflections, restricting to the small ones, as usual in engineering treatments. Nevertheless our treatment peculiarity consists in having taken into account both the weight and the wind, getting a bending moment depending simultaneously on x and $\dfrac{\mathrm{d}y}{\mathrm{d}x}$.

2.5.3 The Analytical Solution of the Third Order ODE

Starting from (2.74) the linearization leads to a beam model consisting of a third order, linear differential equation:

$$\begin{cases} \eta_{\xi\xi\xi} + k^2\xi\eta_\xi = -\beta\xi\left(1-\dfrac{\xi}{2}\right), \\ \eta(1) = 0,\ \eta_\xi(1) = 0,\ \eta_{\xi\xi}(0) = 0 \end{cases} \tag{2.75}$$

whose order can immediately be lowered as

$$\begin{cases} p_{\xi\xi} + k^2\xi p = -\beta\xi\left(1-\dfrac{\xi}{2}\right), \\ p(1) = 0,\ p_\xi(0) = 0 \end{cases} \tag{2.76}$$

Looking for a particular integral like $p^* = b\xi + c$, we find $p^* = \dfrac{\beta}{2k^2}\xi - \dfrac{\beta}{k^2}$ so that the non-homogeneous solution will be $p = p^* + \tilde{p}$, where $\tilde{p}$ is the general solution of the following Equation (2.77):

$$\tilde{p}_{\xi\xi} + k^2\xi\tilde{p} = 0. \tag{2.77}$$

In such a way the general solution to the Airy differential equation appearing in the boundary value problem (2.76) is

$$p(\xi) = \frac{\beta}{k^2}\left(\frac{1}{2}\xi - 1\right) + \frac{\sqrt{\xi}}{3}\left[A\,J_{\frac{1}{3}}\left(\frac{2}{3}k\xi^{\frac{3}{2}}\right) + B\,J_{-\frac{1}{3}}\left(\frac{2}{3}k\xi^{\frac{3}{2}}\right)\right] \tag{2.78}$$

being $J_{\pm 1/3}(x)$ the Bessel ordinary functions of the first kind, see [49], and with the constants A, B to be determined by the boundary conditions in (2.76). Expanding (2.78) in power series in a neighborhood of $\xi = 0$, we can write (2.78) as

$$p(\xi) = \frac{\beta}{k^2}\left(\frac{1}{2}\xi - 1\right) + \frac{B}{\sqrt[3]{9k}\,\Gamma\left(\frac{2}{3}\right)} + \frac{A\,\sqrt[3]{k}\,\xi}{\sqrt[3]{3}\Gamma\left(\frac{1}{3}\right)} - \frac{B\,\sqrt[3]{k^5}\,\xi^3}{6\sqrt[3]{9}\,\Gamma\left(\frac{2}{3}\right)} + \mathrm{o}\left(\xi^3\right)$$

with the usual meaning of the Landau symbol "o". So we can infer that

$$p_\xi(0) = \frac{A\sqrt[3]{k}}{\sqrt[3]{3}\,\Gamma\left(\frac{1}{3}\right)} + \frac{\beta}{2k^2}$$

and now, from the second boundary condition in (2.76) we find

$$A = -\frac{\sqrt[3]{3}}{2}\Gamma\left(\tfrac{1}{3}\right)\frac{\beta}{\sqrt[3]{k^7}} \tag{2.79}$$

Now imposing the first boundary condition in (2.76), we finally obtain

$$B = \frac{\beta\left(3\sqrt[3]{k} + \sqrt[3]{3}\,\Gamma\left(\frac{1}{3}\right)\,J_{\frac{1}{3}}\left(\frac{2k}{3}\right)\right)}{2\sqrt[3]{k^7}\,J_{-\frac{1}{3}}\left(\frac{2k}{3}\right)} \tag{2.80}$$

The formulæ obtained in (2.79) and (2.80) are now plugged in (2.78), leading to the solution to (2.76). For $\beta = 0$ (which means no wind) we have a homogeneous linear differential equation for which the boundary value problem (2.76) has the zero solution only. It will be recalled that the Greenhill criterion for detecting the pole critical condition is founded on the application of $p(1) = 0$ on the (2.78) without the wind, namely $\beta = 0$. In fact for $\xi \to 0^+$, $J_{1/3}$ would be divergent: then A shall be zero. But, in order to get an effective deformed configuration, it shall be $B \neq 0$, so that, in consequence, $J_{-1/3}(2\sqrt{k/3}) = 0$. Recalling that the smallest root of $J_{-1/3}$ is about 1.86635 and $k = \sqrt{q_g\, L^3/EJ}$, one can find the smallest value of L satisfying such a transcendental equation, namely the first eigenvalue, or the critical length for the pole of given EJ and under some k. Should the length of the pole exceed it, then there will be a bend at the upper end: i.e., the pole will buckle. This problem was studied for the first time by Greenhill with some simplifications, see [20]. To end our task on (2.75), we need a further integration of (2.78). Formally the solution of (2.75) is given by the quadrature formula:

$$\eta(\xi) = \int_1^\xi p(\zeta)\mathrm{d}\zeta$$

where two integrals of Bessel functions have to be evaluated. Our integrals are

$$\int_1^\xi \sqrt{\zeta}J_{\pm\frac{1}{3}}\left(\frac{2k}{3}\zeta^{\frac{3}{2}}\right)\mathrm{d}\zeta = \frac{1}{k}\left\{\int_0^{\frac{2k}{3}\xi^{\frac{3}{2}}} J_{\pm\frac{1}{3}}(\zeta)\mathrm{d}\zeta - \int_0^{\frac{2k}{3}} J_{\pm\frac{1}{3}}(\zeta)\mathrm{d}\zeta\right\}$$

Using entry at page 44 of [25], we find out the integral formula:

$$\int_0^z J_\nu(\zeta)\mathrm{d}\zeta = \frac{z^{1\pm\frac{1}{3}}}{2^{\pm\frac{1}{3}}\left(1\pm\frac{1}{3}\right)\Gamma\left(1\pm\frac{1}{3}\right)}\,{}_1\mathrm{F}_2\left(\begin{array}{c}(1\pm 1/3)/2\\ (3\pm 1/3)/2,\ 1\pm 1/3\end{array}\middle| -\frac{z^2}{4}\right)$$

where $\mathrm{Re}(\mu+\nu) > -1$, and ${}_1\mathrm{F}_2$ is a generalized hypergeometric function, see [45] Chapter 2. So that we arrive at the explicit solution of (2.75)

$$\boxed{\eta(\xi) = \frac{1}{12\Gamma\left(\frac{2}{3}\right)\sqrt[3]{k^8}J_{-\frac{1}{3}}\left(\frac{2k}{3}\right)}\mathcal{N}(\xi;\beta,k)} \tag{2.81}$$

where

$$\begin{aligned}\mathcal{N}(\xi;\beta,k) = \beta\Bigg\{ & 2\sqrt[3]{3}\,\xi\left[3\sqrt[3]{k}+\sqrt[3]{3}\,\Gamma\left(\tfrac{1}{3}\right)J_{\frac{1}{3}}\left(\tfrac{2k}{3}\right)\right]{}_1\mathrm{F}_2\left(\begin{matrix}\frac{1}{3}\\ \frac{2}{3},\frac{4}{3}\end{matrix}\middle| -\frac{k^2\xi^3}{9}\right)\\ & -2\left[\sqrt[3]{81k}+\sqrt[3]{9}\,\Gamma\left(\tfrac{1}{3}\right)J_{\frac{1}{3}}\left(\tfrac{2k}{3}\right)\right]{}_1\mathrm{F}_2\left(\begin{matrix}\frac{1}{3}\\ \frac{2}{3},\frac{4}{3}\end{matrix}\middle| -\frac{k^2}{9}\right)+3\sqrt[3]{k^2}\,\Gamma\left(\tfrac{2}{3}\right)J_{-\frac{1}{3}}\left(\tfrac{2k}{3}\right)\\ & \times\left[-\xi^2\,{}_1\mathrm{F}_2\left(\begin{matrix}\frac{2}{3}\\ \frac{4}{3},\frac{5}{3}\end{matrix}\middle| -\frac{k^2\xi^3}{9}\right)+{}_1\mathrm{F}_2\left(\begin{matrix}\frac{2}{3}\\ \frac{4}{3},\frac{5}{3}\end{matrix}\middle| -\frac{k^2}{9}\right)+\xi^2-4\xi+3\right]\Bigg\}\end{aligned}$$

where we take advantage of the capability of Mathematica® which allows full computability, with each being generalized hypergeometric ${}_p\mathrm{F}_q$ implemented on it, so that we can do an *effective computation* of the (β, k) pole's total deflection for any ξ. Let us now provide some plots made by our last formulae (Figures 2.13 and 2.14).

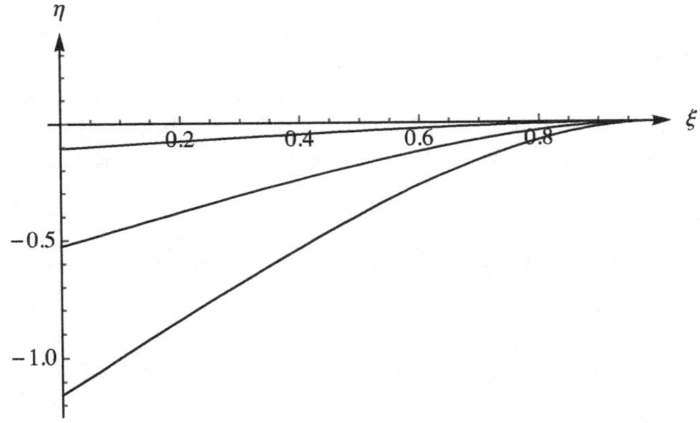

FIGURE 2.13
Pole's elasticæ under increasing winds: $\beta = 1, 3, 5$; $k^2 = 1$.

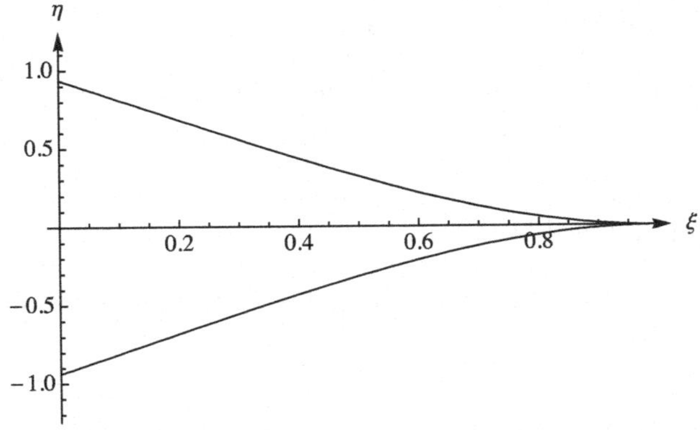

FIGURE 2.14
Pole's elasticæ under opposite winds: $\beta = -5, 5$; $k^2 = 2$.

2.6 Curvature Effects on the Statically Redundant Reactions: The Heavy Cantilever

2.6.1 Statement of the Problem

Much literature deals with the solution of the elastica differential Equation (2.40): we provide the solution by means of elliptic and Lauricella hypergeometric functions (for details see [32]). Nevertheless, such a nonlinear analysis has never been applied, until now, for computing the statically indeterminate unknowns of a beam: in such a way the constraints are popularly credited to exert on them reactions depending on loads and length only, but not on the stiffness EJ. For instance, given a heavy L-rod under a q-weight (N/m) for unity span length, clamped at its (left) side and supported by a roller at the right one: then the roller acts by a vertical upward force of an intensity resulting to be $3qL/8$. A beam, clamped at both ends, will undergo at each of them a bending moment of value $qL^2/12$. All these outcomes, that the reader can find in any textbook on elastic straight beam theory: for instance [50], pages 99 and 100, where the unit weight is called w instead of q, stem from the small strains assumption: if $y(x)$ is the deformed centerline's equation, for each x, it is assumed that

$$y_x^2 \ll 1 \quad \text{with} \quad 0 \leq x \leq L. \tag{2.82}$$

But the above independence is, to some extent, against the physical intuitionism: it is in fact not credible keeping the same q load on a statically indeterminate L-beam whose stiffness (for example) is increasing, and this can be without effects on the constraint reactions. Keeping almost all assumptions upon which the above popular values $3qL/8$ and $qL^2/12$ have been computed and, with this sole difference, founding on the *exact* curvature expression, we will get formulae for the reactions involving EJ too. Therefore the above values are only a first approximation of them. On the basis of these new considerations, we compute the statical indeterminate unknowns through a nonlinear curvature treatment. The nonlinear ODE (2.39) of deflections is employed in advanced applications of the strength of materials in contexts of aerospace or satellites, as in [4]. Our basic assumptions are as follows:

- A_1 the rod is thin, initially straight, homogeneous, with a uniform cross section and flexural stiffness EJ;
- A_2 the rod is always charged by coplanar dead (not follower) loads;
- A_3 linear constitutive law holds: so that the induced curvature is proportional via $(EJ)^{-1}$ to the bending moment.
- A_4 the shear transverse deformation is ignored. As in Section 2.4 the cantilever deformation is assumed isometric.
- A_5 a stationary strain field takes place, and, due to static equilibrium, no rod element undergoes acceleration.

2.6.2 Hypergeometric Tools

In the following we will employ several hypergeometric identities, Theorem 2.5, and reduction formulæ Theorems 2.6 and 2.7 below, which are proved in [33].

Theorem 2.5. *If* $0 < a,\, b < 1$ *then:*

$$\int_0^a \frac{u}{(1-u^2)^{1/2}(1-bu)^{2/3}}\,\mathrm{d}u = \frac{a^2}{2}\mathrm{F}_D^{(3)}\left(\begin{array}{c|c} 2;1/2,1/2,2/3 & \\ & a,-a,ab \\ 3 & \end{array}\right). \tag{2.83}$$

If $0 < a,\, b,\, c$ and $ab < 1$ then:

$$\int_0^a \frac{u}{(1-b^2u^2)^{1/2}(1+cu)^{1/2}}\mathrm{d}u = \frac{a^2}{2}\mathrm{F}_D^{(3)}\left(\begin{array}{c|c} 2;1/2,1/2,1/2 & \\ & ab,-ab,-ac \\ 3 & \end{array}\right). \tag{2.84}$$

Theorem 2.6. *If* $c > a > 0$ *and* $|x| < 1$ *the reduction formula holds true:*

$$\mathrm{F}_1\left(\begin{array}{c|c} a;b,b, & \\ & x,-x \\ c & \end{array}\right) = {}_3\mathrm{F}_2\left(\begin{array}{c|c} \frac{a+1}{2},\frac{a}{2},b & \\ & x^2 \\ \frac{c+1}{2},\frac{c}{2} & \end{array}\right). \tag{2.85}$$

Theorem 2.7. *Assuming that* $c > a + b_3 > 0$, $c > b_3$ *and* $|x| < 1$ *then:*

$$\mathrm{F}_D^{(3)}\left(\begin{array}{c|c} a;b,b,b_3 & \\ & x,-x,1 \\ c & \end{array}\right) = \frac{\Gamma(c)\Gamma(c-a-b_3)}{\Gamma(c-a)\Gamma(c-b_3)}\,{}_3\mathrm{F}_2\left(\begin{array}{c|c} \frac{a+1}{2},\frac{a}{2},b & \\ & x^2 \\ \frac{c-b_3+1}{2},\frac{c-b_3}{2} & \end{array}\right) \tag{2.86}$$

We will also invoke the classical integral representation theorem for the Lauricella $\mathrm{F}_D^{(3)}$ function, see [14], which in the case $n = 3$ reads as if $|x_1| < 1$, $|x_2| < 1$, $|x_3| < 1$ and $c > a > 0$, then

$$\mathrm{F}_D^{(3)}\left(\begin{array}{c|c} a,b_1,b_2,b_3 & \\ & x_1,x_2,x_3 \\ c & \end{array}\right) = \frac{\Gamma(c)}{\Gamma(a)\,\Gamma(c-a)}\int_0^1 \frac{u^{a-1}(1-u)^{c-a-1}}{(1-x_1u)^{b_1}(1-x_2u)^{b_2}(1-x_3u)^{b_3}}\,\mathrm{d}u \tag{2.87}$$

2.6.3 The Statically Indeterminate Heavy Cantilever Supported by a Roller

Let AB be a statically indeterminate rod, Figure 2.15(a), under a uniform continuous load q, whose side A is built-in, while the free tip B is supported by a perfect roller: for such a 1-statically redundant system we assume as unknown the reaction X acted by the roller on the rod. Then the statically equivalent one is that of Figure 2.15(b), whose analysis can be decoupled according to Figure 2.15(c) and (d) and tackled separately. This superposition is allowed first for being within the linear elasticity with the produced curvature proportional through EJ to the bending moment; and furthermore because we are dealing with a isostatic, Figure 2.15(b), loaded beam. Some objection could be raised due to (2.40) nonlinearity: on the purpose, see the relevant discussion on [33].

2.6.3.1 First Subsystem: The Heavy Cantilever

Let us consider the subsystem of Figure 2.15c. The sign conventions are those of use, e.g., the displacements are positive if downwards, and so on. The bending moment is $\mathscr{M}(x) = -(q/2)x^2$. The reality condition of the integrand in (2.41), for this load will imply

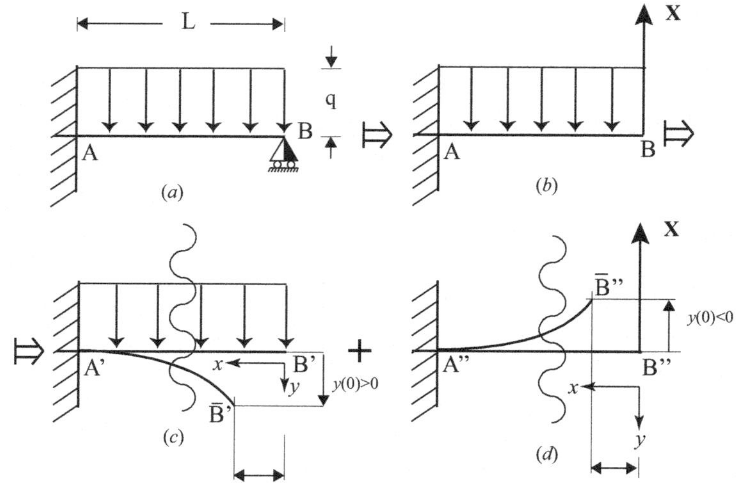

FIGURE 2.15
The heavy cantilever supported by a roller (a), the relevant statically determinate structure (b), and the loads decoupling (c) and (d).

$q < 6EJ/L^3$, namely an upper limit to the q intensity ruled by the flexural rigidity and rod length. Thus (2.41) reads as

$$y(x) = \int_x^L \frac{q(L^3 - s^3)}{6\sqrt{(EJ)^2 - \left(\frac{q}{6}\right)^2 (L^3 - s^3)^2}} \, \mathrm{d}s$$

$$= \frac{2EJ}{qL^2} \int_0^{\frac{q}{6EJ}(L^3 - x^3)} \frac{\sigma}{(1-\sigma^2)^{1/2}\left(1 - \frac{6EJ}{qL^3}\sigma\right)^{2/3}} \, \mathrm{d}\sigma \tag{2.88}$$

where we did the variable change $s^3 = L^3 - (6EJ\sigma)/q$.

Using formula (2.83) of Theorem 2.5 we can evaluate the integral in (2.88), so that

$$y(x) = \frac{q\left(L^3 - x^3\right)^2}{36EJL^2} \mathrm{F}_D^{(3)} \left(\begin{array}{c} 2; 1/2, 1/2, 2/3 \\ 3 \end{array} \middle| \frac{q}{6EJ}(L^3 - x^3), -\frac{q}{6EJ}(L^3 - x^3), 1 - \frac{x^3}{L^3} \right) \tag{2.89}$$

which provides the deformed centerline equation for each x. We are interested to take $x = 0$ in (2.89) getting

$$y(0) = \frac{L^4 q}{36EJ} \mathrm{F}_D^{(3)} \left(\begin{array}{c} 2; 1/2, 1/2, 2/3 \\ 3 \end{array} \middle| \frac{L^3 q}{6EJ}, -\frac{L^3 q}{6EJ}, 1 \right). \tag{2.90}$$

Notice that for $x = 0$ the third argument of Lauricella's becomes 1, so that such a function collapses into an Appell one, and from there it can be reduced to a ${}_3\mathrm{F}_2$ hypergeometric function, see Theorem 2.7, so that we are allowed to describe the elastica by means of a function of only one variable. We can then simplify (2.90) using a the reduction formula (2.87), and therefore we get

$$y(0) = \frac{L^4 q}{8EJ} \, {}_3\mathrm{F}_2 \left(\begin{array}{c} \frac{3}{2}, 1, \frac{1}{2} \\ \frac{5}{3}, \frac{7}{6} \end{array} \middle| \frac{L^6 q^2}{36(EJ)^2} \right) \tag{2.91}$$

which provides the free tip "true" deflection of rod of Figure 2.15c as a function of q, L, E, J.

2.6.3.2 Second Subsystem: The Tip-Sheared Cantilever

We consider now the subsystem of Figure 2.15d. The bending moment is given by $\mathscr{M}(x) = Xx$, so that (2.41) reads as

$$\begin{aligned} y(x) &= \int_x^L \frac{X(s^2 - L^2)}{\sqrt{(2EJ)^2 - X^2(s^2 - L^2)^2}} \, \mathrm{d}s \\ &= \frac{X}{4EJL} \int_{x^2-L^2}^{0} \frac{\sigma}{\left(1 - \left(\frac{X}{2EJ}\sigma\right)^2\right)^{1/2} \left(1 + \frac{1}{L^2}\sigma\right)^{1/2}} \, \mathrm{d}\sigma \end{aligned} \tag{2.92}$$

where we did the change of variable $s = \sqrt{L^2 + \sigma}$. We use the integral representation theorem to evaluate the integral in (2.92), obtaining

$$y(x) = -\frac{\left(x^2 - L^2\right)^2 X}{8EJL} \mathrm{F}_D^{(3)} \left(\begin{array}{c} 2; 1/2, 1/2, 1/2 \\ 3 \end{array} \middle| \frac{(x^2 - L^2)X}{2EJ}, -\frac{(x^2 - L^2)X}{2EJ}, 1 - \frac{x^2}{L^2} \right). \tag{2.93}$$

Taking $x = 0$ in (2.93) too, we get

$$y(0) = -\frac{L^3 X}{8EJ} \mathrm{F}_D^{(3)} \left(\begin{array}{c} 2; 1/2, 1/2, 1/2 \\ 3 \end{array} \middle| -\frac{L^2 X}{2EJ}, \frac{L^2 X}{2EJ}, 1 \right). \tag{2.94}$$

We can simplify (2.94) using Theorem 2.7, formula (2.86) obtaining

$$y(0) = -\frac{L^3 X}{3EJ} \, {}_3\mathrm{F}_2 \left(\begin{array}{c} \frac{1}{2}, 1, \frac{3}{2} \\ \frac{5}{4}, \frac{7}{4} \end{array} \middle| \frac{L^4 X^2}{4(EJ)^2} \right). \tag{2.95}$$

2.6.3.3 Consistence and Detection of the Statically Redundant Unknown

Acting the roller as a perfect constraint, by summing up the displacements given by (2.91) and (2.95), the result shall be zero. Then we arrive at the transcendental equation:

$$3Lq \, {}_3\mathrm{F}_2 \left(\begin{array}{c} \frac{1}{2}, 1, \frac{3}{2} \\ \frac{7}{6}, \frac{5}{3} \end{array} \middle| \frac{L^6 q^2}{36(EJ)^2} \right) - 8X \, {}_3\mathrm{F}_2 \left(\begin{array}{c} \frac{1}{2}, 1, \frac{3}{2} \\ \frac{5}{4}, \frac{7}{4} \end{array} \middle| \frac{L^4 X^2}{4(EJ)^2} \right) = 0 \tag{2.96}$$

whose unknown is the reaction X. Of course, (2.96) cannot be solved analytically, and we will provide an approximate solution. As a check, if we approximate both hypergeometric series only by their first term which is equal to 1, the solution to (2.96) is approximately given by $X = (3/8)\, Lq$ according to the linearized theory. To obtain a better approximation, we proceed through the Lagrange series reversion: for the purpose, in (2.96) we put $X = \left(2EJ/L^2\right) u$ so that (2.96) becomes

$$u \, {}_3\mathrm{F}_2 \left(\begin{array}{c} \frac{1}{2}, 1, \frac{3}{2} \\ \frac{5}{4}, \frac{7}{4} \end{array} \middle| u^2 \right) = \frac{3}{16} \frac{L^3 q}{EJ} \, {}_3\mathrm{F}_2 \left(\begin{array}{c} \frac{1}{2}, 1, \frac{3}{2} \\ \frac{7}{6}, \frac{5}{3} \end{array} \middle| \frac{L^6 q^2}{36(EJ)^2} \right). \tag{2.97}$$

After this, we provide a method to find an approximate u solution of a transcendental equation of the form:

$$f(u) := u \, {}_3\mathrm{F}_2 \left(\begin{array}{c} \frac{1}{2}, 1, \frac{3}{2} \\ \frac{5}{4}, \frac{7}{4} \end{array} \middle| u^2 \right) = z. \tag{2.98}$$

Of course for the application of our interest the real number z is intended as the right-hand side of (2.97).

Lemma 2.2. *Equation* (2.98) *has a unique solution* $f^{-1}(z)$ *for any real number* z *which is approximated by*

$$f^{-1}(z) = z - \frac{12}{35}z^3 + \frac{1952}{13475}z^5 - \frac{1156544}{18393375}z^7 + \frac{4200758784}{152487209875}z^9 - \frac{62541006848}{5157655628125}z^{11} + \cdots \tag{2.99}$$

To achieve an approximate algebraic analytical expression for the solution of Equation (2.96) using the approximation (2.99) we need three more steps: going back to the original unknown X recalling $u = L^2X/(2EJ)$, inserting as z value the right-hand side of (2.97) and eventually expand the hypergeometric function with respect to q. All this leads to the approximate formula, for details, see [33]

$$\boxed{X = \frac{3Lq}{8} - \frac{9L^7q^3}{17920EJ^2} - \frac{16389L^{13}q^5}{2870067200EJ^4} - \frac{3051891L^{19}q^7}{34900017152000EJ^6} - \frac{78942830931L^{25}q^9}{5116621714620416000 0EJ^8} - \cdots} \tag{2.100}$$

In Table 2.1 we presented the value of X obtained with the approximate formula (2.100) as a function of the n number of terms employed there. The linearized solution is obtained for $n = 1$ and clearly overestimates the effective solution; the addition of further terms provides us with a solution which is stable after a reasonable number of terms. We fixed the system parameters as follows: $L = 1$, $q = 1000$, $EJ = 200$. Table 2.1 shows with small step that after ten terms the solution has substantially got a saturation.

Table 2.2 with a greater step ensures the total stationary condition is soundly kept adding whatever terms.

The numerical solution of (2.100) obtained directly using Mathematica® for the selected parameters is $X = 356.2970653$.

2.6.4 Conclusions about the Statically Indeterminate Unknowns

Before going to the conclusions as it concerns the statically indeterminate unknowns, the reader should be aware of the existence of three specific points whose discussion is deferred to [33], namely:

- The nonlinearity of (2.39) and its impliance on the superposition effects.
- About the correctness of the bending moment assumed to depend on x only, while strictly speaking it should also take into account the axial strain of the rod
- Some very particular theorems about the hypergeometric functions to be found as appendix in [33].

TABLE 2.1
Approximation from 1 to 11 Terms with Step of 2

1	3	5	7	9	11
375.	362.4442	358.87525	357.50889	356.90621	356.61728

TABLE 2.2
Approximation from 13 to 49 Terms with Step of 6

13	19	25	31	37	43	49
356.47092	356.32868	356.30355	356.298501	356.2974	356.29715	356.29709

Let us discuss the divergence from the linear approximation as far as it concerns the statically indeterminate unknowns via the elastica method. For a better clarity, we refer to a sample problem shown in Figure 2.15d. Rod of length $L = 1m$ and a square uniform section of side $s = 10^{-2}m$, so that $J = s^4/12 = 10^{-9}m^4$. The material is steel with $E = 2.1 \times 10^{11} N/m^2$, so that each section has a flexural stiffness $EJ \simeq 200Nm^2$. Let us consider first the statically indeterminate unknown relevant to Figure 2.15: according to the approximate linear theory, it is given by $3qL/8$, namely $375N$ for our sample problem. We take a unit continuous load of $q = 10^3 N/m$ which is compliant to the prescription of being $< 6EJ/L^3$. Our solution is provided by a relationship of kind (2.100), namely:

$$X = \sum_{n=0}^{\infty} a_n \frac{L^{6n+1} q^{2n+1}}{(EJ)^{2n}}.$$

The statically indeterminate reaction X of the roller is then computed via a infinite power series involving length, flexural rigidity, unit load, and the relevant coefficients, some of which are shown by (2.100). Therefore we conclude

- C_1 The traditional linearized approach leads to significant errors on the statically indeterminate unknowns. Such deviations would increase for rods that are gradually more flexible. In fact formula (2.100) shows that for $EJ \to 0$ the successive terms' weight is increasing.
- C_2 The above mentioned effects obviously lead to a remarkable impact on the axial bending stresses, say σ s.
- C_3 It is commonly believed and written that the redundant unknowns do not depend on the flexural stiffness, but generally speaking, this is not true.

In fact, our final formula (2.100) shows that this is due to having taken $n = 0$, namely only the first term of the expansion whose subsequent terms hold the reciprocal of powers of EJ. This is against the common textbooks statement, see e.g., [5] page 363, like this:

> Si osservi che quando i vincoli sono rigidi la reazione A non dipende da EJ, mentre ne dipende quando sono cedevoli.[10]

can be accepted *solely within a first order approximation.* The same statement can be found at page 282 of [21]. Mind that we are stating that for a *prismatic* rod the redundant unknowns depend on rigidity. But let us consider a non-prismatic beam on three supports under a continuous load q. The beam has two sections of lengths ℓ_1 and ℓ_2 and rigidities EJ_1 and EJ_2; assuming as a redundant unknown the moment M on the intermediate support, it is well known that

$$M = \frac{\dfrac{q\ell_1^3}{EJ_1} + \dfrac{q\ell_2^3}{EJ_2}}{8\left(\dfrac{\ell_1}{EJ_1} + \dfrac{\ell_2}{EJ_2}\right)}$$

which depends on rigidities. But, for a *prismatic beam* of a constant section, so that $J_1 = J_2$, we get

$$M = \frac{q}{8}(\ell_1^2 + \ell_2^2 - \ell_1\ell_2)$$

which does not depend on flexural rigidity, as currently assumed. Furthermore, note that when the load on the statically indeterminate beam is an inelastic yielding, it is well known

[10] Notice that when the constraints are rigid, the (redundant) reaction does not depend upon EJ, whilst it does when they undergo some yield.

that the redundant, and all, reactions depend not only on the lengths and the imposed loads, but also on EJ. But our investigation—and this is the new thing—has thus revealed the same behavior of the statically indeterminate prismatic rod under non-geometric loads, provided that its deflection is exactly evaluated following the nonlinear curvature. All above and even the methodology that it suggests, namely analytic methods with some numerical help, should be of interest for advanced engineering problems.

References

[1] C. Agostinelli and A. Pignedoli. *Meccanica razionale,* vol I. Zanichelli, Bologna, 1978.

[2] P. Appel and E. Lacour. *Fonctions elliptiques.* Gauthier-Villars, Paris, 1922.

[3] M. Basoco. On the inflexional elastica. *The American Mathematical Monthly,* 48(5):303–309, 1941.

[4] V. V. Beletsky and E. M. Levin. *Dynamics of space tether systems*, vol. 83. Univelt Incorporate, San Diego, 1993.

[5] O. Belluzzi. *Scienza delle costruzioni,* vol. I. Zanichelli, Bologna, 1970.

[6] E. Benvenuto. *La scienza delle costruzioni e il suo sviluppo storico.* Sansoni, Firenze, 1981.

[7] J. Bernoulli. Curvatura laminae elasticae. Ejus identitas cum curvatura lintei a pondere inclusi fluidi expansi. Radii circulorum osculantium in terminis simplicissimis exhibiti una cum novis quibusdam theorematis huc pertinentibus. *Acta Eruditorum,* 13:262–276, 1694.

[8] J. Bernoulli. *Lectiones de methodo integralium, aliisque conscriptae in usum Ill. Marchionis Hospitalii cum auctor Parisiis ageret annis 1691 et 1692.* Bousquet, Lausannæ et Genevæ, 1742.

[9] V. Blasjö. The rectification of quadratures as a central foundational problem for the early leibnizian calculus. *Historia Mathematica,* 39:405–431, 2012.

[10] P. Burgatti. *Teoria matematica dell'elasticità.* Zanichelli, Bologna, 1931.

[11] P. F. Byrd and M. Friedman. *Handbook of elliptic integrals for engineers and scientists.* Springer Verlag, New York, 1971.

[12] S. Chucheepsakul and B. Phungpaigram. Elliptic integral solutions of variable -arc-length elastica under an inclined follower force. *Zeitschrift für Angewandte Mathematik und Mechanik,* 84:29–38, 2004.

[13] R. Dickey and J. Rosemann. Equilibria of the circular elastica under a uniform central force field. *Quarterly of Applied Mathematics,* 51:201–216, 1993.

[14] H. Exton. *Multiple hypergeometric functions and applications.* Hellis Horwood, Chichester, 1976.

[15] C. Fraser. Mathematical technique and physical conception in Euler's investigation of the elastica. *Centaurus,* 34:211–246, 1991.

[16] R. Frisch-Fay. *Flexible bars.* Butterworths, London, 1962.

[17] B. Golley. Large deflections of bars bent through frictionless supports. *The International Journal of Non-Linear Mechanics*, 19:1–9, 1984.

[18] Y. Goto, T. Yoshimitsu, and M. Obata. Elliptic integral solutions of plane elastica with axial and shear deformations. *The International Journal of Non-Linear Mechanics*, 26:375–390, 1990.

[19] I. Gradshteyn and I. Ryzhik. *Table of Integrals, Series and Products*, 6th ed. Academic Press, New York, 2000.

[20] G. Greenhill. Determination of the greatest height consistent with stability that a vertical pole or mast can be made, and of the greatest height to which a tree of given proportion can grow. *Proceedings of the Cambridge Philosophical Society*, IV:65–73, 1881.

[21] E. Guagenti, F. Buccino, E. Garavaglia, and G. Novati. *Statica-Fondamenti di Meccanica Strutturale, Metodi di analisi di strutture reali*, terza edizione. McGraw Hill Education, Milan, 2009.

[22] L. Landau and E. Lifchitz. *Théorie de l'élasticité.* MIR, Moscow, 1967.

[23] T. J. Lardner. A note on the elastica with large loads. *International Journal of Solids and Structures*, 21:21–26, 1985.

[24] A. Love. *A treatise on the mathematical theory of elasticity*, 4th ed., 1927. Reprint Dover, New York, 1944.

[25] Y. Luke. *Integrals of Bessel functions.* McGraw-Hill, Cambridge, MA, 1962.

[26] E. Mathieu. *Dynamique analytique.* Gauthier-Villars, Paris, 1870.

[27] S. Matsutani. Statistical mechanics of elastica on a plane: origin of the mkdv hierarchy. *Journal of Physics A*, 31:2705–2725, 1998.

[28] S. Ming-Gang. Behaviour of a post-buckling riser vibrating due to effects of waves and current. *The International Journal of Non-Linear Mechanics*, 27:437–445, 1992.

[29] G. Mingari Scarpello and D. Ritelli. Elliptic integral solutions of spatial elastica of a thin straight rod bent under concentrated terminal forces. *Meccanica*, 41:519–527, 2006.

[30] G. Mingari Scarpello and D. Ritelli. Elliptic integrals solution to elastica's boundary value problem of a rod bent by axial compression. *Journal of Analysis and Applications*, 5(1):53–69, 2007.

[31] G. Mingari Scarpello and D. Ritelli. Exact curvature elastica of a thin cantilever under terminal loads. *Journal of Geometry and Symmetry in Physics*, 12(1):75–92, 2008.

[32] G. Mingari Scarpello and D. Ritelli. Exact solutions of nonlinear equation of rod deflections involving the Lauricella hypergeometric functions. *International Journal of Mathematics and Mathematical Sciences*, 2011.

[33] G. Mingari Scarpello and D. Ritelli. A hypergeometric treatment to explain the nonlinear true behavior of redundant constraints on a straight elastic rod. *arXiv*, 2015.

[34] S. Navaee and R. Elling. Equilibrium configurations of cantilever beams subjected to inclined end loads. *The Journal of Applied Mechanics*, 59(3–4):572–579, 1992.

[35] D. E. Panayotounakos and P. Theocaris. Exact solution of the non-linear differential equation concerning the elastic line of a straight rod due to terminal loading. *The International Journal of Non-Linear Mechanics*, 17:395–402, 1982.

[36] D. E. Panayotounakos and P. Theocaris. Exact solution for an approximate differential equation of a straight bar under conditions of non-linear equilibrium. *The International Journal of Non-Linear Mechanics*, 21:421–429, 1986.

[37] D. E. Panayotounakos and P. Theocaris. Analytic solutions for nonlinear differential equations describing the elastica of straight bars: theory. *Journal of the Franklin Institute*, 325:621–633, 1988.

[38] P. Payley, L. Shampine, and P. Waltman. *Nonlinear two point boundary value problems.* Academic Press, New York, 1968.

[39] R. Perline. *Proceedings of First Non-Orthodox School on Nonlinearity & Geometry: Luigi Bianchi Days*, September 21–28, 1995, Warsaw, pp. 359–364. Polish Scientific Publishers, Poland, 1998.

[40] B. Rao and V. Rao. Numerical results from large deflection beam and frame problems analysis by means of elliptic integrals. *Zeitschrift für Angewandte Mathematik und Mechanik*, 66(10):507–508, 1986.

[41] L. Saalschütz. *Der belastete Stab unter Entwicklung einer seitlichen Kraft.* Teubner, Leipzig, 1880.

[42] W. Schell. *Theorie der Bewegung und der Kräfte.* Teubner, Leipzig, 1880.

[43] C. Severini. Sopra gl'integrali delle equazioni differenziali ordinarie del secondo ordine, con valori prestabiliti in punti dati. *Atti della Reale accademia delle Scienze di Torino*, 40:1035–1040, 1904–5.

[44] C. Severini. Sopra gl'integrali delle equazioni differenziali ordinarie di ordine superiore al primo, con valori prestabiliti in punti dati. *Atti della Reale accademia delle Scienze di Torino*, 40:853–869, 1904–5.

[45] L. J. Slater. *Generalized hypergeometric functions.* Cambridge University Press, Cambridge, Massachusetts, 1966.

[46] A. B. Sotiropoulou and D. E. Panayotounakos. Exact parametric analytic solutions of the elastica ODEs for bars including effects of the transverse deformation. *The International Journal of Non-Linear Mechanics*, 39(10):1555–1570, 2004.

[47] D. Stump and W. B. Fraser. Multiple solutions for writhed rods: implications for DNA supercooling. *Proceedings of the Royal Society of London A*, 456:455–467, 2000.

[48] F. G. Tricomi. *Funzioni ellittiche.* Zanichelli, Bologna, 1951.

[49] N. Watson. *A treatise on the theory of Bessel functions.* Cambridge University Press, Cambridge, Massachusetts, 1980.

[50] W. C. Young and R. J. Roark. *Formulas for stress and strain*, 5th ed. McGraw Hill, New York, 1984.

3

Second Kind Chebyshev Wavelets for Solving the Variable-Order Space-Time Fractional Telegraph Equation

Mohammad Hossein Heydari
Shiraz University of Technology

Ali Shakiba
Vali-e-Asr University of Rafsanjan

Zakieh Avazzadeh
Xi'an Jiaotong-Liverpool University

Carlo Cattani
University of Tuscia

CONTENTS

3.1 Introduction

One of the most famous hyperbolic partial differential equations (PDEs), which widely arises in many physical and engineering applications, is the telegraph equation:

$$u_{tt}(x,t) + 2\rho u_t(x,t) + \sigma^2 u(x,t) = u_{xx}(x,t) + f(x,t), \tag{3.1}$$

where $\rho > \sigma > 0$ are known constants and $f(x,t)$ is the source term given on a proper domain [1].

Equation (3.1) is referred to as the second-order telegraph equation with constant coefficients and is used in the study of wave propagation of electric signals in a cable transmission line as well as in wave phenomena. The equation is also used in modeling the reaction-diffusion processes in various branches of engineering and biological sciences [1]. Moreover, this equation represents a damped wave motion for $\rho > 0$ and $\sigma = 0$. Recently, it wa s found that the telegraph equation performs better than the ordinary diffusion equation in modeling the reaction-diffusion for these branches of sciences [2]. In recent years, much attention is given to the analysis, development, and implementation of stable methods to find the numerical solution of a second-order hyperbolic equation, especially the telegraph equation, e.g., see [1,2], and the references therein.

The fractional calculus is an interesting topic in applied mathematics dealing with the derivatives and integrals of fractional order [3,4] and plays an important role in engineering and science. The history of this topic is almost as long as the integer calculus; however, it did not attract much attention until recent decades. The modeling of the heat transfer in heterogeneous media [5] and the modeling of the ultracapacitor and beam heating [6] are good applications of the fractional calculus, to mention a few. These applications are mainly due to the fact that many physical systems show fractional order dynamics and their behavior is governed by the fractional differential equations (FDEs) [4]. It is worth to mention that one of the most important advantages of using the FDEs is their non-local property [7], which means that the next state of a dynamical system depends on its current state as well as all of its previous states. This memory effect is one of the main properties that led them to be used in various applications. The reader should note that obtaining analytical solutions for the fractional calculus problems is difficult in general, and in most cases, it is impossible. Therefore, numerical and approximation methods are commonly used to obtain approximate solutions for these kinds of problems. In recent years, operational matrices of the fractional derivative and the integration of many well-known polynomials have successfully applied to solve many different types of the FDEs, e.g., [8–16]. Note that these numerical methods, which are based on the operational matrices of the orthogonal basis functions, lead to approximate solutions with high accuracy for the FDEs.

Though the classical telegraph equation cannot completely describe the abnormal diffusion aspects while the finite long transmit progress, the voltage wave or the current wave may exist [17,18]. Hence, it is essential to study the fractional telegraph equation, including the time and (or) space fractional derivatives. Therefore, based on these reasons and the literature connected with the fractional calculus, Equation (3.1) is successfully generalized by the concept of fractional derivatives in the Caputo sense as

$$ {}_0^cD_t^{\alpha}u(x,t) + 2\rho\, {}_0^cD_t^{\alpha-1}u(x,t) + \sigma^2 u(x,t) = {}_0^cD_x^{\beta}u(x,t) + f(x,t), \tag{3.2} $$

where ${}_0^cD_t^{\alpha}$ and ${}_0^cD_t^{\alpha-1}$ denote the fractional derivatives of orders $1 < \alpha \leq 2$ and $0 < \alpha - 1 \leq 1$ with respect to t, respectively, and ${}_0^cD_x^{\beta}$ denotes the fractional derivative of order $1 < \beta \leq 2$ with respect to x.

Equation (3.2) is usually called the space-time fractional telegraph equation. Considering $\beta = 2$ reduces Equation (3.2) to the time-fractional telegraph equation. Moreover, by considering $\alpha = 2$, this equation is reduced to the space-fractional telegraph equation. The random walks of these particles exist for the suspension flows, e.g., the blood cells moving in the body's vessels or the tunnels of an unnatural organ or lung, the fruit juice pulpy material which is processed or drank using a straw, and pieces of coal in a coal-oil slurry fed into a burner in an electrical power generating plant. In many cases, such phenomena can be quite modeled using hyperbolic equations like the telegraph equation [17]. Moreover, the fractional telegraph equation is extensively studied for a better understanding of the random walks for the suspension flows.

As the order of the fractional derivatives and integrals in the fractional calculus can be an arbitrary value, the non-constant fractional order of the derivatives and integrals is another interesting extension, which is called the variable-order fractional calculus. This generalization of the classical fractional calculus can be used wherever the orders of the fractional derivatives and integrals are dependent on the time and (or) space. With this approach, the non-local property is more evident and has wide applications in Physics [19–21] and Mechanics [22] to model the linear and nonlinear viscoelasticity oscillators [23,24], signal processing [25], and optimization control [26]. Whenever the memory properties vary with spatial locations and/or time, the variable-order fractional derivatives are very useful [27,28]. The authors of [29–31] showed that many complex models in Physics are expressible by the variable-order fractional derivatives with a great success. A comparison of employing variable-order fractional derivatives with the constant-order fractional derivatives is given in [32]. The FDEs involving variable-order, usually called variable-order fractional differential equations (V-FDEs), has attracted many researchers in recent years. In [33–35], the existence and uniqueness of the solutions for a class of generalized V-FDEs are investigated. An important issue that often arises in variable-order fractional functional equations is the fundamental difficulty in finding an analytical solution for them. Therefore, it is usually impossible to get exact solutions for these problems. So, one needs to try finding approximate numerical solutions for such equations. Hence, proposing efficient and accurate approximate/numerical methods to find numerical solutions for such kinds of problems is very important in practice. Some of the recent numerical methods for solving the V-FDEs can be found in [36–48]. Nowadays, finite difference methods are popular to solve various classes of the V-FDEs. It should be noted that the main challenge in finite difference methods is their limited accuracy interwoven with their inherent local character, while the variable-order fractional derivatives are essentially global operators. Hence, global techniques such as spectral schemes may be more suitable for handling the variable-order operators. Recently, spectral methods have been efficiently developed for various types of the V-FDEs, due to their flexible implementation over infinite and finite domains and their exponential convergence, e.g., [49–53]. Moreover, these methods lead to a high level of accuracy in practice. This property of spectral methods is often called the "spectral accuracy" [54]. Indeed, the research on the variable-order fractional PDEs is relatively current, and the development of numerical approximation schemes for these equations is at its early stages.

The theory of wavelets is a fresh and fast developing area of mathematics [55]. It is deeply rooted in the engineering and physics and has an effective role in time-frequency analysis of complex signals, image/signal compression, acoustic and seismic signal processing, and functional and signal analysis through a derived basis using dilations and translations of the mother wavelet, which is a localized wave function [55]. Recently, orthogonal wavelets are widely used to solve various constant-order fractional differential equations such as [56–66]. The main reason of using orthogonal wavelets to solve functional equations is that the problem under study is transferred to a system of linear/nonlinear algebraic equations [67] at first and thereby is solved to obtain an approximated solution. Usually, this is carried by truncating series of orthogonal basis wavelets including unknown coefficients and using the operational matrices of integration and the derivative. Moreover, wavelets are useful basis functions with many advantages compared with other basis functions, e.g., in [67], some of these advantages are given.

The second kind of the Chebyshev wavelets (SKCWs) are a kind of orthonormal wavelets which have the properties of the second kind Chebyshev polynomials, which are orthogonally and spectrally accurate, as well as useful properties of wavelets. Therefore, they are recently used to solve various differential equations successfully, e.g., [68–74]. In [68], a Galerkin method based on the SKCWs is proposed to solve a class of stochastic heat equations by deriving a new stochastic operational matrix for the SKCWs. The SKCWs and their integration and stochastic integration operational matrices are used to transform the problem under study to a linear system of algebraic equations that solves the solution of the problem. In [69], a spectral method based on the SKCWs is proposed by Abd-Elhameed et al. to solve the linear and nonlinear second-order two-point boundary value problems. Mohammadi [70] proposed an SKCW Galerkin method to solve the stochastic Volterra integral equations. In [71], Gupta and Ray proposed a numerical method based on the SKCWs and their fractional integration and derivative operational matrices to solve the time-fractional fifth-order Sawada–Kotera. equation. Wang and Fan [72] used the SKCWs to solve linear and nonlinear FDEs. In order to solve a class of nonlinear Fredholm integro-differential equations of fractional order, Zhu and Fan in [73] proposed a computational method based on the SKCWs. Recently, Salai Mathi Selvi and Hariharan in [74] proposed an efficient SKCW-based method to solve the nonlinear diffusion equation for the steady-state condition.

In this paper, a generalization of the variable-order space-time fractional telegraph equation (V-S-FTE) is proposed by the concept of the variable-order fractional derivative and replacing the fractional derivative terms in the space-time fractional telegraph equation with variable-order fractional derivatives. Consider the following space-time fractional telegraph (3.2) with the variable-order fractional derivatives, i.e.,

$$
{}_0^cD_t^{\alpha(x,t)}u(x,t) + 2\rho\,{}_0^cD_t^{\alpha(x,t)-1}u(x,t) + \sigma^2 u(x,t) = {}_0^cD_x^{\beta(x,t)}u(x,t) + f(x,t), \qquad (x,t) \in \Omega, \tag{3.3}
$$

where $\Omega = [0,1] \times [0,1]$, subject to the initial-boundary conditions:

$$
\begin{aligned}
u(x,0) &= g_0(x), \qquad & u(0,t) &= h_0(t), \\
u_t(x,0) &= g_1(x), \qquad & u(1,t) &= h_1(t).
\end{aligned} \tag{3.4}
$$

Here, ${}_0^cD_t^{\alpha(x,t)}$ and ${}_0^cD_t^{\alpha(x,t)-1}$ denote the variable-order fractional derivatives of orders $1 < \alpha(x,t) \leq 2$ and $0 < \alpha(x,t) - 1 \leq 1$ with respect to the time variable t, respectively, and ${}_0^cD_x^{\beta(x,t)}$ denotes the variable-order fractional derivative of order $1 < \beta(x,t) \leq 2$ with respect to space variable x. The details on variable-order fractional calculus are given in the next section.

To the best of authors' knowledge, no computational methods are developed for the V-S-FTE (3.3) up to date. This motivates our interest to introduce an efficient and accurate computational method for solving V-S-FTE (3.3) subject to the initial-boundary conditions (3.4). The proposed method is based on the SKCWs in conjunction with their operational matrix of variable-order fractional derivative (OMV-FD), which are described in the Caputo sense. To this end, a new OMV-FD is derived for SKCWs at first and then used to compute an approximate solution for the problems at hand. So, a new class of basis functions is introduced and used to form this matrix. In our proposed method, we expand the solution to the problem under study in terms of the SKCWs. Then, we use a combination of the collocation and tau methods to transform the problem to a system of linear algebraic equations with unknown coefficients. It is worth noting that the proposed method is a computational scheme which does not need discretization. Moreover, this method can also be used to solve other kinds of variable-order fractional problems.

Note that the family of the second kind Chebyshev polynomials can be used to approximate solutions of V-FDEs since the integrals in Equations (3.5) and (3.6) are easy to compute and it maintains the property of Equation (3.7). Besides, these basis polynomials have the property of spectral accuracy. However, since the SKCWs maintain the properties of the second kind Chebyshev polynomials as well as the properties of the wavelets, we believe that these basis functions are appropriate to approximate solutions for various variable-order fractional functional equations, especially the V-S-FTE (3.3). Furthermore, there is no work in the literature which uses the wavelets to solve the V-S-FTE (3.3), and this motivated our interest in the method.

A brief outline of the remainder of the paper is as follows: the definition and mathematical preliminary of variable-order fractional derivatives are reviewed in Section 3.2. In Section 3.3, SKCWs and their properties, including function approximation, convergence, and error analysis, and OMV-FD are expressed. In Section 3.4, the proposed method is explained for V-S-FTE (3.3). Section 3.5 contains some illustrative numerical examples which are solved by our proposed method. Finally, a conclusion is drawn in Section 3.6.

3.2 Definitions and Mathematical Preliminaries

By developing variable-order fractional calculus, many new definitions were drafted. Here, we give the Caputo type definition of the variable-order fractional derivative which is used in this paper.

Definition 3.1. The variable-order fractional derivative of order $\bar{m}-1<\alpha(x,t)\leq\bar{m}$ of the function $u(x,t)$ with respect to the variable t in the Caputo type is defined in [41,42] by

$$ {}_0^cD_t^{\alpha(x,t)}u(x,t)=\begin{cases}\dfrac{1}{\Gamma(\bar{m}-\alpha(x,t))}\displaystyle\int_0^t(t-\tau)^{\bar{m}-\alpha(x,t)-1}\frac{\partial^{\bar{m}}u(x,\tau)}{\partial\tau^{\bar{m}}}d\tau, & \bar{m}-1<\alpha(x,t)<\bar{m},\\ \dfrac{\partial^{\bar{m}}u(x,t)}{\partial t^{\bar{m}}}, & \alpha(x,t)=\bar{m},\end{cases} \tag{3.5}$$

where $\bar{m}\in\mathbb{N}$ and $\Gamma(.)$ is the gamma function.

Similarly, the variable-order fractional derivative of order $\bar{n}-1<\beta(x,t)\leq\bar{n}$ of the function $u(x,t)$ with respect to the variable x in the Caputo type is as follows:

$$ {}_0^cD_x^{\beta(x,t)}u(x,t)=\begin{cases}\dfrac{1}{\Gamma(\bar{n}-\beta(x,t))}\displaystyle\int_0^x(x-s)^{\bar{n}-\beta(x,t)-1}\frac{\partial^{\bar{n}}u(s,t)}{\partial s^{\bar{n}}}ds, & \bar{n}-1<\beta(x,t)<\bar{n},\\ \dfrac{\partial^{\bar{n}}u(x,t)}{\partial x^{\bar{n}}}, & \beta(x,t)=\bar{n},\end{cases} \tag{3.6}$$

where $\bar{n}\in\mathbb{N}$.

Corollary 3.1. Based on the Caputo type definition, we have the following property

$$ {}_0^cD_t^{\alpha(x,t)}t^k=\begin{cases}\dfrac{\Gamma(k+1)}{\Gamma(k-\alpha(x,t)+1)}t^{k-\alpha(x,t)}, & \text{for } k\in\mathbb{N}_0 \text{ and } k\geq\bar{m},\\ 0, & \text{for } k\in\mathbb{N}_0 \text{ and } k<\bar{m},\end{cases} \tag{3.7}$$

where $\bar{m}-1<\alpha(x,t)\leq\bar{m}$ and $\mathbb{N}_0=\mathbb{N}\cup\{0\}$.

3.3 The SKCWs and Their Properties

In this section, the SKCWs are briefly introduced and some of their properties are presented.

3.3.1 Wavelets and the SKCWs

Wavelets consist of a class of functions which are constructed by the translation and dilation of a single function $\psi(t)$, which is called the mother wavelet. If the translation parameter b and the dilation parameter a vary continuously, the following family of continuous wavelets is obtained [75]:

$$\psi_{ab}(t) = |a|^{-\frac{1}{2}}\psi\left(\frac{t-b}{a}\right), \qquad a, b \in \mathbb{R}, \ a \neq 0. \tag{3.8}$$

By restricting the parameters a and b to discrete values of $a = a_0^{-k}$ and $b = nb_0a_0^{-k}$ for $a_0 > 1$, $b_0 > 0$, respectively, the following class of discrete wavelets is obtained:

$$\psi_{kn}(t) = |a_0|^{\frac{k}{2}}\psi\left(a_0^k t - nb_0\right), \qquad k, n \in \mathbb{Z}, \tag{3.9}$$

where $\psi_{kn}(t)$ is a wavelet basis for $L^2(\mathbb{R})$.

Particularly, $\psi_{kn}(t)$ is an orthonormal basis when $a_0 = 2$ and $b_0 = 1$ [75].

The SKCWs are defined on $[0,1]$ in [68,69] as

$$\psi_{nm}(t) = \begin{cases} \dfrac{2}{\sqrt{\pi}}2^{\frac{k}{2}}U_m^*\left(2^k t - n\right), & t \in \left[\dfrac{n}{2^k}, \dfrac{n+1}{2^k}\right], \\ 0, & \text{otherwise}, \end{cases} \tag{3.10}$$

for $n = 0, 1, \ldots, 2^k - 1$, $m = 0, 1, \ldots, M-1$, $(k, M) \in \mathbb{N}$, where $U_m^*(t)$ denotes the shifted second kind Chebyshev polynomials which are orthogonal with respect to the weight function $w^*(t) = \sqrt{1-(2t-1)^2}$ on the interval $[0,1]$ and is defined by

$$U_m^*(t) = \sum_{i=0}^{m}(-1)^{m-i}\frac{2^{2i}(m+i+1)!}{(m-i)!(2i+1)!}t^i, \qquad m = 0, 1, \ldots \tag{3.11}$$

The SKCWs form an orthonormal basis for $L^2_{w_n}[0,1]$ (the subscript denotes weighted orthogonal), i.e.,

$$\left(\psi_{nm}(t), \psi_{n'm'}(t)\right)_{w_n} = \int_0^1 \psi_{nm}(t)\psi_{n'm'}(t)w_n(t)dt = \begin{cases} 1, & (m,n) = (n',m'), \\ 0, & (m,n) \neq (n',m'), \end{cases} \tag{3.12}$$

where

$$w_n(t) = \begin{cases} \sqrt{1-\left(2^{k+1}t - 2n - 1\right)^2}, & t \in \left[\dfrac{n}{2^k}, \dfrac{n+1}{2^k}\right], \\ 0, & \text{otherwise}. \end{cases} \tag{3.13}$$

An interested reader is advised to refer to [68] for a construction scheme for the SKCWs on the interval $[0,1]$ with the recursive wavelet construction.

3.3.2 Function Approximation

An arbitrary function $u(t)$ defined over $[0, 1]$ can be expanded by the SKCWs as

$$u(t) = \sum_{n=0}^{\infty} \sum_{m=0}^{\infty} c_{nm} \psi_{nm}(t), \tag{3.14}$$

where $c_{nm} = (u(t), \psi_{nm}(t))_{w_n(t)}$ and $(.,.)_{w_n(t)}$ indicates the inner product in $L^2_{w_n}[0, 1]$.

We can approximate $u(t)$ in terms of the SKCWs by truncating the infinite series in Equation (3.14) as follows:

$$u(t) \simeq \sum_{n=0}^{2^k-1} \sum_{m=0}^{M-1} c_{nm} \psi_{nm}(t) = C^T \Psi(t), \tag{3.15}$$

where T denotes the matrix transposition, C and $\Psi(t)$ are column vectors with $\hat{m} = 2^k M$ elements.

Equation (3.15) can be written simply as

$$u(t) \simeq \sum_{i=1}^{\hat{m}} c_i \psi_i(t) = C^T \Psi(t), \tag{3.16}$$

where $\psi_i(t) = \psi_{nm}(t)$ and $c_i = c_{nm}$, and i is the index determined as $i = Mn + m + 1$.

So, we have

$$\begin{aligned} C &\triangleq [c_1, c_2, \dots, c_{\hat{m}}]^T, \\ \Psi(t) &\triangleq [\psi_1(t), \psi_2(t), \dots, \psi_{\hat{m}}(t)]^T. \end{aligned} \tag{3.17}$$

Similarly, an arbitrary function of two variables $u(x, t)$ over $L^2_w([0, 1] \times [0, 1])$ may be expanded by the SKCWs as

$$u(x, t) \simeq \sum_{i=1}^{\hat{m}} \sum_{j=1}^{\hat{m}} u_{ij} \psi_i(x) \psi_j(t) = \Psi(x)^T U \Psi(t), \tag{3.18}$$

where $U = [u_{ij}]$ and $u_{ij} = \left(\psi_i(x), (u(x, t), \psi_j(t))_{w_{n'}(t)}\right)_{w_n(x)}$ for $i = 1, 2, \dots, \hat{m}$ and $j = 1, 2, \dots, \hat{m}$.

3.3.3 Convergence and Error Analysis

In this section, we investigate the convergence of the SKCWs in two dimensions and then obtain an upper bound for the truncation error of their series expansion. To this end, we first determine an upper bound for the absolute error of the double shifted second kind Chebyshev polynomials using Lagrange interpolation polynomials and then employ the obtained result for our main goal.

To analytically express the error norm with respect to the best approximation of a smooth function $u(x, t)$, where $(x, t) \in \Omega$, in terms of the double shifted second kind Chebyshev polynomials, we use the following definition:

$$\varPi_M = \text{span}\left\{U_i^*(x) U_j^*(t), \quad i = 0, 1, \dots, M-1, \ j = 0, 1, \dots, M-1\right\}. \tag{3.19}$$

In the sequel, $\widetilde{u}(x, t) \in \varPi_M$ denotes the best approximation for $u(x, t)$. So, by the definition we have

$$\|u(x, t) - \widetilde{u}(x, t)\| \leq \|u(x, t) - v(x, t)\|, \quad \forall\, v(x, t) \in \varPi_M. \tag{3.20}$$

It should be noted that the previous inequality also holds if $v(x,t)$ denotes the interpolating polynomial for $u(x,t)$ at points (x_i,t_j) where x_i and t_j for $i,j=0,1,\dots,M-1$ are the roots of $U_M^*(x)$. So, similar to the procedures in [76,77], we have

$$u(x,t)-v(x,t)=\frac{\partial^M u(\eta,t)}{\partial x^M M!}\prod_{i=0}^{M-1}(x-x_i)+\frac{\partial^M u(x,\xi)}{\partial t^M M!}\prod_{j=0}^{M-1}(t-t_j)$$
$$-\frac{\partial^{2M}u(\eta',\xi')}{\partial x^M\partial t^M(M!)^2}\prod_{i=0}^{M-1}(x-x_i)\prod_{j=0}^{M-1}(t-t_j), \tag{3.21}$$

where $\eta,\eta',\xi,\xi'\in[0,1]$, and then, we obtain

$$\begin{aligned}\|u(x,t)-v(x,t)\|_\infty \leq &\frac{1}{M!}\max_{(x,t)\in\Omega}\left|\frac{\partial^M u(\eta,t)}{\partial x^M}\right|\left\|\prod_{i=0}^{M-1}(x-x_i)\right\|_\infty\\ &+\frac{1}{M!}\max_{(x,t)\in\Omega}\left|\frac{\partial^M u(x,\xi)}{\partial t^M}\right|\left\|\prod_{j=0}^{M-1}(t-t_j)\right\|_\infty\\ &+\frac{1}{(M!)^2}\max_{(x,t)\in\Omega}\left|\frac{\partial^{2M} u(\eta',\xi')}{\partial x^M\partial t^M}\right|\left\|\prod_{i=0}^{M-1}(x-x_i)\right\|_\infty\left\|\prod_{j=0}^{M-1}(t-t_j)\right\|_\infty.\end{aligned} \tag{3.22}$$

By the assumption that $u(x,t)$ is a smooth function on Ω, the constants L_1, L_2, and L_3 exist such that

$$\max_{(x,t)\in\Omega}\left|\frac{\partial^M u(x,t)}{\partial x^M}\right|\leq L_1,\quad \max_{(x,t)\in\Omega}\left|\frac{\partial^M u(x,t)}{\partial t^M}\right|\leq L_2,\quad \max_{(x,t)\in\Omega}\left|\frac{\partial^{2M} u(x,t)}{\partial x^M\partial t^M}\right|\leq L_3. \tag{3.23}$$

In order to minimize the factor $\left\|\prod_{i=0}^{M-1}(x-x_i)\right\|_\infty$ by the one-to-one mapping $x=\frac{1}{2}(z+1)$ from the interval $[-1,1]$ to the interval $[0,1]$ like [78], we have

$$\begin{aligned}\min_{x_i\in[0,1]}\max_{x\in[0,1]}\left|\prod_{i=0}^{M-1}(x-x_i)\right| &= \min_{z_i\in[-1,1]}\max_{z\in[-1,1]}\left|\prod_{i=0}^{M-1}\frac{1}{2}(z-z_i)\right|\\ &=\frac{1}{2^M}\min_{z_i\in[-1,1]}\max_{z\in[-1,1]}\left|\prod_{i=0}^{M-1}(z-z_i)\right|\\ &=\frac{1}{2^M}\min_{z_i\in[-1,1]}\max_{z\in[-1,1]}\left|\frac{U_M(z)}{\varpi_M}\right|,\end{aligned} \tag{3.24}$$

where $\varpi_M=2^M$ is the leading coefficient of $U_M(z)$ (the second kind Chebyshev polynomial of degree M) and z_i are the roots of $U_M(z)$. Moreover, by the known equation of

$$\max_{z\in[-1,1]}|U_M(z)|=U_M(1)=M+1, \tag{3.25}$$

and Equations (3.23) and (3.24), we have the following result:

$$\|u(x,t)-\widetilde{u}(x,t)\|_\infty\leq(L_1+L_2)\frac{M+1}{4^M M!}+L_3\frac{(M+1)^2}{4^{2M}(M!)^2}. \tag{3.26}$$

Hence, the absolute error of the double shifted second kind Chebyshev polynomials is bounded from the above. In the sequel, we employ the obtained result to establish an upper bound for the truncation error of the SKCWs expansion.

Theorem 3.1. *Suppose that $u(x,t)$ is a smooth function over Ω with*

$$\max_{(x,t)\in\Omega}\left|\frac{\partial^M u(x,t)}{\partial x^M}\right| \le L_1, \quad \max_{(x,t)\in\Omega}\left|\frac{\partial^M u(x,t)}{\partial t^M}\right| \le L_2, \quad \max_{(x,t)\in\Omega}\left|\frac{\partial^{2M} u(x,t)}{\partial x^M \partial t^M}\right| \le L_3, \tag{3.27}$$

and $\Psi(x)^T U\Psi(t) = \sum_{i=1}^{\hat{m}}\sum_{j=1}^{\hat{m}} u_{ij}\psi_i(x)\psi_j(t)$ be its SKCWs expansion. Then, the error bound is bounded by

$$\left\|u(x,t) - \Psi(x)^T U\Psi(t)\right\|_{w^*} \le (L_1+L_2)\,\frac{\pi(M+1)}{2^{kM}4^{M+1}M!} + L_3\,\frac{\pi(M+1)^2}{2^{2kM}4^{2M+1}(M!)^2}. \tag{3.28}$$

Proof. By the definition of the norm in an inner product space, we have

$$\left\|u(x,t) - \Psi(x)^T U\Psi(t)\right\|_{w^*}^2 = \int_0^1\int_0^1 w^*(x)w^*(t)\left(u(x,t) - \Psi(x)^T U\Psi(t)\right)^2 dxdt. \tag{3.29}$$

Since the domain Ω is divided into 2^{2k} subdomains $\Omega_{n,n'} = \left[\frac{n}{2^k}, \frac{n+1}{2^k}\right] \times \left[\frac{n'}{2^k}, \frac{n'+1}{2^k}\right]$ $(n,\, n' = 0, 1, \ldots, 2^k-1)$, which the function $u(x,t)$ is approximated on them by the SKCWs as a polynomial of $(M-1)$-th degree, then we have

$$\begin{aligned}\left\|u(x,t) - \Psi(x)^T U\Psi(t)\right\|_{w^*}^2 &= \sum_{n=0}^{2^k-1}\sum_{n'=0}^{2^k-1}\int_{\frac{n}{2^k}}^{\frac{n+1}{2^k}}\int_{\frac{n'}{2^k}}^{\frac{n'+1}{2^k}} w_n(x)w_{n'}(t)\\ &\quad\times\left(u(x,t) - \Psi(x)^T U\Psi(t)\right)^2 dxdt\\ &\le \sum_{n=0}^{2^k-1}\sum_{n'=0}^{2^k-1}\int_{\frac{n}{2^k}}^{\frac{n+1}{2^k}}\int_{\frac{n'}{2^k}}^{\frac{n'+1}{2^k}} w_n(x)w_{n'}(t)\left(u(x,t) - u^*(x,t)\right)^2 dxdt,\end{aligned} \tag{3.30}$$

where $u^*(x,t)$ is the interpolating polynomial of degree $(M-1)$ agreeing with $u(x,t)$ at the shifted second kind Chebyshev nodes of order M on $\Omega_{n,n'}$, and by Equation (3.26) satisfies the error bound for interpolation:

$$\left\|u(x,t) - u^*(x,t)\right\|_\infty |_{\Omega_{n,n'}} \le (L_1+L_2)\,\frac{M+1}{2^{kM}4^M M!} + L_3\,\frac{(M+1)^2}{2^{2kM}4^{2M}(M!)^2} \triangleq \mathcal{B}(M, L_1, L_2, L_3). \tag{3.31}$$

Therefore, using Equations (3.30) and (3.31), we have

$$\begin{aligned}\left\|u(x,t) - \Psi(x)^T U\Psi(t)\right\|_{w^*}^2 &\le \mathcal{B}(M, L_1, L_2, L_3)^2 \sum_{n=0}^{2^k-1}\sum_{n'=0}^{2^k-1}\int_{\frac{n}{2^k}}^{\frac{n+1}{2^k}}\int_{\frac{n'}{2^k}}^{\frac{n'+1}{2^k}} w_n(x)w_{n'}(t)dxdt\\ &= \mathcal{B}(M, L_1, L_2, L_3)^2\,\frac{\pi^2}{16},\end{aligned} \tag{3.32}$$

which completes the proof.

Corollary 3.2. Suppose that $u(x,t)$ be a continuous function defined over Ω and the above assumptions are satisfied. By putting $w^*(x) = w^*(t) = 1$, we have the following L_2 error bound

$$\left\|u(x,t) - \Psi(x)^T U\Psi(t)\right\|_2 \le (L_1+L_2)\,\frac{M+1}{2^{kM}4^M M!} + L_3\,\frac{(M+1)^2}{2^{2kM}4^{2M}(M!)^2}. \tag{3.33}$$

Proof. The proof is clear.

3.3.4 The Operational Matrix of Variable-Order Fractional Derivative (OMV-FD)

The variable-order fractional derivative of order $(q-1) < \vartheta(x,t) \leq q$, $q \in \mathbb{N}$, of the vector $\Psi(t)$ in Equation (3.17), is expressible as

$$ {}_0^cD_t^{\vartheta(x,t)}\Psi(t) \simeq \mathbf{Q}_t^{\vartheta(x,t)}\Psi(t), \tag{3.34}$$

where the matrix $\mathbf{Q}_t^{\vartheta(x,t)}$ of dimension $\hat{m} \times \hat{m}$ is called the OMV-FD of order $\vartheta(x,t)$ for SKCWs.

In order to derive an explicit form for this matrix, we propose another class of basis functions as

$$\varphi_i(t) = t^{i-1}, \qquad i = 1, 2, \ldots, \hat{m}. \tag{3.35}$$

Unlike the SKCWs, this family of monomials is not normalized. An $\hat{m}$-set of these monomials are expressible as

$$\Phi(t) \triangleq [\varphi_1(t), \varphi_2(t), \ldots, \varphi_{\hat{m}}(t)]^T. \tag{3.36}$$

The following relation holds between these monomials and the SKCWs

$$\Phi(t) \simeq \mathbf{P}\Psi(t), \tag{3.37}$$

where $\mathbf{p}_{ij} = (\varphi_i(t), \psi_j(t))_{w_n(t)}$.

Lemma 3.1. *Let $\varphi_i(t)$ be as defined in Equation (3.35), and $(q-1) < \vartheta(x,t) \leq q$ be a positive real function defined on $\mathbb{R}^2$. Then*

$$ {}_0^cD_t^{\vartheta(x,t)}\varphi_i(t) = \begin{cases} \dfrac{(i-1)!}{\Gamma(i-\vartheta(x,t))}\, t^{i-\vartheta(x,t)-1}, & i = q+1, q+2, \ldots, \hat{m}, \\ 0, & \text{otherwise}. \end{cases}$$

Proof. By considering Equation (3.7), the proof is obvious. □

Lemma 3.2. *Let $\Phi(t)$ be the monomials vector of Equation (3.36) and $(q-1) < \vartheta(x,t) \leq q$ be a positive real function defined on $\mathbb{R}^2$. The variable-order fractional derivative of order $\vartheta(x,t)$ of $\Phi(t)$ in the Caputo sense is expressible as*

$$ {}_0^cD_t^{\vartheta(x,t)}\Phi(t) = \boldsymbol{T}_t^{\vartheta(x,t)}\Phi(t), \tag{3.38}$$

where $\boldsymbol{T}_t^{\vartheta(x,t)}$ is an $\hat{m} \times \hat{m}$ matrix given by

$$T_t^{\vartheta(x,t)} = \\ t^{-\vartheta(x,t)} \begin{pmatrix} 0 & 0 & \cdots & 0 & 0 & 0 & \cdots & 0 \\ \vdots & \vdots & \vdots & \vdots & \vdots & \vdots & \vdots & \vdots \\ 0 & 0 & 0 & 0 & 0 & 0 & 0 & 0 \\ 0 & \cdots & 0 & \dfrac{q!}{\Gamma(q-\vartheta(x,t)+1)} & 0 & 0 & \cdots & 0 \\ 0 & \cdots & \cdots & 0 & \dfrac{(q+1)!}{\Gamma(q-\vartheta(x,t)+2)} & 0 & 0 & 0 \\ \vdots & \vdots & \vdots & \vdots & \vdots & \ddots & 0 & 0 \\ \vdots & \vdots & \vdots & \vdots & \vdots & \vdots & \dfrac{(\hat{m}-2)!}{\Gamma(\hat{m}-\vartheta(x,t)-1)} & 0 \\ 0 & 0 & \cdots & 0 & 0 & 0 & 0 & \dfrac{(\hat{m}-1)!}{\Gamma(\hat{m}-\vartheta(x,t))} \end{pmatrix}.$$

Proof. It is an immediate consequence of Lemma 3.1.

Remark 3.1. Note that for $\vartheta(x,t) = 1$, we have

$$\mathbf{T}_t^{(1)} = \begin{pmatrix} 0 & 0 & 0 & 0 & 0 & 0 & \dots & 0 \\ 1 & 0 & 0 & 0 & 0 & 0 & \dots & 0 \\ 0 & 2 & 0 & 0 & 0 & 0 & \dots & 0 \\ 0 & 0 & 3 & 0 & 0 & 0 & \dots & 0 \\ 0 & 0 & 0 & 4 & 0 & 0 & \dots & 0 \\ \vdots & \vdots & \vdots & \vdots & \ddots & \vdots & \vdots & \vdots \\ 0 & 0 & \dots & 0 & 0 & \hat{m}-2 & 0 & 0 \\ 0 & 0 & \dots & 0 & 0 & 0 & \hat{m}-1 & 0 \end{pmatrix},$$

Theorem 3.2. *Let $\Psi(t)$ be the SKCWs vector of Equation (3.17) and $(q-1) < \vartheta(x,t) \leq q$ be a positive real function over $[0,1] \times [0,1]$. The variable-order fractional derivative of order $\vartheta(x,t)$ of $\Psi(t)$ in the Caputo sense is expressible as*

$${}_0^C D_t^{\vartheta(x,t)} \Psi(t) \simeq \boldsymbol{Q}_t^{\vartheta(x,t)} \Psi(t) = \left(\boldsymbol{P}^{-1} \boldsymbol{T}_t^{\vartheta(x,t)} \boldsymbol{P} \right) \Psi(t), \tag{3.39}$$

where $\boldsymbol{P}$ is the coefficients matrix obtained in Equation (3.37), $\boldsymbol{T}_t^{\vartheta(x,t)}$ is OMV-FD of order $\vartheta(x,t)$ of the monomials vector defined in Equation (3.36), and $\boldsymbol{Q}_t^{\vartheta(x,t)}$ is called OMV-FD of order $\vartheta(x,t)$ for SKCWs.

Proof. By Lemma 3.2 and Equation (3.37), we have

$$\Psi(t) \simeq \mathbf{P}^{-1} \Phi(t),$$

and then

$${}_0^C D_t^{\vartheta(x,t)} \Psi(t) \simeq \mathbf{P}^{-1} \, {}_0^C D_t^{\vartheta(x,t)} \Phi(t) = \mathbf{P}^{-1} \mathbf{T}_t^{\vartheta(x,t)} \Phi(t) \simeq \left(\mathbf{P}^{-1} \mathbf{T}_t^{\vartheta(x,t)} \mathbf{P} \right) \Psi(t),$$

which concludes the proof.

3.4 The Proposed Method

In this section, we employ the OMV-FD of the SKCWs to solve V-S-FTE (3.3), subject to the initial-boundary conditions (3.4). Therefore, we approximate the unknown function $u(x,t)$ by the SKCWs as

$$u(x,t) \simeq \Psi(x)^T \mathbf{U} \Psi(t), \tag{3.40}$$

where $\mathbf{U} = [u_{ij}]$ is an unknown matrix of size $\hat{m} \times \hat{m}$ that should be computed, and $\Psi(.)$ is the SKCWs vector defined in Equation (3.17).
Using variable-order fractional derivative of orders $\alpha(x,t)$ and $\alpha(x,t) - 1$ of Equation (3.40) with respect to variable t and by Equation (3.39), we have

$${}_0^C D_t^{\alpha(x,t)} u(x,t) \simeq \Psi(x)^T \mathbf{U} \mathbf{Q}_t^{\alpha(x,t)} \Psi(t), \qquad {}_0^C D_t^{\alpha(x,t)-1} u(x,y) \simeq \Psi(x)^T \mathbf{U} \mathbf{Q}_t^{\alpha(x,t)-1} \Psi(t). \tag{3.41}$$

Also, the derivative of Equation (3.40) with respect to variable t yields

$$u_t(x,t) \simeq \Psi(x)^T \mathbf{U}\mathbf{Q}_t^{(1)}\Psi(t). \tag{3.42}$$

Moreover, the variable-order fractional derivative of order $\beta(x,t)$ of Equation (3.40) with respect to the variable x gives

$${}_0^cD_x^{\beta(x,t)}u(x,t) \simeq \Psi(x)^T \left(\mathbf{Q}_x^{\beta(x,t)}\right)^T \mathbf{U}\Psi(t). \tag{3.43}$$

Further, the functions $g_i(x)$ and $h_i(t)$ for $i = 0, 1$ can be approximated by the SKCWs like in Equation (3.15) as

$$\begin{aligned} g_0(x) &\simeq \Psi(x)^T\mathbf{C}_1, \qquad & h_0(t) &\simeq \mathbf{C}_3^T\Psi(t), \\ g_1(x) &\simeq \Psi(x)^T\mathbf{C}_2, \qquad & h_1(t) &\simeq \mathbf{C}_4^T\Psi(t), \end{aligned} \tag{3.44}$$

where $\mathbf{C}_i$ for $i = 1, 2, 3, 4$ are known coefficient vectors.
By applying Equation (3.40) into the initial-boundary conditions in Equation (3.4), we have

$$\begin{aligned} &\Psi(x)^T\left(\mathbf{U}\Psi(0) - \mathbf{C}_1\right) \simeq 0, \qquad & &\left(\Psi(0)^T\mathbf{U} - \mathbf{C}_3^T\right)\Psi(t) \simeq 0, \\ &\Psi(x)^T\left(\mathbf{U}Q_t^{(1)}\Psi(0) - \mathbf{C}_2\right) \simeq 0, \qquad & &\left(\Psi(1)^T\mathbf{U} - \mathbf{C}_4^T\right)\Psi(t) \simeq 0. \end{aligned} \tag{3.45}$$

Since the entries of the vector $\Psi(.)$ are independent, the Equation (3.45) gives

$$\begin{aligned} \Lambda_1 &\triangleq \mathbf{U}\Psi(0) - \mathbf{C}_1 \simeq 0, \qquad & \Lambda_3 &\triangleq \Psi(0)^T\mathbf{U} - \mathbf{C}_3^T \simeq 0, \\ \Lambda_2 &\triangleq \mathbf{U}Q_t^{(1)}\Psi(0) - \mathbf{C}_2 \simeq 0, \qquad & \Lambda_4 &\triangleq \Psi(1)^T\mathbf{U} - \mathbf{C}_4^T \simeq 0. \end{aligned} \tag{3.46}$$

Now, by substituting the Equations (3.41), (3.42) and (3.43) into the V-S-FTE (3.3), we have

$$\Psi(x)^T\left(\mathbf{U}\mathbf{Q}_t^{\alpha(x,t)} + 2\rho\,\mathbf{U}\mathbf{Q}_t^{\alpha(x,t)-1} + \sigma^2\mathbf{U} - \left(\mathbf{Q}_x^{\beta(x,t)}\right)^T\mathbf{U}\right)\Psi(t) - f(x,t) \triangleq \mathbf{R}(x,t) \simeq 0. \tag{3.47}$$

To compute an approximate solution for the problem at hand, we need to find the unknown matrix $\mathbf{U}$. So, we need to construct and solve a linear system of $\hat{m}^2$ equations to determine the unknown matrix $\mathbf{U}$. To this end, we choose $\hat{m}^2 - 4\hat{m} + 4$ algebraic equations using Equation (3.47) as

$$\mathbf{R}\left(\frac{2i-1}{2\hat{m}}, \frac{2j-1}{2\hat{m}}\right) = 0, \qquad i = 2, 3, \ldots, \hat{m}-1,\ j = 2, 3, \ldots, \hat{m}-1. \tag{3.48}$$

Also, by selecting $\hat{m} - 2$ equations of $\Lambda_i = 0\,(i = 1, 2)$, and $\hat{m}$ equations of $\Lambda_i = 0\,(i = 3, 4)$, we get $4\hat{m} - 4$ equations, i.e.,

$$\begin{cases} \Lambda_{ij} = 0, \qquad i = 1, 2, \qquad j = 2, 3, \ldots, \hat{m}-1, \\ \Lambda_{ij} = 0, \qquad i = 3, 4, \qquad j = 1, 2, \ldots, \hat{m}. \end{cases} \tag{3.49}$$

Equations (3.48) and (3.49) give $\hat{m}^2$ equations, which can be solved for $\mathbf{U}$ with the MAPLE or MATLAB® software packages. After determining $\mathbf{U}$, we can determine the approximate solution for $u(x,t)$ by Equation (3.40).

The algorithm of the proposed method is stated in Algorithm 3.1.

Algorithm 3.1: The process of the proposed method.

Input: $M \in \mathbb{N}$, $k \in \mathbb{N} \cup \{0\}; 1 < \alpha(x,t), \beta(x,t) \leq 2$; constants $\rho > \sigma > 0$, and functions $f(x,t)$, $g_0(x)$, $g_1(x)$, $h_0(t)$, and $h_1(t)$.
Output: The approximate solution: $u(x,t) \simeq \Psi(x)^T \mathbf{U}\Psi(t)$.
Step 1: Define the SKCWs, $\psi_{nm}(t)$ by Equation (3.10).
Step 2: Construct the KSCWs vector $\Psi(t)$ by Equation (3.17).
Step 3: Construct the vector $\Phi(t)$ by Equation (3.36).
Step 4: Compute the coefficients matrix $\mathbf{P}$ by Equation (3.37).
Step 5: Define the unknown matrix $U = [u_{ij}]_{\hat{m}\times\hat{m}}$, where $\hat{m} = 2^k M$.
Step 6: Compute the operational matrices $\mathbf{T}_t^{\alpha(x,t)}$, $\mathbf{T}_t^{\alpha(x,t)-1}$, $\mathbf{T}_t^{(1)}$, and $\mathbf{T}_x^{\beta(x,t)}$ by Equation (3.38).
Step 7: Compute the operational matrices $\mathbf{Q}_t^{\alpha(x,t)}$, $\mathbf{Q}_t^{\alpha(x,t)-1}$, $\mathbf{Q}_t^{(1)}$, and $\mathbf{Q}_x^{\beta(x,t)}$ by Equation (3.39).
Step 8: Compute the vectors C_1, C_2, C_3 and C_4 in Equation (3.44) by Equation (3.14).
Step 9: Compute the vectors Λ_i for $i = 1, 2, 3, 4$ by Equation (3.46).
Step 10: Compute the residual function $\mathbf{R}(x,t)$ by Equation (3.47).
Step 11: Compute $\hat{m}^2 - 4\hat{m} + 4$ algebraic equations by Equation (3.48).
Step 12: Compute $4\hat{m} - 4$ algebraic equations by Equation (3.49).
Step 13: Solve the system of linear algebraic equations, including Equations (3.48) and (3.49).

3.5 Illustrative Examples

In this section, we provide several numerical examples to demonstrate the reliability and efficiency of our proposed method. Moreover, we report the absolute errors of the proposed method for some of them in various points $(x_i, t_i) \in \Omega$ as

$$|e(x_i, t_i)| = \left|\Psi(x_i)^T \mathbf{U}\Psi(t_i) - u(x_i, t_i)\right|.$$

Note that all of the numeric computations are performed with MAPLE 17 with 50 decimal digits on a X64-based PC with 4.0 GB of RAM and an Intel (R) CPU Corei 4, 3.10 GHz.

Example 3.1. Consider the V-S-FTE (3.3), subject to the following initial-boundary conditions

$$u(x,0) = x^2, \qquad u(0,t) = t,$$
$$u_t(x,0) = 1, \qquad u(1,t) = 1 + t,$$

and

$$f(x,t) = \frac{2\rho}{\Gamma(3-\alpha(x,t))} t^{2-\alpha(x,t)} + \sigma^2\left(x^2 + t\right) - \frac{2}{\Gamma(3-\beta(x,t))} x^{2-\beta(x,t)}.$$

Its exact solution is $u(x,t) = x^2 + t$. We have solved the problem using our proposed method with $k = 1$ and $M = 3$ for $\alpha(x,t) = 1.7 + 0.3\sin(xt)$, $\beta(x,t) = 1.5 + 0.5\cos(xt)$, $\rho = 2$ and $\sigma = 1$. The space-time plots of the approximate solution and the absolute error are given in Figure 3.1. This figure demonstrates that the proposed method can provide an approximate solution of high accuracy by a few SKCWs.

Example 3.2. Consider V-S-FTE (3.3), subject to the following initial-boundary conditions

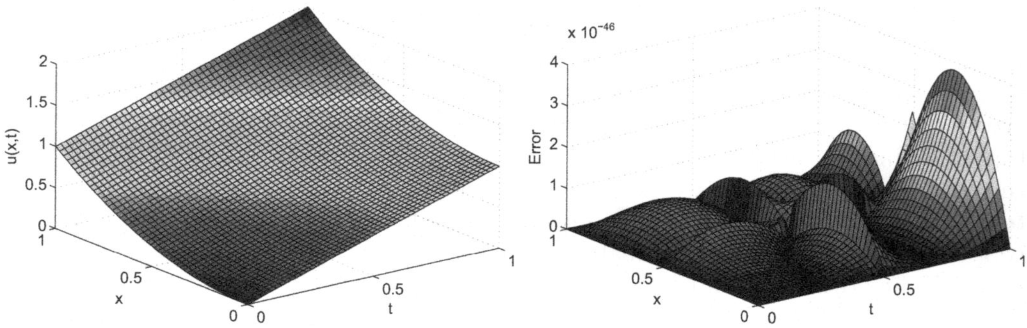

FIGURE 3.1
The graphs of the approximate solution (a) and the absolute error (b) with ($k = 1, M = 3$) for Example 3.1.

$$u(x,0) = x - x^2, \qquad u(0,t) = 0,$$
$$u_t(x,0) = x^2 - x, \qquad u(1,t) = 0,$$

and

$$f(x,t) = \left(x - x^2\right)\left(t^{2-\alpha(x,t)}\mathbf{E}_{1,3-\alpha(x,t)}(-t) - 2\rho\, t^{2-\alpha(x,t)}\mathbf{E}_{1,3-\alpha(x,t)}(-t) + \sigma^2 e^{-t}\right) + \frac{2}{\Gamma\left(3 - \beta(x,t)\right)} x^{2-\beta(x,t)} e^{-t},$$

where $\mathbf{E}_{\theta,\varsigma}(z)$ is the Mittag-Leffler function with two parameters defined in [3] by

$$\mathbf{E}_{\theta,\varsigma}(z) = \sum_{j=0}^{\infty} \frac{z^j}{\Gamma(j\theta + \varsigma)}.$$

Its exact solution is $u(x,t) = \left(x - x^2\right) e^{-t}$. By solving this problem using the proposed method with $k = 1$ and different values of M for $\alpha(x,t) = 1.2 + 0.8\cos(xt)$, $\beta(x,t) = 2 - 0.4|2x - 1|\cos(t)$, $\rho = 4$ and $\sigma = 2$. The plots of the absolute error at $t = 0.6$ (space graph) and $x = 0.6$ (time graph) for $M = 9$ are shown in Figure 3.2. The space-time plots of

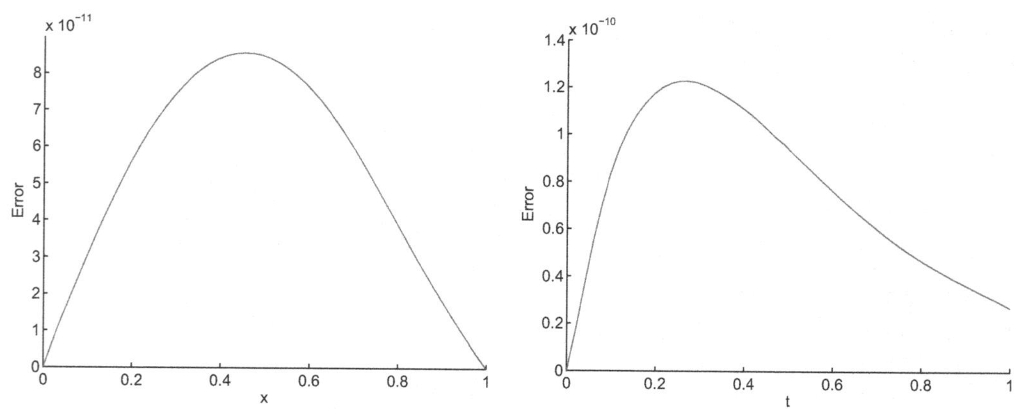

FIGURE 3.2
The plots of the absolute error at $t = 0.6$ (left) and $x = 0.6$ (right) with ($k = 1, M = 9$) for Example 3.2.

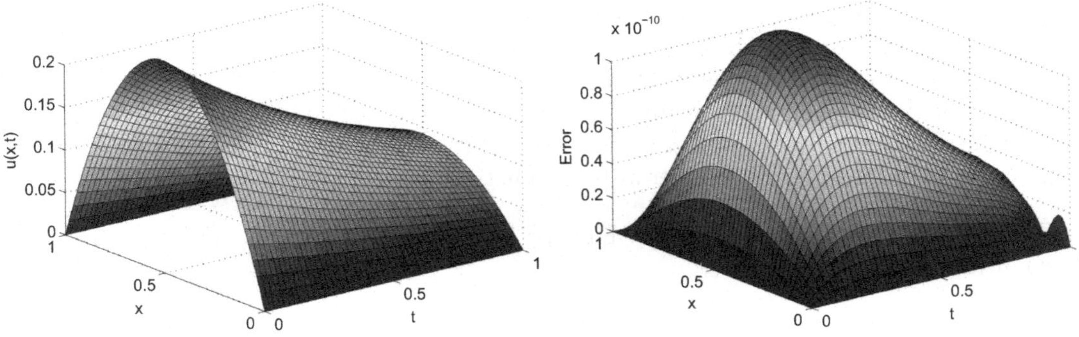

FIGURE 3.3
The graphs of the approximate solution (left) and absolute error (right) with ($k = 1, M = 9$) for Example 3.2.

TABLE 3.1
The Absolute Errors of the Approximate Solutions for $k = 1$ and Different Values of M in Various Points for Example 3.2

(x_i, t_i)	$M = 4$	$M = 5$	$M = 6$	$M = 7$	$M = 8$	$M = 9$
$(0.1, 0.1)$	3.5390E-06	5.5064E-07	5.9363E-08	6.6114E-09	5.3471E-10	3.0207E-11
$(0.2, 0.2)$	5.8150E-07	3.7757E-07	1.5032E-07	1.7163E-08	1.3761E-09	7.8345E-11
$(0.3, 0.3)$	5.6553E-07	3.0003E-07	2.0444E-07	2.4033E-08	1.9316E-09	1.0974E-10
$(0.4, 0.4)$	4.4676E-06	1.0232E-06	2.2893E-07	2.6279E-08	2.0635E-09	1.1668E-10
$(0.5, 0.5)$	8.2261E-07	6.9596E-07	1.9651E-07	2.1545E-08	1.6443E-09	9.0546E-11
$(0.6, 0.6)$	7.7592E-07	7.3540E-07	1.7098E-07	1.8140E-08	1.3904E-09	7.6106E-11
$(0.7, 0.7)$	2.5260E-06	3.7253E-07	1.0197E-07	1.0648E-08	8.5885E-10	4.6826E-11
$(0.8, 0.8)$	1.5737E-06	1.3924E-07	5.2544E-08	5.6426E-09	4.5213E-10	2.4471E-11
$(0.9, 0.9)$	1.1303E-06	4.5356E-08	1.9817E-08	1.8486E-09	1.4435E-10	8.7952E-12

the absolute error and the the approximate solution are shown in Figure 3.3. The absolute errors for the approximate solutions in different points $(x_i, t_i) \in \Omega$ are reported in Table 3.1.

From numerical results obtained in Table 3.1 and Figures 3.2 and 3.3, it can be observed that the proposed method provides high accuracy approximate solutions. Table 3.1 also shows that only a few number of the SKCWs is necessary to achieve an acceptable numerical solution, and the accuracy of the obtained results improves by increasing the number of basis functions.

Example 3.3. Consider the V-S-FTE (3.3), subject to the following initial-boundary conditions

$$u(x,0) = e^{x}, \qquad u(0,t) = e^{-t},$$
$$u_t(x,0) = -e^{x}, \qquad u(1,t) = e^{1-t},$$

and

$$f(x,t) = \left(t^{2-\alpha(x,t)}\mathbf{E}_{1,3-\alpha(x,t)}(-t) - 2\rho\, t^{2-\alpha(x,t)}\mathbf{E}_{1,3-\alpha(x,t)}(-t) + \sigma^2 e^{-t}\right) e^{x} - e^{-t} x^{2-\beta(x,t)}\mathbf{E}_{1,3-\beta(x,t)}(x).$$

Its exact solution is $u(x,t) = e^{x-t}$, and we solve it using our proposed method with $k = 1$ and different values of M for $\alpha(x,t) = 1.7 + 0.2\cos(xt)$, $\beta(x,t) = 1.1 + 0.7e^{-xt}$, $\rho = 3$, and $\sigma = 2$. The plots of the absolute error at $t = 0.6$ and $x = 0.6$ for $M = 9$ are shown in Figure 3.4. The space-time plots of the absolute error and the approximate solution are given in Figure 3.5. The absolute errors for the approximate solutions in different points $(x_i, t_i) \in \Omega$ are reported in Table 3.2.

The obtained results illustrate that our proposed method is very accurate in this example. Moreover, in Table 3.2, one can see that the accuracy of the numerical results improves by increasing the number of basis functions.

Example 3.4. Considering the V-S-FTE (3.3), subject to the following initial-boundary conditions

$$u(x,0) = 0, \qquad u(0,t) = 0,$$
$$u_t(x,0) = 0, \qquad u(1,t) = t^3 \sin(1),$$

and

$$f(x,t) = \left(\frac{6}{\Gamma(4-\alpha(x,t))} t^{3-\alpha(x,t)} + \frac{12\rho}{\Gamma(5-\alpha(x,t))} t^{4-\alpha(x,t)} + \sigma^2 t^3 \right) \sin(x) + t^3 x^{3-\beta(x,t)} \mathbf{E}_{2,4-\beta(x,t)}(-x^2).$$

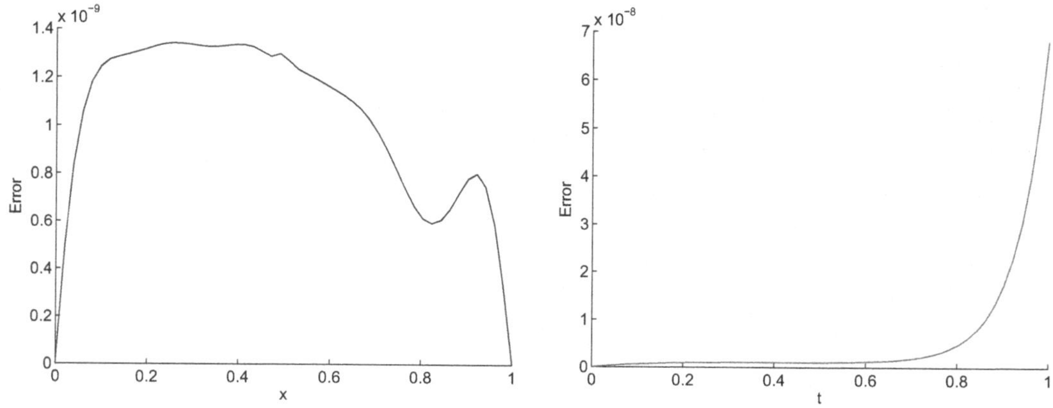

FIGURE 3.4
The plots of the absolute error at $t = 0.6$ (left) and $x = 0.6$ (right) with $(k = 1, M = 9)$ for Example 3.3.

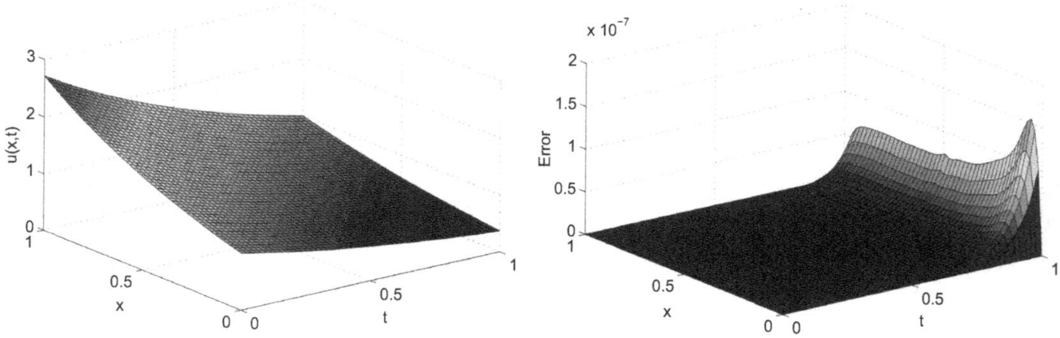

FIGURE 3.5
The plots of the approximate solution (left) and absolute error (right) with $(k = 1, M = 9)$ for Example 3.3.

TABLE 3.2
The Absolute Errors of the Approximate Solutions for $k = 1$ and Different Values of M in Various Points for Example 3.3

(x_i, t_i)	$M = 4$	$M = 5$	$M = 6$	$M = 7$	$M = 8$	$M = 9$
(0.1, 0.1)	1.1610E-05	9.7040E-05	7.2743E-07	1.4810E-08	2.7936E-09	3.0613E-10
(0.2, 0.2)	7.2673E-05	2.6303E-05	6.5561E-07	3.6723E-08	5.7886E-09	6.1891E-10
(0.3, 0.3)	4.6081E-05	1.1722E-05	5.3251E-07	4.2080E-08	8.8373E-09	8.4377E-10
(0.4, 0.4)	2.0960E-05	1.3187E-05	4.8686E-07	4.6853E-08	1.3974E-08	9.9480E-10
(0.5, 0.5)	8.9155E-05	1.0358E-05	7.2069E-08	8.9126E-08	4.1071E-08	1.1116E-09
(0.6, 0.6)	1.2332E-05	5.3837E-06	8.4484E-08	7.9050E-08	6.0129E-08	1.1592E-09
(0.7, 0.7)	5.1108E-06	5.0453E-06	1.0891E-07	1.8814E-08	1.3047E-08	1.5208E-09
(0.8, 0.8)	6.7335E-05	9.7559E-06	1.7438E-06	3.9410E-08	2.7804E-08	4.4809E-09
(0.9, 0.9)	2.5070E-05	1.2881E-05	8.5434E-07	7.5709E-08	4.6799E-08	1.5587E-08

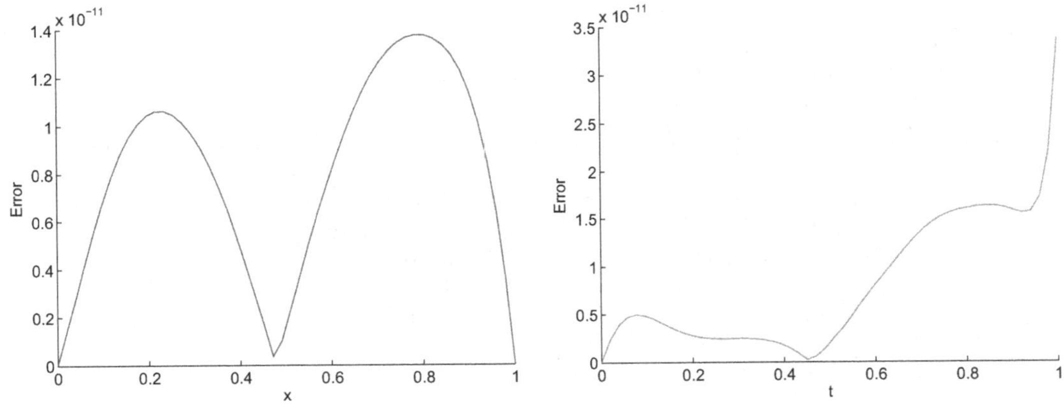

FIGURE 3.6
The plots of the absolute error at $t = 0.6$ (left) and $x = 0.6$ (right) with $(k = 1, M = 9)$ for Example 3.4.

Its exact solution is $u(x,t) = t^3 \sin(x)$ and we solve it by our proposed method with $k = 1$ and different values of M for $\alpha(x,t) = 1.2+0.4\left(x^2+t^2\right)$, $\beta(x,t) = 1.4+0.5\sin(10xt)$, $\rho = 2$, and $\sigma = 1$. The plots of the absolute error at $x = 0.6$ and $t = 0.6$ for $M = 9$ are shown in Figure 3.6. The space-time plots of the absolute error and the approximate solution are given in Figure 3.7. The absolute errors of the approximate solutions in different points $(x_i, t_i) \in \Omega$ are reported in Table 3.3.

The obtained results demonstrate that we can achieve a good approximation with the exact solution by using only a few terms of SKCWs.

3.6 Conclusion

In this paper, the SKCWs are applied to compute an approximate solution for the V-S-FTE. The SKCW's OMV-FD is derived and used in the proposed method. The main advantage of our approach is that it reduces these kinds of problems to those of solving systems of

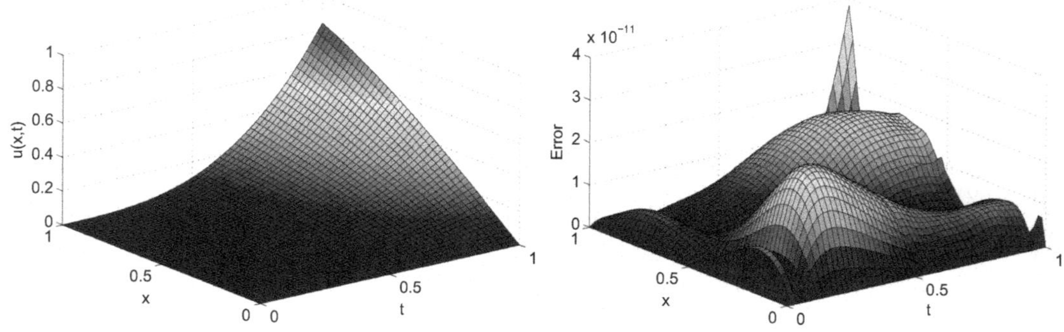

FIGURE 3.7
The plots of the approximate solution (left) and absolute error (right) with ($k = 1, M = 9$) for Example 3.4.

TABLE 3.3
The Absolute Errors of the Approximate Solutions for $k = 1$ and Some Different Values of M in Various Points for Example 3.4

(x_i, t_i)	$M = 4$	$M = 5$	$M = 6$	$M = 7$	$M = 8$	$M = 9$
$(0.1, 0.1)$	1.5692E-07	4.1923E-08	1.0807E-09	3.0635E-10	2.0552E-11	8.7653E-12
$(0.2, 0.2)$	3.0396E-07	1.1576E-08	1.0095E-08	3.0889E-09	1.7947E-10	2.5885E-11
$(0.3, 0.3)$	2.6852E-07	4.4281E-08	1.5156E-08	1.4108E-09	1.5842E-10	2.1540E-11
$(0.4, 0.4)$	2.0566E-07	4.5765E-08	1.0766E-09	3.9439E-10	1.2101E-10	1.4087E-11
$(0.5, 0.5)$	1.6562E-07	4.8719E-08	3.4014E-08	2.1409E-09	6.0796E-11	1.9214E-12
$(0.6, 0.6)$	1.9506E-06	4.1273E-08	2.8006E-09	1.5743E-09	3.1950E-11	8.1002E-12
$(0.7, 0.7)$	1.7719E-06	1.4077E-07	6.5835E-08	3.0487E-09	5.8169E-11	1.7427E-11
$(0.8, 0.8)$	3.0313E-06	2.0613E-07	2.0910E-08	8.3456E-09	1.4924E-10	1.8591E-11
$(0.9, 0.9)$	1.1283E-05	7.7870E-07	1.5742E-08	9.9465E-09	2.1989E-11	4.2337E-12

algebraic equations and simplifiers. Moreover, our approach leads to global approximations of their solutions. Also, the convergence and error analysis of the SKCWs expansion in two dimensions is investigated. A number of examples are provided to illustrate the efficiency and accuracy of our proposed method. The results of these examples illustrated that our proposed method is very accurate and efficient for the problem under study.

References

[1] M. H. Heydari, M. R. Hooshmandasl, and F. M. M. Ghaini, "A new approach of the Chebyshev wavelets method for partial differential equations with boundary conditions of the telegraph type," *Appl. Math. Modell.*, vol. 38, pp. 1597–1606, 2014.

[2] A. Saadatmandi and M. Dehghan, "Numerical solution of hyperbolic telegraph equation using the Chebyshev tau method," *Numer. Methods Partial. Differ. Equations*, vol. 26, pp. 239–252, 2010.

[3] I. Podlubny, *Fractional Differential Equations*. San Diego, CA: Academic Press, 1999.

[4] K. B. Oldham and J. Spanier, *The Fractional Calculus*. New York: Academic Press, 1974.

[5] D. Sierociuk, A. Dzielinski, G. Sarwas, I. Petras, I. Podlubny, and T. Skovranek, "Modelling heat transfer in heterogeneous media using fractional calculus," *Philos. Trans. R. Soc.*, vol. 371, p. 20130146, 2013.

[6] A. Dzielinski, D. Sierociuk, and G. Sarwas, "Some applications of fractional order calculus," *Bull. Pol. Acad. Sci. Tech. Sci.*, vol. 58, pp. 583–592, 2010.

[7] E. Hesameddini, A. Rahimi, and E. Asadollahifard, "On the convergence of a new reliable algorithm for solving multi-order fractional differential equations," *Commun. Nonlinear Sci. Numer. Simul.*, vol. 34, pp. 154–164, 2016.

[8] A. Saadatmandi and M. Dehghan, "A new operational matrix for solving fractional-order differential equations," *Comput. Math. Appl.*, vol. 59, pp. 1326–1336, 2010.

[9] E. H. Doha, A. H. Bhrawy, and S. S. Ezz-Eldien, "A new Jacobi operational matrix: an application for solving fractional differential equations," *Appl. Math. Model.*, vol. 36, pp. 4931–4943, 2012.

[10] E. H. Doha, A. H. Bhrawy, and S. S. Ezz-Eldien, "A Chebyshev spectral method based on operational matrix for initial and boundary value problems of fractional order," *Comput. Math. Appl.*, vol. 62, pp. 2364–2373, 2011.

[11] S. Kazem, S. Abbasbandy, and S. Kumar, "Fractional-order Legendre functions for solving fractional-order differential equations," *Appl. Math. Model.*, vol. 37, pp. 5498–5510, 2013.

[12] A. Ahmadian, M. Suleiman, and S. Salahshour, "An operational matrix based on Legendre polynomials for solving fuzzy fractional-order differential equations," *Abstr. Appl. Anal.*, vol. 2013, 29 p., 2013.

[13] D. Baleanu, A. H. Bhrawy, and T. M. Taha, "Two efficient generalized Laguerre spectral algorithms for fractional initial value problems," *Abstr. Appl. Anal.*, vol. 2013, 2013. doi:10.1155/2013/546502.

[14] M. Ishteva and L. Boyadjiev, "On the c-laguerre functions," *C. R. Acad. Bulg. Sci.*, vol. 58(9), pp. 1019–1024, 2005.

[15] M. Ishteva, L. Boyadjiev, and R. Scherer, "On the Caputo operator of fractional calculus and C-Laguerre functions," *Math. Sci. Res.*, vol. 9(6), pp. 161–170, 2005.

[16] M. H. Heydari, M. R. Hooshmandasl, and F. M. M. Ghaini, "An efficient computational method for solving fractional biharmonic equation," *Comput. Math. Appl.*, vol. 68(9), pp. 269–287, 2014.

[17] Z. Zhao and C. Li, "Fractional difference/finite element approximations for the timespace fractional telegraph equation," *Appl. Math. Comput.*, vol. 219, pp. 2975–2988, 2012.

[18] O. P. Agrawal, "Solution for a fractional difusion-wave equation defined in a bounded domain," *Nonlinear Dyn.*, vol. 29, pp. 145–155, 2002.

[19] H. T. C. Pedro, M. H. Kobayashi, J. M. C. Pereira, and C. F. M. Coimbra, "Variable order modeling of diffusive-convective effects on the oscillatory flow past a sphere," *J. Vib. Control*, vol. 14, pp. 1569–1672, 2008.

[20] L. E. S. Ramirez and C. F. M. Coimbra, "On the selection and meaning of variable order operators for dynamic modelling," *Int. J. Differ. Equation*, vol. 2010, pp. 846107, 16 pp, 2010.

[21] L. E. S. Ramirez and C. Coimbra, "On the variable order dynamics of the nonlinear wake caused by a sedimenting particle," *Physica D*, vol. 240, pp. 1111–1118, 2011.

[22] C. F. M. Coimbra, "Mechanics with variable-order differential operators," *Ann. Phys.*, vol. 12, pp. 692–703, 2003.

[23] C. M. Soon, C. F. M. Coimbra, and M. Kobayashi, "The variable viscoelasticity oscillator," *Ann. Phys.*, vol. 14(6), pp. 378–389, 2005.

[24] L. E. S. Ramirez, C. Coimbra, and M. Kobayashi, "Variable order constitutive relation for viscoelasticity," *Ann. Phys.*, vol. 16, pp. 543–552, 2007.

[25] H. G. Sun, Y. Q. Chen, and W. Chen, "Random-order fractional differential equation models," *Sign. Process.*, vol. 91, pp. 525–530, 2011.

[26] W. K. Zahra and M. M. Hikal, "Non standard finite difference method for solving variable order fractional optimal control problems," *J. Vibr. Control*, pp. 1–11, 2015.

[27] B. Ross and S. Samko, "Fractional integration operator of a variable order in the holder spaces $h(x)$," *Int. J. Math. Math. Sci.*, vol. 18, pp. 777–788, 1995.

[28] S. S. S. Umarov, "Variable order differential equations and diffusion with changing modes," *Z. Anal. Anwend.*, vol. 28, pp. 431–450, 2009.

[29] A. V. Chechkin, R. Gorenflo, and I. M. Sokolov, "Fractional diffusion in inhomogeneous media," *J. Phys. A: Math. Gen.*, vol. 38, pp. 679–684, 2005.

[30] F. Santamaria, S. Wils, E. D. Schutter, and G. J. Augustine, "Anomalous diffusion in purkinje cell dendrites caused by spines," *Neuron*, vol. 52, pp. 635–648, 2006.

[31] H. G. Sun, W. Chen, and Y. Q. Chen, "Variable-order fractional differential operators in anomalous diffusion modeling," *Phys. A*, vol. 388, pp. 4586–4592, 2009.

[32] H. G. Sun, W. Chen, H. Wei, and Y. Chen, "A comparative study of constant-order and variable-order fractional models in characterizing memory property of systems," *Eur. Phys. J. Spec. Top.*, vol. 193, pp. 185–192, 2011.

[33] R. Razminia, A. Dizaji, and V. J. Majd, "Solution existence for non-autonomous variable-order fractional differential equations," *Math. Comput. Modell.*, vol. 55, pp. 1106–1117, 2011.

[34] S. Zhang, "Existence and uniqueness result of solutions to initial value problems of fractional differential equations of variable-order," *J. Frac. Calc. Anal.*, vol. 4(1), pp. 82–98, 2013.

[35] S. Zhang, "Existence result of solutions to differential equations of variable-order with nonlinear boundary value conditions," *Commun. Nonlinear Sci. Numer. Simul.*, vol. 18, pp. 3289–3297, 2013.

[36] Y. Liu, Z. Fang, H. Li, and S. He, "A mixed finite element method for a time-fractional fourth-order partial differential equation," *Appl. Math. Comput.*, vol. 243, pp. 703–717, 2014.

[37] C. F. Lorenzo and T. T. Hartley, "Variable order and distributed order fractional operators," *Nonlinear Dyn.*, vol. 29, pp. 57–98, 2002.

[38] M. M. Meerschaert and C. Tadjeran, "Finite difference approximations for fractional advection dispersion equations," *J. Comput. Appl. Math.*, vol. 172, pp. 65–77, 2004.

[39] Y. Zhang, "A finite difference method for fractional partial differential equation," *Appl. Math. Comput.*, vol. 215, pp. 524–529, 2009.

[40] R. Lin, F. Liu, V. Anh, and I. Turner, "Stability and convergence of a new explicit finite-difference approximation for the variable-order nonlinear fractional diffusion equation," *Appl. Math. Comput.*, vol. 212, pp. 435–445, 2009.

[41] S. Shen, F. Liu, J. Chen, I. Turner, and V. Anh, "Numerical techniques for the variable order time fractional diffusion equation," *Appl. Math. Comput.*, vol. 218, pp. 10861–10870, 2012.

[42] Y. Chen, L. Liu, B. Li, and Y. Sun, "Numerical solution for the variable order linear cable equation with Bernstein polynomials," *Appl. Math. Comput.*, vol. 238, pp. 329–341, 2014.

[43] N. H. Sweilam, M. M. Khader, and H. M. Almarwm, "Numerical studies for the variable-order nonlinear fractional wave equation," *Fract. Calculus Appl. Anal.*, vol. 15, pp. 669–683, 2012.

[44] M. H. Heydari, M. R. Hooshmandasl, C. Cattani, and G. Hariharan, "An optimization wavelet method for multi variable-order fractional differential equations," *Fundam. Inf.*, vol. 151, pp. 255–273, 2017.

[45] M. H. Heydari and Z. Avazzadeh, "A new wavelet method for variable-order fractional optimal control problems," *Asian J. Control*, vol. 20(5), pp. 1–14, 2018.

[46] M. H. Heydari and Z. Avazzadeh, "Legendre wavelets optimization method for variable-order fractional Poisson equation," *Chaos, Solitons Fractals*, vol. 112, pp. 180–190, 2018.

[47] M. H. Heydari, "A new direct method based on the Chebyshev cardinal functions for variable-order fractional optimal control problems," *J. Franklin Inst.*, vol. 335, pp. 4970–4995, 2018.

[48] M. H. Heydari, Z. Avazzadeh, and M. Farzi Haromi, "A wavelet approach for solving multi-term variable-order time fractional diffusion-wave equation," *Appl. Math. Comput.*, vol. 341, pp. 215–228, 2019.

[49] A. H. Bhrawy and M. A. Zaky, "Numerical simulation for two-dimensional variable-order fractional nonlinear cable equation," *Nonlinear Dyn.*, vol. 80(1), pp. 101–116, 2016.

[50] M. Zayernouri and G. E. Karniadakis, "Fractional spectral collocation methods for linear and nonlinear variable order fpdes," *J. Comput. Phys.*, vol. 80(1), pp. 312–338, 2015.

[51] M. A. Abdelkawy, M. A. Zaky, A. H. Bhrawy, and D. Baleanu, "Numerical simulation of time variable fractional order mobile-immobile advection-dispersion model," *Rom. Rep. Phys.*, vol. 67, pp. 773–791, 2015.

[52] Y.-M. Chen, Y.-Q. Wei, D.-Y. Liu, and H. Yu, "Numerical solution for a class of nonlinear variable order fractional differential equations with Legendre wavelets," *Appl. Math. Lett.*, vol. 46, pp. 83–88, 2015.

[53] A. H. Bhrawy and M. A. Zaky, "Numerical algorithm for the variable-order Caputo fractional functional differential equation," *Nonlinear Dyn.*, vol. 1, no. 3, pp. 1–9, 2016.

[54] C. Canuto, M. Hussaini, A. Quarteroni, and T. Zang, *Spectral Methods in Fluid Dynamics*. Berlin: Springer, 1998.

[55] C. K. Chui, *Wavelet Analysis and Its Applications*. San Diego, CA: Academic Press, 1992.

[56] M. H. Heydari, M. R. Hooshmandasl, F. Mohammadi, and C. Cattani, "Wavelets method for solving systems of nonlinear singular fractional Volterra integro-differential equations," *Commun. Nonlinear Sci. Numer. Simul.*, vol. 19, pp. 37–48, 2014.

[57] M. H. Heydari, M. R. Hooshmandasl, and F. Mohammadi, "Legendre wavelets method for solving fractional partial differential equations with Dirichlet boundary conditions," *Appl. Math. Comput.*, vol. 234, pp. 267–276, 2014.

[58] M. H. Heydari, M. R. Hooshmandasl, and F. Mohammadi, "Two-dimensional Legendre wavelets for solving time-fractional telegraph equation," *Adv. Appl. Math. Mech.*, vol. 6(2), pp. 247–260, 2014.

[59] M. H. Heydari, M. R. Hooshmandasl, F. M. M. Ghaini, and C. Cattani, "Wavelets method for the time fractional diffusion-wave equation," *Phys. Lett. A*, vol. 379, pp. 71–76, 2015.

[60] M. H. Heydari, M. R. Hooshmandasl, F. M. M. Ghaini, and C. Cattani, "Wavelets method for solving fractional optimal control problems," *Appl. Math. Comput.*, vol. 286, pp. 139–154, 2016.

[61] L. Zhu and Q. Fan, "Solving fractional nonlinear Fredholm integro-differential equations by the second kind Chebyshev wavelet," *Commun. Nonlinear Sci. Numer. Simul.*, vol. 17(6), pp. 2333–2341, 2012.

[62] Y. Wang and Q. Fan, "The second kind Chebyshev wavelet method for solving fractional differential equations," *Appl. Math. Comput.*, vol. 218(17), pp. 8592–8601, 2012.

[63] A. K. Gupta and S. S. Ray, "Numerical treatment for the solution of fractional fifth-order Sawada-Kotera equation using second kind Chebyshev wavelet method," *Appl. Math. Model*, vol. 39(19), pp. 5121–5130, 2015.

[64] Y. L. Li and W. W. Zhao, "Haar wavelet operational matrix of fractional order integration and its applications in solving the fractional order differential equations," *Appl. Math. Comput.*, vol. 216, pp. 2276–2285, 2010.

[65] Y. L. Li, "Solving a nonlinear fractional differential equation using Chebyshev wavelets," *Commun. Nonlinear Sci. Numer. Simul.*, vol. 15, pp. 2284–2292, 2010.

[66] H. Saeedi, "A cas wavelet method for solving nonlinear Fredholm integro-differential equation of fractional order," *Commun. Nonlinear Sci. Numer. Simul.*, vol. 16, pp. 1154–1163, 2011.

[67] M. H. Heydari, M. R. Hooshmandasl, F. M. M. Ghaini, and F. Fereidouni, "Two-dimensional Legendre wavelets for solving fractional Poisson equation with Dirichlet boundary conditions," *Eng. Anal. Boundary Elem.*, vol. 37, no. 11, pp. 1331–1338, 2013.

[68] M. H. Heydari, M. R. Hooshmandasl, G. B. Loghmania, and C. Cattani, "Wavelets Galerkin method for solving stochastic heat equation," *Int. J. Comput. Math.*, vol. 1, no. 3, pp. 1–18, 2015.

[69] E. H. D. W. M. Abd-Elhameed and Y. H. Youssri, "New spectral second kind Chebyshev wavelets algorithms for solving linear and nonlinear second order differential equation involving singular and Bratu type equations," *Abstr. Appl. Anal.*, vol. 2013, 9 p., 2013.

[70] F. Mohammadi, "Second kind Chebyshev wavelet Galerkin method for stochastic Ito-Volterra integral equations," *Mediterr. J. Math.*, vol. 2, no. 1, pp. 53–69, 2016.

[71] A. K. Gupta and S. S. Ray, "Numerical treatment for the solution of fractional fifth-order Sawada-Kotera equation using second kind Chebyshev wavelet method," *Appl. Math. Model*, vol. 39(19), pp. 5121–5130, 2015.

[72] Y. Wang and Q. Fan, "The second kind Chebyshev wavelet method for solving fractional differential equations," *Appl. Math. Comput.*, vol. 218(17), pp. 8592–8601, 2012.

[73] L. Zhu and Q. Fan, "Solving fractional nonlinear Fredholm integro-differential equations by the second kind Chebyshev wavelet," *Commun. Nonlinear Sci. Numer. Simul.*, vol. 17(6), pp. 2333–2341, 2012.

[74] M. S. M. Selvi and G. Hariharan, "Wavelet-based analytical algorithm for solving steady-state concentration in immobilized glucose isomerase of packed-bed reactor model," *J. Membr. Biol.*, vol. 2, no. 4, pp. 47–59, 2018.

[75] J. S. Gu and W. S. Jiang, "The Haar wavelets operational matrix of integration," *Int. J. Syst. Sci.*, vol. 27, pp. 623–628, 1996.

[76] A. H. Bhrawya and M. A. Zakyc, "A method based on the Jacobi tau approximation for solving multi-term time-space fractional partial differential equations," *J. Comput. Phys.*, vol. 281, pp. 876–895, 2015.

[77] M. Gasca and T. Sauer, "On the history of multivariate polynomial interpolation," *J. Comput. Appl. Math.*, vol. 122, pp. 23–35, 2000.

[78] J. de Villiers, *Mathematics of Approximation.* Paris: Atlantis Press, 2012.

4

Hyers–Ulam–Rassias Stabilities of Some Classes of Fractional Differential Equations

J. E. Restrepo
Regional Mathematical Center of Southern Federal University
Nazarbayev University

R. A. Higuita
University of Antioquia

Shilpi Jain
Poornima College of Engineering India

CONTENTS

4.1 Introduction

In 1940, S. Ulam worked on the problem of the stability of group homomorphims, which led him to arrive at another new problem about the stability of functional equations. In 1941, D. H. Hyers gave the first and significant approximation to solve this problem in the case of additive mapping between Banach spaces [23]. Almost 30 years later, T. M. Rassias proved a significant result on the stability of linear mapping by means of an extension of Hyers's theorem [34]. Further, there was a great interest on this problem which led to several approaches and generalizations that have been published, see e.g. [5,16]. Also different stabilities of functional equations have been considered for the researchers; nevertheless, Hyers–Ulam and Hyers–Ulam–Rassias stabilities are now very popular and useful than the other ones [7,8,10–22,25,26,28–33,35,36,39,40].

Some years ago, the Hyers–Ulam and Hyers–Ulam–Rassias stabilities have been studied and analyzed in the case of fractional integro-differential equations. Besides, some new expected and unexpected results have been obtained in this direction, and many types of fractional derivatives have been studied in different systems of integro-differential equations for a great variety of applications such as engineering, physics, and mathematics, see e.g. [1,6,27,41–45] and the references therein.

Here, we will use the recent Caputo fractional derivative of a function with respect to another function introduced by Almeida in 2016 [4] to define a new class of fractional differential equation and study the Hyers–Ulam and Hylers–Ulam–Rassias stabilities. We also consider the generalized Riemann–Liouville fractional derivative to define another class of fractional differential Equation [38]. The main idea is to study and analyze the stabilities of the considered Caputo and Riemann–Liouville fractional equations. In the end, the advantages and disadvantages are obvious for each fractional operator involved in the considered equations.

The chapter is organized as follows: In Section 4.2 we give definitions and preliminary results on the generalized Riemann–Liouville fractional integro-differentiation operators and the Caputo fractional derivative. We also introduce the classes of Riemann–Liouville and Caputo fractional Volterra type integro-differential equations. Moreover, we introduce the stabilities of Hyers–Ulam and Hyler–Ulam–Rassias type for the considered fractional integro-differential equations. In Sections 4.3 and 4.4 we present sufficient conditions for the Hyers–Ulam–Rassias and Hylers–Ulam stabilities, respectively, of the considered fractional integro-differential equations in a finite interval. Finally, Section 4.5 is devoted to study the Hyers–Ulam–Rassias stabilities of the fractional integro-differential equations in an infinite interval.

4.2 Preliminaries Results

Here we recollect some definitions and auxiliary results. We first recall some definitions and properties of the generalized Riemann–Liouville fractional integro-differentiation operators, see e.g. [38].

Definition 4.1. Let $\alpha > 0$, $I = [a, b]$ $(-\infty \leqslant a < b \leqslant \infty)$ be a finite or infinite interval, f an integrable function defined on I, and $\phi(x) \in C^1(I)$ an increasing function such that $\phi'(t) \neq 0$, for all $t \in I$. The left-sided Riemann–Liouville fractional integral of a function f with respect to another function ϕ is defined as [38]

$$I_{a+}^{\alpha,\phi} f(x) = \frac{1}{\Gamma(\alpha)} \int_a^x \phi'(t)(\phi(x) - \phi(t))^{\alpha-1} f(t)dt. \tag{4.1}$$

Definition 4.2. Let $\alpha > 0$, $I = [a, b]$ $(-\infty \leqslant a < b \leqslant \infty)$ be a finite or infinite interval, f an integrable function defined on I, and $\phi(x) \in C^1(I)$ an increasing function such that $\phi'(t) \neq 0$, for all $t \in I$. The left-sided Riemann–Liouville fractional derivative of a function f with respect to another function ϕ is defined as [38]

$$D_{a+}^{\alpha;\phi} f(x) = \left(\frac{1}{\phi'(x)} \frac{d}{dx}\right)^n (I_{a+}^{n-\alpha;\phi} f)(x), \tag{4.2}$$

where $n = \lfloor \alpha \rfloor + 1$ and $\lfloor \cdot \rfloor$ is the floor function.

The above definitions are generalizations of the well-known Riemann–Liouville fractional integral and derivative operators. Hence, analogous to the properties of the Riemann–Liouville operators, there exists the corresponding properties for the generalized Riemann–Liouville fractional operators. Below we recall some useful results on the reciprocal operations between the generalized fractional integro-differentiation operators, so that it can be proved in the same manner as [38, Theorem 2.4]. Notice that in [38], the parameter α is a complex number with $\mathrm{Re}\,\alpha > 0$, but, in this paper, we just treat the case $\alpha > 0$.

Theorem 4.1. *Let $\alpha > 0$. Then the equality*

$$D_{a+}^{\alpha,\phi} I_{a+}^{\alpha,\phi} f(x) = f(x),$$

is valid for any summable function $f(x)$.

Theorem 4.2. *Let $\alpha > 0$, $n = \lfloor \alpha \rfloor + 1$, and $f_{n-\alpha}(x) = I_{a+}^{n-\alpha,\phi} f(x)$. If $f(x) \in L_1(a,b)$ and $f_{n-\alpha} \in AC^n[a,b]$, then the equality*

$$(I_{a+}^{\alpha,\phi} D_{a+}^{\alpha,\phi} f)(x) = f(x) - \sum_{j=0}^{n-1} \frac{f_{n-\alpha,\phi}^{[n-j-1]}(a)}{\Gamma(\alpha - j)} (\phi(x) - \phi(a))^{\alpha-j-1},$$

holds almost everywhere on $[a,b]$, where

$$f_{n-\alpha,\phi}^{[n-j-1]}(x) = \left(\frac{1}{\phi'(x)} \frac{d}{dx} \right)^{n-j-1} f_{n-\alpha}(x).$$

The following lemma can be found in a general form in [27, Property 2.20]. But, it is not so clear when the result of the property 2.20 is identically zero. Hence we give a proof.

Lemma 4.1. *Let $\alpha > 0$, $n = \lfloor \alpha \rfloor + 1$, $[a,b]$ $(-\infty \leqslant a < b \leqslant \infty)$ be a finite or infinite interval, and $\phi(x) \in C^1([a,b])$ an increasing function such that $\phi'(t) \neq 0$, for all $t \in [a,b]$. Then*

$$D_{a+}^{\alpha,\phi} (\phi(x) - \phi(a))^{\alpha-j-1} = 0, \quad j = 0, \ldots, n-1, \quad x \in [a,b].$$

Proof. We have

$$\begin{aligned} D_{a+}^{\alpha,\phi} (\phi(x) - \phi(a))^{\alpha-j-1} &= \left(\frac{1}{\phi'(x)} \frac{d}{dx} \right)^n \left(I_{a+}^{n-\alpha;\phi} (\phi(x) - \phi(a))^{\alpha-j-1} \right) \\ &= \left(\frac{1}{\phi'(x)} \frac{d}{dx} \right)^n \frac{1}{\Gamma(n-\alpha)} \int_a^x \phi'(t)(\phi(x) - \phi(t))^{\alpha-1} (\phi(t) - \phi(a))^{\alpha-j-1} dt. \end{aligned}$$

Making the substitution $u = \frac{\phi(t)-\phi(a)}{\phi(x)-\phi(a)}$, it follows that $1 - u = \frac{\phi(x)-\phi(t)}{\phi(x)-\phi(a)}$. Hence

$$\begin{aligned} D_{a+}^{\alpha,\phi} (\phi(x) - \phi(a))^{\alpha-j-1} &= \frac{1}{\Gamma(n-\alpha)} \int_0^1 u^{\alpha-j-1} (1-u)^{\alpha-1} du \\ &\quad \times \left(\frac{1}{\phi'(x)} \frac{d}{dx} \right)^n (\phi(x) - \phi(a))^{n-j-1} = 0, \end{aligned}$$

for any $j = 0, \ldots, n-1$. □

Now we recall some facts on the Caputo fractional derivative.

Definition 4.3. Let $\alpha > 0$, $[a,b]$ be a finite or infinite interval$(-\infty \leqslant a < b \leqslant \infty)$, $f, \phi \in C^n([a,b])$ two functions such that ϕ is increasing and $\phi'(t) \neq 0$, for all $t \in [a,b]$. The left ϕ-Caputo fractional derivative of f of order α is given by [4]

$$^C D_{a+}^{\alpha,\phi} f(x) := I_{a+}^{n-\alpha,\phi} \left(\frac{1}{\phi'(x)} \frac{d}{dx} \right)^n f(x), \tag{4.3}$$

where $n = \lfloor \alpha \rfloor + 1$ for $\alpha \notin \mathbb{N}$ and $n = \alpha$ for $\alpha \in \mathbb{N}$.

The following results can be found in [4], see Theorems 4.4 and 4.5.

Theorem 4.3. *Given a function $f \in C^n[a,b]$ and $\alpha > 0$, we have*

$$I_{a+}^{\alpha,\phi}\, {}^C D_{a+}^{\alpha,\phi} f(x) = f(x) - \sum_{j=0}^{n-1} \frac{f_\phi^{[j]}(a)}{j!} \left(\phi(x) - \phi(a)\right)^j,$$

where

$$f_\phi^{[j]}(x) = \left(\frac{1}{\phi'(x)} \frac{d}{dx}\right)^j f(x).$$

Proof. The result follows by using the semi-group law and the integration by parts formula repeatedly. □

Theorem 4.4. *Given a function $f \in C^1[a,b]$ and $\alpha > 0$, we have*

$${}^C D_{a+}^{\alpha,\phi} I_{a+}^{\alpha,\phi} f(x) = f(x).$$

Lemma 4.2. *If $\alpha > 0$, $n = \lfloor \alpha \rfloor + 1$, and $n > j \in \mathbb{N}$, then*

$${}^C D_{a+}^{\alpha,\phi} \left(\phi(x) - \phi(a)\right)^j = 0.$$

A very important tool in this paper will be the well-known Banach Fixed Point Theorem on a generalized metric:

Theorem 4.5. *Let (X, d) be a generalized complete metric space and $T : X \to X$ a strictly contractive operator with a Lipschitz constant $L < 1$. If there exists a non-negative integer k such that $d(T^{k+1}x, T^k x) < \infty$ for some $x \in X$, then:*

I) the sequence $(T^n x)_{n\in\mathbb{N}}$ converges to a fixed point x^* of T;

II) x^* is the unique fixed point of T in $X^* = \{y \in X : d(T^k x, y) < \infty\}$;

III) if $y \in X^*$, then

$$d(y, x^*) \leqslant \frac{1}{1-L} d(Ty, y). \tag{4.4}$$

In this paper we consider the following classes of Riemann–Liouville and Caputo fractional Volterra type integro-differential equations, respectively, as

$$D_{a+}^{\alpha,\phi} y(x) = f\left(x, y(x), \int_a^x k(x, \tau, y(\tau), y(\theta(\tau)))d\tau\right), \quad x \in [a,b], \tag{4.5}$$

along with the conditions

$$\left(\left(\frac{1}{\phi'(x)} \frac{d}{dx}\right)^{n-j-1} I_{a+}^{n-\alpha,\phi} f\left(x, y(x), \int_a^x k(x, \tau, y(\tau), y(\theta(\tau)))d\tau\right)\right)(a) = \gamma_j, \tag{4.6}$$

and

$${}^C D_{a+}^{\alpha,\phi} y(x) = f\left(x, y(x), \int_a^x k(x, \tau, y(\tau), y(\theta(\tau)))d\tau\right), \quad x \in [a,b], \tag{4.7}$$

along with the conditions

$$\left(\left(\frac{1}{\phi'(x)} \frac{d}{dx}\right)^{j} f\left(x, y(x), \int_a^x k(x, \tau, y(\tau), y(\theta(\tau)))d\tau\right)\right)(a) = \mu_j, \tag{4.8}$$

where a, b are some fixed real numbers, γ_j, μ_j are some constants for $j = 0, \ldots, n-1$, y is a function in $C^{n+1}([a,b])$, $f : [a,b] \times \mathbb{C} \times \mathbb{C} \to \mathbb{C}$ and $k : [a,b] \times [a,b] \times \mathbb{C} \times \mathbb{C} \to \mathbb{C}$ are functions in $C^n([a,b])$, and $\theta : [a,b] \to [a,b]$ is a continuous delay function (i.e. $\theta(t) \leqslant t$ for any $t \in [a,b]$). The above classes of fractional differential equations are the fractional version of that class of differential equations considered recently in [11].

The main aim of this chapter is to analyze the Hyers–Ulam stability and the Hyler–Ulam–Rassias stability of Equations (4.5) and (4.7). Hence we introduce the stabilities for the considered fractional integro-differential equations.

If for each function y which satisfy

$$\left| D_{a+}^{\alpha,\phi} y(x) - f\left(x, y(x), \int_a^x k(x,\tau,y(\tau),y(\theta(\tau)))d\tau\right)\right| \leqslant \theta, \quad x \in [a,b], \tag{4.9}$$

where $\theta \geqslant 0$, there is a solution y_0 of the fractional integro-differential Equation (4.5) and a constant $C > 0$ independent of y and y_0 such that

$$|y(x) - y_0(x)| \leqslant C\theta, \quad \text{for any} \quad x \in [a,b], \tag{4.10}$$

then we say that the fractional integro-differential Equation (4.5) has the Hyers–Ulam stability.

Similarly, if for each function y which satisfy

$$\left| {}^C D_{a+}^{\alpha,\phi} y(x) - f\left(x, y(x), \int_a^x k(x,\tau,y(\tau),y(\theta(\tau)))d\tau\right)\right| \leqslant \theta, \quad x \in [a,b], \tag{4.11}$$

where $\theta \geqslant 0$, there is a solution y_0 of the fractional integro-differential Equation (4.7) and a constant $C > 0$ independent of y and y_0 such that condition (4.10) holds, then we say that the fractional integro-differential Equation (4.7) has the Hyers–Ulam stability.

If for each function y which satisfy

$$\left| D_{a+}^{\alpha,\phi} y(x) - f\left(x, y(x), \int_a^x k(x,\tau,y(\tau),y(\theta(\tau)))d\tau\right)\right| \leqslant \sigma(x), \quad x \in [a,b], \tag{4.12}$$

where σ is a non-negative function, there is a solution y_0 of the fractional integro-differential Equation (4.5) and a constant $C > 0$ independent of y and y_0 such that

$$|y(x) - y_0(x)| \leqslant C\sigma(x), \quad \text{for any} \quad x \in [a,b], \tag{4.13}$$

then we say that the fractional integro-differential Equation (4.5) has the Hyers–Ulam–Rassias stability.

Analogously, if for each function y which satisfy

$$\left| {}^C D_{a+}^{\alpha,\phi} y(x) - f\left(x, y(x), \int_a^x k(x,\tau,y(\tau),y(\theta(\tau)))d\tau\right)\right| \leqslant \sigma(x), \quad x \in [a,b], \tag{4.14}$$

where σ is a non-negative function, there is a solution y_0 of the fractional integro-differential Equation (4.7) and a constant $C > 0$ independent of y and y_0 such that condition (4.13) holds, then we say that the fractional integro-differential Equation (4.7) has the Hyers–Ulam–Rassias stability.

In this chapter we consider the space of functions $C^{n+1}[a,b]$ on $[a,b]$ endowed with a generalization of the Bielecki's metric

$$d(u,v) = \sup_{x \in [a,b]} \frac{|u(x) - v(x)|}{\sigma(x)}, \tag{4.15}$$

where $\sigma : [a,b] \to (0,\infty)$ is a non-decreasing continuous function. For more details of this metric, see e.g. [11]. It is known that $(C[a,b],d)$ and $(C^1[a,b],d)$ are complete metric spaces [11,12]. Hence it can be proved that $(C^{n+1}[a,b],d)$ is a complete metric space.

4.3 Hyers–Ulam–Rassias Stability in a Finite Interval

This section is devoted to present the sufficient conditions for the Hyers–Ulam–Rassias stability of the fractional integro-differential Equations (4.5) and (4.7).

We begin with the case of the generalized Riemann–Liouville fractional differential Equation (4.5).

Theorem 4.6. *Let $\theta : [a,b] \to [a,b]$ be a continuos delay function and $\sigma : [a,b] \to (0,\infty)$ a non-decreasing continuous function. We also assume the following conditions:*

i) There exists $\beta \in [0,1)$ such that

$$\int_a^x \sigma(\tau)d\tau \leqslant \beta\sigma(x), \quad \textit{for any} \quad x \in [a,b]. \tag{4.16}$$

ii) There exists $\lambda \in [0,1)$ such that

$$\int_a^x \phi'(\tau)(\phi(x)-\phi(\tau))^{\alpha-1}\sigma(\tau)d\tau \leqslant \lambda\sigma(x), \quad \textit{for any} \quad x \in [a,b]. \tag{4.17}$$

iii) The function $f : [a,b] \times \mathbb{C} \times \mathbb{C} \to \mathbb{C}$ is in $C^n([a,b])$ and satisfies the Lipschitz condition for some $M > 0$, i.e.

$$\big|f(x,u(x),g(x)) - f(x,v(x),h(x))\big| \leqslant M\big(|u(x)-v(x)| + |g(x)-h(x)|\big). \tag{4.18}$$

iv) The kernel $k : [a,b] \times [a,b] \times \mathbb{C} \times \mathbb{C} \to \mathbb{C}$ is in $C^n([a,b])$ and satisfies the Lipschitz condition for some $L > 0$, i.e.

$$\big|k(x,t,u(t),u(\theta(t))) - k(x,t,v(t),v(\theta(t)))\big| \leqslant L\big|u(\theta(t)) - v(\theta(t))\big|. \tag{4.19}$$

If y is a function in $C^{n+1}([a,b])$ which satisfies condition (4.12) and $M\lambda(1+L\beta) < \Gamma(\alpha)$, then there is a unique function y_0 in $C^{n+1}([a,b])$ satisfying the Equation (4.5) and

$$\big|y(x) - y_0(x)\big| \leqslant \frac{\lambda(\Gamma(\alpha))^2}{\Gamma(\alpha) - M\lambda(1+L\beta)}\sigma(x), \quad \textit{for any} \quad x \in [a,b]. \tag{4.20}$$

Proof. By Theorem 4.2, we see that Equation (4.5) becomes

$$y(x) = r(x) + I_{a+}^{\alpha,\phi} f\left(x, y(x), \int_a^x k(x,\tau,y(\tau),y(\theta(\tau)))d\tau\right),$$

where $r(x) = \sum_{j=0}^{n-1} \frac{f_{n-\alpha,\phi}^{[n-j-1]}(a)}{\Gamma(\alpha-j)}(\phi(x)-\phi(a))^{\alpha-j-1}$.

We now define the mapping $T : C^{n+1}([a,b]) \to C^{n+1}([a,b])$ as

$$(Tu)(x) = r(x) + I_{a+}^{\alpha,\phi} f\left(x, u(x), \int_a^x k(x,\tau,u(\tau),u(\theta(\tau)))d\tau\right) dt,$$

for any $x \in [a,b]$ and any function u in $C^{n+1}([a,b])$. Notice first that if u is in $C^{n+1}([a,b])$ then Tu is in $C^{n+1}([a,b])$. Indeed, making integration by parts to the second term of Tu it can be seen easily that the integral belongs to $C^{n+1}([a,b])$. Below we prove that the mapping T is strictly contractive with respect to the metric (4.15). Let us take any two functions u, v in $C^{n+1}([a,b])$. Notice first that the terms $r_u(x)$ and $r_v(x)$ that appear in the representation of $Tu(x)$ and $Tv(x)$ are the same because of the initial conditions (4.6) imposed over Equation (4.5). Hence by the conditions (4.6), (4.18), (4.19), (4.17), and (4.16), respectively, we obtain that

$$\begin{aligned}
d(Tu, Tv) =& \sup_{x\in[a,b]} \frac{|(Tu)(x) - (Tv)(x)|}{\sigma(x)} \\
\leqslant& M \sup_{x\in[a,b]} \frac{1}{\Gamma(\alpha)\sigma(x)} \int_a^x \phi'(s)(\phi(x)-\phi(s))^{\alpha-1}|u(s)-v(s)|ds \\
&+ M \sup_{x\in[a,b]} \frac{1}{\Gamma(\alpha)\sigma(x)} \int_a^x \phi'(s)(\phi(x)-\phi(s))^{\alpha-1} \times \\
&\qquad \times \int_a^s |k(s,\tau,u(\tau),u(\theta(\tau))) - k(s,\tau,v(\tau),v(\theta(\tau)))|d\tau ds \\
\leqslant& M \sup_{x\in[a,b]} \frac{1}{\Gamma(\alpha)\sigma(x)} \int_a^x \phi'(s)(\phi(x)-\phi(s))^{\alpha-1}|u(s)-v(s)|ds \\
&+ ML \sup_{x\in[a,b]} \frac{1}{\Gamma(\alpha)\sigma(x)} \int_a^x \phi'(s)(\phi(x)-\phi(s))^{\alpha-1} \int_a^s |u(\theta(\tau)) - v(\theta(\tau))|d\tau ds \\
=& M \sup_{x\in[a,b]} \frac{1}{\Gamma(\alpha)\sigma(x)} \int_a^x \phi'(s)(\phi(x)-\phi(s))^{\alpha-1}\sigma(s)\frac{|u(s)-v(s)|}{\sigma(s)}ds \\
&+ ML \sup_{x\in[a,b]} \frac{1}{\Gamma(\alpha)\sigma(x)} \int_a^x \phi'(s)(\phi(x)-\phi(s))^{\alpha-1} \int_a^s \sigma(\tau)\frac{|u(\theta(\tau)) - v(\theta(\tau))|}{\sigma(\tau)}d\tau ds \\
\leqslant& M \sup_{s\in[a,b]} \frac{|u(s)-v(s)|}{\sigma(s)} \sup_{x\in[a,b]} \frac{1}{\Gamma(\alpha)\sigma(x)} \int_a^x \phi'(s)(\phi(x)-\phi(s))^{\alpha-1}\sigma(s)ds \\
&+ ML \sup_{\tau\in[a,b]} \frac{|u(\tau)-v(\tau)|}{\sigma(\tau)} \sup_{x\in[a,b]} \frac{1}{\Gamma(\alpha)\sigma(x)} \int_a^x \phi'(s)(\phi(x)-\phi(s))^{\alpha-1} \int_a^s \sigma(\tau)d\tau ds \\
\leqslant& \frac{M}{\Gamma(\alpha)} d(u,v)\lambda + \frac{ML}{\Gamma(\alpha)} d(u,v)\beta\lambda = \frac{M\lambda(1+L\beta)}{\Gamma(\alpha)} d(u,v).
\end{aligned}$$

Since $M\lambda(1 + L\beta) < \Gamma(\alpha)$ it follows that T is strictly contractive. Hence by the Banach Fixed Point Theorem, there exists y_0 in $C^{n+1}[a,b]$ such that $(Ty_0)(x) = y_0(x)$, or what is equivalent

$$\begin{aligned}
y_0(x) =& \sum_{j=0}^{n-1} \frac{f_{n-\alpha,\phi}^{[n-j-1]}(a)}{\Gamma(\alpha-j)} (\phi(x)-\phi(a))^{\alpha-j-1} \\
&+ I_{a+}^{\alpha,\phi} f\left(x, y_0(x), \int_a^x k(x,\tau,y_0(\tau),y_0(\theta(\tau)))d\tau\right).
\end{aligned}$$

Applying the operator $D_{a+}^{\alpha,\phi}$ to both sides of the above equality and taking into account Lemma 4.1 and Theorem 4.1, we obtain that y_0 satisfies Equation (4.5). Moreover, by the Banach Fixed Point Theorem, we also have that

$$d(y, y_0) \leqslant \frac{\Gamma(\alpha)}{\Gamma(\alpha) - M\lambda(1+L\beta)} d(Ty, y). \tag{4.21}$$

Notice now that, by the application of the fractional integral $I_{a+}^{\alpha,\phi}$ over the condition (4.12) and Theorem 4.2, we get that

$$\left|y(x) - r(x) - I_{a+}^{\alpha,\phi} f\left(x, y(x), \int_a^x k(x,\tau,y(\tau),y(\theta(\tau)))d\tau\right)\right| \leqslant I_{a+}^{\alpha,\phi}\sigma(x), \quad x \in [a,b].$$

By inequality (4.21), the definition of the metric d, the above inequality, and condition (4.17), we obtain

$$d(y, y_0) = \sup_{x\in[a,b]} \frac{|y(x) - y_0(x)|}{\sigma(x)} \leqslant \frac{\Gamma(\alpha)}{\Gamma(\alpha) - M\lambda(1+L\beta)} d(Ty, y)$$

$$\leqslant \frac{\Gamma(\alpha)}{\Gamma(\alpha) - M\lambda(1+L\beta)} \sup_{x\in[a,b]} \frac{I_{a+}^{\alpha,\phi}\sigma(x)}{\sigma(x)} \leqslant \frac{\lambda(\Gamma(\alpha))^2}{\Gamma(\alpha) - M\lambda(1+L\beta)},$$

hence it follows the desired inequality (4.20). □

We now continue with the case of Caputo fractional differential Equation (4.7).

Theorem 4.7. *Let $\theta : [a,b] \to [a,b]$ be a continuous delay function and $\sigma : [a,b] \to (0,\infty)$ a non-decreasing continuous function. Moreover, we assume that the conditions i), ii), iii), and iv) from Theorem 4.6 hold. If y is a function in $C^{n+1}([a,b])$ which satisfies condition* (4.14) *and $M\lambda(1+L\beta) < \Gamma(\alpha)$, then there is a unique function y_0 in $C^{n+1}([a,b])$ satisfying Equation* (4.7) *and condition* (4.20).

Proof. Since Theorem 4.3, we have that Equation (4.7) is equivalent to

$$y(x) = c(x) + I_{a+}^{\alpha,\phi} f\left(x, y(x), \int_a^x k(x,\tau,y(\tau),y(\theta(\tau)))d\tau\right),$$

where $c(x) = \sum_{k=0}^{n-1} \frac{f_\phi^{[k]}(a)}{k!} (\phi(x) - \phi(a))^k$.

Let us take the function $D : C^{n+1}([a,b]) \to C^{n+1}([a,b])$ defined as

$$(Du)(x) = c(x) + I_{a+}^{\alpha,\phi} f\left(x, u(x), \int_a^x k(x,\tau,u(\tau),u(\theta(\tau)))d\tau\right),$$

for any $x \in [a,b]$ and any function u in $C^{n+1}([a,b])$. It can be proved in the same manner as was done with the function T in Theorem 4.6 that D is contractive with the constant $\frac{M\lambda(1+L\beta)}{\Gamma(\alpha)}$. For proving this, it is necessary to take into account the conditions i), ii), iii), and iv) from Theorem 4.6 and the initial conditions (4.8) of Equation (4.7). Hence by the Banach Fixed Point Theorem, there exists y_0 in $C^{n+1}[a,b]$ such that $(Dy_0)(x) = y_0(x)$, i.e.

$$y_0(x) = \sum_{k=0}^{n-1} \frac{f_\phi^{[k]}(a)}{k!} (\phi(x) - \phi(a))^k + I_{a+}^{\alpha,\phi} f\left(x, y_0(x), \int_a^x k(x,\tau,y_0(\tau),y_0(\theta(\tau)))d\tau\right).$$

Applying the operator $D_{a+}^{\alpha,\phi}$ to both sides of the above equality, by Lemma 4.2 and Theorem 4.4, it follows that y_0 satisfies the Equation (4.7). Again, by the Banach Fixed Point Theorem, we know that

$$d(y, y_0) \leqslant \frac{\Gamma(\alpha)}{\Gamma(\alpha) - M\lambda(1+L\beta)} d(Dy, y). \tag{4.22}$$

By the application of $I_{a+}^{\alpha,\phi}$ over condition (4.14) and Theorem 4.3, we get that

$$\left|y(x) - c(x) - I_{a+}^{\alpha,\phi} f\left(x, y(x), \int_a^x k(x,\tau,y(\tau),y(\theta(\tau)))d\tau\right)\right| \leqslant I_{a+}^{\alpha,\phi}\sigma(x), \quad x \in [a,b].$$

By inequality (4.22), the definition of the metric d, the above inequality, and condition (4.17), we finally arrive at

$$d(y, y_0) \leqslant \frac{\Gamma(\alpha)}{\Gamma(\alpha) - M\lambda(1 + L\beta)} d(Dy, y) \leqslant \frac{\lambda(\Gamma(\alpha))^2}{\Gamma(\alpha) - M\lambda(1 + L\beta)}.$$

From here we get inequality (4.20). □

Remark 4.1. *Notice that the conditions* (4.16) *and* (4.17) *are in general independent. It can be expected in many cases that the condition* (4.17) *implies the condition* (4.16), *but, we now show that this is not true in general. Indeed, suppose that* $\phi(x) = \sqrt{x}$, $\sigma(x) = ln(x)$, $\alpha = 2/3$, *and* $[a, b] = [e, 4]$. *Further we prove that condition* (4.17) *holds under these functions. It follows for any* $x \in [e, 4]$ *that*

$$\int_e^x \frac{\ln(t)}{2\sqrt{t}\sqrt[3]{\sqrt{x} - \sqrt{t}}} dt \leqslant \ln(x) \int_e^x \frac{dt}{2\sqrt{t}\sqrt[3]{\sqrt{x} - \sqrt{t}}} \leqslant \frac{3\ln(x)}{2}\left(\sqrt{4} - \sqrt{e}\right)^{\frac{2}{3}} \leqslant 0.8\ln(x).$$

On the other hand, we now prove that there is no $\beta \in [0, 1)$ *such that condition* (4.16) *holds for any* $x \in [e, 4]$. *We have that the condition* (4.16) *is equivalent to*

$$x(\ln(x) - 1) \leqslant \beta \ln(x) \quad x \in [e, 4].$$

And it can be proved that at the point $x = 4$ *we have the converse inequality*

$$4(\ln(4) - 1) > \ln(4).$$

4.4 Hyers–Ulam Stability in a Finite Interval

In this section we present two results on the sufficient conditions for the Hyers–Ulam stability of the Riemann–Liouville and Caputo fractional integro-differential Equations (4.5) and (4.7), respectively.

Theorem 4.8. *Let* $\theta : [a, b] \to [a, b]$ *be a continuous delay function and* $\sigma : [a, b] \to (0, \infty)$ *a non-decreasing continuous function. Suppose also that the conditions i), ii), iii), and iv) from Theorem 4.6 are fulfilled. If* y *is a function in* $C^{n+1}([a, b])$ *which satisfies the condition* (4.9) *and* $M\lambda(1 + L\beta) < \Gamma(\alpha)$, *then there is a unique function* y_0 *in* $C^{n+1}([a, b])$ *satisfying Equation* (4.5) *and*

$$|y(x) - y_0(x)| \leqslant \frac{(\phi(b) - \phi(a))^\alpha \sigma(b)}{\alpha(\Gamma(\alpha) - M\lambda(1 + L\beta))\sigma(a)} \theta, \quad \textit{for any} \quad x \in [a, b]. \tag{4.23}$$

Proof. The proof is similar to the proof of Theorem 4.6. Hence we just present the proof. Again we can consider the mapping $T : C^{n+1}([a, b]) \to C^{n+1}([a, b])$ defined by

$$(Tu)(x) = c(x) + I_{a+}^{\alpha;\phi} f\left(x, u(x), \int_a^x k(x, \tau, u(\tau), u(\theta(\tau)))d\tau\right) dt,$$

for any $x \in [a, b]$ and any function u in $C^{n+1}([a, b])$. Besides, under the conditions of the theorem, we can prove that T is strictly contractive (with the contractive constant

$M\lambda(1 + L\beta)/\Gamma(\alpha))$ with respect to the metric (4.15), since $M\lambda(1 + L\beta) < \Gamma(\alpha)$. Hence by the Banach Fixed Point Theorem, it follows that there exists y_0 in $C^{n+1}[a,b]$ satisfying Equation (4.5). By the same theorem we know that

$$d(y, y_0) \leqslant \frac{\Gamma(\alpha)}{\Gamma(\alpha) - M\lambda(1 + L\beta)} d(Ty, y). \tag{4.24}$$

Notice now that by inequality (4.9) that

$$-\theta \leqslant D_{a+}^{\alpha,\phi} y(x) - f\left(x, y(x), \int_a^x k(x, \tau, y(\tau), y(\theta(\tau)))d\tau\right) \leqslant \theta, \quad x \in [a,b].$$

By the application of $I_{a+}^{\alpha,\phi}$ to the above inequalities and Theorem 4.2, we have that

$$\left| y(x) - r(x) - I_{a+}^{\alpha,\phi} f\left(x, y(x), \int_a^x k(x, \tau, y(\tau), y(\theta(\tau)))d\tau\right)\right|$$
$$\leqslant \frac{1}{\Gamma(\alpha)} \int_a^x \phi'(s)(\phi(x) - \phi(s))^{\alpha-1}\theta ds, \quad \text{for any} \quad x \in [a,b],$$

this is equivalent to

$$|y(x) - (Ty)(x)| \leqslant \frac{(\phi(b) - \phi(a))^\alpha}{\alpha\Gamma(\alpha)}\theta, \quad \text{for any} \quad x \in [a,b]. \tag{4.25}$$

Then by the inequalities (4.24) and (4.25), it follows inequality (4.23). □

Theorem 4.9. *Let $\theta : [a,b] \to [a,b]$ be a continuous delay function and $\sigma : [a,b] \to (0,\infty)$ a non-decreasing continuous function. Moreover, we assume that the conditions i), ii), iii), and iv) from Theorem 4.6 hold. If y is a function in $C^{n+1}([a,b])$ which satisfies the condition* (4.11) *and $M\lambda(1+L\beta) < \Gamma(\alpha)$, then there is a unique function y_0 in $C^{n+1}([a,b])$ satisfying Equation* (4.7) *and inequality* (4.23).

Proof. The proof is similar to the proof of Theorem 4.7. Below we give the main steps of the proof. Again we consider the function $T : C^{n+1}([a,b]) \to C^{n+1}([a,b])$ as

$$(Tu)(x) = c(x) + I_{a+}^{\alpha,\phi} f\left(x, u(x), \int_a^x k(x, \tau, u(\tau), u(\theta(\tau)))d\tau\right) dt,$$

for any $x \in [a,b]$ and any function u in $C^{n+1}([a,b])$. By the initial conditions of the theorem, we can prove that T is strictly contractive (with a contractive constant $M\lambda(1+L\beta)/\Gamma(\alpha))$ with respect to the metric (4.15), since $M\lambda(1 + L\beta) < \Gamma(\alpha)$. By the Banach Fixed Point Theorem, there exists y_0 in $C^{n+1}[a,b]$ satisfying the Equation (4.7). Also, it follows inequality (4.24). Besides, by inequality (4.11), we obtain

$$-\theta \leqslant {}^C D_{a+}^{\alpha,\phi} y(x) - f\left(x, y(x), \int_a^x k(x, \tau, y(\tau), y(\theta(\tau)))d\tau\right) \leqslant \theta, \quad x \in [a,b].$$

Hence by Theorem 4.3 we get

$$\left| y(x) - c(x) - I_{a+}^{\alpha,\phi} f\left(x, y(x), \int_a^x k(x, \tau, y(\tau), y(\theta(\tau)))d\tau\right) ds\right|$$
$$\leqslant \frac{1}{\Gamma(\alpha)} \int_a^x \phi'(s)(\phi(x) - \phi(s))^{\alpha-1}\theta ds, \quad \text{for any} \quad x \in [a,b],$$

then

$$|y(x) - (Ty)(x)| \leqslant \frac{(\phi(b) - \phi(a))^\alpha}{\alpha\Gamma(\alpha)}\theta, \quad \text{for any} \quad x \in [a,b]. \tag{4.26}$$

Then by the inequalities (4.24) and (4.26), it follows inequality (4.23). □

4.5 Hyers–Ulam–Rassias Stability in an Infinite Interval

In this section, we study the Hyers–Ulam–Rassias stabilities of the Riemann–Liouville and Caputo fractional integro-differential Equations (4.5) and (4.7), in an infinite interval $[a,\infty)$, for some fixed $a \in \mathbb{R}$. Below we introduce the new fractional integro-differential equations in the case of an infinite interval $[a,\infty)$. From now on, we will treat the following equations:

$$D_{a+}^{\alpha,\phi} y(x) = f\left(x, y(x), \int_a^x k(x,\tau,y(\tau),y(\theta(\tau)))d\tau\right), \quad x \in [a,\infty), \tag{4.27}$$

along with the conditions

$$\left(\left(\frac{1}{\phi'(x)}\frac{d}{dx}\right)^{n-j-1} I_{a+}^{n-\alpha,\phi} f\left(x, y(x), \int_a^x k(x,\tau,y(\tau),y(\theta(\tau)))d\tau\right)\right)(a) = \gamma_j, \tag{4.28}$$

and

$$^{C}D_{a+}^{\alpha,\phi} y(x) = f\left(x, y(x), \int_a^x k(x,\tau,y(\tau),y(\theta(\tau)))d\tau\right), \quad x \in [a,\infty), \tag{4.29}$$

along with the conditions

$$\left(\left(\frac{1}{\phi'(x)}\frac{d}{dx}\right)^{j} f\left(x, y(x), \int_a^x k(x,\tau,y(\tau),y(\theta(\tau)))d\tau\right)\right)(a) = \mu_j, \tag{4.30}$$

where a is a fixed real number, γ_j, μ_j are some constants for $j = 0,\ldots,n-1$, y is a function in $C^{n+1}([a,\infty))$, $f : [a,\infty) \times \mathbb{C} \times \mathbb{C} \to \mathbb{C}$ and $k : [a,\infty) \times [a,\infty) \times \mathbb{C} \times \mathbb{C} \to \mathbb{C}$ are functions in $C^n([a,\infty))$, and $\theta : [a,\infty) \to [a,\infty)$ is a continuous delay function.

Below, by $C_b^{n+1}([a,\infty))$ we denote the space of those bounded functions that belong to $C^{n+1}([a,\infty))$. Besides, we endowed the space $C_b^{n+1}([a,\infty))$ with the metric

$$d_b(u,v) = \sup_{x\in[a,\infty)} \frac{|u(x)-v(x)|}{\sigma(x)}, \tag{4.31}$$

where $\sigma : [a,\infty) \to (c,d)$ is a non-decreasing continuous function, for some $c,d > 0$.

Theorem 4.10. *Let $\theta : [a,\infty) \to [a,\infty)$ be a continuous delay function and $\sigma : [a,\infty) \to (c,d)$ a non-decreasing continuous function. We assume the following conditions:*

i) There exists $\beta \in [0,1)$ such that

$$\int_a^x \sigma(\tau)d\tau \leqslant \beta\sigma(x), \quad \text{for any} \quad x \in [a,\infty). \tag{4.32}$$

ii) There exists $\lambda \in [0,1)$ such that

$$\int_a^x \phi'(\tau)(\phi(x)-\phi(\tau))^{\alpha-1}\sigma(\tau)d\tau \leqslant \lambda\sigma(x), \quad \text{for any} \quad x \in [a,\infty). \tag{4.33}$$

iii) The function $f : [a,\infty) \times \mathbb{C} \times \mathbb{C} \to \mathbb{C}$ is in $C^n([a,\infty))$ and satisfies the Lipschitz condition for some $M > 0$, i.e.

$$\big|f(x,u(x),g(x)) - f(x,v(x),h(x))\big| \leqslant M\big(|u(x)-v(x)| + |g(x)-h(x)|\big). \tag{4.34}$$

iv) *The kernel $k : [a,\infty) \times [a,\infty) \times \mathbb{C} \times \mathbb{C} \to \mathbb{C}$ is in $C^n([a,\infty))$ which satisfy that $\int_a^x k(x,\tau,q(\tau),q(\theta(\tau)))d\tau$ is bounded for any bounded function q in $C^n([a,\infty))$. Also, k satisfies the Lipschitz condition for some $L > 0$, i.e.*

$$\big|k(x,t,u(t),u(\theta(t))) - k(x,t,v(t),v(\theta(t)))\big| \leqslant L\big|u(\theta(t)) - v(\theta(t))\big|. \tag{4.35}$$

If y is a function in $C_b^{n+1}([a,\infty))$ which satisfies

$$\left|D_{a+}^{\alpha,\phi}y(x) - f\left(x, y(x), \int_a^x k(x,\tau,y(\tau),y(\theta(\tau)))d\tau\right)\right| \leqslant \sigma(x), \quad x \in [a,\infty), \tag{4.36}$$

and $M\lambda(1+L\beta) < \Gamma(\alpha)$, then there is a unique function y_0 in $C_b^{n+1}([a,\infty))$ satisfying Equation (4.27) and

$$\big|y(x) - y_0(x)\big| \leqslant \frac{\lambda\big(\Gamma(\alpha)\big)^2}{\Gamma(\alpha) - M\lambda(1+L\beta)}\sigma(x), \quad \textit{for any} \quad x \in [a,\infty). \tag{4.37}$$

Proof. For any $n \in \mathbb{N}$, we consider the interval $I_n = [a, a+n]$. As y is a function in $C_b^{n+1}([a,\infty))$ satisfying (4.36), by Theorem 4.6, there exists a unique $y_{0,n} : I_n \to \mathbb{C}$ in $C_b^{n+1}(I_n)$ such that

$$y_{0,n}(x) = r(x) + I_{a+}^{\alpha,\phi} f\left(x, y_{0,n}(x), \int_a^x k(x,\tau,y_{0,n}(\tau),y_{0,n}(\theta(\tau)))d\tau\right) \tag{4.38}$$

and

$$\big|y(x) - y_{0,n}(x)\big| \leqslant \frac{\lambda(\Gamma(\alpha))^2}{\Gamma(\alpha) - M\lambda(1+L\beta)}\sigma(x) \tag{4.39}$$

for any $x \in I_n$. By the uniqueness of the function $y_{0,n}$ it follows that

$$y_{0,n}(x) = y_{0,n+1}(x) = y_{0,n+2}(x) = \cdots, \quad \text{for any} \quad x \in I_n. \tag{4.40}$$

For any $x \in [a,\infty)$, we define $n(x) \in \mathbb{N}$ as $n(x) = \min\{n \in \mathbb{N} : x \in I_n\}$. We also define the function $y_0 : [a,\infty) \to \mathbb{C}$ by

$$y_0(x) = y_{0,n(x)}(x). \tag{4.41}$$

For any $x_1 \in [a,\infty)$, let $n_1 = n(x_1)$. Then $x \in \operatorname{Int}(I_{n_1+1})$ and there exists an $\epsilon > 0$ such that $y_0(x) = y_{0,n_1+1}(x)$ for any $x \in (x_1 - \epsilon, x_1 + \epsilon)$. By Theorem 4.6, y_{0,n_1+1} is unique and belongs to $C_b^{n+1}(I_{n_1+1})$, hence y_0 is in $C_b^{n+1}(I_{n_1+1})$ at any point $x_1 \in [a,\infty)$. It remains to prove that y_0 satisfies Equation (4.27) and inequality (4.37). Indeed, for any arbitrary $x \in [a,\infty)$, there exists $n(x)$ such that $x \in I_{n(x)}$. By (4.41), (4.38), and (4.40), we have

$$\begin{aligned} y_0(x) = y_{0,n(x)}(x) &= r(x) + I_{a+}^{\alpha,\phi} f\left(x, y_{0,n(x)}(x), \int_a^x k(x,\tau,y_{0,n(x)}(\tau),y_{0,n(x)}(\theta(\tau)))d\tau\right) \\ &= r(x) + I_{a+}^{\alpha,\phi} f\left(x, y_0(x), \int_a^x k(x,\tau,y_0(\tau),y_0(\theta(\tau)))d\tau\right). \end{aligned} \tag{4.42}$$

Notice that $n(\tau) \leqslant n(x)$, for any $\tau \in I_{n(x)}$, hence by (4.40) we get that $y_0(\tau) = y_{0,n(\tau)}(\tau) = y_{0,n(x)}(\tau)$. Thus, the last inequality in (4.42) holds. Hence y_0 satisfies Equation (4.27).

Inequality (4.37) follows equality (4.41) and inequality (4.39):

$$\big|y(x) - y_0(x)\big| = \big|y(x) - y_{0,n(x)}(x)\big| \leqslant \frac{\lambda(\Gamma(\alpha))^2}{\Gamma(\alpha) - M\lambda(1+L\beta)}\sigma(x), \quad x \in [a,\infty).$$

Further we prove the uniqueness of y_0. Hence, let y_1 in $C_b^{n+1}([a,\infty))$ satisfy Equation (4.27) and inequality (4.37). By the uniqueness of the solution on $I_{n(x)}$ for any $n(x) \in \mathbb{N}$ we have that $y_0(x)\big|_{I_{n(x)}} = y_{0,n(x)}(x)$ and $y_1(x)\big|_{I_{n(x)}}$ satisfies (4.37) and (4.38) for any $x \in I_{n(x)}$, then $y_0(x) = y_0(x)\big|_{I_{n(x)}} = y_1(x)\big|_{I_{n(x)}} = y_1(x)$. □

A similar result for the case of the Caputo fractional integro-differential Equation (4.29) is given later.

Theorem 4.11. *Let $\theta : [a,\infty) \to [a,\infty)$ be a continuous delay function and $\sigma : [a,\infty) \to (c,d)$ a non-decreasing continuous function. We also assume the conditions i), ii), iii), and iv) given in Theorem 4.10. If y is a function in $C_b^{n+1}([a,\infty)$ which satisfies*

$$\left| {}^C D_{a+}^{\alpha,\phi} y(x) - f\left(x, y(x), \int_a^x k(x,\tau,y(\tau),y(\theta(\tau)))d\tau\right)\right| \leqslant \sigma(x), \quad x \in [a,\infty), \tag{4.43}$$

and $M\lambda(1+L\beta) < \Gamma(\alpha)$, then there is a unique function y_0 in $C_b^{n+1}([a,\infty)$ satisfying Equation (4.29) *and inequality* (4.37).

Proof. The proof follows the same lines of the proof of Theorem 4.10 with the corresponding changes. For any $n \in \mathbb{N}$, $I_n = [a, a+n]$. By Theorem 4.7, there exists a unique bounded and $C^{n+1}(I_n)$, $y_{0,n} : I_n \to \mathbb{C}$ which satisfy that

$$y_{0,n}(x) = c(x) + I_{a+}^{\alpha,\phi} f\left(x, y_{0,n}(x), \int_a^x k(x,\tau,y_{0,n}(\tau),y_{0,n}(\theta(\tau)))d\tau\right), \tag{4.44}$$

and inequality (4.39) for any $x \in I_n$. By the uniqueness of $y_{0,n}$ we obtain that

$$y_{0,n}(x) = y_{0,n+1}(x) = y_{0,n+2}(x) = \cdots, \quad \text{for any} \quad x \in I_n. \tag{4.45}$$

Again, for any $x \in [a,\infty)$, we define $n(x) \in \mathbb{N}$ as $n(x) = \min\{n \in \mathbb{N} : x \in I_n\}$ and the function $y_0 : [a,\infty) \to \mathbb{C}$ by $y_0(x) = y_{0,n(x)}(x)$. The rest of the proof follows in the same form as was done in Theorem 4.10 for the function $y_{0,n(x)}(x)$, where it was proved that y_0 belongs to $C_b^{n+1}([a,\infty))$, satisfies inequality (4.37), and it is unique. □

Remark 4.2. *The Hyers–Ulam–Rassias stabilities of the Riemann–Liouville and Caputo fractional integro-differential Equations* (4.27) *and* (4.29), *in an infinite interval of the form $(-\infty, b]$ or $(-\infty,+\infty)$, where b is a fixed real number, it can be just proved in the same form as those theorems above.*

4.6 Conclusions

The study of the Hyers–Ulam stability and the Hyler–Ulam–Rassias stability of some good-enough classes of integro-differentiation fractional equations leads to generalize and extend some classical results. Moreover, the tools to prove it are the same as usual, with just some additional conditions appearing natural in the fractional case. The key of this chapter was to use the reciprocal operations between the fractional integro-differentiation operators to recover the function.

On the other hand, it is important to say that the idea of this paper started in June (2018), and we did not at this moment think that a pre-print of the published paper [9] exists. Some months later, we saw the work and we can say that the tools and techniques are similar, but the mathematical objects, we mean fractional operators and the classes of fractional equations that we consider are very different than those considered in [9], even, the initial conditions of the main theorems contain other important and necessary additions as those independent conditions given in (4.16) and (4.17) (see Remark 4.1 to take into account the independence between them). For instance, the indispensable condition (4.17) was not considered in [9].

In the future, we will consider more general fractional operators, as those introduced in [2,3,24,37]. With them, we can introduce a very general class of integro-differentiation fractional equations that leads us to obtain rare and more unexpected results than the known ones.

Acknowledgments

J.E. Restrepo was partially supported by the Nazarbayev University Program 091019CRP2120. No new data was collected or generated during the course of this research. Shilpi Jain is thankful for SERB for their research grant supported by SERB Project Number: MTR/2017/000194.

References

[1] S. Abbas, M. Benchohra, J.E. Lagreg, A. Alsaedi, Y. Zhou: Existence and Ulam stability for fractional differential equations of Hilfer-Hadamard type. *Adv. Diff. Equa.* **2017**, 180 (2017).

[2] P. Agarwal, A.M. Jerbashian, J.E. Restrepo: Weighted integral inequalities in terms of Omega-Fractional integro-differentiation. In: Agarwal P., Dragomir S., Jleli M., Samet B. (eds) *Advances in Mathematical Inequalities and Applications*, Trends in Mathematics, Birkhäuser, Singapore (2018) 199–218.

[3] P. Agarwal, J.E. Restrepo: An extension by means of ω-weighted classes of the generalized Riemann-Liouville k-fractional integral inequalities. *J. Math Inequal.* **14** (1) (2020) 35–46.

[4] R. Almeida: A Caputo fractional derivative of a function with respect to another function, *Commun. Nonlinear Sci. Numer. Simul.* **44** (2017) 460–481.

[5] T. Aoki: On the stability of the linear transformation in Banach spaces. *J. Math. Soc. Japan* **2** (1950) 64–66.

[6] M. Benchohra, J.E. Lazreg: Existence and Ulam stability for nonlinear implicit fractional differential equations with Hadamard derivative. *Stud. Univ. Babes-Bolyai Math.* **62** (1) (2017) 27–38.

[7] W.W. Breckner, T. Trif: *Convex Functions and Related Functional Equations*, Selected Topics, Cluj University Press, Cluj (2008).

[8] J. Brzdek, L. Cădariu, K. Cieplińsk: Fixed point theory and the Ulam stability. *J. Funct. Space* **2014** (2014) 16.

[9] E. Capelas de Oliveira, J.V.C. Sousa: Ulam-Hyers-Rassias stability for a class of fractional integro-differential equations. *Results Math.* **73** (2018) 111.

[10] L.P. Castro, A.M. Simões: Hyers–Ulam and Hyers–Ulam–Rassias stability of a class of Hammerstein integral equations. *AIP Conf. Proc.* **1798** (2017) 10.

[11] L.P. Castro, A.M. Simões: Hyers–Ulam and Hyers–Ulam–Rassias stability for a class of integro-differential equations. In: Taş K., Baleanu D., Machado J. (eds) *Mathematical Methods in Engineering*, Nonlinear Systems and Complexity, Vol **23**. Springer, Cham (2019).

[12] L. Cădariu, L. Găvruta, P. Găvruta: Weighted space method for the stability of some nonlinear equations. *Appl. Anal. Discrete Math.* **6** (1) (2012) 126–139.

[13] Y.J. Cho, Th.M. Rassias, R. Saadati: *Stability of Functional Equations in Random Normed Spaces*, Springer, New York (2013).

[14] J. Chung: Hyers–Ulam–Rassias stability of Cauchy equation in the space of Schwartz distributions. *J. Math. Anal. Appl.* **300** (2) (2004) 343–350.

[15] S. Czerwik: *Functional Equations and Inequalities in Several Variables*, World Scientific Publishing Co, Singapore (2002).

[16] J.B. Diaz, B. Margolis: A fixed point theorem of the alternative, for contractions on a generalized complete metric space. *Bull. Am. Math. Soc.* **74** (2) (1968) 305–309.

[17] W.S. Du: A generalization of Diaz-Margoliss fixed point theorem and its application to the stability of generalized Volterra integral equations. *J. Inequal. Appl.* **407** (2015) 15.

[18] P. Gavruta: A generalization of the Hyers–Ulam–Rassias stability of approximately additive mappings. *J. Math. Anal. Appl.* **184** (1994) 431–436.

[19] P. Gavruta, L. Gavruta: A new method for the generalized Hyers–Ulam–Rassias stability. *Int. J. Nonlinear Anal. Appl.* **1** (2) (2010) 6.

[20] A.M. Hassan, E. Karapinar, H.H. Alsulami: Ulam-Hyers stability for MKC mappings via fixed point theory. *J. Funct. Space* **2016** (2016) 11.

[21] M. Hazewinkel (ed): Hyers–Ulam–Rassias stability. In: *Encyclopaedia of Mathematics*, Supplement III, Kluwer Academic Publishers, Dordrecht 2001 194–196.

[22] J. Huang, Y. Li: Hyers–Ulam stability of delay differential equations of first order. *Math. Nachr.* **289** (1) (2016) 60–66.

[23] D.H. Hyers: On the stability of linear functional equation. *Proc. Natl. Acad. Sci.* **27** (4) (1941) 222–224.

[24] A.M. Jerbashian, J.E. Restrepo: On some classes of harmonic functions with nonnegative harmonic majorants in the half-plane. *J. Contemp. Math. Anal.* **51** (2016) 79–89.

[25] S.M. Jung: *Hyers–Ulam–Rassias Stability of Functional Equations in Mathematical Analysis*, Hadronic Press, Palm Harbor (2001).

[26] S.M. Jung: Hyers–Ulam–Rassias stability of Jensen's equation and its application. *Proc. Am. Math. Soc.* **126** (1998) 3137–3143.

[27] A.A. Kilbas, H.M. Srivastava, J.J. Trujillo: *Theory and Applications of Fractional Differential Equations*, North-Holland Mathematics Studies, vol. 204. Elsevier Science B.V., Amsterdam (2006).

[28] G.H. Kim: A generalization of Hyers–Ulam–Rassias stability of the G-functional equation. *Math. Inequal. Appl.* **10** (2007) 351–358.

[29] Y.H. Lee, K.W. Jun: A generalization of the Hyers–Ulam–Rassias stability of the pexider equation. *J. Math. Anal. Appl.* **246** (2000) 627–638.

[30] L. Maligranda: A result of Tosio Aoki about a generalization of Hyers–Ulam stability of additive functionsa question of priority. *Aequationes Math.* **75** (2008) 289–296.

[31] T. Miura, S.E. Takahasi, G. Hirasawa: Hyers–Ulam–Rassias stability of Jordan homomorphisms on Banach algebras. *J. Inequal. Appl.* **4** (2005) 435–441.

[32] A. Najati, C. Park: HyersUlam-Rassias stability of homomorphisms in quasi-Banach algebras associated to the Pexiderized Cauchy functional equation. *J. Math. Anal. Appl.* **335** (2007) 763–778.

[33] D. Popa, I. Rasa.: On the best constant in Hyers–Ulam stability of some positive linear operators. *J. Math. Anal. Appl.* **412** (1) (2014) 103–108.

[34] Th.M. Rassias: On the stability of the linear mapping in Banach spaces. *Proc. Am. Math. Soc.* **72** (1978) 297–300.

[35] Th.M. Rassias: On the stability of functional equations and a problem of Ulam. *Acta Appl. Math.* **62** (1) (2000) 23–130.

[36] Th.M. Rassias, J. Brzdek: *Functional Equations in Mathematical Analysis*, Springer, New York (2012).

[37] J.E. Restrepo, A. Kilicman, P. Agarwal, O. Altun: Weighted hypergeometric functions and fractional derivative. *Adv. Differ. Equ.* **2017** 105 (2017).

[38] S.G. Samko, A.A. Kilbas, O.I. Marichev: *Fractional Integrals and Derivatives, Translated from the 1987 Russian Original*, Gordon and Breach, Yverdon (1993).

[39] S. Sevgin, H. Sevli: Stability of a nonlinear Volterra integro-differential equation via a fixed point approach. *J. Nonlinear Sci. Appl.* **9** (2016) 200–207.

[40] T. Trif: Hyers–Ulam–Rassias stability of a Jensen type functional equation. *J. Math. Anal. Appl.* **250** (2000) 579–588.

[41] J. Vanterler da C. Sousa, E. Capelas de Oliveira: Ulam-Hyers stability of a nonlinear fractional Volterra integro-differential equation. *Appl. Math. Lett.* **81** (2018) 50–56.

[42] J. Wang, M. Fĕckan, Y. Zhou: A survey on impulsive fractional differential equations. *Fract. Cal. Appl. Anal.* **19** (4) (2016) 806–831.

[43] J. Wang, M. Fĕckan, Y. Tian: Stability analysis for a general class of non-instantaneous impulsive differential equations. *Mediterr. J. Math.* **14** (2) (2017) 46.

[44] Z. Yong, w. Jinrong, Z. Lu: *Basic Theory of Fractional Differential Equations*, World Scientific Publishing Company, Singapore, New Jersey, London and Hong Kong (2014).

[45] Y. Zhou: Existence and uniqueness of solutions for a system of fractional differential equations. *J. Fract. Calc. Appl. Anal.* **12** (2) (2009) 195–204.

5

Applications of Fractional Derivatives to Heat Transfer in Channel Flow of Nanofluids

Muhammad Saqib
Universiti Teknologi Malaysia

Ilyas Khan
Ton DucThang University
Majmaah University

Sharidan Shafie
Universiti Teknologi Malaysia

CONTENTS

5.1 Introduction

The literature of nanofluids is rich enough, but the memory effect is neglected using partial differential equations with integer order. The fractionalized nanofluid models are more general and better in the description of memory effect and to analyze the influence of flow parameters. It is well accepted that non-integer order derivatives are used to describe various physical phenomena which include viscoelastic materials, rheology, fluid flow, diffusive transport, probability, electrical networks, electromagnetic theory, etc. [1–4]. It is worth mentioning here that various physical and real-world phenomena follow three dominant mathematical functions, i.e., the exponential decay function, power-law function, and the generalized Mittag-Leffler [5]. In the existing literature, the mathematical definitions of different fractional operators are based on functions such as Riemann–Liouville fractional derivative (RLFD), and Caputo fractional derivative (CFD) is based on the power law functions. The RLFD and CFD have been in practice for a long time and applied to various practical problems successfully [6,7]. However, RLFD and CFD have some limitations. The kernel of CFD is singular, and the derivative of a constant is not zero in case of RLFD [8,9].

The Caputo–Fabrizio fractional derivative (CFFD) is based on the exponential kernel which overcomes the limitations in RLFD and CFD. But CFFD is criticized by many researchers due to the locality in the kernel [10,11]. Recently, Atangana and Baleanu introduced new fractional derivatives in Riemann–Liouville and Caputo sense [12]. The kernel Atangana–Baleanu fractional derivative (ABFD) is based on the generalized Mittag-Leffler function without singularity and locality. ABFDs are suitable to describe the physical and real-world phenomena. The non-singularity and non-locality of the kernel gives a better description of the memory within the structure with different scales [13]. Furthermore, the Atangana–Baleanu operators satisfy all the mathematical principles under the scope of fractional calculus [14].

Fractional models or generalized models in fluid dynamics are rarely studied in the literature. The flow of different types of fractionalized nanofluids over an infinite plate is studied by Saqib et al. [15] using a CFD approach. They established fractional results for velocity profile, temperature distribution, entropy generation, and Bejan number. Sheikh et al. [16,17] considered Casson fluid along with heat and mass transfer over an infinite plate. The exact solution was generated using the Laplace transform technique. They compared the results of ABFD and CFFD in their studies. In another paper, Sheikh et al. [18] carried out a comparative study of ABFD and CFFD to the application of nanofluid in the solar collector. Jan et al. [19] analyzed engine oil-based molybdenum disulfide nanofluid using ABFD approach. They concluded that fractional velocity and temperature distributions of nanofluid show a decreasing trend for a smaller value of a fractional parameter. The unsteady flow fractional nanofluid over a moving plate was studied by Azhar et al. [20]. They reported that the fractional parameter allows to establishing a strong agreement between theoretical and experimental work. Fetecau et al. [21] developed the closed form solutions for fractional nanofluid flowing over isothermal vertical plate taking into consideration the radiation effect. As discussed above, many known mathematicians, physicists, and engineers contributed to the theory of fractional derivatives, yet the idea of fractional derivatives, particularly ABFD, needs to be further explored.

On the other hand, carbon nanotubes (CNTs) form a cylinder of carbon atoms possessing significant electrical and mechanical thermal performance. It is a valuable material in nanotechnology because of its structural backdrop, baby admeasurement and mass, stronger, higher electrical and thermal conductivity, and so on [22]. In CNTs, the carbon atoms form a hexagonal cylindrical network of 1 nm in bore and 100 nm in length. There are three kinds of CNTs: Single-wall CNTs (SWCNTs), double-wall CNTs (DWCNTs), and multi-wall CNTs (MWCNTs). Murshed et al. [23] reported that CNT-based nanofluids have six times higher thermal conductivity when compared with other material nanofluids at room temperature. The main applications of CNTs include nanolithography, lithium-battery anodes, additives in polymers, supercapacitor, drug discovery, hydrogen storage, electromagnetic wave absorption, and shielding gas-discharge tubes in telecom networks [24]. Aman et al. [25] considered CNT's nanofluids with three different types of base fluids in their study. They found that the heat transfer in all three types of base fluids is higher for SWCNTs, and the nanofluid velocity decreases for greater values of the volume fraction of nanoparticles. Safaei et al. [26] studied the heat transfer flow of water-based fractionalized MWCNTs (FMWCNTs) nanofluid. They concluded that by increasing the concentration of FMWCNTs the boundary layer thickness increases which causes a decrease in the fluid motion. Haq et al. [27] analyzed the magnetohydrodynamic (MHD) flow of engine oil-based CNT's nanofluid in two concentric cylinders. Rahmati and Reiszadeh [28] experimentally studied the effect of MWCNTs in water and ethylene glycol (EG) base fluid with indirect heater gas reducing stations. Their study shows that the rate of heat transfer (Nusselt number), thermal conductivity coefficient, and viscosity ratios increased with an increase in volume fraction of nanoparticles.

TABLE 5.1
Thermophysical Properties of Base Fluids and Nanoparticles

	Base Fluid	Nanoparticles	
Material	**CMC**	**MWCNT**	**SWCNT**
$\rho\,(\mathrm{kg/m^3})$	997	1600	2600
$C_p\,(\mathrm{J/kg\,K})$	4179	796	425
$K\,(\mathrm{W/m\,K})$	0.613	3000	6600
Pr	6.2		

Based on the above-discussed literature, this chapter focused on the convective flow of carboxy-methyl-cellulose (CMC)-based-CNT's nanofluid in a vertical microchannel embedded in a porous medium. The ABFDs approach is used in the mathematical modeling. The exact analytical results are obtained via the Laplace transform technique for velocity and temperature profiles. The solutions are presented in graphs and discussed in detail.

5.2 Mathematical Modeling

Let us consider the MHD flow of electrically conducting CMC-based-CNT's nanofluid through a porous medium in a vertical microchannel. The channel consists of two parallel plates at a distance d. The fluid flows in x-direction, whereas the y-axis is chosen perpendicular to the flow direction. A uniform magnetic field of strength B_0 is chosen transversely to the flow direction. At the time, $t = 0$, the fluid is at rest with ambient temperature T_0. After $t = 0^+$, the temperature of the plate at $y = d$ is raised from T_0 to T_W which enhances the buoyancy force and the fluid starts its motion in the x-direction.

Based on the above assumptions, the flow of incompressible CMC-based-CNT's nanofluid through a porous medium, under the influence of a uniform magnetic field along with heat transfer, is governed by [16]

$$\rho_{nf}\frac{\partial u\,(y,t)}{\partial t} = \mu_{nf}\frac{\partial^2 u\,(y,t)}{\partial y^2} - \sigma_{nf}B_0^2 u\,(y,t) - \frac{\mu_{nf}}{K}u\,(y,t) + g\,(\rho\beta_T)_{nf}\,(T\,(y,t) - T_0)\,, \tag{5.1}$$

$$(\rho C_p)_{nf}\frac{\partial T\,(y,t)}{\partial t} = k_{nf}\frac{\partial^2 T\,(y,t)}{\partial y^2}, \tag{5.2}$$

associated with the following physical initial and boundary conditions:

$$u\,(y,0) = 0,\ T\,(y,0) = T_0,\ \forall\ y \geq 0, \tag{5.3}$$

$$\left.\begin{aligned} &u\,(0,t) = 0,\ T\,(0,t) = T_0 \\ &u\,(d,t) = 0,\ T\,(d,t) = T_W, \quad t > 0, \end{aligned}\right\}. \tag{5.4}$$

The thermophysical properties of nanoparticles and base fluid are presented in Table 5.1, and the mathematical expressions are defined as follows [26,27]:

$$\mu_{nf} = \frac{\mu_f}{(1-\phi)^{2.5}}, \tag{5.5}$$

$$\rho_{nf} = (1-\phi)\,\rho_f + \phi\rho_{\mathrm{CNT}}, \tag{5.6}$$

$$\frac{\sigma_{nf}}{\sigma_f} = 1 + \frac{3\left(\frac{\sigma_{\mathrm{CNT}}}{\sigma_f} - 1\right)\phi}{\left(\frac{\sigma_{\mathrm{CNT}}}{\sigma_f} + 2\right) - 3\left(\frac{\sigma_{\mathrm{CNT}}}{\sigma_f} - 1\right)\phi}, \tag{5.7}$$

$$(\rho\beta_T)_{nf} = (1-\phi)\,\rho_f + \phi\,(\rho\beta_T)_{\mathrm{CNT}}, \tag{5.8}$$

$$(\rho C_p)_{nf} = (1-\phi)\,(\rho C_p)_f + \phi\,(\rho C_p)_{\mathrm{CNT}}, \tag{5.9}$$

$$\frac{K_{nf}}{K_f} = \frac{(1-\phi) + 2\phi\frac{K_{\mathrm{CNT}}}{K_{\mathrm{CNT}}-K_f}\ln\frac{K_{\mathrm{CNT}}-K_f}{2K_f}}{(1-\phi) + 2\phi\frac{K_{\mathrm{CNT}}}{K_{\mathrm{CNT}}-K_f}\ln\frac{K_{\mathrm{CNT}}+K_f}{2K_f}}. \tag{5.10}$$

Introducing the following dimensionless variables

$$v = \frac{ud}{\nu_f},\ \tau = \frac{t\nu_f}{d},\ \xi = \frac{y}{d},\ \theta = \frac{T-T_0}{T_W-T_0},$$

into Equations (5.1)–(5.4) and using Equations (5.5)–(5.10) the following dimensionless system is obtained:

$$\left((1-\phi) + \phi\frac{\rho_{\mathrm{CNT}}}{\rho_f}\right)\frac{\partial u(\xi,\tau)}{\partial\tau} = \frac{1}{(1-\phi)^{2.5}}\frac{\partial^2 u(\xi,\tau)}{\partial\xi^2} - K_{eff}v$$
$$+ \left((1-\phi) + \phi\frac{(\rho\beta_T)_{\mathrm{CNT}}}{(\rho\beta_T)_f}\right) Gr\theta(\xi,\tau), \tag{5.11}$$

$$\left((1-\phi) + \phi\frac{(\rho C_p)_{\mathrm{CNT}}}{(\rho C_p)_f}\right)\frac{\partial\Phi(\xi,\tau)}{\partial\tau} = \frac{1}{\mathrm{Pr}}\left(\frac{K_{nf}}{K_f}\right)\frac{\partial^2\Phi(\xi,\tau)}{\partial\xi^2}. \tag{5.12}$$

$$v(\xi,0) = 0,\ \theta(\xi,0) = 0,\ \forall\,\xi \geq 0, \tag{5.13}$$

$$\left.\begin{array}{l} v(0,\tau) = 0,\ \theta(0,t) = 0 \\ u(1,\tau) = 0,\ \theta(d,t) = 1,\ \tau > 0, \end{array}\right\}, \tag{5.14}$$

with

$$M = \frac{\sigma_f B_0^2 d^2}{\mu_f},\ K_1 = \frac{K}{d^2},\ Gr = \frac{g(\beta_T)_f d^3 (T_W - T_0)}{\nu_f^2},$$

$$\mathrm{Pr} = \frac{(\mu C_p)_f}{k_f},\ M_0 = \frac{\sigma_{nf}}{\sigma_f}M,\ K_{\mathrm{eff}} = M_0 + \frac{a_1}{K_1}$$

Here M is a magnetic parameter called Hartman number, K_1 is the permeability of porous medium, Gr is the thermal Grashof number, Pr is the Prandtl number, M_0 and K_{eff} are constants.

The Atangana–Baleanu fractional model of CMC-based-CNT's nanofluid coupled with energy equations is given by [29]

$$a_0 D_\tau^\alpha v(\xi,\tau) = a_1\frac{\partial^2 v(\xi,\tau)}{\partial\xi^2} - K_{\mathrm{eff}}v(\xi,\tau) + a_2 Gr\theta(\xi,\tau), \tag{5.15}$$

$$a_3 D_\tau^\alpha \theta(\xi,\tau) = \frac{\lambda_{nf}}{\mathrm{Pr}}\frac{\partial^2\theta(\xi,\tau)}{\partial\xi^2}, \tag{5.16}$$

together with the initial and boundary conditions defined in (5.13) and (5.14). The constants appearing in Equations (5.15) and (5.16) are as follow:

$$a_0 = (1-\phi) + \phi\frac{\rho_{\text{CNT}}}{\rho_f},\ a_1 = \frac{1}{(1-\phi)^{2.5}},\ \lambda_{nf} = \frac{K_{nf}}{K_f}$$
$$a_2 = (1-\phi) + \phi\frac{(\rho\beta_T)_{\text{CNT}}}{(\rho\beta_T)_f},\ a_3 = (1-\phi) + \phi\frac{(\rho C_p)_{\text{CNT}}}{(\rho C_p)_f}$$

The Atangana–Baleanu fractional operator "$D_\tau^\alpha(.,.)$" is based on Mittag-Leffler function without singular and local kernel, which is defined as [12]

$$D_\tau^\alpha f(\xi,\tau) = \frac{N(\alpha)}{1-\alpha}\int_0^\tau E_\alpha\left\{\frac{-\alpha(\tau-t)^\alpha}{1-\alpha}\right\} f'(\xi,\tau)\,dt, \tag{5.17}$$

where $N(\alpha)$ is a normalization function such that $N(1) = N(0) = 1$ and $E_\alpha(-t^\alpha) = \sum_{k=0}^{\infty}\frac{(-t)^{\alpha k}}{\Gamma(\alpha k+1)}$ is the Mittag-Leffler function. The Laplace transform of Equation (5.17) is obtained as [16]

$$L\{f(\xi,\tau);q\} = \frac{q^\alpha}{(1-\alpha)q^\alpha+\alpha}\overline{f}(\xi,q). \tag{5.18}$$

5.3 Solution of the Problem

Here the Laplace transformation is applied to obtain analytical solutions for velocity and temperature distributions in the transformed domain.

5.3.1 Solution of Energy Equation

Applying the fractional Laplace transformation to Equation (5.15) using the corresponding initial condition from Equation (5.13) and boundary conditions from (5.14) we get the following:

$$\frac{d^2\overline{\theta}(\xi,q)}{d\xi^2} - \frac{b_0 b_3 q^\alpha}{q^\alpha+b_1}\overline{\theta}(\xi,\tau) = 0, \tag{5.19}$$

and

$$\overline{\theta}(0,q) = 0,\ \overline{\theta}(1,q) = \frac{1}{q}, \tag{5.20}$$

with $b_0 = \frac{1}{1-\alpha}$, $b_1 = b_0\alpha$ and $b_3 = \frac{a_3 \Pr}{\lambda_{nf}}$.

The solution of Equation (5.19) using the corresponding boundary conditions from (5.20) is given as follows:

$$\overline{\theta}(\xi,\tau) = \frac{1}{q}\frac{\sinh \xi\sqrt{\frac{b_0 b_3 q^\alpha}{q^\alpha+b_1}}}{\sinh\sqrt{\frac{b_0 b_3 q^\alpha}{q^\alpha+b_1}}} \tag{5.21}$$

Equation (5.21) is further simplified as

$$\overline{\theta}(\xi,q) = \frac{1}{q^{1-\alpha}}\left[\begin{array}{c}\sum\limits_{n=0}^{\infty}\frac{1}{q^\alpha}\exp\left\{-(1+2n-\xi)\sqrt{\frac{b_0 b_3 q^\alpha}{q^\alpha+b_1}}\right\} \\ -\sum\limits_{n=0}^{\infty}\frac{1}{q^\alpha}\exp\left\{-(1+2n+\xi)\sqrt{\frac{b_0 b_3 q^\alpha}{q^\alpha+b_1}}\right\}\end{array}\right]. \tag{5.22}$$

Upon taking the inverse Laplace transform, Equation (5.22) takes the following form:

$$\theta(\xi,\tau) = h(\tau) * \left\{ \begin{array}{c} \sum_{n=0}^{\infty} \Phi\left((1+2n-\xi), \tau; b_0 b_3, b_1\right) \\ - \sum_{n=0}^{\infty} \Phi\left((1+2n+\xi), \tau; b_0 b_3, b_1\right) \end{array} \right\}, \tag{5.23}$$

here "*" represents convolution product, and the functions $h(\tau)$ and $\Phi(\eta, t, a, b)$ are presented in Appendix 5.A (5.A.1)–(5.A.3).

5.3.2 Solution of Momentum Equation

Applying the fractional Laplace transformation to Equation (5.15) using the corresponding initial condition from Equation (5.13) and boundary conditions from (5.14) we get the following:

$$\frac{d^2\overline{v}(\xi,q)}{d\xi^2} - \frac{1}{a_1}\left(\frac{b_4 q^\alpha + b_5}{q^\alpha + b_1}\right)\overline{v}(\xi,\tau) = -\frac{b_6}{q}\frac{\sinh \xi\sqrt{\frac{b_0 b_3 q^\alpha}{q^\alpha + b_1}}}{\sinh\sqrt{\frac{b_0 b_3 q^\alpha}{q^\alpha + b_1}}}, \tag{5.24}$$

$$\overline{v}(0,q) = 0,\ \overline{v}(1,q) = 0, \tag{5.25}$$

with $b_4 = a_0 b_0 + K_{\text{eff}}$, $b_5 = K_{\text{eff}} b_1$ and $b_6 = \frac{a_2 Gr}{a_1}$.

The solution of Equation (5.24) using the boundary conditions (5.25) is given as follows:

$$\begin{aligned}\overline{v}(\xi,q) = {} & \frac{b_6\left(q^\alpha + b_1\right)}{b_7 q^\alpha - b_8}\frac{1}{q}\frac{\sinh \xi\sqrt{\frac{1}{a_1}\left(\frac{b_4 q^\alpha + b_5}{q^\alpha + a_1}\right)}}{\sinh\sqrt{\frac{1}{a_1}\left(\frac{b_4 q^\alpha + b_5}{q^\alpha + a_1}\right)}} \\ & - \frac{b_6\left(q^\alpha + b_1\right)}{b_7 q^\alpha - b_8}\frac{1}{q}\frac{\sinh \xi\sqrt{\frac{b_3 q^\alpha}{q^\alpha + b_1}}}{\sinh\sqrt{\frac{b_3 q^\alpha}{q^\alpha + b_1}}},\end{aligned} \tag{5.26}$$

here $b_7 = \frac{a_1 b_0 b_3 - b4}{a_1}$, $b_8 = \frac{b_5}{a_1}$.

Equivalently:

$$\begin{aligned}\overline{v}(\xi,q) = {} & b_9\left(\frac{q^\alpha}{q^\alpha - b_{10}} + \frac{b1}{q^\alpha + b_{10}}\right)\frac{1}{q^{1-\alpha}} \\ & \times\left[\begin{array}{l} \sum_{n=0}^{\infty}\frac{1}{q^\alpha}\exp\left\{-(1+2n-\xi)\sqrt{\frac{1}{a_1}\left(\frac{b_4 q^\alpha + b_5}{q^\alpha + a_1}\right)}\right\} \\ - \sum_{n=0}^{\infty}\frac{1}{q^\alpha}\exp\left\{-(1+2n+\xi)\sqrt{\frac{1}{a_1}\left(\frac{b_4 q^\alpha + b_5}{q^\alpha + a_1}\right)}\right\} \end{array}\right] \\ & - b_9\left(\frac{q^\alpha}{q^\alpha + b_{10}} + \frac{b1}{q^\alpha - b_{10}}\right)\frac{1}{q^{1-\alpha}} \\ & \times\left[\begin{array}{l} \sum_{n=0}^{\infty}\frac{1}{q^\alpha}\exp\left\{-(1+2n-\xi)\sqrt{\frac{b_0 b_3 q^\alpha}{q^\alpha + b_1}}\right\} \\ - \sum_{n=0}^{\infty}\frac{1}{q^\alpha}\exp\left\{-(1+2n+\xi)\sqrt{\frac{b_0 b_3 q^\alpha}{q^\alpha + b_1}}\right\} \end{array}\right]\end{aligned} \tag{5.27}$$

with $b_9 = \frac{b_6}{b_7}$, $b_{10} = \frac{b_8}{b_7}$

Inverting the Laplace transform, Equation (5.27) takes a simplified and suitable form as

$$
\begin{aligned}
v(\xi,\tau) = {} & h(\tau) * \{b_9 (R_{\alpha,\alpha}(b_{10},\tau) + b_1 F_\alpha (b_{10},\tau))\} \\
& * \left\{ \begin{array}{l} \sum_{n=0}^{\infty} \Psi\left((1+2n-\xi)\frac{1}{\sqrt{a_1}}, \tau; b_4, b_5, a_1\right) \\ -\sum_{n=0}^{\infty} \Psi\left((1+2n+\xi)\frac{1}{\sqrt{a_1}}, \tau; b_4, b_5, a_1\right) \end{array} \right\} \\
& - h(\tau) * (R_{\alpha,\alpha}(b_{10},\tau) + b_9 F_\alpha (b_{10},\tau)) \\
& * \left\{ \begin{array}{l} \sum_{n=0}^{\infty} \Phi\left((1+2n-\xi), \tau; b_0 b_3, b_1\right) \\ -\sum_{n=0}^{\infty} \Phi\left((1+2n+\xi), \tau; b_0 b_3, b_1\right) \end{array} \right\}
\end{aligned} \tag{5.28}
$$

The functions $h(\tau)$ and $\Phi(\eta, t, a, b)$ are previously defined, whereas the new function $\Psi(\eta, t; a, b, c)$ is presented in Appendix 5.A (5.A.4) and (5.A.5). Here $R_{\alpha,v}(.,.)$ and $F_\alpha(.,.)$ the Lorenzo and Hartleys' and Robotnov and Hartleys' functions, respectively. These special functions can be defined as [30]

$$
R_{\alpha,v}(-m,t) = \left(\frac{q^v}{q^\alpha + m}\right) = \sum_{n=0}^{\infty} \frac{(-m)^n t^{(n+1)\alpha-1-v}}{\Gamma\{(n+1)\alpha - v\}} \tag{5.29}
$$

$$
F_\alpha(-m,t) = \left(\frac{1}{q^\alpha + m}\right) = \sum_{n=0}^{\infty} \frac{(-m)^n t^{(n+1)\alpha-1}}{\Gamma\{(n+1)\alpha\}}. \tag{5.30}
$$

It is worth mentioning here that Equations (5.23) and (5.28) clearly satisfy all the imposed initial and boundary conditions which validate our results.

5.4 Parametric Studies

In this chapter, the convection flow of CMC-based CNT's nanofluid in a porous medium is studied. The newly introduced ABFD approach is used in the problem formulation. The Laplace transform is used to obtain exact analytical solutions for velocity and temperature distributions. To explore the impact of pertinent parameters, the solutions are brought into light through numerical computations and various graphs are presented. The velocity and temperature distributions are compared for MWCNT-CMC and SWCNT-CMC.

Figures (5.1) and (5.2) depict the influence of the fractional parameter α on velocity and temperature distributions. These distributions show an identical trend when α is varied. The velocity and temperature distributions show an increasing behavior with increasing values of α. For $\alpha = 1$, the fractional velocity and temperature distributions reduced to ordinary (classical) velocity and temperature profiles. In case of ordinary velocity and temperature profiles, the momentum and thermal boundary layers are the thickest and gradually decrease with a decrease in α, and as a result the velocity and temperature distributions decrease. In other words, when α is increased, the thickness of the thermal and momentum boundary layers is increased which tend to increase the temperature and velocity profiles. The same behavior for the fractional parameter on the velocity and temperature profiles is reported in [20,21].

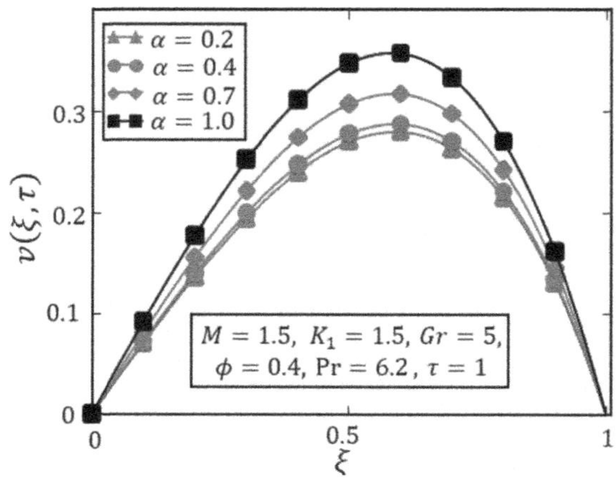

FIGURE 5.1
Variation in velocity distributions due to fractional parameter α where the other parameters are kept constant.

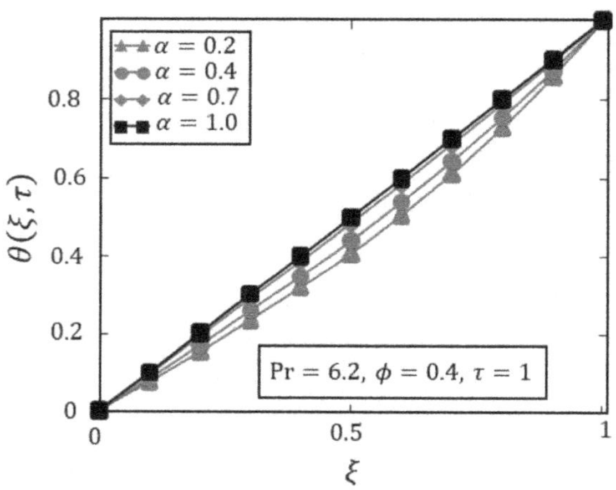

FIGURE 5.2
Variation in temperature distributions due to fractional parameter α where the other parameters are kept constant.

Figures (5.3) and (5.4) are plotted to show the influence of ϕ on velocity and temperature distributions. The velocity and temperature distributions show an opposite trend for increasing values of ϕ. The velocity profile decreases with an increase in ϕ, whereas the temperature distributions increases with an increase in ϕ. Increasing values of ϕ, increase the volume fraction of CNTs in the base fluid CMC which tends to increase the viscosity of nanofluid. As a result, the viscous forces became stronger which tends to decrease the fluid velocity. On the other hand, the temperature profile shows an increasing trend due to the effective thermal conductivity defined in Equation (5.10). The effective thermal conductivity is responsible for higher conduction of heat, which leads to an increase in temperature distribution.

In Figures (5.5) and (5.6) the velocity and temperature profiles are compared for MWCNT-CMC and SWCNT-CMC. It is found that the velocity of MWCNT-CMC is

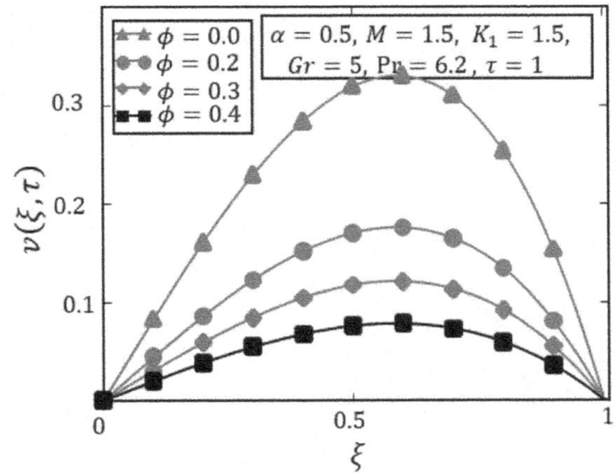

FIGURE 5.3
Variation in velocity distributions due to volume concentration of nanoparticles ϕ where the other parameters are kept constant.

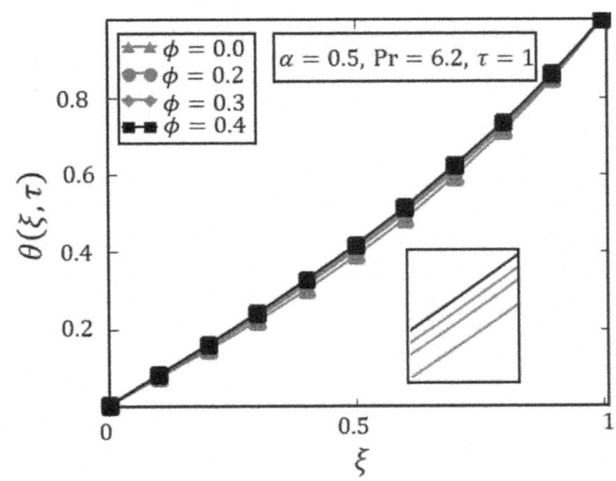

FIGURE 5.4
Variation in temperature distributions due to volume concentration of nanoparticles ϕ where the other parameters are kept constant.

comparatively less than the velocity of SWCNT-CMC because the density of MWCNT is greater than the density of SWCNT. When MWCNT nanoparticles are dispersed in CMC the resulting MWCNT-CMC nanofluid became denser than the SWCNT-CMC nanofluid. The higher density of MWCNT is a significant factor which reduces the velocity of MWCNT-CMC nanofluid. However, the trend of the temperature profile is in contrast with that of the velocity profile as MWCNTs have higher thermal conductivity than SWCNTs. Consequently, MWCNTs conduct more heat than SWCNTs; consequently, the temperature distributions of MWCNTs are higher than SWCNTs. Aman et al. [25] and Haq et al. [27] reported the same trend for MWCNT and SWCNT nanofluids in their studies.

The effect M on the velocity profile is shown in Figure (5.7). The velocity profile decreases with an increase in M. Physically, this trend corresponds to the fact that the

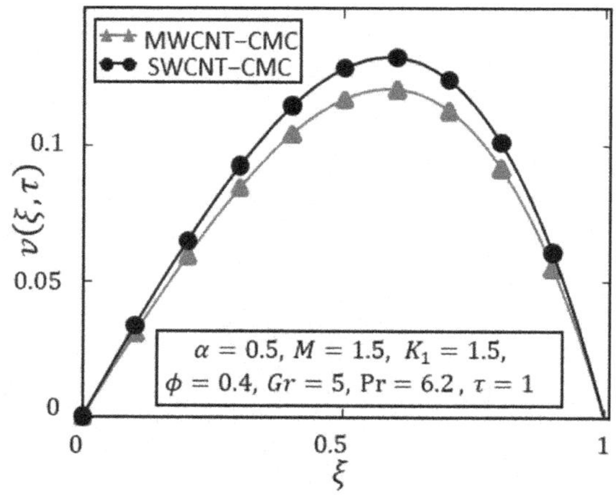

FIGURE 5.5
Comparison of velocity distributions for MWCNT-CMC and SWCNT-CMC nanofluids.

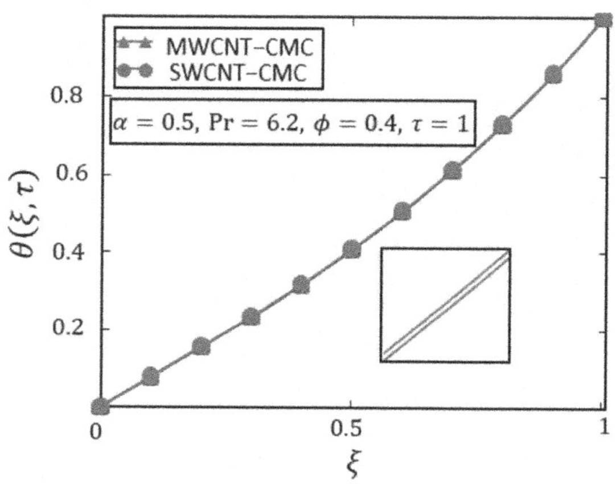

FIGURE 5.6
Comparison of temperature distributions for MWCNT-CMC and SWCNT-CMC nanofluids.

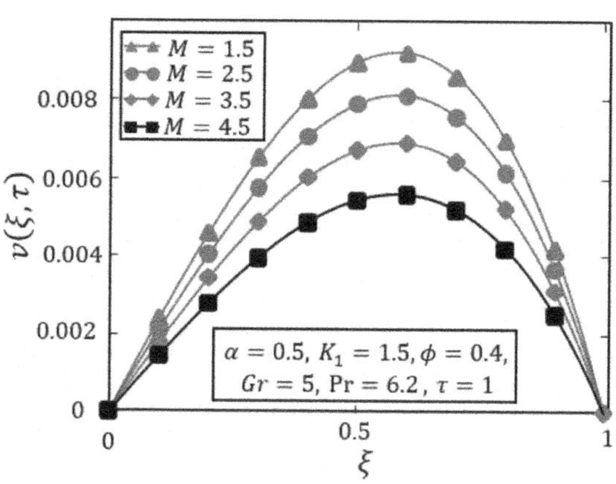

FIGURE 5.7
Variation in velocity distributions due to magnetic parameter M where the other parameters are kept constant.

effect of the transverse magnetic field on the electrically conducting nanofluid gives rise to the resistive type of forces called Lorentz force which is similar to the drag forces. Upon increasing the values of M the velocity distributions decrease.

Figure (5.8) depicts the influence of K_1 on the nanofluid velocity. It is found that the velocity increases with an increase in K_1. This can be physically justified, as when K_1 is increased the porous medium permits more fluid to flow due to a decline in the resistive force and correspondingly the fluid velocity raises. Figure (5.9) shows the influence of Gr on the velocity distribution. The nanofluid velocity increases with an increase in Gr. This tendency can be physically justifying as greater values of Gr strengthening the buoyancy forces due to which the nanofluid velocity increases.

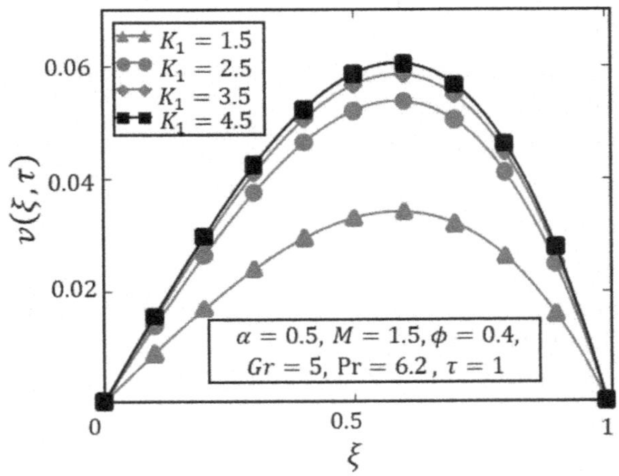

FIGURE 5.8
Variation in velocity distributions due to permeability of porous medium K_1 where the other parameters are kept constant.

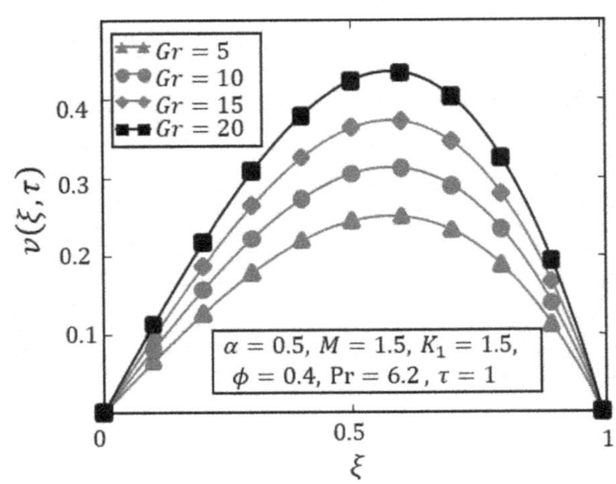

FIGURE 5.9
Variation in velocity distributions due to thermal Grashof number Gr where the other parameters are kept constant.

5.5 Concluding Remarks

In this chapter, the time fractional model of the convective flow of CMC-based-CNT's nanofluid through a porous media is studied in a microchannel. The ordinary (classical model) is generalized using Atangana–Baleanu fractional approach. The Laplace transform is utilized to obtain exact analytical solutions. Finally, the solutions are plotted in the different figures to analyze the influence of pertinent parameters. The major key points extracted from this study are as follows:

- The fractional solutions are more general than classical solutions.
- For $\alpha = 1$ the fractional solutions reduced to the classical solutions.
- The velocity and temperature profiles show a decreasing trend for smaller values of α and are maximum for $\alpha = 1$.
- The viscosity of the nanofluid increases with an increase in ϕ which causes a decrease in the velocity profile.
- The temperature distribution increases with an increase in ϕ because of the improvement of thermal conductivity of nanofluid.
- MWCNTs are more efficient in the enhancement of thermal conductivity of nanofluids when compared with SWCNTs.
- The nanofluid velocity decreases with an increase in M because of the enhancement in the resistive type forces.
- The fluid velocity increases with an increase in K_1 due to a decrease in resistive forces of the porous medium.
- The velocity distribution shows an increasing trend for increasing values of Gr.

Acknowledgment

The authors would like to acknowledge Ministry of Education (MOE) and Research Management Centre-UTM, Universiti Teknologi Malaysia (UTM) for the financial support through vote numbers 5F004, 07G70, 07G72, 07G76, 07G77 and 08G33 for this research.

Appendix 5.A

$$\Phi_1(\eta, t; a, b) = L^{-1}\left\{\frac{1}{q}\exp\left(-\eta\left(\sqrt{\frac{aq}{q+b}}\right)\right)\right\} = 1 - \frac{2a}{\pi}\int_0^\infty \frac{\sin(\eta s)}{s(a+s^2)}\exp\left(-\frac{bts^2}{a+s^2}\right)ds \tag{5.A.1}$$

$$\begin{aligned}\Phi(\eta, t; a, b) &= L^{-1}\left\{\frac{1}{q^\alpha}\exp\left(-\eta\left(\sqrt{\frac{aq^\alpha}{q^\alpha+b}}\right)\right)\right\} \\ &= \frac{1}{\pi}\int_0^\infty\int_0^\infty \Phi_1(\eta, u; a, b)\exp\left(-\tau r - ur^\alpha\cos\alpha\pi\right)\left(ur^\alpha\sin\alpha\pi\right)drdu \end{aligned}\tag{5.A.2}$$

$$h(\tau) = L^{-1}\left(\frac{1}{q^{1-\alpha}}\right) = \frac{1}{\tau^{\alpha}\Gamma(1-\alpha)} \tag{5.A.3}$$

$$\begin{aligned}\Psi_1(\eta,t;a,b,c) &= L^{-1}\left\{\frac{1}{q}\exp\left(-\eta\sqrt{\frac{aq+b}{q+c}}\right)\right\} \\ &= \exp\left(-\eta\sqrt{a}\right) - \frac{\eta\sqrt{b-ac}}{2\sqrt{\pi}}\int_0^{\infty}\int_0^{\tau}\frac{1}{\sqrt{t}}\exp\left(-ct-\frac{\eta}{4u}-au\right)I_1\left(2\sqrt{(b-ac)\,ut}\right)dtdu\end{aligned} \tag{5.A.4}$$

$$\begin{aligned}\Psi(\eta,t;a,b,c) &= L^{-1}\left\{\frac{1}{q^{\alpha}}\exp\left(-\eta\sqrt{\frac{aq^{\alpha}+b}{q^{\alpha}+c}}\right)\right\} \\ &= \frac{1}{\pi}\int_0^{\infty}\int_0^{\infty}\Psi_1(\eta,u;a,b,c)\exp\left(-\tau r-ur^{\alpha}\cos\alpha\pi\right)\left(ur^{\alpha}\sin\alpha\pi\right)drdu\end{aligned} \tag{5.A.5}$$

References

[1] Ali, F., et al., Application of Caputo-Fabrizio derivatives to MHD free convection flow of generalized Walters'-B fluid model. *The European Physical Journal Plus*, 2016. **131**(10): p. 377.

[2] Ali, F., et al., Solutions with Wright function for time fractional free convection flow of Casson fluid. *Arabian Journal for Science and Engineering*, 2017. **42**(6): pp. 2565–2572.

[3] Alkahtani, B.S.T. and A. Atangana, Modeling the potential energy field caused by mass density distribution with Eton approach. *Open Physics*, 2016. **14**(1): pp. 106–113.

[4] Singh, J., et al., A fractional epidemiological model for computer viruses pertaining to a new fractional derivative. *Applied Mathematics and Computation*, 2018. **316**: pp. 504–515.

[5] Caputo, M., Linear models of dissipation whose Q is almost frequency independent. *Annals of Geophysics*, 1966. **19**(4): pp. 383–393.

[6] Atangana, A., *Derivative with a New Parameter: Theory, Methods and Applications.* 2015: Academic Press, London.

[7] Caputo, M., Linear models of dissipation whose Q is almost frequency independent—II. *Geophysical Journal International*, 1967. **13**(5): pp. 529–539.

[8] Caputo, M. and M. Fabrizio, A new definition of fractional derivative without singular kernel. *Progress in Fractional Differentiation and Applications*, 2015. **1**(2): pp. 1–13.

[9] Saqib, M., et al., Exact solutions for free convection flow of generalized Jeffrey fluid: A Caputo-Fabrizio fractional model. *Alexandria Engineering Journal*, 2017. **57**(3): pp. 1849–1858.

[10] Atangana, A. and J.F. Gómez-Aguilar, Hyperchaotic behaviour obtained via a nonlocal operator with exponential decay and Mittag-Leffler laws. *Chaos, Solitons and Fractals*, 2017. **102**: pp. 285–294.

[11] Atangana, A. and J.F. Gómez-Aguilar, A new derivative with normal distribution kernel: Theory, methods and applications. *Physica A: Statistical Mechanics and Its Applications*, 2017. **476**: pp. 1–14.

[12] Atangana, A. and D. Baleanu, New fractional derivatives with nonlocal and non-singular kernel: Theory and application to heat transfer model. *Thermal Science*, 2016. **2**(20): pp. 763–769.

[13] Atangana, A. and D. Baleanu, Caputo-Fabrizio derivative applied to groundwater flow within confined aquifer. *Journal of Engineering Mechanics*, 2017. **143**(5): p. D4016005.

[14] Atangana, A., Fractal-fractional differentiation and integration: Connecting fractal calculus and fractional calculus to predict complex system. *Chaos, Solitons and Fractals*, 2017. **102**: pp. 396–406.

[15] Saqib, M., et al., Entropy generation in different types of fractionalized nanofluids. *Arabian Journal for Science and Engineering*, 2019. **44**(1): pp. 531–540.

[16] Sheikh, N.A., et al., A comparative study of Atangana-Baleanu and Caputo-Fabrizio fractional derivatives to the convective flow of a generalized Casson fluid. *The European Physical Journal Plus*, 2017. **132**(1): p. 54.

[17] Sheikh, N.A., et al., Comparison and analysis of the Atangana–Baleanu and Caputo–Fabrizio fractional derivatives for generalized Casson fluid model with heat generation and chemical reaction. *Results in Physics*, 2017. **7**: pp. 789–800.

[18] Sheikh, N.A., et al., On the applications of nanofluids to enhance the performance of solar collectors: A comparative analysis of Atangana-Baleanu and Caputo-Fabrizio fractional models. *The European Physical Journal Plus*, 2017. **132**(12): p. 540.

[19] Jan, S.A.A., et al., Engine oil based generalized brinkman-type nano-liquid with molybdenum disulphide nanoparticles of spherical shape: Atangana-Baleanu fractional model. *Numerical Methods for Partial Differential Equations*, 2017. **34**(5): pp. 1472–1488.

[20] Azhar, W.A., D. Vieru, and C. Fetecau, Free convection flow of some fractional nanofluids over a moving vertical plate with uniform heat flux and heat source. *Physics of Fluids*, 2017. **29**(8): p. 082001.

[21] Fetecau, C., D. Vieru, and W.A. Azhar, Natural convection flow of fractional nanofluids over an isothermal vertical plate with thermal radiation. *Applied Sciences*, 2017. **7**(3): p. 247.

[22] Iijima, S., Helical microtubules of graphitic carbon. *Nature*, 1991. **354**(6348): p. 56.

[23] Murshed, S.M.S., et al., A review of boiling and convective heat transfer with nanofluids. *Renewable and Sustainable Energy Reviews*, 2011. **15**(5): pp. 2342–2354.

[24] Lin, Y., L. Zheng, and X. Zhang, Radiation effects on Marangoni convection flow and heat transfer in pseudo-plastic non-Newtonian nanofluids with variable thermal conductivity. *International Journal of Heat and Mass Transfer*, 2014. **77**: pp. 708–716.

[25] Aman, S., et al., Magnetic field effect on Poiseuille flow and heat transfer of carbon nanotubes along a vertical channel filled with Casson fluid. *AIP Advances*, 2017. **7**(1): p. 015036.

[26] Safaei, M.R., et al., Boundary layer flow and heat transfer of FMWCNT/water nanofluids over a flat plate. *Fluids*, 2016. **1**(4): p. 31.

[27] Haq, R.U., F. Shahzad, and Q.M. Al-Mdallal, MHD pulsatile flow of engine oil based carbon nanotubes between two concentric cylinders. *Results in Physics*, 2017. **7**: pp. 57–68.

[28] Rahmati, A.R. and M. Reiszadeh, An experimental study on the effects of the use of multi-walled carbon nanotubes in ethylene glycol/water-based fluid with indirect heaters in gas pressure reducing stations. *Applied Thermal Engineering*, 2018. **134**: pp. 107–117.

[29] Saqib, M., et al., Convection in ethylene glycol-based molybdenum disulfide nanofluid. *Journal of Thermal Analysis and Calorimetry*, 2019. **135**: pp. 1–10.

[30] Ali, F., et al., Magnetic field effect on blood flow of Casson fluid in axisymmetric cylindrical tube: A fractional model. *Journal of Magnetism and Magnetic Materials*, 2017. **423**: pp. 327–336.

6

The Hyperbolic Maximum Principle Approach to the Construction of Generalized Convolutions

Rúben Sousa

Universidade do Porto

Manuel Guerra

Universidade de Lisboa

Semyon Yakubovich

Universidade do Porto

CONTENTS

6.1 Introduction

Given a Sturm–Liouville operator on an interval of the real line, it is well known that its eigenfunction expansion gives rise to an integral transform which shares many properties with the ordinary Fourier transform [18,52]. Since various standard special functions are solutions of Sturm–Liouville equations, the class of integral transforms of Sturm–Liouville type includes, as particular cases, many common integral transforms (Hankel, Kontorovich–Lebedev, Mehler–Fock, Jacobi, Laguerre, etc.).

The Fourier transform lies at the heart of the classical theory of harmonic analysis. This naturally raises a question: *is it possible to generalize the main facts of harmonic analysis to integral transforms of Sturm–Liouville type?*

Starting from the seminal works of Delsarte [17] and Levitan [33] it was noticed that the key ingredient for developing such a generalized harmonic analysis is the so-called product formula. We say that an indexed family of complex-valued functions $\{w_\lambda\}$ on an interval $I \subset \mathbb{R}$ has a *product formula* if for each $x, y \in I$ there exists a complex Borel measure $\boldsymbol{\nu}_{x,y}$ (independent of λ) such that

$$w_\lambda(x)\, w_\lambda(y) = \int_I w_\lambda \, d\boldsymbol{\nu}_{x,y} \qquad (\lambda \in \Lambda). \tag{6.1}$$

Product formulas naturally lead to generalized convolution operators. To fix ideas, let $\ell(u) = \frac{1}{r}\big[-(pu')' + qu\big]$ be a usual Sturm–Liouville differential expression defined on the interval I, and let $(\mathcal{F}h)(\lambda) := \int_I h(x)\, w_\lambda(x)\, dm(x)$ be a Sturm–Liouville type integral transform, where the w_λ are solutions of $\ell(w) = \lambda w$ $(\lambda \in \mathbb{C})$. If $\{w_\lambda\}$ has a product formula, then we can define a generalized (Sturm–Liouville type) convolution operator $*$ by

$$(f * g)(x) := \int_I \left(\int_I f \, d\boldsymbol{\nu}_{x,y} \right) g(y)\, dm(y). \tag{6.2}$$

It is not difficult to show that, under reasonable assumptions, the property $\mathcal{F}(f * g) = (\mathcal{F}f)\cdot(\mathcal{F}g)$ holds for this convolution operator; this means that the analogue of one of the basic identities in harmonic analysis—the Fourier convolution theorem—is satisfied by the generalized convolution.

Consider now the associated hyperbolic partial differential equation

$$\frac{1}{r(x)}\Big\{-\partial_x\big[p(x)\,\partial_x f(x,y)\big] + q(x)f(x,y)\Big\} = \frac{1}{r(y)}\Big\{-\partial_y\big[p(y)\,\partial_y f(x,y)\big] + q(y)f(x,y)\Big\}. \tag{6.3}$$

If the kernel of the Sturm–Liouville transform is defined via some initial condition $w_\lambda(a) = 1$, then the product $f(x,y) = w_\lambda(x)w_\lambda(y)$ is a solution of (6.3) satisfying the boundary condition $f(x,a) = w_\lambda(x)$. Studying the properties of the associated hyperbolic equation is therefore a natural strategy for proving the existence of a product formula and extracting information about the measure $\boldsymbol{\nu}_{x,y}$.

An especially interesting case is that where $\{\boldsymbol{\nu}_{x,y}\}$ turns out to be a family of probability measures (satisfying appropriate continuity assumptions). Indeed, in this case one can show that the convolution (6.2) gives rise to a Banach algebra structure in the space of finite complex Borel measures in which various probabilistic concepts and properties can be developed in analogy with the classical theory [6,53]. Establishing explicit product formulas, or even proving their existence, has been recognized as a difficult problem [12,15]. Nevertheless, using the maximum principle for hyperbolic equations [60], it was shown by Levitan [34] (and, under weakened assumptions, by Chebli [11] and Zeuner [63]) that this probabilistic property of the product formula holds for a general family of Sturm–Liouville differential expressions on $I = [0, \infty)$ of the form $\ell(u) = -\frac{1}{A}(Au')'$. This family of Sturm–Liouville operators includes, as important particular cases, the generators of the Hankel transform and the (Fourier–)Jacobi transform; these cases are noteworthy due to the fact that the explicit expression for the measure in the product formula can be determined using results from the theory of special functions (see Examples 6.3–6.4).

Various examples show that the probabilistic property of the product formula holds only for a restricted class of Sturm–Liouville operators [37,45]; this is connected with the fact that the hyperbolic maximum principle requires rather strong assumptions on the coefficients. Notwithstanding, the recent work [46,47] of the authors on the index Whittaker transform made it apparent that there is room for generalization of the results of [11,34,63]. In fact, the family of Sturm–Liouville operators considered in these works only includes operators for which Equation (6.3) is uniformly hyperbolic on $[0, \infty)^2$; a consequence of this is that, under their assumptions, the support $\mathrm{supp}(\boldsymbol{\nu}_{x,y})$ of the measures in the product formula is always compact. In contrast, the case of the index Whittaker transform provides an example of a product formula whose measures $\boldsymbol{\nu}_{x,y}$ have the probabilistic property and satisfy $\mathrm{supp}(\boldsymbol{\nu}_{x,y}) = [0, \infty)$ for $x, y > 0$; here the associated hyperbolic Equation (6.3) is parabolically degenerate at the boundaries $x = 0$ and $y = 0$. (The index Whittaker transform is generated by the Sturm–Liouville expression $x^2 u'' + (1 + 2(1 - \alpha)x)u'$ on $I = [0, \infty)$, and its product formula, which is known in closed form, is given in Example 6.8.)

The goal of this work is to introduce a unified framework for the construction of Sturm–Liouville type convolution operators associated with possibly degenerate hyperbolic equations. We will consider a Sturm–Liouville differential expression of the form

$$\ell = -\frac{1}{r}\frac{d}{dx}\Bigl(p\,\frac{d}{dx}\Bigr), \qquad x \in (a, b) \tag{6.4}$$

($-\infty \leq a < b \leq \infty$), where p and r are (real-valued) coefficients such that $p(x), r(x) > 0$ for all $x \in (a, b)$, and p, p', r, and r' are locally absolutely continuous on (a, b). Concerning the behavior of the coefficients at the boundaries $x = a$ and $x = b$, we will assume, respectively, that

$$\int_a^c \int_y^c \frac{dx}{p(x)}\, r(y)dy < \infty \tag{6.5}$$

$$\int_c^b \int_y^b \frac{dx}{p(x)}\, r(y)dy = \int_c^b \int_c^y \frac{dx}{p(x)}\, r(y)dy = \infty \tag{6.6}$$

where $c \in (a, b)$ is an arbitrary point. These conditions mean that a is a regular or entrance boundary and b is a natural boundary for the operator ℓ. The notions of regular, entrance, and natural boundary refer to the Feller classification of boundaries, which is recalled in Remark 6.2, where we also give some comments on the role of conditions (6.5)–(6.6).

The point of departure is the study of the Cauchy problem for the possibly degenerate hyperbolic Equation $\frac{1}{r(x)}\partial_x\bigl(p(x)\,\partial_x f(x, y)\bigr) = \frac{1}{r(y)}\partial_y\bigl(p(y)\,\partial_y f(x, y)\bigr)$. Under the assumption that the product $p(x)r(x)$ of the coefficients of (6.4) is an increasing function, we prove an existence and uniqueness theorem for the Cauchy problem which is based on the spectral theory of Sturm–Liouville operators. We then give a sufficient condition for the maximum principle to hold for the hyperbolic equation is given and, as a corollary, the positivity preserving property of the solution of the Cauchy problem is obtained.

Our existence theorem (and the positivity result) covers many hyperbolic equations with initial data on the parabolic line which are outside the scope of the classical theory, and for which the problem of well-posedness of the Cauchy problem was, to the best of our knowledge, open. In fact, given that our results depend heavily on the assumption that the left boundary a is of entrance type (cf. Remark 6.2), they indicate that the well-posedness of the degenerate problem with initial line $y = a$ depends on the Feller boundary classification of ℓ at the endpoint a.

If the maximum principle holds for the hyperbolic equation associated with ℓ, then the solution of the hyperbolic Cauchy problem can be written as $f(x,y) = \int_{[a,b)} h\, d\boldsymbol{\nu}_{x,y}$, where $h(x) = f(x,a)$ is the initial condition and $\{\boldsymbol{\nu}_{x,y}\}$ is a family of finite positive Borel measures on $[a,b)$. Formally, this suggests that the product formula (6.1) should hold for the kernel w_λ of the Sturm–Liouville transform. It turns out that (6.1) indeed holds and that the $\boldsymbol{\nu}_{x,y}$ are probability measures, but the proof requires some effort, especially when the Cauchy problem is parabolically degenerate [48]. We then define the generalized convolution by (6.2), so that the expected convolution theorem $\mathcal{F}(f * g) = (\mathcal{F}f)\cdot(\mathcal{F}g)$ holds. Moreover, the Young inequality for the L_p-spaces with respect to the weighted measure $r(x)dx$ is valid for the convolution (6.2), demonstrating that the mapping properties of the generalized convolution structure resemble those of the ordinary convolution.

A fundamental tool for studying the continuity and mapping properties of the generalized convolution is the extension of the Sturm–Liouville transform to complex measures, defined by $\widehat{\mu}(\lambda) = \int_{[a,b)} w_\lambda(x)\mu(dx)$. Actually, if we define the convolution of two Dirac measures by $\delta_x * \delta_y = \boldsymbol{\nu}_{x,y}$ and then define the convolution $\mu * \nu$ of two complex measures so that $(\mu,\nu) \mapsto \mu * \nu$ is weakly continuous, then the space $\mathcal{M}_{\mathbb{C}}[a,b)$ of finite complex measures on $[a,b)$ becomes a convolution measure algebra for which the Sturm–Liouville transform is a generalized characteristic function, in the sense that the property $\widehat{\mu * \nu} = \widehat{\mu} \cdot \widehat{\nu}$ holds. The algebra $(\mathcal{M}_{\mathbb{C}}[a,b), *)$ is therefore a natural environment for studying notions from probabilistic harmonic analysis, in particular infinite divisibility, Gaussian-type measures, and Lévy-type (additive) stochastic processes. As anticipated above, the study of these concepts leads to analogues of chief results in probability theory, such as the Lévy–Khintchine formula or the contraction property of convolution semigroups.

The class of Lévy-type processes with respect to the convolution measure algebra includes the diffusion process generated by the Sturm–Liouville expression ℓ, as well as many other Markov processes with discontinuous paths. We hope that this work illuminates the role of product formulas and hyperbolic Cauchy problems on a purely probabilistic problem—that of constructing a class of Lévy-type processes which accommodates a given diffusion process—and stimulates further research on this topic.

The remaining sections are organized as follows. In Section 6.2, after introducing the basic properties of the solution of the Sturm–Liouville equation $\ell(w) = \lambda w$, we summarize some key facts from the theory of eigenfunction expansions of Sturm–Liouville operators and from the theory of one-dimensional diffusion processes. Section 6.3 is devoted to the hyperbolic Cauchy problem associated with ℓ: an existence and uniqueness theorem is proved and, under suitable assumptions, it is shown that the unique solution satisfies a weak maximum principle. In Section 6.4, the solution of the hyperbolic Cauchy problem is used to define the generalized convolution of probability measures and the generalized translation of functions; moreover, the Sturm–Liouville transform of finite measures is introduced and an analogue of the Lévy continuity theorem is established, together with some other basic properties. The product formula for the solution of the Sturm–Liouville equation is discussed in Section 6.5. In Section 6.6 we establish the basic properties of the generalized convolution as an operator on weighted L_p-spaces. Section 6.7 explores the probabilistic properties of the convolution, demonstrating that the main concepts and facts from the classical theory of infinitely divisible distributions and convolution semigroups can be developed, in a parallel fashion, in the framework of the generalized convolutions considered here. The concluding Section 6.8 presents several examples and shows that various convolutions associated with standard integral transforms constitute particular cases of the general construction presented here.

6.2 Preliminaries

We use the following standard notations. For a subset $E \subset \mathbb{R}^d$, $\mathrm{C}(E)$ is the space of continuous complex-valued functions on E; $\mathrm{C_b}(E)$, $\mathrm{C_0}(E)$, and $\mathrm{C_c}(E)$ are, respectively, its subspaces of bounded continuous functions, of continuous functions vanishing at infinity and of continuous functions with compact support; $\mathrm{C}^k(E)$ stands for the subspace of k times continuously differentiable functions. $\mathrm{B_b}(E)$ is the space of complex-valued bounded and Borel measurable functions. The corresponding spaces of real-valued functions are denoted by $\mathrm{C}(E,\mathbb{R})$, $\mathrm{C_b}(E,\mathbb{R})$, etc.

$L_p(E;\mu)$ $(1 \leq p \leq \infty)$ denotes the Lebesgue space of complex-valued p-integrable functions with respect to a given measure μ on E. The space of probability (respectively, finite positive, finite complex) Borel measures on E will be denoted by $\mathcal{P}(E)$ (respectively, $\mathcal{M}_+(E)$, $\mathcal{M}_\mathbb{C}(E)$). The total variation of $\mu \in \mathcal{M}_\mathbb{C}(E)$ is denoted by $\|\mu\|$, and δ_x denotes the Dirac measure at a point x.

6.2.1 Solutions of the Sturm–Liouville Equation

We begin by collecting some properties of the solutions of the Sturm–Liouville equation $\ell(u) = \lambda u$ $(\lambda \in \mathbb{C})$, where ℓ is of the form (6.4) and satisfies the boundary condition (6.5). We shall write $f^{[1]} = pf'$ and $\mathfrak{s}(x) = \int_c^x \frac{d\xi}{p(\xi)}$ (this is the so-called *scale function*, cf. [8]).

If the Sturm–Liouville equation is regular at the left endpoint a, it is well known that there is an entire solution $w_\lambda(x)$ of $\ell(u) = \lambda u$ satisfying the initial conditions $w_\lambda(a) = \cos\theta$, $w_\lambda^{[1]}(a) = \sin\theta$ $(0 \leq \theta < \pi)$. When we only require that (6.5) holds (so that a may be an entrance boundary), the following lemma ensures that the same continues to hold for the boundary condition with vanishing derivative $(\theta = 0)$:

Lemma 6.1. *For each $\lambda \in \mathbb{C}$, there exists a unique solution $w_\lambda(\cdot)$ of the boundary value problem*

$$\ell(w) = \lambda w \quad (a < x < b), \qquad w(a) = 1, \qquad w^{[1]}(a) = 0. \tag{6.7}$$

Moreover, $\lambda \mapsto w_\lambda(x)$ is, for each fixed x, an entire function of exponential type.

Proof. The proof is similar to [28, Lemma 3], but for completeness, we give a sketch here. Let

$$\eta_0(x) = 1, \qquad \eta_j(x) = \int_a^x \bigl(\mathfrak{s}(x) - \mathfrak{s}(\xi)\bigr)\eta_{j-1}(\xi)r(\xi)d\xi \quad (j = 1, 2, \ldots). \tag{6.8}$$

Pick an arbitrary $\beta \in (a,b)$ and define $\mathcal{S}(x) = \int_a^x \bigl(\mathfrak{s}(\beta) - \mathfrak{s}(\xi)\bigr)r(\xi)d\xi$. From the boundary assumption (6.5) it follows that $0 \leq \mathcal{S}(x) \leq \mathcal{S}(\beta) < \infty$ for $x \in (a,\beta]$. Furthermore, it is easy to show (using induction) that $|\eta_j(x)| \leq \frac{1}{j!}(\mathcal{S}(x))^j$ for all j. Therefore, the function

$$w_\lambda(x) = \sum_{j=0}^{\infty} (-\lambda)^j \eta_j(x) \qquad (a < x \leq \beta,\ \lambda \in \mathbb{C})$$

is well-defined as an absolutely convergent series. The estimate

$$|w_\lambda(x)| \leq \sum_{j=0}^{\infty} |\lambda|^j \frac{(\mathcal{S}(x))^j}{j!} = e^{|\lambda|\mathcal{S}(x)} \leq e^{|\lambda|\mathcal{S}(\beta)} \qquad (a < x \leq \beta)$$

shows that $\lambda \mapsto w_\lambda(x)$ is entire and of exponential type. In addition, for $a < x \le \beta$ we have

$$\begin{aligned}
1 - \lambda \int_a^x \frac{1}{p(y)} \int_a^y w_\lambda(\xi)\, r(\xi) d\xi\, dy &= 1 - \lambda \int_a^x (\mathfrak{s}(x) - \mathfrak{s}(\xi)) w_\lambda(\xi)\, r(\xi) d\xi \\
&= 1 - \lambda \int_a^x (\mathfrak{s}(x) - \mathfrak{s}(\xi)) \Big(\sum_{j=0}^\infty (-\lambda)^j \eta_j(\xi) \Big) r(\xi) d\xi \\
&= 1 + \sum_{j=0}^\infty (-\lambda)^{j+1} \int_a^x (\mathfrak{s}(x) - \mathfrak{s}(\xi)) \eta_j(\xi)\, r(\xi) d\xi \\
&= 1 + \sum_{j=0}^\infty (-\lambda)^{j+1} \eta_{j+1}(x) \;=\; w_\lambda(x),
\end{aligned}$$

i.e., $w_\lambda(x)$ satisfies

$$w_\lambda(x) = 1 - \lambda \int_a^x \frac{1}{p(y)} \int_a^y w_\lambda(\xi)\, r(\xi) d\xi\, dy$$

This integral equation is equivalent to (6.7), so the proof is complete. □

Throughout this work, $\{a_m\}_{m\in\mathbb{N}}$ will denote a sequence $b > a_1 > a_2 > \ldots$ with $\lim a_m = a$. Next we verify that the solution w_λ for the Sturm–Liouville equation on the interval (a, b) is approximated by the corresponding solutions on the intervals (a_m, b):

Lemma 6.2. *For $m \in \mathbb{N}$, let $w_{\lambda,m}(x)$ be the unique solution of the boundary value problem*

$$\ell(w) = \lambda w \quad (a_m < x < b), \qquad w(a_m) = 1, \qquad w^{[1]}(a_m) = 0. \tag{6.9}$$

Then

$$\lim_{m\to\infty} w_{\lambda,m}(x) = w_\lambda(x) \quad \textit{pointwise for each } a < x < b \textit{ and } \lambda \in \mathbb{C}.$$

Proof. In the same way as in the proof of Lemma 6.1, we can check that the solution of (6.9) is given by

$$w_{\lambda,m}(x) = \sum_{j=0}^\infty (-\lambda)^j \eta_{j,m}(x) \qquad (a_m < x < b,\ \lambda \in \mathbb{C})$$

where $\eta_{0,m}(x) = 1$ and $\eta_{j,m}(x) = \int_{a_m}^x (\mathfrak{s}(x) - \mathfrak{s}(\xi)) \eta_{j-1,m}(\xi) r(\xi) d\xi$. As before we have $|\eta_{j,m}(x)| \le \frac{1}{j!}(\mathcal{S}(x))^j$ for $a_m < x \le \beta$ (where $\mathcal{S}$ is the function from the proof of Lemma 6.1). Using this estimate and induction on j, it is easy to see that $\eta_{j,m}(x) \to \eta_j(x)$ as $m \to \infty$ ($a < x \le \beta$, $j = 0, 1, \ldots$). Noting that the estimate on $|\eta_{j,m}(x)|$ allows us to take the limit under the summation sign, we conclude that $w_{\lambda,m}(x) \to w_\lambda(x)$ as $m \to \infty$ ($a < x \le \beta$). □

The following lemma provides a sufficient condition for the solution $w_\lambda(\cdot)$ to be uniformly bounded in the variables $x \in (a, b)$ and $\lambda \ge 0$:

Lemma 6.3. *If $x \mapsto p(x)r(x)$ is an increasing function, then the solution of (6.7) is bounded:*

$$|w_\lambda(x)| \le 1 \qquad \textit{for all } a < x < b,\ \lambda \ge 0.$$

Proof. Let us start by assuming that $p(a)r(a) > 0$. For $\lambda = 0$ the result is trivial because $w_0(x) \equiv 1$. Fix $\lambda > 0$. Multiplying both sides of the differential equation $\ell(w_\lambda) = \lambda w_\lambda$ by $2w_\lambda^{[1]}$, we obtain $-\frac{1}{pr}[(w_\lambda^{[1]})^2]' = \lambda(w_\lambda^2)'$. Integrating the differential equation and then using integration by parts, we get

$$\begin{aligned}\lambda(1-w_\lambda(x)^2) &= \int_a^x \frac{1}{p(\xi)r(\xi)}(w_\lambda^{[1]}(\xi)^2)'d\xi \\ &= \frac{w_\lambda^{[1]}(x)^2}{p(x)r(x)} + \int_a^x (p(\xi)r(\xi))'\Big(\frac{w_\lambda^{[1]}(\xi)}{p(\xi)r(\xi)}\Big)^2 d\xi, \qquad a<x<b\end{aligned}$$

where we also used the fact that $w_\lambda^{[1]}(a) = 0$ and the assumption that $p(a)r(a) > 0$. The right-hand side is nonnegative, because $x \mapsto p(x)r(x)$ is increasing and therefore $(p(\xi)r(\xi))' \geq 0$. Given that $\lambda > 0$, it follows that $1 - w_\lambda(x)^2 \geq 0$, so that $|w_\lambda(x)| \leq 1$.

If $p(a)r(a) = 0$, the above proof can be used to show that the solution of (6.9) is such that $|w_{\lambda,m}(x)| \leq 1$ for all $a < x < b$, $\lambda \geq 0$ and $m \in \mathbb{N}$; then Lemma 6.2 yields the desired result. □

Remark 6.1. *We shall make extensive use of the fact that the differential expression* (6.4) *can be transformed into the standard form*

$$\widetilde{\ell} = -\frac{1}{A}\frac{d}{d\xi}\Big(A\frac{d}{d\xi}\Big) = -\frac{d^2}{d\xi^2} - \frac{A'}{A}\frac{d}{d\xi}.$$

This is achieved by setting

$$A(\xi) := \sqrt{p(\gamma^{-1}(\xi))\, r(\gamma^{-1}(\xi))}, \tag{6.10}$$

where γ^{-1} is the inverse of the increasing function

$$\gamma(x) = \int_c^x \sqrt{\frac{r(y)}{p(y)}}dy,$$

$c \in (a,b)$ being a fixed point (if $\sqrt{\frac{r(y)}{p(y)}}$ is integrable near a, we may also take $c = a$). Indeed, it is straightforward to check that a given function $\omega_\lambda : (a,b) \to \mathbb{C}$ satisfies $\ell(\omega_\lambda) = \lambda\omega_\lambda$ if and only if $\widetilde{\omega}_\lambda(\xi) := \omega_\lambda(\gamma^{-1}(\xi))$ satisfies $\widetilde{\ell}(\widetilde{\omega}_\lambda) = \lambda\widetilde{\omega}_\lambda$.

It is interesting to note that the assumption of the previous lemma ($x \mapsto p(x)r(x)$ is increasing) is equivalent to requiring that the first-order coefficient $\frac{A'}{A}$ of the transformed operator $\widetilde{\ell}$ is nonnegative. We also observe that if this assumption holds then we have $\gamma(b) = \infty$ (otherwise the left-hand side integral in (6.6) *would be finite, contradicting that b is a natural boundary). We have $\gamma(a) > -\infty$ if a is a regular endpoint (Remark 6.2); if a is entrance, $\gamma(a)$ can be either finite or infinite.*

6.2.2 Sturm–Liouville Type Transforms

For simplicity, we shall write $L_p(r) := L_p\big((a,b); r(x)dx\big)$ $(1 \leq p < \infty)$, and the norm of this space will be denoted by $\|\cdot\|_p$.

It follows from the boundary conditions (6.5)–(6.6) that one obtains a self-adjoint realization of ℓ in the Hilbert space $L_2(r)$ by imposing the Neumann boundary condition $\lim_{x\downarrow a} u^{[1]}(x) = 0$ at the left endpoint a. We state this well-known fact (cf. [35,40]) as a lemma:

Lemma 6.4. *The operator*

$$\mathcal{L} : \mathcal{D}_{\mathcal{L}}^{(2)} \subset L_2(r) \longrightarrow L_2(r), \qquad \mathcal{L}u = \ell(u)$$

where

$$\mathcal{D}_{\mathcal{L}}^{(2)} := \Big\{u \in L_2(r) \;\Big|\; u \text{ and } u' \text{ locally abs. continuous on } (a,b),\ \ell(u) \in L_2(r), \lim_{x\downarrow a} u^{[1]}(x) = 0\Big\} \tag{6.11}$$

is self-adjoint.

The self-adjoint realization $\mathcal{L}$ gives rise to an integral transform, which we will call the *$\mathcal{L}$-transform*, given by

$$(\mathcal{F}h)(\lambda) := \int_a^b h(x)\, w_\lambda(x)\, r(x)dx \qquad (h \in L_1(r),\ \lambda \geq 0) \tag{6.12}$$

(this is also known as the generalized Fourier transform or the Sturm–Liouville transform). The $\mathcal{L}$-transform is an isometry with an inverse which can be written as an integral with respect to the so-called *spectral measure* $\boldsymbol{\rho}_{\mathcal{L}}$:

Proposition 6.1. *There exists a unique locally finite positive Borel measure $\boldsymbol{\rho}_{\mathcal{L}}$ on $\mathbb{R}$ such that the map $h \mapsto \mathcal{F}h$ induces an isometric isomorphism $\mathcal{F} : L_2(r) \longrightarrow L_2(\mathbb{R}; \boldsymbol{\rho}_{\mathcal{L}})$ whose inverse is given by*

$$(\mathcal{F}^{-1}\varphi)(x) = \int_{\mathbb{R}} \varphi(\lambda)\, w_\lambda(x)\, \boldsymbol{\rho}_{\mathcal{L}}(d\lambda),$$

the convergence of the latter integral being understood with respect to the norm of $L_2(r)$. The spectral measure $\boldsymbol{\rho}_{\mathcal{L}}$ is supported on $[0, \infty)$. Moreover, the differential operator $\mathcal{L}$ is connected with the transform (6.12) via the identity

$$[\mathcal{F}(\mathcal{L}h)](\lambda) = \lambda \cdot (\mathcal{F}h)(\lambda), \qquad h \in \mathcal{D}_{\mathcal{L}}^{(2)} \tag{6.13}$$

and the domain $\mathcal{D}_{\mathcal{L}}^{(2)}$ defined by (6.11) can be written as

$$\mathcal{D}_{\mathcal{L}}^{(2)} = \Bigl\{ u \in L_2(r) \Bigm| \lambda \cdot (\mathcal{F}f)(\lambda) \in L_2\bigl([0, \infty); \boldsymbol{\rho}_{\mathcal{L}}\bigr) \Bigr\}. \tag{6.14}$$

Proof. The existence of a generalized Fourier transform associated with the operator $\mathcal{L}$ is a consequence of the standard Weyl-Titchmarsh-Kodaira theory of eigenfunction expansions of Sturm–Liouville operators (see [49, Section 3.1] and [59, Section 8]).

In the general case the eigenfunction expansion is written in terms of two linearly independent eigenfunctions and a 2×2 matrix measure. However, from the regular/entrance boundary assumption (6.5), it follows that the function $w_\lambda(x)$ is square-integrable near $x = 0$ with respect to the measure $r(x)dx$; moreover, by Lemma 6.1, $w_\lambda(x)$ is (for fixed x) an entire function of λ. Therefore, the possibility of writing the expansion in terms of the eigenfunction $w_\lambda(x)$ only follows from the results of [19, Sections 9 and 10]. □

It is worth pointing out that the transformation of the Sturm–Liouville operator ℓ into its standard form $\widetilde{\ell}$ (Remark 6.1) leaves the spectral measure unchanged: indeed, it is easily verified that the operator $\widetilde{\mathcal{L}} : \mathcal{D}_{\widetilde{\mathcal{L}}}^{(2)} \subset L_2(A) \longrightarrow L_2(A)$, $\widetilde{\mathcal{L}}u = \widetilde{\ell}(u)$ is unitarily equivalent to the operator $\mathcal{L}$ and, consequently, $\boldsymbol{\rho}_{\widetilde{\mathcal{L}}} = \boldsymbol{\rho}_{\mathcal{L}}$.

The following lemma gives a sufficient condition for the inversion integral of the $\mathcal{L}$-transform to be absolutely convergent.

Lemma 6.5. *(a) For each $\mu \in \mathbb{C} \setminus \mathbb{R}$, the integrals*

$$\int_{[0,\infty)} \frac{w_\lambda(x)\, w_\lambda(y)}{|\lambda - \mu|^2} \boldsymbol{\rho}_{\mathcal{L}}(d\lambda) \qquad \text{and} \qquad \int_{[0,\infty)} \frac{w_\lambda^{[1]}(x)\, w_\lambda^{[1]}(y)}{|\lambda - \mu|^2} \boldsymbol{\rho}_{\mathcal{L}}(d\lambda) \tag{6.15}$$

converge uniformly on compact squares in $(a, b)^2$.

(b) If $h \in \mathcal{D}_{\mathcal{L}}^{(2)}$, then

$$h(x) = \int_{[0,\infty)} (\mathcal{F}h)(\lambda)\, w_\lambda(x)\, \boldsymbol{\rho}_{\mathcal{L}}(d\lambda) \tag{6.16}$$

$$h^{[1]}(x) = \int_{[0,\infty)} (\mathcal{F}h)(\lambda)\, w_\lambda^{[1]}(x)\, \boldsymbol{\rho}_{\mathcal{L}}(d\lambda) \tag{6.17}$$

where the right-hand side integrals converge absolutely and uniformly on compact subsets of (a,b).

Proof. **(a)** By [19, Lemma 10.6] and [51, p. 229],

$$\int_{[0,\infty)} \frac{w_\lambda(x)w_\lambda(y)}{|\lambda-\mu|^2}\boldsymbol{\rho}_{\mathcal{L}}(d\lambda) = \int_a^b G(x,\xi,\mu)G(y,\xi,\mu)\,r(\xi)d\xi = \frac{1}{\mathrm{Im}(\mu)}\,\mathrm{Im}\big(G(x,y,\mu)\big)$$

where $G(x,y,\mu)$ is the resolvent kernel (or Green function) of the operator $(\mathcal{L},\mathcal{D}_{\mathcal{L}}^{(2)})$. Moreover, according to [19, Theorems 8.3 and 9.6], the resolvent kernel is given by

$$G(x,y,\mu) = \begin{cases} w_\mu(x)\vartheta_\mu(y), & x<y \\ w_\mu(y)\vartheta_\mu(x), & x\geq y \end{cases}$$

where $\vartheta_\lambda(\cdot)$ is a solution of $\ell(u) = \lambda u$ which is square-integrable near b with respect to the measure $r(x)dx$ and verifies the identity $w_\lambda(x)\vartheta_\lambda^{[1]}(x) - w_\lambda^{[1]}(x)\vartheta_\lambda(x) \equiv 1$. It is easily seen (cf. [41, p. 125]) that the functions $\mathrm{Im}\big(G(x,y,\mu)\big)$ and $\partial_x^{[1]}\partial_y^{[1]}\mathrm{Im}\big(G(x,y,\mu)\big)$ are continuous in $0<x,y<\infty$. Essentially the same proof as that of [41, Corollary 3] now yields that

$$\int_{[0,\infty)} \frac{w_\lambda^{[1]}(x)\,w_\lambda^{[1]}(y)}{|\lambda-\mu|^2}\boldsymbol{\rho}_{\mathcal{L}}(d\lambda) = \frac{1}{\mathrm{Im}(\mu)}\,\partial_x^{[1]}\partial_y^{[1]}\mathrm{Im}\big(G(x,y,\mu)\big)$$

and that the integrals (6.15) converge uniformly for x,y in compacts.

(b) By Proposition 6.1 and the classical theorem on differentiation under the integral sign for Riemann–Stieltjes integrals, to prove (6.16)–(6.17) it only remains to justify the absolute and uniform convergence of the integrals in the right-hand side.

Recall from Proposition 6.1 that the condition $h\in\mathcal{D}_{\mathcal{L}}^{(2)}$ implies that $\mathcal{F}h\in L_2\big([0,\infty);\boldsymbol{\rho}_{\mathcal{L}}\big)$ and also $\lambda\,(\mathcal{F}h)(\lambda)\in L_2\big([0,\infty);\boldsymbol{\rho}_{\mathcal{L}}\big)$. As a consequence, we obtain

$$\begin{aligned}
&\int_{[0,\infty)} \big|(\mathcal{F}h)(\lambda)w_\lambda(x)\big|\boldsymbol{\rho}_{\mathcal{L}}(d\lambda)\\
&\qquad\leq \int_{[0,\infty)} \lambda\,\big|(\mathcal{F}h)(\lambda)\big|\left|\frac{w_\lambda(x)}{\lambda+i}\right|\boldsymbol{\rho}_{\mathcal{L}}(d\lambda) + \int_{[0,\infty)} \big|(\mathcal{F}h)(\lambda)\big|\left|\frac{w_\lambda(x)}{\lambda+i}\right|\boldsymbol{\rho}_{\mathcal{L}}(d\lambda)\\
&\qquad\leq \big(\|\lambda\,(\mathcal{F}h)(\lambda)\|_\rho + \|(\mathcal{F}h)(\lambda)\|_\rho\big)\left\|\frac{w_\lambda(x)}{\lambda+i}\right\|_\rho\\
&\qquad<\infty
\end{aligned}$$

where $\|\cdot\|_\rho$ denotes the norm of the space $L_2\big(\mathbb{R};\boldsymbol{\rho}_{\mathcal{L}}\big)$, and similarly

$$\int_{[0,\infty)} \big|(\mathcal{F}h)(\lambda)\,w_\lambda^{[1]}(x)\big|\boldsymbol{\rho}_{\mathcal{L}}(d\lambda) \leq \big(\|\lambda\,(\mathcal{F}h)(\lambda)\|_\rho + \|(\mathcal{F}h)(\lambda)\|_\rho\big)\left\|\frac{w_\lambda^{[1]}(x)}{\lambda+i}\right\|_\rho < \infty.$$

We know from part (a) that the integrals which define $\|\frac{w_\lambda(x)}{\lambda+i}\|_\rho$ and $\|\frac{w_\lambda^{[1]}(x)}{\lambda+i}\|_\rho$ converge uniformly; hence, the integrals in (6.16)–(6.17) converge absolutely and uniformly for x in compact subsets. □

6.2.3 Diffusion Processes

In what follows we write P_{x_0} for the distribution of a given time-homogeneous Markov process started at the point x_0 and $\mathbb{E}_{x_0}$ for the associated expectation operator.

By an *irreducible diffusion process* X on an interval $I \subset \mathbb{R}$, we mean a continuous strong Markov process $\{X_t\}_{t\geq 0}$ with state space I such that

$$P_x(\tau_y < \infty) > 0 \text{ for any } x \in \operatorname{int} I \text{ and } y \in I, \qquad \text{where } \tau_y = \inf\{t \geq 0 \mid X_t = y\}.$$

The *resolvent* $\{\mathcal{R}_\eta\}_{\eta>0}$ of such a diffusion (or of a general Feller process) X is defined by $\mathcal{R}_\eta u = \int_0^\infty e^{-\eta t} T_t u\, dt$, $u \in \mathrm{C_b}(I,\mathbb{R})$, where $(T_t u)(x) = \mathbb{E}_x[u(X_t)]$ is the transition semigroup of the process X. The $\mathrm{C_b}$-*generator* $(\mathcal{G}, \mathcal{D}(\mathcal{G}))$ of X is the operator with domain $\mathcal{D}(\mathcal{G}) = \mathcal{R}_\eta\big(\mathrm{C_b}(I,\mathbb{R})\big)$ $(\eta > 0)$ and defined by

$$(\mathcal{G}u)(x) = \eta u(x) - g(x) \qquad \text{for } u = \mathcal{R}_\eta g,\ g \in \mathrm{C_b}(I,\mathbb{R}),\ x \in I$$

($\mathcal{G}$ is independent of η, cf. [24, p. 295]). A *Feller semigroup* is a family $\{T_t\}_{t\geq 0}$ of operators $T_t : \mathrm{C_b}(I,\mathbb{R}) \longrightarrow \mathrm{C_b}(I,\mathbb{R})$ satisfying

(i) $T_t T_s = T_{t+s}$ for all $t, s \geq 0$;

(ii) $T_t\big(\mathrm{C_0}(I,\mathbb{R})\big) \subset \mathrm{C_0}(I,\mathbb{R})$ for all $t \geq 0$;

(iii) If $h \in \mathrm{C_b}(I,\mathbb{R})$ satisfies $0 \leq h \leq 1$, then $0 \leq T_t h \leq 1$;

(iv) $\lim_{t\downarrow 0} \|T_t h - h\|_\infty = 0$ for each $h \in \mathrm{C_0}(I,\mathbb{R})$.

The Feller semigroup is said to be *conservative* if $T_t \mathbb{1} = \mathbb{1}$ (here $\mathbb{1}$ denotes the function identically equal to one). A *Feller process* is a time-homogeneous Markov process $\{X_t\}_{t\geq 0}$ whose transition semigroup is a Feller semigroup. For further background on the theory of Markov diffusion processes and Feller semigroups, we refer to [8] and references therein.

We now recall a known fact from the theory of (one-dimensional) diffusion processes, namely that the negative of the Sturm–Liouville differential operator (6.4) generates a diffusion process which is conservative and has the Feller property. The proof can be found in [24, Sections 4 and 6] (see also [39, Section II.5]).

Lemma 6.6. *The operator*

$$\mathcal{L}^{(\mathrm{b})} : \mathcal{D}_{\mathcal{L}}^{(\mathrm{b})} \subset \mathrm{C_b}([a,b),\mathbb{R}) \longrightarrow \mathrm{C_b}([a,b),\mathbb{R}), \qquad \mathcal{L}^{(\mathrm{b})} u = -\ell(u)$$

with domain

$$\mathcal{D}_{\mathcal{L}}^{(\mathrm{b})} = \big\{u \in \mathrm{C_b}([a,b),\mathbb{R}) \;\big|\; \ell(u) \in \mathrm{C_b}([a,b),\mathbb{R}),\ \lim_{x\downarrow a} u^{[1]}(x) = 0\big\}$$

is the $\mathrm{C_b}$*-generator of a one-dimensional irreducible diffusion process* $X = \{X_t\}_{t\geq 0}$ *with state space* $[a,b)$ *whose transition semigroup* $\{T_t\}_{t\geq 0}$ *defines a conservative Feller semigroup on* $\mathrm{C_0}([a,b),\mathbb{R})$.

The transition probabilities of the one-dimensional diffusion process from the previous lemma admits an explicit representation as the inverse $\mathcal{L}$-transform of the function $e^{-t\lambda} w_\lambda(x)$:

Lemma 6.7. *The transition semigroup admits the representation*

$$(T_t h)(x) = \int_a^b h(y)\, p(t,x,y)\, r(y) dy \qquad \big(h \in \mathrm{B_b}\big([a,b),\mathbb{R}\big),\ t > 0,\ a < x < b\big) \tag{6.18}$$

where $p(t,x,y)$ is a nonnegative function which is called the fundamental solution *for the parabolic equation $\frac{\partial u}{\partial t} = -\ell_x u$ (the subscript indicates the variable in which the operator ℓ acts). The fundamental solution and its derivatives are explicitly given by*

$$(\partial_t^n p)(t,x,y) = \int_{[0,\infty)} \lambda^n e^{-t\lambda}\, w_\lambda(x)\, w_\lambda(y)\, \boldsymbol{\rho}_{\mathcal{L}}(d\lambda)$$

$$(\partial_x^{[1]}\partial_t^n p)(t,x,y) = \int_{[0,\infty)} \lambda^n e^{-t\lambda}\, w_\lambda^{[1]}(x)\, w_\lambda(y)\, \boldsymbol{\rho}_{\mathcal{L}}(d\lambda)$$

for $n \in \mathbb{N}_0$, where, for fixed $t > 0$, the integrals converge absolutely and uniformly on compact squares of $(a,b) \times (a,b)$. If $h \in L_2(r) \cap \mathrm{B_b}([a,b),\mathbb{R})$, then (6.18) *can also be written as*

$$(T_t h)(x) = \int_{[0,\infty)} e^{-t\lambda}\, w_\lambda(x)\, (\mathcal{F}h)(\lambda)\, \boldsymbol{\rho}_{\mathcal{L}}(d\lambda) \qquad (t > 0,\ a < x < b) \tag{6.19}$$

where the integral converges with respect to the norm of $L_2(r)$.

Proof. These assertions are a consequence of the results of [35, Sections 2–3] and [40, Section 4]. □

Remark 6.2. *Let X be a one-dimensional diffusion process on an interval with endpoints a and b, whose $\mathrm{C_b}$-generator is of the form* (6.4). *Let*

$$I_a = \int_a^c \int_a^y \frac{dx}{p(x)}\, r(y)dy, \qquad J_a = \int_a^c \int_y^c \frac{dx}{p(x)}\, r(y)dy$$

According to Feller's boundary classification for the diffusion X, the left endpoint a is called

regular	*if* $I_a < \infty$, $J_a < \infty$;		*entrance*	*if* $I_a = \infty$, $J_a < \infty$;	
exit	*if* $I_a < \infty$, $J_a = \infty$;		*natural*	*if* $I_a = \infty$, $J_a = \infty$.	

(the right endpoint is classified in a similar way).

The probabilistic meaning of this classification is the following [8, Chapter II]: an irreducible diffusion can be started from the boundary a if and only if a is regular or entrance; the boundary a is reached from $x_0 \in (a,b)$ with positive probability by an irreducible diffusion if and only if a is regular or exit.

Our standing assumption (6.5) *on the coefficients of the Sturm–Liouville operator means that a is a regular or an entrance boundary for the diffusion process X generated by ℓ. It is clear from the preceding remarks that Lemma 6.6 relies crucially on this assumption. The same is true for some of the results of the previous subsections: in fact, Lemma 6.1 fails if a is exit or natural [26, Sections 5.13–5.14], and the boundary conditions defining $\mathcal{D}_{\mathcal{L}}^{(2)}$ differ from those in* (6.11) *when a is exit or natural [40]. In turn, the assumption* (6.6) *means that b is a natural boundary for the diffusion X. Since one can show that* (6.6) *is automatically satisfied whenever Assumption MP below holds [48, Proposition 3.5], this boundary assumption at b yields no loss of generality on our results concerning product formulas and generalized convolutions.*

6.3 The Hyperbolic Equation $\ell_x f = \ell_y f$

In this section we investigate the (possibly degenerate) hyperbolic Cauchy problem

$$(\ell_x f)(x,y) = (\ell_y f)(x,y) \quad (x,y \in (a,b)), \qquad f(x,a) = h(x), \qquad (\partial_y^{[1]} f)(x,a) = 0 \tag{6.20}$$

where $\partial^{[1]}u = pu'$, ℓ is the Sturm–Liouville operator (6.4), and the subscripts indicate the variable in which the operators act.

Since $\ell_y - \ell_x = \frac{p(x)}{r(x)}\frac{\partial^2}{\partial x^2} - \frac{p(y)}{r(y)}\frac{\partial^2}{\partial y^2}$ + lower-order terms, the equation $\ell_x f = \ell_y f$ is hyperbolic at the line $y = a$ if $\frac{p(a)}{r(a)} > 0$; otherwise, the initial conditions of the Cauchy problem are given at a line of parabolic degeneracy. If $\gamma(a) = -\int_a^c \sqrt{\frac{r(y)}{p(y)}}dy > -\infty$, then we can remove the degeneracy via the change of variables $x = \gamma(\xi)$, $y = \gamma(\zeta)$ (cf. Remark 6.1), through which the partial differential equation is transformed to the standard form $\widetilde{\ell}_\xi u = \widetilde{\ell}_\zeta u$, with initial condition at the line $\zeta = \gamma(a)$. In the case $\gamma(a) = -\infty$, the standard form of the equation is also parabolically degenerate in the sense that its initial line is $\zeta = -\infty$.

6.3.1 Existence and Uniqueness of Solution

We start by proving a result which not only assures the existence of solution for Cauchy problems with well-behaved initial conditions but also provides an explicit representation for the solution as an inverse $\mathcal{L}$-transform:

Theorem 6.1 (Existence of solution). *Suppose that $x \mapsto p(x)r(x)$ is an increasing function. If $h \in \mathcal{D}_\mathcal{L}^{(2)}$ and $\ell(h) \in \mathcal{D}_\mathcal{L}^{(2)}$, then the function*

$$f_h(x,y) := \int_{[0,\infty)} w_\lambda(x)\, w_\lambda(y)\, (\mathcal{F}h)(\lambda)\, \boldsymbol{\rho}_\mathcal{L}(d\lambda) \tag{6.21}$$

solves the Cauchy problem (6.20).

For ease of notation, unless necessary, we drop the dependence in h and denote (6.21) by $f(x,y)$.

Proof. Let us begin by justifying that $\ell_x f$ can be computed via differentiation under the integral sign. It follows from (6.7) that $w_\lambda^{[1]}(x) = -\lambda \int_a^x w_\lambda(\xi)\, r(\xi) d\xi$ and therefore (by Lemma 6.3) $|w_\lambda^{[1]}(x)| \leq \lambda \int_a^x r(\xi)d\xi$. Hence

$$\int_{[0,\infty)} \bigl|(\mathcal{F}h)(\lambda)\, w_\lambda^{[1]}(x)\, w_\lambda(y)\bigr| \boldsymbol{\rho}_\mathcal{L}(d\lambda) \leq \int_a^x r(\xi)d\xi \cdot \int_{[0,\infty)} \lambda \bigl|(\mathcal{F}h)(\lambda)\, w_\lambda(y)\bigr| \boldsymbol{\rho}_\mathcal{L}(d\lambda) < \infty, \tag{6.22}$$

where the convergence (which is uniform on compacts) follows from (6.13) and Lemma 6.5(b). From the convergence of the differentiated integral we conclude that $\partial_x^{[1]} f(x,y) = \int_{[0,\infty)} (\mathcal{F}h)(\lambda)\, w_\lambda^{[1]}(x)\, w_\lambda(y)\, \boldsymbol{\rho}_\mathcal{L}(d\lambda)$. Since $(\ell w_\lambda)(x) = \lambda w_\lambda(x)$, in the same way we check that $\int_{[0,\infty)} (\mathcal{F}h)(\lambda)\, (\ell w_\lambda)(x)\, w_\lambda(y)\, \boldsymbol{\rho}_\mathcal{L}(d\lambda)$ converges absolutely and uniformly on compacts and is therefore equal to $(\ell_x f)(x,y)$. Consequently,

$$(\ell_x f)(x,y) = (\ell_y f)(x,y) = \int_{[0,\infty)} \lambda\, (\mathcal{F}h)(\lambda)\, w_\lambda(x)\, w_\lambda(y)\, \boldsymbol{\rho}_\mathcal{L}(d\lambda). \tag{6.23}$$

Concerning the boundary conditions, Lemma 6.5(b) together with the fact that $w_\lambda(a) = 1$ imply that $f(x,a) = h(x)$, and from (6.22) we easily see that $\lim_{y \downarrow a} \partial_y^{[1]} f(x,y) = 0$. This shows that f is a solution of the Cauchy problem (6.20). □

Under the assumptions of the theorem, the solution (6.21) of the hyperbolic Cauchy problem satisfies

$$f(\cdot,y) \in \mathcal{D}_{\mathcal{L}}^{(2)} \qquad \text{for all } a<y<b, \tag{6.24}$$

$$\mathcal{F}[\ell_y f(\cdot,y)](\lambda) = \ell_y[\mathcal{F}f(\cdot,y)](\lambda) \qquad \text{for all } a<y<b, \tag{6.25}$$

$$\lim_{y\downarrow a}[\mathcal{F}f(\cdot,y)](\lambda) = (\mathcal{F}h)(\lambda), \tag{6.26}$$

$$\lim_{y\downarrow a} \partial_y^{[1]}[\mathcal{F}f(\cdot,y)](\lambda) = 0. \tag{6.27}$$

Indeed, by Proposition 6.1 we have $[\mathcal{F}f(\cdot,y)](\lambda) = (\mathcal{F}h)(\lambda)\, w_\lambda(y)$ for all $\lambda \in \operatorname{supp}(\boldsymbol{\rho}_{\mathcal{L}})$ and $a<y<b$. Since $h \in \mathcal{D}_{\mathcal{L}}^{(2)}$ and $|w_\lambda(\cdot)| \leq 1$ (Lemma 6.3), it is clear from (6.14) that $f(x,y)$ satisfies (6.24). Moreover, it follows from (6.23) that $\mathcal{F}[\ell_y f_j(\cdot,y)](\lambda) = \lambda\,(\mathcal{F}h)(\lambda)\, w_\lambda(y) = \ell_y[\mathcal{F}f_j(\cdot,y)](\lambda)$, hence (6.25) holds. The properties (6.26)–(6.27) follow immediately from Lemma 6.1.

Next we show that the solution from the above existence theorem is the unique solution satisfying the conditions (6.24)–(6.27):

Theorem 6.2 (Uniqueness). *Let $h \in \mathcal{D}_{\mathcal{L}}^{(2)}$. Let $f_1, f_2 \in \mathrm{C}^2\big((a,b)^2\big)$ be two solutions of $(\ell_x f)(x,y) = (\ell_y f)(x,y)$. For $f \in \{f_1, f_2\}$, suppose that* (6.24) *holds and that there exists a zero $\boldsymbol{\rho}_{\mathcal{L}}$-measure set $\Lambda_0 \subset [0,\infty)$ such that* (6.25)–(6.27) *hold for each $\lambda \in [0,\infty) \setminus \Lambda_0$. Then*

$$f_1(x,y) \equiv f_2(x,y) \qquad \textit{for all } x,y \in (a,b). \tag{6.28}$$

Proof. Fix $\lambda \in [0,\infty)\setminus\Lambda_0$ and let $\Psi_j(y,\lambda) := [\mathcal{F}f_j(\cdot,y)](\lambda)$. We have

$$\ell_y \Psi_j(y,\lambda) = \mathcal{F}[\ell_y f_j(\cdot,y)](\lambda) = \mathcal{F}[\ell_x f_j(\cdot,y)](\lambda) = \lambda \Psi_j(y,\lambda), \qquad a<y<b$$

where the first equality is due to (6.25) and the last step follows from (6.13). Moreover,

$$\lim_{y\downarrow a} \Psi_j(y,\lambda) = (\mathcal{F}h)(\lambda) \quad \text{and} \quad \lim_{y\downarrow a} \partial_y^{[1]} \Psi_j(y,\lambda) = 0$$

by (6.26) and (6.27), respectively. It thus follows from Lemma 6.1 that

$$[\mathcal{F}f_j(\cdot,y)](\lambda) = \Psi_j(y,\lambda) = (\mathcal{F}h)(\lambda)\, w_\lambda(y), \qquad a<y<b.$$

This equality holds for $\boldsymbol{\rho}_{\mathcal{L}}$-almost every λ, so the isometric property of $\mathcal{F}$ gives $f_1(\cdot,y) = f_2(\cdot,y)$ Lebesgue-almost everywhere; since the f_j are continuous, we conclude that (6.28) holds. □

If the hyperbolic equation $\ell_x f = \ell_y f$ (or the transformed equation $\widetilde{\ell}_\xi u = \widetilde{\ell}_y u$) is uniformly hyperbolic, the existence and uniqueness of solution for this Cauchy problem is a standard result which follows from the classical theory of hyperbolic problems in two variables (see e.g., [16, Chapter V]); in fact, the existence and uniqueness holds under much weaker restrictions on the initial condition. However, our existence and uniqueness result becomes nontrivial in the presence of a (nonremovable) parabolic degeneracy at the initial line.

Indeed, even though many authors have addressed Cauchy problems for degenerate hyperbolic equations in two variables, most studies are restricted to equations where the $\frac{\partial^2}{\partial x^2}$ term vanishes at an initial line $y = y_0$ (we refer to [5, Section 2.3], [43, Section 5.4] and references therein). Much less is known for hyperbolic equations whose $\frac{\partial^2}{\partial y^2}$ term vanishes at the same initial line: it is known that the Cauchy problem is, in general, not well-posed, and the relevance of determining sufficient conditions for its well-posedness has long been pointed out [5, Section 2.4], but as far as we are aware, little progress has been made on this problem (for related work see [38]). The application of spectral techniques to hyperbolic Cauchy problems associated with Sturm–Liouville operators is by no means new, see e.g., [10,11] and references therein; however, it seems that such techniques had never been applied to degenerate cases.

It is helpful to know that an existence theorem analogous to Theorem 6.1 holds when the initial line is shifted away from the degeneracy, because this will allow us to justify that the solution of the degenerate Cauchy problem is the pointwise limit of solutions of nondegenerate problems.

Proposition 6.2. *Suppose that $x \mapsto p(x)r(x)$ is an increasing function, and let $m \in \mathbb{N}$. If $h \in \mathcal{D}_{\mathcal{L}}^{(2)}$ and $\ell(h) \in \mathcal{D}_{\mathcal{L}}^{(2)}$, then the function*

$$f_m(x,y) = \int_{[0,\infty)} w_\lambda(x)\, w_{\lambda,m}(y)\, (\mathcal{F}h)(\lambda)\, \boldsymbol{\rho}_{\mathcal{L}}(d\lambda) \tag{6.29}$$

is a solution of the Cauchy problem

$$\begin{aligned} (\ell_x f_m)(x,y) &= (\ell_y f_m)(x,y), && a < x < b,\ a_m < y < b \\ f_m(x,a_m) &= h(x), && a < x < b \\ (\partial_y^{[1]} f_m)(x,a_m) &= 0, && a < x < b. \end{aligned} \tag{6.30}$$

Proof. Let us begin by justifying that $\partial_x^{[1]} f_m(x,y)$ and $(\ell_x f_m)(x,y)$ can be computed via differentiation under the integral sign. The differentiated integrals are given by

$$\int_{[0,\infty)} w_\lambda^{[1]}(x)\, w_{\lambda,m}(y)\, (\mathcal{F}h)(\lambda)\, \boldsymbol{\rho}_{\mathcal{L}}(d\lambda) \tag{6.31}$$

$$\int_{[0,\infty)} w_\lambda(x)\, w_{\lambda,m}(y)\, [\mathcal{F}(\ell(h))](\lambda)\, \boldsymbol{\rho}_{\mathcal{L}}(d\lambda) \tag{6.32}$$

(for the latter, we used the identities $(\ell w_\lambda)(x) = \lambda w_\lambda(x)$ and (6.13)), and their absolute and uniform convergence on compacts follows from the fact that $h, \ell(h) \in \mathcal{D}_{\mathcal{L}}^{(2)}$, together with Lemma 6.5(b) and the inequality $|w_{\lambda,m}(\cdot)| \leq 1$ (which follows from Lemma 6.3 if we replace a by a_m). This justifies that $\partial_x^{[1]} f_m(x,y)$ and $(\ell_x f_m)(x,y)$ are given by (6.31) and (6.32), respectively.

We also need to ensure that $\partial_y^{[1]} f_m(x,y)$ and $(\ell_y f_m)(x,y)$ are given by the corresponding differentiated integrals, and to that end, we must check that

$$\int_{[0,\infty)} w_\lambda(x)\, w_{\lambda,m}^{[1]}(y)\, (\mathcal{F}h)(\lambda)\, \boldsymbol{\rho}_{\mathcal{L}}(d\lambda)$$

converges absolutely and uniformly. Indeed, it follows from (6.9) that for $y \geq a_m$ we have $w_{\lambda,m}^{[1]}(y) = \lambda \int_{a_m}^y w_{\lambda,m}(\xi)\, r(\xi) d\xi$ and consequently $|w_{\lambda,m}^{[1]}(y)| \leq \lambda \int_{a_m}^y r(\xi) d\xi$; hence

$$\int_{[0,\infty)} \bigl| w_\lambda(x)\, w_{\lambda,m}^{[1]}(y)\, (\mathcal{F}h)(\lambda) \bigr| \boldsymbol{\rho}_{\mathcal{L}}(d\lambda) \leq \int_{a_m}^y r(\xi) d\xi \cdot \int_{[0,\infty)} \lambda \bigl| w_\lambda(x) (\mathcal{F}h)(\lambda) \bigr| \boldsymbol{\rho}_{\mathcal{L}}(d\lambda) \tag{6.33}$$

and the uniform convergence in compacts follows from (6.13) and Lemma 6.5(b).

The verification of the boundary conditions is straightforward: Lemma 6.5(b) together with the fact that $w_{\lambda,m}(a_m) = 1$ imply that $f_m(x,a_m) = h(x)$, and from (6.33) we easily see that $\partial_y^{[1]} f_m(x,a_m) = 0$. This shows that f_m is a solution of the Cauchy problem (6.30). □

Corollary 6.1. *Suppose that $x \mapsto p(x)r(x)$ is an increasing function. Let $h \in \mathcal{D}_{\mathcal{L}}^{(2)}$ and consider the functions f_m, f defined by (6.29), (6.21). Then*

$$\lim_{m\to\infty} f_m(x,y) = f(x,y) \qquad \textit{pointwise for each } x,y \in (a,b).$$

Proof. Since $w_{\lambda,m}(y) \to w_\lambda(y)$ pointwise as $m \to \infty$ (Lemma 6.2), the conclusion follows from the dominated convergence theorem (which is applicable due to Lemmas 6.3 and 6.5(b)). □

6.3.2 Maximum Principle and Positivity of Solution

After having shown that the Cauchy problem is well-posed whenever the function $x \mapsto p(x)r(x)$ is increasing, we introduce a stronger assumption on the coefficients which will be seen to be sufficient for a maximum principle to hold for the hyperbolic equation $\ell_x f = \ell_y f$ and, in consequence, for the solution of the Cauchy problem (6.20) to preserve positivity and boundedness of its initial condition. We shall rely on the transformation of ℓ into the standard form (Remark 6.1); in the following assumption, A is the function defined in (6.10).

Assumption MP (Maximum Principle). There exists $\eta \in \mathrm{C}^1(\gamma(a),\infty)$ such that $\eta \geq 0$, $\phi_\eta := \frac{A'}{A} - \eta \geq 0$, and the functions ϕ_η and $\psi_\eta := \frac{1}{2}\eta' - \frac{1}{4}\eta^2 + \frac{A'}{2A}\cdot\eta$ are both decreasing on $(\gamma(a),\infty)$.

Observe that Assumption MP allows for $\gamma(a) = -\infty$ (this will enable us to treat the case of nonremovable degeneracy), and it does not include the left endpoint in the interval where the conditions on η are imposed. This assumption is therefore a generalization of an assumption introduced by Zeuner, cf. Example 6.8.

The proof of the maximum principle presented in the sequel is based on [63, Proposition 3.7] and on the maximum principles of [60]. The key tool is the integral identity which we now state:

Lemma 6.8. *Let $\boldsymbol{\ell}^B$ be the differential expression $\boldsymbol{\ell}^B v := -v'' - \phi_\eta v' + \psi_\eta v$. For $\gamma(a) < c \leq y \leq x$, consider the triangle $\Delta_{c,x,y} := \{(\xi,\zeta) \in \mathbb{R}^2 \mid \zeta \geq c,\, \xi+\zeta \leq x+y,\, \xi-\zeta \geq x-y\}$, and let $v \in \mathrm{C}^2(\Delta_{c,x,y})$. Write $B(x) := \exp(\frac{1}{2}\int_\beta^x \eta(\xi)d\xi)$ (with $\beta > \gamma(a)$ arbitrary) and $A_B(x) = \frac{A(x)}{B(x)^2}$. Then the following integral equation holds:*

$$A_B(x)A_B(y)\,v(x,y) = H + I_0 + I_1 + I_2 + I_3 - I_4 \tag{6.34}$$

where

$$\begin{aligned}
H &:= \tfrac{1}{2}A_B(c)\bigl[A_B(x-y+c)\,v(x-y+c,c) + A_B(x+y-c)\,v(x+y-c,c)\bigr]\\
I_0 &:= \tfrac{1}{2}A_B(c)\int_{x-y+c}^{x+y-c} A_B(s)(\partial_y v)(s,c)\,ds\\
I_1 &:= \tfrac{1}{2}\int_c^y A_B(s)A_B(x-y+s)\bigl[\phi_\eta(s)+\phi_\eta(x-y+s)\bigr]v(x-y+s,s)\,ds\\
I_2 &:= \tfrac{1}{2}\int_c^y A_B(s)A_B(x+y-s)\bigl[\phi_\eta(s)-\phi_\eta(x+y-s)\bigr]v(x+y-s,s)\,ds\\
I_3 &:= \tfrac{1}{2}\int_{\Delta_{c,x,y}} A_B(\xi)A_B(\zeta)\bigl[\psi_\eta(\zeta)-\psi_\eta(\xi)\bigr]v(\xi,\zeta)\,d\xi d\zeta\\
I_4 &:= \tfrac{1}{2}\int_{\Delta_{c,x,y}} A_B(\xi)A_B(\zeta)\,(\boldsymbol{\ell}_\zeta^B v - \boldsymbol{\ell}_\xi^B v)(\xi,\zeta)\,d\xi d\zeta.
\end{aligned}$$

Proof. Just compute

$$\begin{aligned} I_4 - I_3 &= \tfrac{1}{2}\int_{\Delta_{c,x,y}} \left(\frac{\partial}{\partial\xi}\big[A_B(\xi)A_B(\zeta)\,(\partial_\xi v)(\xi,\zeta)\big] - \frac{\partial}{\partial\zeta}\big[A_B(\xi)A_B(\zeta)\,(\partial_\zeta v)(\xi,\zeta)\big] \right) d\xi d\zeta \\ &= I_0 - \tfrac{1}{2}\int_0^y A_B(s)A_B(x-y+s)\,(\partial_\zeta v + \partial_\xi v)(x-y+s,s)\,ds \\ &\quad - \tfrac{1}{2}\int_0^y A_B(s)A_B(x+y-s)\,(\partial_\zeta v - \partial_\xi v)(x+y-s,s)\,ds \\ &= I_0 + I_1 - \int_c^y \frac{d}{ds}\big[A_B(s)A_B(x-y+s)\,v(x-y+s,s)\big]ds \\ &\quad + I_2 - \int_c^y \frac{d}{ds}\big[A_B(s)A_B(x+y-s)\,v(x+y-s,s)\big]ds \end{aligned}$$

where in the second equality we used Green's theorem, and the third equality follows easily from the fact that $(A_B)' = \phi_\eta A_B$. □

Theorem 6.3 (Weak maximum principle). *Suppose Assumption MP holds, and let* $\gamma(a) < c \le y_0 \le x_0$. *If* $u \in \mathrm{C}^2(\Delta_{c,x_0,y_0})$ *satisfies*

$$\begin{aligned} (\widetilde{\ell}_y u - \widetilde{\ell}_x u)(x,y) &\le 0, && (x,y) \in \Delta_{c,x_0,y_0} \\ u(x,c) &\ge 0, && x \in [x_0 - y_0 + c, x_0 + y_0 - c] \\ (\partial_y u)(x,c) + \tfrac{1}{2}\eta(c)u(x,c) &\ge 0, && x \in [x_0 - y_0 + c, x_0 + y_0 - c] \end{aligned} \tag{6.35}$$

then $u \ge 0$ *in* Δ_{c,x_0,y_0}.

Proof. Pick a function $\omega \in \mathrm{C}^2[c,\infty)$ such that $\boldsymbol{\ell}^B\omega < 0$, $\omega(c) > 0$ and $\omega'(c) \ge 0$. Clearly, it is enough to show that for all $\delta > 0$ we have $v(x,y) := B(x)B(y)u(x,y) + \delta\omega(y) > 0$ for $(x,y) \in \Delta_{c,x_0,y_0}$.

Assume by contradiction that there exist $\delta > 0$, $(x,y) \in \Delta_{c,x_0,y_0}$ for which we have $v(x,y) = 0$ and $v(\xi,\zeta) \ge 0$ for all $(\xi,\zeta) \in \Delta_{c,x,y} \subset \Delta_{c,x_0,y_0}$. It is clear from the choice of ω that $v(\cdot,c) > 0$, thus we have $H \ge 0$ in the right-hand side of (6.34). Similarly, $(\partial_y v)(\cdot,c) = B(x)B(y)\big[(\partial_y u)(\cdot,c) + \frac{1}{2}\eta(c)u(\cdot,c)\big] + \delta\omega'(c) \ge 0$, hence $I_0 \ge 0$. Since ϕ_η is positive and decreasing and ψ_η is decreasing (cf. Assumption MP) and we are assuming that $u \ge 0$ on $\Delta_{c,x,y}$, it follows that $I_1 \ge 0$, $I_2 \ge 0$ and $I_3 \ge 0$. In addition, $I_4 < 0$ because $(\boldsymbol{\ell}_\zeta^B v - \boldsymbol{\ell}_\xi^B v)(\xi,\zeta) = B(x)B(y)(\widetilde{\ell}_\zeta u - \widetilde{\ell}_\xi u)(\xi,\zeta) + (\boldsymbol{\ell}^B\omega)(\zeta) < 0$. Consequently, (6.34) yields $0 = A_B(x)A_B(y)v(x,y) \ge -I_4 > 0$. This contradiction shows that $v(x,y) > 0$ for all $(x,y) \in \Delta_{c,x_0,y_0}$. □

Naturally, this weak maximum principle can be restated in terms of the operator $\ell = -\frac{1}{r}\frac{d}{dx}(p\,\frac{d}{dx})$; this is left to the reader. As anticipated above, the positivity-preserving property of the Cauchy problem is a by-product of the maximum principle.

Proposition 6.3. *Suppose Assumption MP holds, let* $m \in \mathbb{N}$. *If* $h \in \mathcal{D}_{\mathcal{L}}^{(2)}$, $\ell(h) \in \mathcal{D}_{\mathcal{L}}^{(2)}$ *and* $h \ge 0$, *then the function* f_m *given by* (6.29) *is such that*

$$f_m(x,y) \ge 0 \qquad \textit{for } x \ge y > a_m. \tag{6.36}$$

If, in addition, $h \le C$ *(where* C *is a constant), then* $f_m(x,y) \le C$ *for* $x \ge y > a_m$.

Proof. It follows from Proposition 6.2 that the function $u_m(x,y) := f_m(\gamma^{-1}(x),\gamma^{-1}(y))$ is a solution of the Cauchy problem

$$(\widetilde{\ell}_x u_m)(x,y) = (\widetilde{\ell}_y u_m)(x,y), \qquad x,y > \tilde{a}_m \tag{6.37}$$

$$u_m(x,\tilde{a}_m) = h(\gamma^{-1}(x)), \qquad x > \tilde{a}_m \tag{6.38}$$

$$(\partial_y u_m)(x,\tilde{a}_m) = 0, \qquad x > \tilde{a}_m \tag{6.39}$$

where $\tilde{a}_m = \gamma(a_m)$. Clearly, u_m satisfies the inequalities (6.35) for arbitrary $x_0 \geq y_0 \geq \tilde{a}_m$ (here $c = \tilde{a}_m$). By Theorem 6.3, $u_m(x_0, y_0) \geq 0$ for all $x_0 \geq y_0 > \tilde{a}_m$; consequently, (6.36) holds.

The proof of the last statement is straightforward: if we have $h \leq C$, then $\widetilde{u}_m(x,y) = C - u_m(x,y)$ is a solution of (6.37) with initial conditions $\widetilde{u}_m(x, \tilde{a}_m) = C - h(\gamma^{-1}(x)) \geq 0$ and (6.39), thus the reasoning of the previous paragraph yields that $C - u_m \geq 0$ for $x \geq y > \tilde{a}_m$. □

The previous result gives the positivity-preservingness for the solution of the nondegenerate Cauchy problem (6.30). The extension of this property to the possibly degenerate problem (6.20) is an immediate consequence of the pointwise convergence result of Corollary 6.1:

Corollary 6.2. *Suppose Assumption MP holds. If $h \in \mathcal{D}_{\mathcal{L}}^{(2)}$, $\ell(h) \in \mathcal{D}_{\mathcal{L}}^{(2)}$ and $h \geq 0$, then the function f given by (6.21) is such that*

$$f(x,y) \geq 0 \qquad \text{for } x, y \in (a,b).$$

If, in addition, $h \leq C$, then $f(x,y) \leq C$ for $x, y \in (a,b)$.

Note that the conclusion holds for all $x, y \in (a,b)$ because the function $f(x,y)$ is symmetric.

6.4 Sturm–Liouville Translation and Convolution

Assumption MP will always be in force throughout this and the subsequent sections.

6.4.1 Definition and First Properties

In view of the comments made in the Introduction, it is natural to define the $\mathcal{L}$-convolution $\mu * \nu$ ($\mu, \nu \in \mathcal{M}_{\mathrm{C}}[a,b)$) in order that, for sufficiently well-behaved initial conditions, the integral $\int_{[a,b)} h(\xi)\, (\delta_x * \delta_y)(d\xi)$ coincides with the solution (6.21) of the hyperbolic Cauchy problem.

Having this in mind, the first step is to confirm that the solution of the hyperbolic Cauchy problem can be represented as an integral with respect to a family of positive measures. This is done in the next result for initial conditions belonging to the linear space

$$\mathfrak{X} := \bigcup_{t>0} T_t\big(\mathrm{C}_0[a,b) \cap L_2(r)\big)$$

where $\{T_t\}$ is the Feller semigroup generated by ℓ (Lemma 6.6). One should keep in mind that $\mathfrak{X}$ is a dense linear subspace of $\{h \in \mathcal{D}_{\mathcal{L}}^{(2)} \mid \ell(h) \in \mathcal{D}_{\mathcal{L}}^{(2)}\}$ (in the norm of $L_2(r)$) and also of $\mathrm{C}_0[a,b)$ (in the uniform norm). The first of these two facts is easily checked using Proposition 6.1 combined with the identity (6.19), while the second follows from the Feller property of the semigroup.

Proposition 6.4. *Fix $x, y \in [a,b)$. There exists a subprobability measure $\boldsymbol{\nu}_{x,y} \in \mathcal{M}_+[a,b)$ such that, for all initial conditions $h \in \mathfrak{X}$, the solution (6.21) of the hyperbolic Cauchy problem (6.20) can be written as*

$$f_h(x,y) = \int_{[a,b)} h(\xi)\, \boldsymbol{\nu}_{x,y}(d\xi) \qquad (h \in \mathfrak{X}). \tag{6.40}$$

Proof. For each fixed $x, y \in [a,b)$, the right-hand side of (6.21) defines a linear functional $\mathfrak{X} \ni h \mapsto f_h(x,y) \in \mathbb{C}$. By Corollary 6.2, $|f_h(x,y)| \leq \|h\|_\infty$ for $h \in \mathfrak{X}$. Thus it follows from the Hahn-Banach theorem that this functional admits a linear extension $\mathcal{T}_{x,y} : \mathrm{C}_0[a,b) \to \mathbb{C}$ such that $|\mathcal{T}_{x,y} h| \leq \|h\|_\infty$ for all $h \in \mathrm{C}_0[a,b)$. According to the Riesz representation theorem (cf. [14, Theorem 7.3.6]), $\mathcal{M}_\mathbb{C}[a,b)$ is the dual of $\mathrm{C}_0[a,b)$; we thus have $\mathcal{T}_{x,y}h = \int_{[a,b)} h(\xi)\boldsymbol{\nu}_{x,y}(d\xi)$, where $\boldsymbol{\nu}_{x,y}$ is a finite complex measure with $\|\boldsymbol{\nu}_{x,y}\| \leq 1$. Finally, the fact that $\int_{[a,b)} h(\xi)\boldsymbol{\nu}_{x,y}(d\xi) \equiv f_h(x,y) \geq 0$ for all $h \in \mathfrak{X}$, $h \geq 0$ (Corollary 6.2) yields that $\boldsymbol{\nu}_{x,y} \in \mathcal{M}_+[a,b)$ is a subprobability measure. □

Definition 6.1. Let $\mu, \nu \in \mathcal{M}_\mathbb{C}[a,b)$. The measure

$$(\mu * \nu)(\cdot) = \int_{[a,b)} \int_{[a,b)} \boldsymbol{\nu}_{x,y}(\cdot)\, \mu(dx)\, \nu(dy)$$

is called the *$\mathcal{L}$-convolution* of the measures μ and ν. The *$\mathcal{L}$-translation* of a function $h \in \mathrm{B_b}[a,b)$ is defined as

$$(\mathcal{T}^y h)(x) = \int_{[a,b)} h(\xi)\, \boldsymbol{\nu}_{x,y}(d\xi) \equiv \int_{[a,b)} h(\xi)\, (\delta_x * \delta_y)(d\xi), \qquad x, y \in [a,b).$$

It follows from this definition, together with (6.21), that the $\mathcal{L}$-convolution is such that (for $\mu_1, \mu_2, \nu, \pi \in \mathcal{M}_\mathbb{C}[a,b)$ and $p_1, p_2 \in \mathbb{C}$):

(i) $\mu * \nu = \nu * \mu$ (Commutativity);

(ii) $(\mu * \nu) * \pi = \mu * (\nu * \pi)$ (Associativity);

(iii) $(p_1\mu_1 + p_2\mu_2) * \nu = p_1(\mu_1 * \nu) + p_2(\mu_2 * \nu)$ (Bilinearity);

(iv) $\|\mu * \nu\| \leq \|\mu\| \cdot \|\nu\|$ (Submultiplicativity);

(v) If $\mu, \nu \in \mathcal{M}_+[a,b)$, then $\mu * \nu \in \mathcal{M}_+[a,b)$ (Positivity).

Summarizing this, we have

Proposition 6.5. *The space $(\mathcal{M}_\mathbb{C}[a,b), *)$, equipped with the total variation norm, is a commutative Banach algebra over $\mathbb{C}$ whose identity element is the Dirac measure δ_a.*

Moreover, $\mathcal{M}_+[a,b)$ is an algebra cone (i.e., it is closed under $\mathcal{L}$-convolution, addition and multiplication by positive scalars, and it contains the identity element).

Remark 6.3. *Given a measure $\mu \in \mathcal{M}_\mathbb{C}[a,b)$, it is natural to define the $\mathcal{L}$-translation by μ as*

$$(\mathcal{T}^\mu h)(x) := \int_{[a,b)} (\mathcal{T}^y h)(x)\, \mu(dy) \equiv \int_{[a,b)} h(\xi)\, (\delta_x * \mu)(d\xi) \qquad (h \in \mathrm{B_b}[a,b))$$

(so that $\mathcal{T}^x \equiv \mathcal{T}^{\delta_x}$ for $a \leq x < b$). It is easy to see that $\|\mathcal{T}^\mu h\|_\infty \leq \|\mu\| \cdot \|h\|_\infty$ for all $h \in \mathrm{B_b}[a,b)$ and $\mu \in \mathcal{M}_\mathbb{C}[a,b)$. Observe also that for $h \in \mathfrak{X}$ we can write (by (6.21) and (6.40))

$$(\mathcal{T}^\mu h)(x) = \int_{[0,\infty)} (\mathcal{F}h)(\lambda)\, w_\lambda(x)\, \widehat{\mu}(\lambda)\, \boldsymbol{\rho}_\mathcal{L}(d\lambda) \qquad (h \in \mathfrak{X}) \tag{6.41}$$

or equivalently (cf. Proposition 6.1)

$$\big(\mathcal{F}(\mathcal{T}^\mu h)\big)(\lambda) = \widehat{\mu}(\lambda)(\mathcal{F}h)(\lambda) \quad \textit{for } \boldsymbol{\rho}_\mathcal{L}\textit{-almost every } \lambda \qquad (h \in \mathfrak{X}). \tag{6.42}$$

Due to Lemma 6.5, the integral in the right-hand side of (6.41) converges absolutely and uniformly for x on compact subsets of (a,b). It is easy to see that the identity (6.41) extends (by continuity) to all $h \in \mathcal{D}_\mathcal{L}^{(2)} \cap \mathrm{C}_0[a,b)$.

6.4.2 Sturm–Liouville Transform of Measures

An important tool for the subsequent analysis is the extension of the $\mathcal{L}$-transform (6.12) to finite complex measures, defined as follows:

Definition 6.2. Let $\mu \in \mathcal{M}_{\mathbb{C}}[a,b)$. The *$\mathcal{L}$-transform* of the measure μ is the function defined by the integral

$$\widehat{\mu}(\lambda) = \int_{[a,b)} w_\lambda(x)\, \mu(dx), \qquad \lambda \geq 0.$$

The next proposition contains some basic properties of the $\mathcal{L}$-transform of measures which, as one would expect, resemble those of the ordinary Fourier transform (or characteristic function) of finite measures. We recall that, by definition, the complex measures μ_n converge weakly to $\mu \in \mathcal{M}_{\mathbb{C}}[a,b)$ if $\lim_n \int_{[a,b)} g(\xi)\mu_n(d\xi) = \int_{[a,b)} g(\xi)\mu(d\xi)$ for all $g \in \mathrm{C_b}[a,b)$. We also recall that a family $\{\mu_j\} \subset \mathcal{M}_{\mathbb{C}}[a,b)$ is said to be uniformly bounded if $\sup_j \|\mu_j\| < \infty$, and $\{\mu_j\}$ is said to be tight if for each $\varepsilon > 0$ there exists a compact $K_\varepsilon \subset [a,b)$ such that $\sup_j |\mu_j|([a,b) \setminus K_\varepsilon) < \varepsilon$. (These definitions are taken from [7].) In the sequel, the notation $\mu_n \xrightarrow{w} \mu$ denotes weak convergence of measures.

Proposition 6.6. *The $\mathcal{L}$-transform $\widehat{\mu}$ of $\mu \in \mathcal{M}_{\mathbb{C}}[a,b)$ has the following properties:*

(a) $\widehat{\mu}$ is continuous on $[0,\infty)$. Moreover, if a family of measures $\{\mu_j\} \subset \mathcal{M}_{\mathbb{C}}[a,b)$ is tight and uniformly bounded, then $\{\widehat{\mu_j}\}$ is equicontinuous on $[0,\infty)$.

(b) Each measure $\mu \in \mathcal{M}_{\mathbb{C}}[a,b)$ is uniquely determined by $\widehat{\mu}$. In particular, each $f \in L_1(r)$ is uniquely determined by $\mathcal{F}f \equiv \widehat{\mu_f}$, where $\mu_f \in \mathcal{M}_{\mathbb{C}}[a,b)$ is defined by $\mu_f(dx) = f(x)r(x)dx$.

(c) If $\{\mu_n\}$ is a sequence of measures belonging to $\mathcal{M}_+[a,b)$, $\mu \in \mathcal{M}_+[a,b)$, and $\mu_n \xrightarrow{w} \mu$, then

$$\widehat{\mu_n} \xrightarrow[n\to\infty]{} \widehat{\mu} \qquad \textit{uniformly for } \lambda \textit{ in compact sets.}$$

(d) Suppose that $\lim_{x\uparrow b} w_\lambda(x) = 0$ for all $\lambda > 0$. If $\{\mu_n\}$ is a sequence of measures belonging to $\mathcal{M}_+[a,b)$ whose $\mathcal{L}$-transforms are such that

$$\widehat{\mu_n}(\lambda) \xrightarrow[n\to\infty]{} f(\lambda) \qquad \textit{pointwise in } \lambda \geq 0 \tag{6.43}$$

for some real-valued function f which is continuous at a neighborhood of zero, then $\mu_n \xrightarrow{w} \mu$ for some measure $\mu \in \mathcal{M}_+[a,b)$ such that $\widehat{\mu} \equiv f$.

Proof. **(a)** Let us prove the second statement, which implies the first. Set $C = \sup_j \|\mu_j\|$. Fix $\lambda_0 \geq 0$ and $\varepsilon > 0$. By the tightness assumption, we can choose $\beta \in (a,b)$ such that $|\mu_j|(\beta,b) < \varepsilon$ for all j. Since the family $\{w_{(\cdot)}(x)\}_{x\in(a,\beta]}$ is equicontinuous on $[0,\infty)$ (this follows easily from the power series representation of $w_{(\cdot)}(x)$, cf. proof of Lemma 6.1), we can choose $\delta > 0$ such that

$$|\lambda - \lambda_0| < \delta \quad \Longrightarrow \quad |w_\lambda(x) - w_{\lambda_0}(x)| < \varepsilon \; \text{ for all } a < x \leq \beta.$$

Consequently,

$$\begin{aligned} \bigl|\widehat{\mu_j}(\lambda) - \widehat{\mu_j}(\lambda_0)\bigr| &= \left|\int_{(a,b)} \bigl(w_\lambda(x) - w_{\lambda_0}(x)\bigr)\mu_j(dx)\right| \\ &\leq \int_{(\beta,b)} \bigl|w_\lambda(x) - w_{\lambda_0}(x)\bigr||\mu_j|(dx) + \int_{(a,\beta]} \bigl|w_\lambda(x) - w_{\lambda_0}(x)\bigr||\mu_j|(dx) \\ &\leq 2\varepsilon + C\varepsilon = (C+2)\varepsilon \end{aligned}$$

for all j, provided that $|\lambda - \lambda_0| < \delta$, which means that $\{\widehat{\mu_j}\}$ is equicontinuous at λ_0.

(b) Let $\mu \in \mathcal{M}_{\mathbb{C}}[a,b)$ be such that $\widehat{\mu}(\lambda) = 0$ for all $\lambda \geq 0$. We need to show that μ is the zero measure. For each $h \in \mathfrak{X}$, by (6.41) we have

$$(\mathcal{T}^{\mu}h)(x) = \int_{[0,\infty)} (\mathcal{F}h)(\lambda)\, w_\lambda(x)\, \widehat{\mu}(\lambda)\, \boldsymbol{\rho}_{\mathcal{L}}(d\lambda) \;=\; 0.$$

Since $h \in \mathfrak{X}$, Theorem 6.1 assures that $\lim_{x\downarrow a}(\mathcal{T}^y h)(x) = h(y)$ for $y \geq a$; therefore, by dominated convergence (which is applicable because $\|\mathcal{T}^y h\|_\infty \leq \|h\|_\infty < \infty$),

$$0 = \lim_{x\downarrow a}(\mathcal{T}^{\mu}h)(x) = \lim_{x\downarrow a}\int_{[a,b)} (\mathcal{T}^y h)(x)\, \mu(dy) = \int_{[a,b)} h(y)\, \mu(dy)$$

This shows that $\int_{[a,b)} h(y)\,\mu(dy) = 0$ for all $h \in \mathfrak{X}$ and, consequently, μ is the zero measure.

(c) Since $w_\lambda(\cdot)$ is continuous and bounded, the pointwise convergence $\widehat{\mu_n}(\lambda) \to \widehat{\mu}(\lambda)$ follows from the definition of weak convergence of measures. By Prokhorov's theorem [7, Theorem 8.6.2], $\{\mu_n\}$ is tight and uniformly bounded, thus (by part (i)) $\{\widehat{\mu_n}\}$ is equicontinuous on $[0,\infty)$. Invoking [31, Lemma 15.22], we conclude that the convergence $\widehat{\mu_n} \to \widehat{\mu}$ is uniform for λ in compact sets.

(d) We only need to show that the sequence $\{\mu_n\}$ is tight and uniformly bounded. Indeed, if $\{\mu_n\}$ is tight and uniformly bounded, then Prokhorov's theorem yields that for any subsequence $\{\mu_{n_k}\}$ there exists a further subsequence $\{\mu_{n_{k_j}}\}$ and a measure $\mu \in \mathcal{M}_+[a,b)$ such that $\mu_{n_{k_j}} \xrightarrow{w} \mu$. Then, due to part (iii) and to (6.43), we have $\widehat{\mu}(\lambda) = f(\lambda)$ for all $\lambda \geq 0$, which implies (by part (ii)) that all such subsequences have the same weak limit; consequently, the sequence μ_n itself converges weakly to μ.

The uniform boundedness of $\{\mu_n\}$ follows immediately from the fact that $\widehat{\mu_n}(0) = \mu_n[a,b)$ converges. To prove the tightness, take $\varepsilon > 0$. Since f is continuous at a neighborhood of zero, we have $\frac{1}{\delta}\int_0^{2\delta}(f(0) - f(\lambda))d\lambda \longrightarrow 0$ as $\delta \downarrow 0$; therefore, we can choose $\delta > 0$ such that

$$\left|\frac{1}{\delta}\int_0^{2\delta} (f(0) - f(\lambda))d\lambda\right| < \varepsilon.$$

Next we observe that, due to the assumption that $\lim_{x\uparrow b} w_\lambda(x) = 0$ for all $\lambda > 0$, we have $\int_0^{2\delta}(1 - w_\lambda(x))d\lambda \longrightarrow 2\delta$ as $x \uparrow b$, meaning that we can pick $\beta \in (a,b)$ such that

$$\int_0^{2\delta} (1 - w_\lambda(x))d\lambda \geq \delta \qquad \text{for all } \beta < x < b.$$

By our choice of β and Fubini's theorem,

$$\begin{aligned}
\mu_n[\beta,b) &= \frac{1}{\delta}\int_{[\beta,b)} \delta\, \mu_n(dx) \\
&\leq \frac{1}{\delta}\int_{[\beta,b)}\int_0^{2\delta} (1 - w_\lambda(x))d\lambda\, \mu_n(dx) \\
&\leq \frac{1}{\delta}\int_{[a,b)}\int_0^{2\delta} (1 - w_\lambda(x))d\lambda\, \mu_n(dx) \\
&= \frac{1}{\delta}\int_0^{2\delta} (\widehat{\mu_n}(0) - \widehat{\mu_n}(\lambda))d\lambda.
\end{aligned}$$

Hence, using the dominated convergence theorem,

$$\begin{aligned}\limsup_{n\to\infty}\mu_n[\beta,b) &\leq \frac{1}{\delta}\limsup_{n\to\infty}\int_0^{2\delta}\big(\widehat{\mu_n}(0)-\widehat{\mu_n}(\lambda)\big)d\lambda\\ &= \frac{1}{\delta}\int_0^{2\delta}\lim_{n\to\infty}\big(\widehat{\mu_n}(0)-\widehat{\mu_n}(\lambda)\big)d\lambda = \frac{1}{\delta}\int_0^{2\delta}\big(f(0)-f(\lambda)\big)d\lambda < \varepsilon\end{aligned}$$

due to the choice of δ. Since ε is arbitrary, we conclude that $\{\mu_n\}$ is tight, as desired. □

Remark 6.4. ***I.*** *Parts (c) and (d) of the proposition above show that (whenever $\lim_{x\uparrow b} w_\lambda(x) = 0$ for all $\lambda > 0$) the $\mathcal{L}$-transform possesses the following important property:* the $\mathcal{L}$-transform is a topological homeomorphism between $\mathcal{P}[a,b)$ with the weak topology and the set $\widehat{\mathcal{P}}$ of $\mathcal{L}$-transforms of probability measures with the topology of uniform convergence in compact sets.

II. *Recall that, by definition [2, Section 30], the measures μ_n converge vaguely to μ if $\lim_n \int_{[a,b)} g(\xi)\mu_n(d\xi) = \int_{[a,b)} g(\xi)\mu(d\xi)$ for all $g \in \mathrm{C}_0[a,b)$. Much like weak convergence, vague convergence of measures can be formulated via the $\mathcal{L}$-transform, provided that $\lim_{x\uparrow b} w_\lambda(x) = 0$ for all $\lambda > 0$. Indeed, using $\overset{v}{\longrightarrow}$ to denote vague convergence of measures, we have*

II.1 If $\{\mu_n\} \subset \mathcal{M}_+[a,b)$, $\mu \in \mathcal{M}_+[a,b)$, and $\mu_n \overset{v}{\longrightarrow} \mu$, then $\lim \widehat{\mu_n}(\lambda) = \widehat{\mu}(\lambda)$ pointwise for each $\lambda > 0$;

II.2 If $\{\mu_n\} \subset \mathcal{M}_+[a,b)$, $\{\mu_n\}$ is uniformly bounded and $\lim \widehat{\mu_n}(\lambda) = f(\lambda)$ pointwise in $\lambda > 0$ for some function $f \in \mathrm{B_b}(0,\infty)$, then $\mu_n \overset{v}{\longrightarrow} \mu$ for some measure $\mu \in \mathcal{M}_+[a,b)$ such that $\widehat{\mu} \equiv f$.

(The first part is trivial; the second follows from the reasoning in the first paragraph of the proof of (d) in the proposition above, together with the fact that any uniformly bounded sequence of positive measures contains a vaguely convergent subsequence [2, p. 213].)

III. *Concerning the additional assumption in the above remarks, one can state:* a necessary and sufficient condition for the condition $\lim_{x\uparrow b} w_\lambda(x) = 0$ $(\lambda > 0)$ to hold is that $\lim_{x\uparrow b} p(x)r(x) = \infty$. *This fact can be proved using the transformation into the standard form (Remark 6.1) and known results on the asymptotic behavior of solutions of the Sturm–Liouville equation $-u'' - \frac{A'}{A}u' = \lambda u$ (see [23, proof of Lemma 3.7]).*

6.5 The Product Formula

We saw in the previous section that the hyperbolic maximum principle allows us to introduce a convolution measure algebra associated with the Sturm–Liouville operator. The next aims are to develop harmonic analysis on L_p spaces and to study notions such as the continuity of the convolution or the divisibility of measures. However, this requires a fundamental tool, namely the trivialization property $\widehat{\delta_x * \delta_y} = \widehat{\delta_x}\cdot\widehat{\delta_y}$ for the $\mathcal{L}$-transform or, which is the same, the product formula for its kernel.

Theorem 6.4 (Product formula for w_λ). *The product $w_\lambda(x)\,w_\lambda(y)$ admits the integral representation*

$$w_\lambda(x)\,w_\lambda(y) = \int_{[a,b)} w_\lambda(\xi)\,(\delta_x * \delta_y)(d\xi), \qquad x,y \in [a,b),\ \lambda \in \mathbb{C}. \tag{6.44}$$

Here we present the proof only in the special (nondegenerate) case $\gamma(a) > -\infty$, and under the assumption that the function A (defined in (6.10)) belongs to $\mathrm{C}^3(a,b)$. The proof of the general case is longer and relies on a different regularization argument; the details are given in [48].

Proof of Theorem 6.4 for the case $\gamma(a) > -\infty$. Assume first that $\ell = -\frac{1}{A}\frac{d}{dx}(A\frac{d}{dx})$, $0 < x < \infty$, and that Assumption MP holds with $a = \gamma(a) = 0$. Fix $\lambda \in \mathbb{C}$, and let $\{\mathbf{w}_\lambda^{\langle n\rangle}\}_{n\in\mathbb{N}} \subset \mathrm{C}^4[a,b)$ be a sequence of functions such that

$$\mathbf{w}_\lambda^{\langle n\rangle}(x) = w_\lambda(x) \quad \text{for } x \in [0,n], \qquad\qquad \mathbf{w}_\lambda^{\langle n\rangle}(x) = 0 \quad \text{for } x \geq n+1.$$

Let $f^{\langle n\rangle}(x,y) \equiv f_{\mathbf{w}_\lambda^{\langle n\rangle}}(x,y)$ be given by (6.21), with initial condition $h(x) = \mathbf{w}_\lambda^{\langle n\rangle}(x)$. By Theorem 6.1, f is a solution of (6.20). (Note that the assumption $A \in \mathrm{C}^3(a,b)$ assures that $\mathbf{w}_\lambda^{\langle n\rangle}$, $\ell(\mathbf{w}_\lambda^{\langle n\rangle}) \in \mathcal{D}_{\mathcal{L}}^{(2)}$.) Since the family of characteristics for the hyperbolic equation $(\ell_x u)(x,y) = (\ell_y u)(x,y)$ is $x \pm y =$ const., the solution $f^{\langle n\rangle}(x,y)$ depends only on the values of the initial condition on the interval $[|x-y|, x+y]$. Observing that the function $\mathbf{w}_\lambda^{\langle n\rangle}(x)\,\mathbf{w}_\lambda^{\langle n\rangle}(y)$ is a solution of the hyperbolic equation $(\ell_x u)(x,y) = (\ell_y u)(x,y)$ on the square $(x,y) \in [0,n]^2$, we deduce that

$$f^{\langle n\rangle}(x,y) = \mathbf{w}_\lambda^{\langle n\rangle}(x)\,\mathbf{w}_\lambda^{\langle n\rangle}(y) = w_\lambda(x)\,w_\lambda(y), \qquad x,y \in [0,\tfrac{n}{2}].$$

Applying Proposition 6.4 with $h = T_t(\mathbf{w}_\lambda^{\langle n\rangle})$ and then taking the limit as $t \downarrow 0$, we find that

$$w_\lambda(x)\,w_\lambda(y) = \int_{[0,\infty)} w_\lambda(\xi)\,\boldsymbol{\nu}_{x,y}(d\xi), \qquad x,y \in [0,\tfrac{n}{2}]$$

(note that $\operatorname{supp}(\boldsymbol{\nu}_{x,y}) = [|x-y|, x+y]$ because of the domain of dependence of the hyperbolic equation). Since n is arbitrary, the identity holds for all $x,y \in [0,\infty)$, proving that the theorem holds for operators of the form $\ell = -\frac{1}{A}\frac{d}{dx}(A\frac{d}{dx})$, $0 < x < \infty$.

Now, in the general case of an operator ℓ of the form (6.4), note that $\gamma(a) > -\infty$ means that $\sqrt{\frac{r(y)}{p(y)}}$ is integrable near a, so that we may assume that $\gamma(a) = 0$ (otherwise, replace the interior point c by the endpoint a in the definition of the function γ). Applying the first part of the proof to the transformed operator $\widetilde{\ell} = -\frac{1}{A}\frac{d}{d\xi}(A\frac{d}{d\xi})$ defined via (6.10), we find that $\widetilde{w}_\lambda(x)\,\widetilde{w}_\lambda(y) = \int_{[0,\infty)} \widetilde{w}_\lambda(\xi)\,(\delta_x \,\widetilde{*}\, \delta_y)(d\xi)$ for $x,y \in [0,\infty)$, where $\widetilde{w}_\lambda(\xi) := w_\lambda(\gamma^{-1}(\xi))$ and $\widetilde{*}$ is the convolution associated with $\widetilde{\ell}$. We can rewrite this as

$$w_\lambda(x)\,w_\lambda(y) = \int_{[a,b)} w_\lambda(\xi)\big[\gamma^{-1}(\delta_{\gamma(x)} \,\widetilde{*}\, \delta_{\gamma(y)})\big](d\xi), \qquad x,y \in [a,b),\ \lambda \in \mathbb{C}$$

where the measure in the right-hand side is the pushforward of the measure $\delta_{\gamma(x)} \,\widetilde{*}\, \delta_{\gamma(y)}$ under the map $\xi \mapsto \gamma^{-1}(\xi)$. But one can easily check that the convolutions $*$ and $\widetilde{*}$ are connected by $\delta_x * \delta_y = \gamma^{-1}(\delta_{\gamma(x)} \,\widetilde{*}\, \delta_{\gamma(y)})$ (this is a simple consequence of the definition of the convolution and the relation between the operators ℓ and $\widetilde{\ell}$), so we are done. □

Corollary 6.3. *Let* $\mu, \nu, \pi \in \mathcal{M}_{\mathbb{C}}[a,b)$.

(a) We have $\pi = \mu * \nu$ *if and only if*

$$\widehat{\pi}(\lambda) = \widehat{\mu}(\lambda)\,\widehat{\nu}(\lambda) \qquad \textit{for all } \lambda \geq 0.$$

(b) Probability measures are closed under $\mathcal{L}$*-convolution: if* $\mu, \nu \in \mathcal{P}[a,b)$*, then* $\mu * \nu \in \mathcal{P}[a,b)$.

If $\lim_{x\uparrow b} p(x)r(x) = \infty$ *holds (cf. Remark 6.2.III), then the following properties also hold:*

(c) The mapping $(\mu,\nu) \mapsto \mu * \nu$ *is continuous in the weak topology.*

(d) If $h \in \mathrm{C_b}[a,b)$*, then* $\mathcal{T}^\mu h \in \mathrm{C_b}[a,b)$ *for all* $\mu \in \mathcal{M}_\mathbb{C}[a,b)$*.*

(e) If $h \in \mathrm{C_0}[a,b)$*, then* $\mathcal{T}^\mu h \in \mathrm{C_0}[a,b)$ *for all* $\mu \in \mathcal{M}_\mathbb{C}[a,b)$*.*

Proof. ***(a)*** Using (6.44), we compute

$$\begin{aligned}\widehat{\mu * \nu}(\lambda) &= \int_{[a,b)} w_\lambda(x)\,(\mu*\nu)(dx)\\ &= \int_{[a,b)}\int_{[a,b)}\int_{[a,b)} w_\lambda(\xi)\,(\delta_x*\delta_y)(d\xi)\,\mu(dx)\nu(dy)\\ &= \int_{[a,b)}\int_{[a,b)} w_\lambda(x)\,w_\lambda(y)\,\mu(dx)\nu(dy) \;=\; \widehat{\mu}(\lambda)\,\widehat{\nu}(\lambda), \qquad \lambda \geq 0.\end{aligned}$$

This proves the "only if" part, and the converse follows from the uniqueness property in Proposition 6.6(b).

(b) Due to Proposition 6.5, it only remains to prove that $(\mu*\nu)[a,b) = 1$ $(\mu,\nu \in \mathcal{P}[a,b))$. But this follows at once from part (a):

$$(\mu*\nu)[a,b) = \widehat{\mu*\nu}(0) = \widehat{\mu}(0)\cdot\widehat{\nu}(0) = \mu[a,b)\cdot\nu[a,b) = 1.$$

(c) Since $\widehat{\delta_x * \delta_y}(\lambda) = w_\lambda(x)w_\lambda(y)$, Proposition 6.6(d) yields that $(x,y) \mapsto \delta_x * \delta_y$ is continuous in the weak topology. Therefore, for $h \in \mathrm{C_b}[a,b)$ and $\mu_n, \nu_n \in \mathcal{M}_\mathbb{C}[a,b)$ with $\mu_n \xrightarrow{w} \mu$ and $\nu_n \xrightarrow{w} \nu$, we have

$$\begin{aligned}\lim_n \int_{[a,b)} h(\xi)(\mu_n*\nu_n)(d\xi) &= \lim_n \int_{[a,b)}\int_{[a,b)}\biggl(\int_{[a,b)} h(\xi)\,(\delta_x*\delta_y)(d\xi)\biggr)\mu_n(dx)\nu_n(dy)\\ &= \int_{[a,b)}\int_{[a,b)}\biggl(\int_{[a,b)} h(\xi)\,(\delta_x*\delta_y)(d\xi)\biggr)\mu(dx)\nu(dy)\\ &= \int_{[a,b)} h(\xi)(\mu*\nu)(d\xi)\end{aligned}$$

due to the continuity of the function in parenthesis.

(d) Since $(\mathcal{T}^\mu h)(x) = \int_{[a,b)} h(\xi)\,(\delta_x * \mu)(d\xi)$, this follows immediately from part (c)

(e) It remains to show that $(\mathcal{T}^\mu h)(x) \to 0$ as $x \uparrow b$. Since $w_\lambda(x)\widehat{\mu}(\lambda) \to 0$ as $x \uparrow b$ $(\lambda > 0)$, it follows from Remark 6.2.II that $\delta_x * \mu \xrightarrow{v} \mathbf{0}$ as $x \uparrow b$, where $\mathbf{0}$ denotes the zero measure; this means that for each $h \in \mathrm{C_0}[a,b)$ we have

$$(\mathcal{T}^\mu h)(x) = \int_{[a,b)} h(\xi)(\delta_x*\mu)(d\xi) \longrightarrow \int_{[a,b)} h(\xi)\,\mathbf{0}(d\xi) = 0 \qquad \text{as } x \uparrow b$$

showing that $\mathcal{T}^\mu h \in \mathrm{C_0}[a,b)$. □

6.6 Harmonic Analysis on L_p Spaces

For the remainder of this work, the coefficients of ℓ will be assumed to satisfy $\lim_{x\uparrow b} p(x)r(x) = \infty$ (cf. Remark 6.2.III), and Assumption MP continues to be in place.

In this section, we turn our attention to the basic mapping properties of the $\mathcal{L}$-translation and convolution on the Lebesgue spaces $L_p(r)$ $(1 \leq p \leq \infty)$. The first result, whose proof depends on the continuity of the mapping $(\mu,\nu) \mapsto \mu * \nu$, ensures that the $\mathcal{L}$-translation defines a linear contraction on $L_p(r)$:

Proposition 6.7. *Let $1 \leq p \leq \infty$ and $\mu \in \mathcal{M}_+[a,b)$. The $\mathcal{L}$-translation $(\mathcal{T}^\mu h)(x) = \int_{[a,b)} h(\xi)\,(\delta_x * \mu)(d\xi)$ is, for each $h \in L_p(r)$, a Borel measurable function of x, and we have*

$$\|\mathcal{T}^\mu h\|_p \leq \|\mu\| \cdot \|h\|_p \qquad \textit{for all } h \in L_p(r) \tag{6.45}$$

(consequently, $\mathcal{T}^\mu\bigl(L_p(r)\bigr) \subset L_p(r)\bigr)$.

Proof. It suffices to prove the result for nonnegative $h \in L_p(r)$, $1 \leq p \leq \infty$.

The map $\nu \mapsto \mu * \nu$ is weakly continuous (Corollary 6.3(c)) and takes $\mathcal{M}_+[a,b)$ into itself. According to [27, Section 2.3], this implies that, for each Borel measurable $h \geq 0$, the function $x \mapsto (\mathcal{T}^\mu h)(x)$ is Borel measurable. It follows that $\int_{[a,b)} g(x)(\mu * r)(dx) := \int_a^b (\mathcal{T}^\mu g)(x) r(x) dx$ $(g \in \mathrm{C_c}[a,b))$ defines a positive Borel measure. For $a \leq c_1 < c_2 < b$, let $\mathbb{1}_{[c_1,c_2)}$ be the indicator function of $[c_1,c_2)$, let $h_n \in \mathcal{D}_{\mathcal{L}}^{(2)} \cap \mathrm{C_c}[a,b)$ be a sequence of nonnegative functions such that $h_n \to \mathbb{1}_{[c_1,c_2)}$ pointwise, and write $\mathfrak{C} = \{g \in \mathrm{C}_\mathrm{c}^\infty(a,b) \mid 0 \leq g \leq 1\}$. We compute

$$\begin{aligned}
(\mu * r)[c_1,c_2) &= \lim_n \int_{[a,b)} h_n(x)(\mu * r)(dx) \\
&= \lim_n \sup_{g \in \mathfrak{C}} \int_a^b (\mathcal{T}^\mu h_n)(x)\, g(x)\, r(x) dx \\
&= \lim_n \sup_{g \in \mathfrak{C}} \int_{[0,\infty)} (\mathcal{F}h_n)(\lambda)\, (\mathcal{F}g)(\lambda)\, \widehat{\mu}(\lambda)\, \boldsymbol{\rho}_{\mathcal{L}}(d\lambda) \\
&= \lim_n \sup_{g \in \mathfrak{C}} \int_a^b h_n(x)\, (\mathcal{T}^\mu g)(x)\, r(x) dx \\
&\leq \|\mu\| \cdot \lim_n \int_{[a,b)} h_n(x)\, r(x) dx \;=\; \|\mu\| \cdot \int_{[c_1,c_2)} r(x) dx
\end{aligned}$$

where the third and fourth equalities follow from (6.41) and the isometric property of $\mathcal{F}$ (Proposition 6.1), and the inequality holds because $\|\mathcal{T}^\mu g\|_\infty \leq \|\mu\| {\cdot} \|g\|_\infty \leq \|\mu\|$. Therefore, $\|\mathcal{T}^\mu h\|_1 = \|h\|_{L_1([a,b),\mu * r)} \leq \|\mu\| \cdot \|h\|_1$ for each Borel measurable $h \geq 0$. Since $\delta_x * \mu \in \mathcal{M}_+[a,b)$, Hölder's inequality yields that $\|\mathcal{T}^\mu h\|_p \leq \|\mu\|^{1/q} \cdot \|\mathcal{T}^\mu |h|^p\|_1^{1/p} \leq \|\mu\| \cdot \|h\|_p$ for $1 < p < \infty$.

Finally, if $h \in L_\infty(r)$, $h \geq 0$ then $h = h_\mathbf{b} + h_\mathbf{0}$, where $0 \leq h_\mathbf{b} \leq \|h\|_\infty$ and $h_\mathbf{0} = 0$ Lebesgue-almost everywhere. Since $\|\mathcal{T}^\mu h_\mathbf{0}\|_1 \leq \|\mu\| {\cdot} \|h_\mathbf{0}\|_1 = 0$, we have $\mathcal{T}^y h_\mathbf{0} = 0$ Lebesgue-almost everywhere, and therefore $\|\mathcal{T}^y h\|_\infty = \|\mathcal{T}^y h_\mathbf{b}\|_\infty \leq \|\mu\| \cdot \|h\|_\infty$. □

It is natural to define the $\mathcal{L}$-convolution of functions so that the fundamental identity $\mathcal{F}(h * g) = (\mathcal{F}h) {\cdot} (\mathcal{F}g)$ holds (where $\mathcal{F}$ denotes the $\mathcal{L}$-transform (6.12)):

Definition 6.3. Let $h, g : [a,b) \longrightarrow \mathbb{C}$. If the integral

$$(h * g)(x) = \int_a^b (\mathcal{T}^y h)(x)\, g(y)\, r(y) dy = \int_a^b \int_{[a,b)} h(\xi)\, (\delta_x * \delta_y)(d\xi)\, g(y)\, r(y) dy$$

exists for almost every $x \in [a,b)$, then we call it the $\mathcal{L}$-*convolution* of the functions h and g.

Proposition 6.8. *If $h \in \mathcal{D}_{\mathcal{L}}^{(2)} \cap \mathrm{C_c}[a,b)$ and $g \in L_1(r)$, then $\bigl(\mathcal{F}(h * g)\bigr)(\lambda) = (\mathcal{F}h)(\lambda)(\mathcal{F}g)(\lambda)$ for all $\lambda \geq 0$.*

Proof. For $h \in \mathcal{D}_{\mathcal{L}}^{(2)} \cap \mathrm{C_c}[a,b)$ and $g \in L_1(r)$ we have

$$\begin{aligned}\big(\mathcal{F}(h * g)\big)(\lambda) &= \int_a^b \int_a^b (\mathcal{T}^\xi h)(x) g(\xi)\, r(\xi) d\xi\, w_\lambda(x) r(x) dx \\ &= \int_a^b \big(\mathcal{F}(\mathcal{T}^\xi h)\big)(\lambda)\, g(\xi) r(\xi) d\xi \\ &= (\mathcal{F}h)(\lambda) \int_a^b g(\xi) w_\lambda(\xi) r(\xi) d\xi \;=\; (\mathcal{F}h)(\lambda)(\mathcal{F}g)(\lambda)\end{aligned}$$

where we have used Fubini's theorem and the identity (6.42). □

Proposition 6.9 (Young convolution inequality). *Let $p_1, p_2 \in [1,\infty]$ such that $\frac{1}{p_1} + \frac{1}{p_2} \geq 1$. For $h \in L_{p_1}(r)$ and $g \in L_{p_2}(r)$, the $\mathcal{L}$-convolution $h * g$ is well- defined and, for $s \in [1,\infty]$ defined by $\frac{1}{s} = \frac{1}{p_1} + \frac{1}{p_2} - 1$, it satisfies*

$$\|h * g\|_s \leq \|h\|_{p_1} \|g\|_{p_2}$$

(in particular, $h * g \in L_s(r)$). Consequently, the $\mathcal{L}$-convolution is a continuous bilinear operator from $L_{p_1}(r) \times L_{p_2}(r)$ into $L_s(r)$.

The proof is given for completeness; it is analogous to that of the Young inequality for the ordinary convolution.

Proof. Define $\frac{1}{t_1} = \frac{1}{p_1} - \frac{1}{s}$ and $\frac{1}{t_2} = \frac{1}{p_2} - \frac{1}{s}$. Observe that

$$|(\mathcal{T}^x h)(y)|\, |g(y)| \leq |(\mathcal{T}^x h)(y)|^{p_1/t_1}\, |g(y)|^{p_2/t_2} \big[|(\mathcal{T}^x h)(y)|^{p_1}\, |g(y)|^{p_2}\big]^{1/s}.$$

Since $\frac{1}{s} + \frac{1}{t_1} + \frac{1}{t_2} = 1$, we have by Hölder's inequality and (6.45)

$$\begin{aligned}\int_a^b |(\mathcal{T}^x h)(y)|\, |g(y)| r(y) dy &\leq \bigg(\int_a^b |(\mathcal{T}^x h)(y)|^{p_1} r(y) dy\bigg)^{1/t_1} \\ &\quad \times \bigg(\int_a^b |g(y)|^{p_2} r(y) dy\bigg)^{1/t_2} \bigg(\int_a^b |(\mathcal{T}^x h)(y)|^{p_1}\, |g(y)|^{p_2} r(y) dy\bigg)^{1/s} \\ &= \|h\|_{p_1}^{p_1/t_1} \|g\|_{p_2}^{p_2/t_2} \bigg(\int_a^b |(\mathcal{T}^x h)(y)|^{p_1}\, |g(y)|^{p_2} r(y) dy\bigg)^{1/s}.\end{aligned}$$

Using again (6.45) we conclude that

$$\|h * g\|_s \leq \|h\|_{p_1}^{p_1/t_1}\, \|g\|_{p_2}^{p_2/t_2}\, \|h\|_{p_1}^{p_1/s}\, \|g\|_{p_2}^{p_2/s} = \|h\|_{p_1} \|g\|_{p_2}.$$ □

A consequence of the Young convolution inequality is that the fundamental identity $\big(\mathcal{F}(h * g)\big)(\lambda) = (\mathcal{F}h)(\lambda)(\mathcal{F}g)(\lambda)$ (Proposition 6.8) extends, by continuity, to $h \in L_1(r) \cup L_2(r)$ and $g \in L_1(r)$. Another consequence is the Banach algebra property of the space $L_1(r)$:

Corollary 6.4. *The Banach space $L_1(r)$, equipped with the convolution multiplication $h \cdot g \equiv h * g$, is a commutative Banach algebra without identity element.*

Proof. The Young convolution inequality shows that the $\mathcal{L}$-convolution defines a binary operation on $L_1(r)$ for which the norm is submultiplicative. The commutativity and associativity of the $\mathcal{L}$-convolution are a consequence of the identity $\mathcal{F}(h * g) = (\mathcal{F}h) \cdot (\mathcal{F}g)$.

Suppose now that there exists $\mathrm{e} \in L_1(r)$ such that $h * \mathrm{e} = h$ for all $h \in L_1(r)$. Then

$$(\mathcal{F}h)(\lambda)(\mathcal{F}\mathrm{e})(\lambda) = \big(\mathcal{F}(h * \mathrm{e})\big)(\lambda) = (\mathcal{F}h)(\lambda) \qquad \text{for all } h \in L_1(r) \text{ and } \lambda \geq 0.$$

Clearly, this implies that $(\mathcal{F}\mathrm{e})(\lambda) = 1$ for all $\lambda \geq 0$. But we know that $\widehat{\delta_a} \equiv 1$, so it follows from Proposition 6.6(b) that $\mathrm{e}(x)r(x)dx = \delta_a(dx)$, which is absurd. This shows that the Banach algebra has no identity element. □

6.7 Applications to Probability Theory

6.7.1 Infinite Divisibility of Measures and the Lévy–Khintchine Representation

The set $\mathcal{P}_{\mathrm{id}}$ of *$\mathcal{L}$-infinitely divisible measures* (or *$\mathcal{L}$-infinitely divisible distributions*) is defined in the obvious way:

$$\mathcal{P}_{\mathrm{id}} = \big\{\mu \in \mathcal{P}[a,b) \;\big|\; \text{for all } n \in \mathbb{N} \text{ there exists } \nu_n \in \mathcal{P}[a,b) \text{ such that } \mu = \nu_n^{*n}\big\}$$

where ν_n^{*n} denotes the n-fold $\mathcal{L}$-convolution of ν_n with itself.

It is a simple exercise to show that the $\mathcal{L}$-transform of $\mu \in \mathcal{P}_{\mathrm{id}}$ is of the form

$$\widehat{\mu}(\lambda) = e^{-\psi_\mu(\lambda)}$$

where ψ_μ is continuous, nonnegative, and $\psi_\mu(0) = 0$. The function ψ_μ is called the *$\mathcal{L}$-exponent* of $\mu \in \mathcal{P}_{\mathrm{id}}$.

As we will see, the exponents of $\mathcal{L}$-infinitely divisible measures admit a representation which is analogous to the well-known Lévy–Khintchine formula for infinitely divisible measures with respect to the ordinary Fourier transform. In the present context, the relevant notions of Poisson and Gaussian measures are defined as follows:

Definition 6.4. Let $\mu \in \mathcal{M}_+[a,b)$. The measure $\mathbf{e}(\mu) \in \mathcal{P}[a,b)$ defined by

$$\mathbf{e}(\mu) = e^{-\|\mu\|} \sum_{k=0}^{\infty} \frac{\mu^{*k}}{k!}$$

(the infinite sum converging in the weak topology) is said to be the *$\mathcal{L}$-compound Poisson measure* associated with μ.

The $\mathcal{L}$-transform of $\mathbf{e}(\mu)$ can be easily deduced using Corollary 6.3(a):

$$\widehat{\mathbf{e}(\mu)}(\lambda) = e^{-\|\mu\|} \sum_{k=0}^{\infty} \frac{\widehat{\mu^{*k}}(\lambda)}{k!} = e^{-\|\mu\|} \sum_{k=0}^{\infty} \frac{\big(\widehat{\mu}(\lambda)\big)^k}{k!} = \exp\big(\widehat{\mu}(\lambda) - \|\mu\|\big).$$

Since $\mathbf{e}(\mu_1 + \mu_2) = \mathbf{e}(\mu_1) * \mathbf{e}(\mu_2)$ $(\mu_1, \mu_2 \in \mathcal{M}_+[a,b))$, every $*$-compound Poisson measure belongs to $\mathcal{P}_{\mathrm{id}}$.

Definition 6.5. A measure $\mu \in \mathcal{P}_{\mathrm{id}}$ is called an *$\mathcal{L}$-Gaussian measure* if

$$\mu = \mathbf{e}(\nu) * \vartheta \quad \big(a > 0,\, \nu \in \mathcal{M}_+[a,b),\, \vartheta \in \mathcal{P}_{\mathrm{id}}\big) \qquad \Longrightarrow \qquad \mathbf{e}(\nu) = \delta_a.$$

We observe that the above definition is analogous with the classical notion of a Gaussian measure on $\mathbb{R}^d$ by the following result: an infinitely divisible probability measure on $\mathbb{R}^d$ is Gaussian if and only if it has no nontrivial divisors of the form $\mathfrak{e}(\nu)$, where ν is a finite positive measure on $\mathbb{R}^d$ and $\mathfrak{e}(\nu)$ denotes the (ordinary) compound Poisson measure associated with ν. (This fact follows from classical results in probability theory, see e.g., [31, Theorem 16.17] and [36, Section III.1].)

We are now ready to state the analogue of the Lévy–Khintchine representation for infinite divisibility with respect to the $\mathcal{L}$-convolution.

Theorem 6.5 (Lévy–Khintchine type formula). *The $\mathcal{L}$-exponent of a measure $\mu \in \mathcal{P}_{\text{id}}$ can be represented in the form*

$$\psi_\mu(\lambda) = \psi_\alpha(\lambda) + \int_{(a,b)} (1 - w_\lambda(x))\nu(dx) \tag{6.46}$$

where ν is a σ-finite measure on (a,b) which is finite on the complement of any neighborhood of a such that

$$\int_{(a,b)} (1 - w_\lambda(x))\nu(dx) < \infty$$

and α is an $\mathcal{L}$-Gaussian measure with $\mathcal{L}$-exponent $\psi_\alpha(\lambda)$. Conversely, each function of the form (6.46) is an $\mathcal{L}$-exponent of some $\mu \in \mathcal{P}_{\text{id}}$.

Proof. We only give a sketch of the proof, and refer to [56] for details.

Let $\mu \in \mathcal{P}_{\text{id}}$, let $b > a_1 > a_2 > \ldots$ with $\lim a_n = a$, and let $I_n = [a, a_n)$, $J_n = [a_n, b)$. Consider the set $\mathcal{Q}$ of all divisors of μ of the form $\mathbf{e}(\pi)$ such that $\pi(I_1) = 0$. One can prove that the set $\mathrm{D}(\mathfrak{P})$ of all divisors (with respect to the $\mathcal{L}$-convolution) of measures $\nu \in \mathfrak{P}$ is relatively compact whenever $\mathfrak{P} \subset \mathcal{P}[a,b)$ is relatively compact (see [57, Corollary 1]); using this fact, it can be shown that $\sup_{\mu=\mathbf{e}(\pi)\in\mathcal{Q}} \bigl[\int_{[a,b)} (1 - w_\lambda(x))\pi(dx)\bigr] < \infty$ and, consequently, there exists a divisor $\mu_1 = \mathbf{e}(\pi_1) \in \mathcal{Q}$ such that $\pi_1(J_1)$ is maximal among all elements of $\mathcal{Q}$. Write $\mu = \mu_1 * \alpha_1$ $(\alpha_1 \in \mathcal{P}_{\text{id}})$. Applying the same reasoning to α_1 with I_1 replaced by I_2, we get $\alpha_1 = \mu_2 * \alpha_2 = \mathbf{e}(\pi_2) * \alpha_2$. If we perform this successively, we get

$$\mu = \alpha_n * \beta_n, \qquad \text{where } \beta_n = \mu_1 * \mu_2 * \ldots \mu_n, \qquad \mu_k = \mathbf{e}(\pi_k)$$

with $\pi_k(I_k) = 0$ and $\pi_k(J_k)$ having the specified maximality property. The sequences $\{\alpha_n\}$ and $\{\beta_n\}$ are relatively compact; letting α and β be limit points, we have

$$\mu = \alpha * \beta \qquad (\alpha, \beta \in \mathcal{P}_{\text{id}}).$$

Suppose, by contradiction, that α is not $\mathcal{L}$-Gaussian, and let $\mathbf{e}(\eta)$, with $\eta \neq \delta_a$, be a divisor of α. Clearly $\eta(J_k) > 0$ for some k; given that each α_n divides α_{n-1}, we have $\alpha_k = \mathbf{e}(\eta) * \nu$ $(\nu \in \mathcal{P}_{\text{id}})$. If we let $\widetilde{\eta}$ be the restriction of η to the interval J_k, then

$$\alpha_{k-1} = \mathbf{e}(\pi_k + \widetilde{\eta}) * \mathbf{e}(\eta - \widetilde{\eta}) * \nu$$

which is absurd (because $(\pi_k + \widetilde{\eta})(J_k) > \pi_k(J_k)$, contradicting the maximality property which defines π_k). To determine the $\mathcal{L}$-exponent of β, note that $\beta_n = \mathbf{e}(\Pi_n)$ is the $\mathcal{L}$-compound Poisson measure associated with $\Pi_n := \sum_{k=1}^n \pi_k$, thus $\psi_{\beta_n}(\lambda) = \int_{(a,b)} (1 - w_\lambda(x))\Pi_n(dx)$. Since $\{\Pi_n\}$ is an increasing sequence of measures and each $\mathbf{e}(\Pi_n)$ dividing μ, there exists a σ-finite measure ν such that

$$\psi_\beta(\lambda) = \lim_n \int_{(a,b)} (1 - w_\lambda(x))\Pi_n(dx) = \lim_n \int_{(a,b)} (1 - w_\lambda(x))\nu(dx) < \infty$$

($\mu \in \mathcal{P}_{\text{id}}$ ensures the finiteness of the integral); from the relative compactness of $\mathrm{D}(\{\mu\})$ it is possible to conclude that $\nu(J_k) < \infty$ for all k.

For the converse, let ν_n be the restriction of ν to the interval J_n defined as above. It is verified without difficulty that the right-hand side of (6.46) is continuous at zero, hence by Proposition 6.6(d) $\alpha * \mathbf{e}(\nu_n) \overset{w}{\longrightarrow} \mu \in \mathcal{P}[a,b)$, and $\mu \in \mathcal{P}_{\text{id}}$ because $\mathcal{P}_{\text{id}}$ is closed under weak convergence of measures. □

6.7.2 Convolution Semigroups and Their Contraction Properties

Definition 6.6. A family $\{\mu_t\}_{t\geq 0} \subset \mathcal{P}[a,b)$ is called an *$\mathcal{L}$-convolution semigroup* if it satisfies the conditions

- $\mu_s * \mu_t = \mu_{s+t}$ for all $s,t \geq 0$;
- $\mu_0 = \delta_a$;
- $\mu_t \xrightarrow{w} \delta_a$ as $t \downarrow 0$.

A direct consequence of this definition is that

$$\{\mu_t\} \longmapsto \mu_1 \in \mathcal{P}_{\mathrm{id}} \tag{6.47}$$

defines a one-to-one correspondence holds between the set of $\mathcal{L}$-convolution semigroups and the set of $\mathcal{L}$-infinitely divisible measures. Indeed, if $\{\mu_t\}$ is an $\mathcal{L}$-convolution semigroup, then it is clear that each μ_t is $\mathcal{L}$-infinitely divisible; and if $\mu \in \mathcal{P}_{\mathrm{id}}$ has exponent $\psi_\mu(\lambda)$, then $\widehat{\mu_t}(\lambda) = \exp(-t\,\psi_\mu(\lambda))$ defines the unique $\mathcal{L}$-convolution semigroup such that $\mu_1 = \mu$ (the proof of this is analogous to that for the classical convolution, cf. [1, Theorem 29.6]).

Proposition 6.10. *Let $\{\mu_t\}$ be an $\mathcal{L}$-convolution semigroup. Then*

$$(T_t h)(x) := (\mathcal{T}^{\mu_t} h)(x) = \int_{[a,b)} h(\xi)(\delta_x * \mu_t)(d\xi)$$

defines a strongly continuous Markovian contraction semigroup $\{T_t\}_{t\geq 0}$ on $\mathrm{C}_0[a,b)$ and on the spaces $L_p(r)$ $(1 \leq p < \infty)$, i.e., the following properties hold:

(i) $T_t T_s = T_{t+s}$ *for all* $t,s \geq 0$;

(ii) $T_t\big(\mathrm{C}_0[0,\infty)\big) \subset \mathrm{C}_0[0,\infty)$ *for all* $t \geq 0$;

(ii') $T_t\big(L_p(r)\big) \subset L_p(r)$ *for all* $t \geq 0$ $\quad (1 \leq p < \infty)$;

(iii) $T_t \mathbb{1} = \mathbb{1}$ *for all* $t \geq 0$, *and if* $f \in \mathrm{C_b}[0,\infty)$ *satisfies* $0 \leq h \leq 1$, *then* $0 \leq T_t h \leq 1$;

(iv) $\lim_{t\downarrow 0} \|T_t h - h\|_\infty = 0$ *for each* $h \in \mathrm{C}_0[0,\infty)$;

(iv') $\lim_{t\downarrow 0} \|T_t h - h\|_p = 0$ *for each* $h \in L_p(r)$ $\quad (1 \leq p < \infty)$.

Moreover, $\{T_t\}$ is translation-invariant: $T_t \mathcal{T}^\nu f = \mathcal{T}^\nu T_t f$ for all $t \geq 0$ and $\nu \in \mathcal{M}_{\mathbb{C}}[a,b)$.

Proof. Parts (ii), (ii'), and (iii) follow at once from Corollary 6.3 and Proposition 6.7. Concerning part (i) and the translation invariance property, notice that by (6.42) we have

$$\mathcal{F}(\mathcal{T}^\mu(\mathcal{T}^\nu h)) = \widehat{\mu} \cdot \mathcal{F}(\mathcal{T}^\nu h) = \widehat{\mu} \cdot \widehat{\nu} \cdot \mathcal{F}h = \widehat{\mu * \nu} \cdot \mathcal{F}h = \mathcal{F}(\mathcal{T}^{\mu*\nu} h) \qquad (h \in \mathcal{D}_{\mathcal{L}}^{(2)} \cap \mathrm{C_c}[a,b))$$

so that $\mathcal{T}^\mu(\mathcal{T}^\nu h) = \mathcal{T}^{\mu*\nu} h$ first for $h \in \mathcal{D}_{\mathcal{L}}^{(2)} \cap \mathrm{C_c}[a,b)$ and then, by continuity, for $h \in \mathrm{C}_0[a,b)$ and $h \in L_p(r)$ $(1 \leq p < \infty)$.

To prove part (iv) we just need to show that $\lim_{t\downarrow 0}(T_t h)(x) = h(x)$ for all $h \in \mathrm{C}_0[a,b)$ and $x \in [a,b)$, because it is well known from the theory of Feller semigroups that for a semigroup satisfying (ii) and (iii) this weak continuity property implies the strong continuity of the semigroup (see e.g., [9, Lemma 1.4]). But for $h \in \mathrm{C}_0[a,b)$ and $x \in [a,b)$, we clearly have

$$\lim_{t\downarrow 0}\big((T_t h)(x) - h(x)\big) = \lim_{t\downarrow 0} \int_{[a,b)} \big((\mathcal{T}^y h)(x) - h(x)\big)\mu_t(dy) = \int_{[a,b)} \big((\mathcal{T}^y h)(x) - h(x)\big)\delta_a(dy) = 0$$

showing that (iv) holds.

For part (iv'), let $h \in L_p(r)$, $\varepsilon > 0$ and choose $g \in \mathrm{C}_\mathrm{c}^\infty(a,b)$ such that $\|h - g\|_p \leq \varepsilon$. Then it follows from (6.45) and part (iv) that

$$\begin{aligned}
\limsup_{t\downarrow 0} \|T_t h - h\|_p &\leq \limsup_{t\downarrow 0}\Big(\|T_t h - T_t g\|_p + \|h - g\|_p + \|T_t g - g\|_p\Big) \\
&\leq 2\varepsilon + C \cdot \limsup_{t\downarrow 0} \|T_t g - g\|_\infty \\
&= 2\varepsilon
\end{aligned}$$

where $C = [\int_{\text{supp}(g)} r(x)dx]^{1/p}$ ($C < \infty$ because the support $\text{supp}(g) \subset (a,b)$ is compact). Since ε is arbitrary, (iv') holds. □

The result for the space $\mathrm{C}_0[a,b)$ means that $\{T_t\}$ is an $\mathcal{L}$-translation-invariant conservative Feller semigroup. This semigroup is also symmetric with the measure $r(x)dx$, that is, $\int_a^b (T_t h)(x)g(x)r(x)dx = \int_a^b h(x)(T_t g)(x)r(x)dx$ for $h, g \in \mathrm{C_c}[a,b)$. Any such symmetric Feller semigroup extends to a strongly continuous Markovian contraction semigroup $\{T_t^{(p)}\}_{t\geq 0}$ on $L_p(r)$, $1 \leq p < \infty$ [9, Lemma 1.45]. However, the conclusion of Proposition 6.10 is stronger: it also states that the integral with respect to the Feller transition function is well-defined for all $h \in \cup_{1\leq p<\infty} L_p(r)$ and, accordingly, the extensions $T_t^{(p)}$ are also given by $h \mapsto (\mathcal{T}^{\mu_t} h)(x) = \int_{[a,b)} h(\xi)(\delta_x * \mu_t)(d\xi)$.

On the Hilbert space $L_2(r)$, we can take advantage of the $\mathcal{L}$-transform to obtain a characterization of the generator for the L_2-Markovian semigroup $T_t^{(2)} \equiv T_t : L_2(r) \longrightarrow L_2(r)$:

Proposition 6.11. *Let $\{\mu_t\}$ be an $\mathcal{L}$-convolution semigroup with exponent ψ. Then the infinitesimal generator $(\mathcal{A}^{(2)}, \mathcal{D}_{\mathcal{A}^{(2)}})$ of the L_2-Markovian semigroup $\{T_t^{(2)}\}$ is given by*

$$\mathcal{F}(\mathcal{A}^{(2)} h) = -\psi \cdot (\mathcal{F}h), \qquad h \in \mathcal{D}_{\mathcal{A}^{(2)}}$$

where

$$\mathcal{D}_{\mathcal{A}^{(2)}} = \biggl\{ h \in L_2(r) \biggm| \int_{[0,\infty)} |\psi(\lambda)|^2 \bigl|(\mathcal{F}h)(\lambda)\bigr|^2 \boldsymbol{\rho}_{\mathcal{L}}(d\lambda) < \infty \biggr\}.$$

Proof. We give a proof which follows closely that of the corresponding result for the ordinary convolution, as given in [4, Theorem 12.16].

Let $h \in \mathcal{D}_{\mathcal{A}^{(2)}}$, so that $L_2\text{-}\lim_{t\downarrow 0} \frac{1}{t}(T_t h - h) = \mathcal{A}^{(2)} h \in L_2(\mathrm{m})$. Recalling that (by (6.42)) $\mathcal{F}(T_t h) = \widehat{\mu_t} \cdot (\mathcal{F}h) = e^{-t\psi} \cdot (\mathcal{F}h)$ for all $h \in L_2(r)$, we see that

$$L_2\text{-}\lim_{t\downarrow 0} \frac{1}{t}\bigl(e^{-t\psi} - 1\bigr) \cdot (\mathcal{F}h) = \mathcal{F}(\mathcal{A}^{(2)} h)$$

The convergence holds almost everywhere along a sequence $\{t_n\}_{n\in\mathbb{N}}$ such that $t_n \to 0$, so we conclude that $\mathcal{F}(\mathcal{A}^{(2)} h) = -\psi \cdot (\mathcal{F}h) \in L_2(\mathbb{R}; \boldsymbol{\rho}_{\mathcal{L}})$.

Conversely, if we let $h \in L_2(r)$ with $-\psi \cdot (\mathcal{F}h) \in L_2(\mathbb{R}; \boldsymbol{\rho}_{\mathcal{L}})$, then we have

$$L_2\text{-}\lim_{t\downarrow 0} \frac{1}{t}\bigl(\mathcal{F}(T_t h) - \mathcal{F}h\bigr) = -\psi \cdot (\mathcal{F}h) \in L_2(\mathbb{R}; \boldsymbol{\rho}_{\mathcal{L}})$$

and the isometry gives that $L_2\text{-}\lim_{t\downarrow 0} \frac{1}{t}(T_t h - h) \in L_2(\mathrm{m})$, meaning that $h \in \mathcal{D}_{\mathcal{A}^{(2)}}$. □

6.7.3 Additive and Lévy Processes

Definition 6.7. An $[a,b)$-valued Markov chain $\{S_n\}_{n\in\mathbb{N}_0}$ is said to be $\mathcal{L}$-*additive* if there exist measures $\mu_n \in \mathcal{P}[a,b)$ such that

$$P[S_n \in B | S_{n-1} = x] = (\mu_n * \delta_x)(B), \qquad n \in \mathbb{N},\ a \leq x < b,\ B \text{ a Borel subset of } [a,b). \tag{6.48}$$

If $\mu_n = \mu$ for all n, then $\{S_n\}$ is said to be an $\mathcal{L}$-*random walk*.

An explicit construction can be given for $\mathcal{L}$-additive Markov chains, based on the following lemma:

Lemma 6.9. *There exists a Borel measurable $\Phi : [a,b) \times [a,b) \times [0,1] \longrightarrow [a,b)$ such that*

$$(\delta_x * \delta_y)(B) = \mathfrak{m}\{\Phi(x,y,\cdot) \in B\}, \qquad x, y \in [a,b),\ B \textit{ a Borel subset of } [a,b)$$

where $\mathfrak{m}$ denotes Lebesgue measure on $[0,1]$.

Proof. Let $\Phi(x,y,\xi) = \max\bigl(a, \sup\{z \in [a,b) : (\delta_x * \delta_y)[a,z] < \xi\}\bigr)$. Using the continuity of the $\mathcal{L}$-convolution, one can show that Φ is Borel measurable, see [6, Theorem 7.1.3]. It is straightforward that $\mathfrak{m}\{\Phi(x,y,\cdot) \in [a,c]\} = \mathfrak{m}\{(\delta_x * \delta_y)[a,c] \geq \xi\} = (\delta_x * \delta_y)[a,c]$. □

Let X_1, U_1, X_2, U_2, ... be a sequence of independent random variables (on a given probability space $(\Omega, \mathfrak{A}, \pi)$) where the X_n have distribution $P_{X_n} = \mu_n \in \mathcal{P}[a,b)$ and each of the (auxiliary) random variables U_n has the uniform distribution on $[0,1]$. Set

$$S_0 = 0, \qquad S_n = S_{n-1} \oplus_{U_n} X_n \tag{6.49}$$

where $X \oplus_U Y := \Phi(X,Y,U)$. Then we have $P_{S_n} = P_{S_{n-1}} * \mu_n$ ($n \in \mathbb{N}_0$) and, consequently, $\{S_n\}_{n \in \mathbb{N}_0}$ is an $\mathcal{L}$-additive Markov chain satisfying (6.48). The identity $P_{S_n} = P_{S_{n-1}} * \mu_n$ is easily checked:

$$\begin{aligned} P_{S_n}(B) = P\bigl[\Phi(S_{n-1}, X_n, U_n) \in B\bigr] &= \int_{[a,b)} \int_{[a,b)} \mathfrak{m}\{\Phi(x,y,\cdot) \in B\} P_{S_{n-1}}(dx) P_{X_n}(dy) \\ &= \int_{[a,b)} \int_{[a,b)} (\delta_x * \delta_y)(B) P_{S_{n-1}}(dx) P_{X_n}(dy) \\ &= (P_{S_{n-1}} * \mu_n)(B). \end{aligned}$$

We now define the continuous-time analogue of $\mathcal{L}$-random walks:

Definition 6.8. An $[a,b)$-valued Markov process $Y = \{Y_t\}_{t \geq 0}$ is said to be an *$\mathcal{L}$-Lévy process* if there exists an $\mathcal{L}$-convolution semigroup $\{\mu_t\}_{t \geq 0}$ such that the transition probabilities of Y are given by

$$P\bigl[Y_t \in B | Y_s = x\bigr] = (\mu_{t-s} * \delta_x)(B), \qquad 0 \leq s \leq t,\ a \leq x < b,\ B \text{ a Borel subset of } [a,b).$$

The notion of an $\mathcal{L}$-Lévy process coincides with that of a Feller process associated with the Feller semigroup $T_t f = \mathcal{T}^{\mu_t} f$. Consequently, the general connection between Feller semigroups and Feller processes (see e.g., [9, Section 1.2]) ensures that for each (initial) distribution $\nu \in \mathcal{P}[a,b)$ and $\mathcal{L}$-convolution semigroup $\{\mu_t\}_{t \geq 0}$ there exists an $\mathcal{L}$-Lévy process Y associated with $\{\mu_t\}_{t \geq 0}$ such that $P_{Y_0} = \nu$. Any $\mathcal{L}$-Lévy process has the following properties:

- It is stochastically continuous: $Y_s \to Y_t$ in probability as $s \to t$, for each $t \geq 0$;
- It has a càdlàg modification: there exists an $\mathcal{L}$-Lévy process $\{\widetilde{Y}_t\}$ with almost surely (a.s.) right-continuous paths and satisfying $P\bigl[Y_t = \widetilde{Y}_t\bigr] = 1$ for all $t \geq 0$.

(These properties hold for all Feller processes, cf. [9, Section 1.2].)

An analogue of the well-known theorem on appoximation of Lévy processes by triangular arrays holds for $\mathcal{L}$-Lévy processes (below the notation $\xrightarrow{d}$ stands for convergence in distribution):

Proposition 6.12. *Let X be an $[a,b)$-valued random variable. The following assertions are equivalent:*

(i) $X = Y_1$ *for some $\mathcal{L}$-Lévy process* $Y = \{Y_t\}_{t \geq 0}$.

(ii) The distribution of X is $\mathcal{L}$-infinitely divisible;

(iii) $S_{m_n}^n \xrightarrow{d} X$ *for some sequence of $\mathcal{L}$-random walks* $S^1, S^2, \ldots$ *(with* $S_0^j = a$*) and some integers* $m_n \to \infty$.

Proof. The equivalence between (i) and (ii) is a restatement of the one-to-one correspondence (6.47) between $\mathcal{L}$-infinitely divisible measures and $\mathcal{L}$-convolution semigroups. It is obvious that (i) implies (iii): simply let $m_n = n$ and S^n the random walk whose step distribution is the law of $Y_{1/n}$.

Suppose that (iii) holds, let π_n, μ be the distributions of S_j^n, X, respectively. Choose $\varepsilon > 0$ small enough so that $\widehat{\mu}(\lambda) > C_\varepsilon > 0$ for $\lambda \in [0, \varepsilon]$, where $C_\varepsilon > 0$ is a constant. By (iii) and Proposition 6.6(c), $\widehat{\pi_n}(\lambda)^{m_n} \to \widehat{\mu}(\lambda)$ uniformly on compacts, which implies that $\widehat{\pi_n}(\lambda) \to 1$ for all $\lambda \in [0, \varepsilon]$ and, therefore, by Proposition 6.6(d) $\pi_n \xrightarrow{w} \delta_a$. Now let $k \in \mathbb{N}$ be arbitrary. Since $\pi_n \xrightarrow{w} \delta_a$, we can assume that each m_n is a multiple of k. Write $\nu_n = \pi_n^{*(m_n/k)}$, so that $\nu_n^{*k} \xrightarrow{w} \mu$. By relative compactness of $\mathrm{D}(\{\pi_n^{*m_n}\})$ (see the proof of Theorem 6.5), the sequence $\{\nu_n\}_{n\in\mathbb{N}}$ has a weakly convergent subsequence, say $\nu_{n_j} \xrightarrow{w} \mu_k$ as $j \to \infty$, and from this it clearly follows that $\mu_k^{*k} = \mu$. Consequently, (ii) holds. □

As one would expect, the diffusion process generated by the Sturm–Liouville operator (6.4) (cf. Lemma 6.6) is an $\mathcal{L}$-Lévy process:

Proposition 6.13. *The irreducible diffusion process X generated by $(\mathcal{L}^{(\mathrm{b})}, \mathcal{D}_\mathcal{L}^{(\mathrm{b})})$ is an $\mathcal{L}$-Lévy process.*

Proof. For $t \geq 0$, $a \leq x < b$ let us write $p_{t,x}(dy) \equiv P_x[X_t \in dy]$. Recall from Lemma 6.7 that

$$p_{t,x}(dy) \equiv p(t,x,y)r(y)dy = \int_{[0,\infty)} e^{-t\lambda} w_\lambda(x)\, w_\lambda(y)\, \boldsymbol{\rho}_\mathcal{L}(d\lambda)\, r(y)dy, \qquad t > 0,\ a < x < b$$

where the integral converges absolutely. Consequently, by Proposition 6.1,

$$\widehat{p_{t,x}}(\lambda) = e^{-t\lambda} w_\lambda(x), \qquad t \geq 0,\ a \leq x < b$$

(the weak continuity of $p_{t,x}$ justifies that the equality also holds for $t = 0$ and for $x = a$). This shows that $p_{t,x} = p_{t,a} * \delta_x$ where $\widehat{p_{t,a}}(\lambda) = e^{-t\lambda}$. It is clear from the properties of the $\mathcal{L}$-transform that $\{p_{t,a}\}_{t\geq 0}$ is an $\mathcal{L}$-convolution semigroup; therefore, X is an $\mathcal{L}$-Lévy process. □

An $\mathcal{L}$-convolution semigroup $\{\mu_t\}_{t\geq 0}$ such that μ_1 is an $\mathcal{L}$-Gaussian measure is said to be an *$\mathcal{L}$-Gaussian convolution semigroup*, and an $\mathcal{L}$-Lévy process associated with an $\mathcal{L}$-Gaussian convolution semigroup is called an *$\mathcal{L}$-Gaussian process.*

It actually turns out that the diffusion X generated by $(\mathcal{L}^{(\mathrm{b})}, \mathcal{D}_\mathcal{L}^{(\mathrm{b})})$ is an $\mathcal{L}$-Gaussian process. This is a consequence of the following characterization of $\mathcal{L}$-Gaussian measures:

Proposition 6.14. *Suppose that $A \in \mathrm{C}^3(a,b)$ (where A is the function defined in* (6.10)*). Let $Y = \{Y_t\}_{t\geq 0}$ be an $\mathcal{L}$-Lévy process, let $\{\mu_t\}_{t\geq 0}$ be the associated $\mathcal{L}$-convolution semigroup and let $(\mathcal{G}, \mathcal{D}(\mathcal{G}))$ be the $\mathrm{C_b}$-generator of the process Y. The following conditions are equivalent:*

- *(i) μ_1 is a Gaussian measure;*
- *(ii) $\lim_{t\downarrow 0} \frac{1}{t}\mu_t\big([a,b) \setminus \mathcal{V}_a\big) = 0$ for every neighborhood $\mathcal{V}_a$ of the point a;*
- *(iii) $\lim_{t\downarrow 0} \frac{1}{t}(\mu_t * \delta_x)\big([a,b) \setminus \mathcal{V}_x\big) = 0$ for every $x \in [a,b)$ and every neighborhood $\mathcal{V}_x$ of the point x;*
- *(iv) Y has a modification whose paths are a.s. continuous.*

If any of these conditions hold then the $\mathrm{C_b}$-generator of Y is a local operator, i.e., $(\mathcal{G}h)(x) = (\mathcal{G}g)(x)$ whenever $h, g \in \mathcal{D}(\mathcal{G})$ and $h = g$ on some neighborhood of $x \in [a,b)$.

Proof. (i) $\Longrightarrow$ (ii): Let $\{t_n\}_{n\in\mathbb{N}}$ be a sequence such that $t_n \to 0$ as $n \to \infty$, and let $\nu_n = \mathbf{e}(\frac{1}{t_n}\mu_{t_n})$. We have

$$\lim_{n\to\infty} \widehat{\nu_n}(\lambda) = \lim_{n\to\infty} \exp\left[\frac{1}{t_n}\big(\widehat{\mu_1}(\lambda)^{t_n} - 1\big)\right] = \widehat{\mu_1}(\lambda), \qquad \lambda > 0 \tag{6.50}$$

and therefore, by Proposition 6.6(d), $\nu_n \xrightarrow{w} \mu_1$ as $n \to \infty$. From this it follows, cf. [56], that if π_n denotes the restriction of $\frac{1}{t_n}\mu_{t_n}$ to $[a,b) \setminus \mathcal{V}_a$, then $\{\pi_n\}$ is relatively compact; if π is a limit point, then $\mathbf{e}(\pi)$ is a divisor of μ_1. Since μ_1 is Gaussian, $\mathbf{e}(\pi) = \delta_a$, hence π must be the zero measure, showing that (ii) holds.

(ii)$\Longrightarrow$ (i): As in (6.50),

$$\widehat{\mu_1}(\lambda) = \lim_{n\to\infty} \exp\left[\frac{1}{t_n}\int_{[a,b)} (w_\lambda(x) - 1)\mu_{t_n}(dx)\right] = \lim_{n\to\infty} \exp\left[\frac{1}{t_n}\int_{\mathcal{V}_a} (w_\lambda(x) - 1)\mu_{t_n}(dx)\right], \qquad \lambda > 0$$

where the second equality is due to (ii), noting that $\frac{1}{t_n}\int_{[a,b)\setminus\mathcal{V}_a}(w_\lambda(x) - 1)\mu_{t_n}(dx) \leq \frac{2}{t}\mu_{t_n}\big([a,b)\setminus\mathcal{V}_a\big)$. Given that $\nu_n = \mathbf{e}(\frac{1}{t_n}\mu_{t_n}) \xrightarrow{w} \mu_1$, we have (again, see [56])

$$\widehat{\mu}_1(\lambda) = \exp\left[\int_{(a,b)} (w_\lambda(x) - 1)\, \eta(dx)\right], \qquad \lambda > 0$$

for some σ-finite measure η on (a,b) which, by the above, vanishes on the complement of any neighborhood of the point a. Therefore, μ_1 is Gaussian.

(ii) $\Longleftrightarrow$ (iii): To prove the nontrivial direction, assume that (ii) holds, and fix $x \in (a,b)$. Let $\mathcal{V}_x$ be a neighborhood of the point x not containing a and write $E_x = [a,b)\setminus\mathcal{V}_x$. Pick a function $h \in \mathrm{C}^4$ such that $0 \leq h \leq 1$, $h = 0$ on E_x and $h = 1$ on some smaller neighborhood $\mathcal{U}_x \subset \mathcal{V}_x$ of the point x. (Using the assumption $A \in \mathrm{C}^3(a,b)$, it is easy to check that $h,\, \ell(h) \in \mathcal{D}_{\mathcal{L}}^{(2)}$.)

We begin by showing that

$$\lim_{y\downarrow a} \frac{1 - (\mathcal{T}^x h)(y)}{1 - w_\lambda(y)} = 0 \qquad \text{for each } \lambda > 0. \tag{6.51}$$

Indeed, it follows from Theorem 6.1 that $\lim_{y\downarrow a}(\mathcal{T}^x h)(y) = 1$, $\lim_{y\downarrow a}\partial_y^{[1]}(\mathcal{T}^x h)(y) = 0$ and

$$\ell_y(\mathcal{T}^x h)(y) = \int_{[0,\infty)} \lambda\,(\mathcal{F}h)(\lambda)\, w_\lambda(x)\, w_\lambda(y)\, \boldsymbol{\rho}_{\mathcal{L}}(d\lambda) = \big(\mathcal{T}^x \ell(h)\big)(y) \xrightarrow[y\downarrow a]{} \ell(h)(x) = 0,$$

hence using L'Hôpital's rule twice, we find that $\lim_{y\downarrow a}\frac{1-(\mathcal{T}^x h)(y)}{1-w_\lambda(y)} = \lim_{y\downarrow a}\frac{\ell_y(\mathcal{T}^x h)(y)}{\lambda\, w_\lambda(y)} = 0$ $(\lambda > 0)$.

By (6.51), for each $\lambda > 0$ there exists $a_\lambda > a$ such that $(\mathcal{T}^x \mathbb{1}_{E_x})(y) \leq \big(\mathcal{T}^x(\mathbb{1} - h)\big)(y) \leq 1 - w_\lambda(y)$ for all $y \in [a, a_\lambda)$ (here $\mathbb{1}_{E_x}$ denotes the indicator function of E_x). We then estimate

$$\begin{aligned}
\frac{1}{t}(\mu_t * \delta_x)(E_x) &= \frac{1}{t}\int_{[a,b)} (\mathcal{T}^x \mathbb{1}_{E_x})(y)\mu_t(dy) \\
&\leq \frac{1}{t}\int_{[a,a_\lambda)} \big(1 - w_\lambda(y)\big)\mu_t(dy) + \frac{1}{t}\mu_t[a_\lambda, b) \\
&\leq \frac{1}{t}\int_{[a,b)} \big(1 - w_\lambda(y)\big)\mu_t(dy) + \frac{1}{t}\mu_t[a_\lambda, b) \\
&= \frac{1}{t}\big(1 - \widehat{\mu_t}(\lambda)\big) + \frac{1}{t}\mu_t[a_\lambda, b).
\end{aligned}$$

Given that we are assuming that (ii) holds and, by the $\mathcal{L}$-semigroup property, $\lim_{t\downarrow 0}\frac{1}{t}\big(1-\widehat{\mu_t}(\lambda)\big)=\lim_{t\downarrow 0}\frac{1}{t}\big(1-\widehat{\mu_1}(\lambda)^t\big)=-\log\widehat{\mu_1}(\lambda)$, the above inequality gives

$$\limsup_{t\downarrow 0}\frac{1}{t}(\mu_t*\delta_x)(E_x)\leq -\log\widehat{\mu_1}(\lambda).$$

This holds for arbitrary $\lambda>0$. Since the right-hand side is continuous and vanishes for $\lambda=0$, we conclude that $\lim_{t\downarrow 0}\frac{1}{t}(\mu_t*\delta_x)(E_x)=0$, as desired.

***(iii)* $\Longrightarrow$ *(iv)*:** This follows from a general result in the theory of Feller processes [21, Chapter 4, Proposition 2.9] according to which $\lim_{t\downarrow 0}\frac{1}{t}P_x[Y_t\in[a,b)\setminus\mathcal{V}_x]=0$ is a sufficient condition for a given $[a,b)$-valued Feller process Y to have continuous paths.

***(iv)* $\Longrightarrow$ *(iii)*:** This is a consequence of Ray's theorem for one-dimensional Markov processes, which is stated and proved in [26, Theorem 5.2.1].

Finally, it is well known that Markov processes with continuous paths have local generators (see e.g., [26, Theorem 5.1.1]), thus the last assertion holds. □

To finish this section, it is worth mentioning that analogues of the classical limit theorems—such as laws of large numbers or central limit theorems—can be established for the $\mathcal{L}$-convolution measure algebra. As in the setting of hypergroup convolution structures (cf. Example 6.8), solutions $\{\varphi_k\}_{k\in\mathbb{N}}$ of the functional equation

$$(\mathcal{T}^y\varphi_k)(x)=\sum_{j=0}^{k}\binom{k}{j}\varphi_j(x)\varphi_{k-j}(y)\qquad(x,y\in[a,b)),\qquad\qquad\varphi_0=0,$$

which are called $\mathcal{L}$-*moment functions*, play a role similar to that of the monomials under the ordinary convolution.

For the sake of illustration, let us state some strong laws of large numbers which hold true for the $\mathcal{L}$-convolution: let $\{S_n\}$ be an $\mathcal{L}$-additive Markov chain constructed as in (6.49), and define the $\mathcal{L}$-moment functions of first and second order by $\varphi_1(x)=\kappa\eta_1(x)$, $\varphi_2(x)=2[\kappa\eta_2(x)+\eta_1(x)]$, respectively, where $\kappa:=\lim_{\xi\to\infty}\frac{A'(\xi)}{A(\xi)}=\lim_{x\uparrow b}\frac{[(pr)^{1/2}]'(x)}{2r(x)}$ and the η_j are given by (6.8). Then:

7.13.I. *If $\{r_n\}_{n\in\mathbb{N}}$ is a sequence of positive numbers such that $\lim_n r_n=\infty$ and $\sum_{n=1}^{\infty}\frac{1}{r_n}\big(\mathbb{E}[\varphi_2(X_n)]-\mathbb{E}[\varphi_1(X_n)]^2\big)<\infty$, then*

$$\lim_n\frac{1}{\sqrt{r_n}}\big(\varphi_1(S_n)-\mathbb{E}[\varphi_1(S_n)]\big)=0\qquad\pi\text{-}a.s.$$

7.13.II. *If $\{S_n\}$ is an $\mathcal{L}$-random walk such that $\mathbb{E}[\varphi_2(X_1)^{\theta/2}]<\infty$ for some $1\leq\theta<2$, then $\mathbb{E}[\varphi_1(X_1)]<\infty$ and*

$$\lim_n\frac{1}{n^{1/\theta}}\big(\varphi_1(S_n)-n\mathbb{E}[\varphi_1(X_1)]\big)=0\qquad\pi\text{-}a.s.$$

7.13.III. *Suppose that $\varphi_1\equiv 0$. If $\{r_n\}_{n\in\mathbb{N}}$ is a sequence of positive numbers such that $\lim_n r_n=\infty$ and $\sum_{n=1}^{\infty}\frac{1}{r_n}\mathbb{E}[\varphi_2(X_n)]<\infty$, then*

$$\lim_n\frac{1}{r_n}\varphi_2(S_n)=0\qquad\pi\text{-}a.s.$$

7.13.IV. *Suppose that $\varphi_1\equiv 0$. If $\{S_n\}$ is an $\mathcal{L}$-random walk such that $\mathbb{E}[\varphi_2(X_1)^{\theta}]<\infty$ for some $0<\theta<1$, then*

$$\lim_n\frac{1}{n^{1/\theta}}\varphi_2(S_n)=0\qquad\pi\text{-}a.s.$$

The above assertions can be proved exactly as in the hypergroup framework: the reader is referred to [63, Section 7].

6.8 Examples

We begin with two simple examples where the Sturm–Liouville operator is regular and nondegenerate, and the kernel of the $\mathcal{L}$-transform can be written in terms of elementary functions.

Example 6.1 (Cosine Fourier transform). Consider the Sturm–Liouville operator

$$\ell = -\frac{d^2}{dx^2}, \qquad 0 < x < \infty$$

which is obtained by setting $p = r = \mathbb{1}$ and $(a, b) = (0, \infty)$. This operator trivially satisfies assumption MP. Since the solution of the Sturm–Liouville boundary value problem (6.7) is $w_\lambda(x) = \cos(\tau x)$ (where $\lambda = \tau^2$), the $\mathcal{L}$-transform is simply the cosine Fourier transform $(\mathcal{F}h)(\tau) = \int_0^\infty h(x)\cos(\tau x)dx$. By elementary trigonometric identities, $w_\tau(x)w_\tau(y) = \frac{1}{2}[w_\tau(|x-y|) + w_\tau(x+y)]$, hence the $\mathcal{L}$-convolution is given by

$$\delta_x * \delta_y = \frac{1}{2}(\delta_{|x-y|} + \delta_{x+y}), \qquad x, y \geq 0.$$

In other words, $*$ is (up to identification) the ordinary convolution of symmetric measures.

Example 6.2. If we let $p(x) = r(x) = (1+x)^2$ and $(a, b) = (0, \infty)$, we obtain the differential operator

$$\ell = -\frac{d^2}{dx^2} - \frac{2}{1+x}\frac{d}{dx}, \qquad 0 < x < \infty,$$

which satisfies Assumption MP with $\eta(x) = \frac{2}{1+x}$. The function

$$w_\lambda(x) = \begin{cases} \frac{1}{1+x}[\cos(\tau x) + \frac{1}{\tau}\sin(\tau x)], & \tau > 0 \\ 1, & \tau = 0 \end{cases} \qquad (\lambda = \tau^2)$$

is the solution of the boundary value problem (6.7), thus the $\mathcal{L}$-transform can be expressed as a sum of cosine and sine Fourier transforms. A straightforward computation [63, Example 4.10] shows that the product formula $w_\lambda(x)\, w_\lambda(y) = \int_{[a,b)} w_\lambda\, d(\delta_x * \delta_y)$ holds for $\delta_x * \delta_y$, defined by

$$\begin{aligned}(\delta_x * \delta_y)(d\xi) = \frac{1}{2(1+x)(1+y)}\big[&(1+|x-y|)\delta_{x-y}(d\xi) + (1+x+y)\delta_{x+y}(d\xi) \\ &+ (1+\xi)\mathbb{1}_{[|x-y|,x+y]}(\xi)d\xi\big]\end{aligned} \tag{6.52}$$

and therefore (by the uniqueness property, Proposition 6.6(b)) the $\mathcal{L}$-convolution is given by (6.52). This example, which was introduced in [63, Example 4.10], illustrates that, in general, convolutions associated with regular Sturm–Liouville operators have both discrete and absolute continuous components.

Next we present the chief example of a convolution associated with a singular Sturm–Liouville operator:

Example 6.3 (Hankel transform). Let $\alpha \geq -\frac{1}{2}$. The Bessel operator

$$\ell = -\frac{d^2}{dx^2} - \frac{2\alpha+1}{x}\frac{d}{dx}, \qquad 0 < x < \infty$$

has coefficients $p(x) = r(x) = x^{2\alpha+1}$. Clearly, Assumption MP holds with $\eta = 0$. Here the kernel of the $\mathcal{L}$-transform is

$$w_\lambda(x) = \boldsymbol{J}_\alpha(\tau x) := 2^\alpha \Gamma(\alpha+1)(\tau x)^{-\alpha} J_\alpha(\tau x) \qquad (\lambda = \tau^2)$$

where J_α is the Bessel function of the first kind (this is easily checked using the basic properties of the Bessel function, cf. [42, Chapter 10]). The Sturm–Liouville type transform associated with the Bessel operator is the *Hankel transform*, $(\mathcal{F}h)(\tau) = \int_0^\infty h(x)\, \boldsymbol{J}_\alpha(\tau x)\, x^{2\alpha+1} dx$. It follows from classical integration formulae for the Bessel function [58, p. 411] that $\boldsymbol{J}_\alpha(\tau x)\, \boldsymbol{J}_\alpha(\tau y) = \int_0^\infty \boldsymbol{J}_\alpha(\tau \xi)\, (\delta_x *_\alpha \delta_y)(d\xi)$, where

$$\begin{aligned}(\delta_x *_\alpha \delta_y)(d\xi) &= \frac{2^{1-2\alpha}\Gamma(\alpha+1)}{\sqrt{\pi}\,\Gamma(\alpha+\frac{1}{2})}(xy\xi)^{-2\alpha} \\ &\quad \times \bigl[(\xi^2-(x-y)^2)((x+y)^2-\xi^2)\bigr]^{\alpha-1/2} \mathbb{1}_{[|x-y|,x+y]}(\xi)\, r(\xi) d\xi\end{aligned}$$

for $x, y > 0$; this convolution is known as the *Hankel* [13,25] or *Kingman convolution* [30,54].

This example has motivated the development of the theory of generalized translation and convolution operators back since the pioneering work of Delsarte [17]. It plays a special role in the context of the Sturm–Liouville hypergroups in Example 6.8; in particular, it appears as the limit distribution in central limit theorems on hypergroups [6, Section 7.5]. Moreover, since the diffusion ($\mathcal{L}$-Lévy) process generated by ℓ is the Bessel process—a fundamental continuous-time stochastic process [8], which in the case $\alpha = \frac{d}{2} - 1$ ($d \in \mathbb{N}$) can be defined as the radial part of a d-dimensional Brownian motion—the Hankel convolution is a useful tool for the study of the Bessel process, cf. e.g., [44,55].

The Jacobi operator provides another example of a singular Sturm–Liouville operator whose product formula and convolution can be written in terms of standard special functions.

Example 6.4 (Jacobi transform). The coefficients $p(x) = r(x) = (\sinh x)^{2\alpha+1}(\cosh x)^{2\beta+1}$ $(\alpha \geq \beta \geq -\frac{1}{2},\ \alpha \neq \frac{1}{2})$ give rise to the Jacobi operator

$$\ell = -\frac{d^2}{dx^2} - [(2\alpha+1)\coth x + (2\beta+1)\tanh x]\frac{d}{dx}, \qquad 0 < x < \infty.$$

As in the previous example, Assumption MP holds with $\eta = 0$. The so-called Jacobi function

$$w_\lambda(x) = \phi_\tau^{(\alpha,\beta)}(x) := {}_2F_1\Bigl(\tfrac{1}{2}(\sigma - i\tau), \tfrac{1}{2}(\sigma + i\tau); \alpha+1; -(\sinh x)^2\Bigr)$$

($\sigma = \alpha + \beta + 1$, $\lambda = \tau^2 + \sigma^2$) where ${}_2F_1$ denotes the hypergeometric function [42, Chapter 15] can be shown to be the unique solution of the Sturm–Liouville problem (6.7). The associated integral transform is the *(Fourier-)Jacobi transform*, $(\mathcal{F}h)(\tau) = \int_0^\infty h(x)\, \phi_\tau^{(\alpha,\beta)}(x)\, (\sinh x)^{2\alpha+1}(\cosh x)^{2\beta+1} dx$ (this transformation is also known as Olevskii transform, index hypergeometric transform or, in the case $\alpha = \beta$, generalized Mehler–Fock transform [62]). By a deep result of Koornwinder [22,32], the product formula $\phi_\tau^{(\alpha,\beta)}(x)\, \phi_\tau^{(\alpha,\beta)}(y) = \int_0^\infty \phi_\tau^{(\alpha,\beta)} d(\delta_x *_{\alpha,\beta} \delta_y)$ holds for the *Jacobi convolution*, defined by

$$(\delta_x *_{\alpha,\beta} \delta_y)(d\xi) = \frac{2^{-2\sigma}\Gamma(\alpha+1)(\cosh x\, \cosh y\, \cosh\xi)^{\alpha-\beta-1}}{\sqrt{\pi}\,\Gamma(\alpha+\frac{1}{2})(\sinh x\, \sinh y\, \sinh\xi)^{2\alpha}} \times (1-Z^2)^{\alpha-1/2}\,{}_2F_1\Bigl(\alpha+\beta,\alpha-\beta;\alpha+\tfrac{1}{2};\tfrac{1}{2}(1-Z)\Bigr)\mathbb{1}_{[|x-y|,x+y]}(\xi)r(\xi)d\xi$$

where $Z := \frac{(\cosh x)^2+(\cosh y)^2+(\cosh\xi)^2-1}{2\cosh x\, \cosh y\, \cosh\xi}$.

For half-integer values of the parameters α, β, the Jacobi transform and convolution have various group theoretic interpretations; in particular, they are related with harmonic analysis on rank one Riemannian symmetric spaces [32]. Moreover, a remarkable property of the Jacobi transform is that it admits a positive dual convolution structure, that is, there exists a family $\{\theta_{\tau_1,\tau_2}\}$ of finite positive measures such that the dual product formula $\phi_{\tau_1}^{(\alpha,\beta)}(x)\,\phi_{\tau_2}^{(\alpha,\beta)}(x) = \int_0^\infty \phi_{\tau_3}^{(\alpha,\beta)}(x)\,\theta_{\tau_1,\tau_2}(d\tau_3)$ holds, and this permits the construction of a generalized convolution which trivializes the inverse Jacobi transform [3].

All the examples presented so far belong to the class of Sturm–Liouville hypergroup convolutions which was introduced by Zeuner [63] as follows:

Example 6.5 (Sturm–Liouville hypergroups). Consider a Sturm–Liouville operator on the positive half-line with coefficients $p = r = A$,

$$\ell = -\frac{d^2}{dx^2} - \frac{A'(x)}{A(x)}\frac{d}{dx}, \qquad 0 < x < \infty,$$

where the function A satisfies the following conditions:

SL0 $A \in \mathrm{C}[0,\infty) \cap \mathrm{C}^1(0,\infty)$ and $A(x) > 0$ for $x > 0$.

SL1 One of the following assertions holds:

- **SL1.1** $A(0) = 0$ and $\frac{A'(x)}{A(x)} = \frac{\alpha_0}{x} + \alpha_1(x)$ for x in a neighborhood of 0, where $\alpha_0 > 0$ and $\alpha_1 \in \mathrm{C}^\infty(\mathbb{R})$ is an odd function;
- **SL1.2** $A(0) > 0$ and $A \in \mathrm{C}^1[0,\infty)$.

SL2 There exists $\eta \in \mathrm{C}^1[0,\infty)$ such that $\eta \geq 0$, $\boldsymbol{\phi}_\eta \geq 0$ and the functions $\boldsymbol{\phi}_\eta$, $\boldsymbol{\psi}_\eta$ are both decreasing on $(0,\infty)$ ($\boldsymbol{\phi}_\eta$, $\boldsymbol{\psi}_\eta$ are defined as in Assumption MP).

The last condition ensures that A satisfies Assumption MP; hence, this is a particular case of the general family of Sturm–Liouville operators considered in the previous sections. It was proved by Zeuner [63] that the convolution measure algebra $(\mathcal{M}_\mathbb{C}[0,\infty),*)$ is a commutative hypergroup with identity involution; this means that the Banach algebra property of Proposition 6.5 and properties (b)–(c) of Corollary 6.3 hold, as well as the following axioms:

- $(x,y) \mapsto \mathrm{supp}(\delta_x * \delta_y)$ is continuous from $[0,\infty)\times[0,\infty)$ into the space of compact subsets of $[0,\infty)$ (endowed with the Michael topology, see [27]);
- $0 \in \mathrm{supp}(\delta_x * \delta_y)$ if and only if $x = y$.

Observe that the Sturm–Liouville operator $\ell = -\frac{d^2}{dx^2} - \frac{A'}{A}\frac{d}{dx}$ is either singular or regular, depending on whether the function A satisfies condition SL1.1 or SL1.2. In any event, the associated hyperbolic equation $\ell_x f = \ell_y f$ is uniformly hyperbolic on $[0,\infty)^2$. The construction of the product formula and convolution presented in the previous sections generalizes that of Zeuner because it is also applicable to parabolically degenerate operators.

The next example shows that the two hypergroup axioms on the (compact) support of $\delta_x * \delta_y$ are generally false for operators associated with degenerate hyperbolic equations:

Example 6.6 (Index Whittaker transform). The choice $p(x) = x^{2-2\alpha}e^{-1/x}$ and $r(x) = x^{-2\alpha}e^{-1/x}$, with $\alpha < \frac{1}{2}$, leads to the normalized Whittaker operator

$$\ell = -x^2\frac{d^2}{dx^2} - (1+2(1-\alpha)x)\frac{d}{dx}, \qquad 0 < x < \infty.$$

The standard form of this differential operator (Remark 6.1) is $\widetilde{\ell} = -\frac{d^2}{dz^2} - (e^{-z}+1-2\alpha)\frac{d}{dz}$, where $z = \log x \in \mathbb{R}$, and it is apparent that Assumption MP holds with $\eta = 0$. As pointed out in Section 6.3, the fact that the operator $\widetilde{\ell}$ is defined on the whole real line means that the hyperbolic partial differential equation associated with the normalized Whittaker operator has a nonremovable parabolic degeneracy at the initial line. The unique solution of the boundary value problem (6.7) turns out to be given by

$$w_\lambda(x) = \boldsymbol{W}_{\alpha,i\tau}(x) := x^\alpha e^{\frac{1}{2x}} W_{\alpha,i\tau}(\tfrac{1}{x}) \qquad (\lambda = \tau^2 + (\tfrac{1}{2}-\alpha)^2)$$

where $W_{\alpha,i\tau}$ is the Whittaker function of the second kind of parameters α and $i\tau$ [42, Chapter 13]. The eigenfunction expansion of the normalized Whittaker operator yields the *index Whittaker transform* [49,50] $(\mathcal{F}h)(\tau) = \int_0^\infty h(x)\boldsymbol{W}_{\alpha,i\tau}(x)x^{-2\alpha}e^{-1/x}dx$. The product formula for the kernel $\boldsymbol{W}_{\alpha,i\tau}$ has recently been established by the authors [46,47] using techniques from classical analysis and known facts in the theory of special functions; it is given by $\boldsymbol{W}_{\alpha,i\tau}(x)\boldsymbol{W}_{\alpha,i\tau}(y) = \int_0^\infty \boldsymbol{W}_{\alpha,i\tau}\, d(\delta_x *_\alpha \delta_y)$, where $*_\alpha$ is the *Whittaker convolution*, defined by

$$(\delta_x *_\alpha \delta_y)(d\xi) = \frac{2^{-1-\alpha}}{\sqrt{\pi}}(xy\xi)^{-\frac{1}{2}+\alpha}\exp\Bigl(\frac{1}{x}+\frac{1}{y}+\frac{1}{\xi}-\frac{(x+y+\xi)^2}{8xy\xi}\Bigr)D_{2\alpha}\Bigl(\frac{x+y+\xi}{\sqrt{2xy\xi}}\Bigr)r(\xi)d\xi$$

for $x, y > 0$, with D_μ denoting the parabolic cylinder function [20, Chapter VIII]. Notice in particular that $\operatorname{supp}(\delta_x *_\alpha \delta_y) = [0,\infty)$ for every $x, y > 0$.

The particular case $\alpha = 0$ is worthy of special mention, because in this case, the index Whittaker transform reduces to $(\mathcal{F}h)(\tau) = \pi^{-1/2}\int_0^\infty h(x)K_{i\tau}(\frac{1}{2x})x^{-1/2}e^{-\frac{1}{2x}}dx$, which is (a normalized form of) the Kontorovich–Lebedev transform; here $K_{i\tau}$ is the modified Bessel function of the second kind with parameter $i\tau$ [42, Chapter 10]. The Kontorovich–Lebedev transform plays a central role in the theory of index-type integral transforms [61]. The Whittaker convolution of parameter $\alpha = 0$, which can be written in the simplified form

$$(\delta_x *_0 \delta_y)(d\xi) = \frac{1}{2\sqrt{\pi xy\xi}}\exp\Bigl(\frac{1}{x}+\frac{1}{y}-\frac{(x+y+\xi)^2}{4xy\xi}\Bigr)d\xi,$$

is identical (up to an elementary change of variables) to the Kontorovich–Lebedev convolution, which was introduced by Kakichev in [29] and has been extensively studied, cf. [61] and references therein.

Our final example illustrates that the (degenerate) hyperbolic equation approach allows us to generalize the results on the Whittaker product formula and convolution to a much larger class of degenerate operators:

Example 6.7. Let $\boldsymbol{\zeta} \in \mathrm{C}^1(0,\infty)$ be a nonnegative decreasing function such that $\int_1^\infty \boldsymbol{\zeta}(y)\frac{dy}{y} = \infty$, and let $\kappa > 0$. The differential expression

$$\ell = -x^2\frac{d^2}{dx^2} - \bigl[\kappa + x\bigl(1+\boldsymbol{\zeta}(x)\bigr)\bigr]\frac{d}{dx}, \qquad 0 < x < \infty$$

is a particular case of (6.4), obtained by considering $p(x) = xe^{-\kappa/x+I_{\boldsymbol{\zeta}}(x)}$ and $r(x) = \frac{1}{x}e^{-\kappa/x+I_{\boldsymbol{\zeta}}(x)}$, where $I_{\boldsymbol{\zeta}}(x) = \int_1^x \boldsymbol{\zeta}(y)\frac{dy}{y}$. (If $\kappa = 1$ and $\boldsymbol{\zeta}(x) = 1-2\alpha > 0$, we recover the normalized Whittaker operator from Example 6.8.) The change of variable $z = \log x \in \mathbb{R}$ transforms ℓ into the standard form $\widetilde{\ell} = -\frac{d^2}{dz^2} - \frac{A'(z)}{A(z)}\frac{d}{dz}$, where $\frac{A'(z)}{A(z)} = \kappa e^{-\kappa z} + \zeta(e^z)$. It is clear that ℓ satisfies Assumption MP with $\eta = 0$, and the additional assumption $\lim_{x\uparrow b} p(x)r(x) = \infty$ holds because $I_\zeta(\infty) = \infty$. Therefore, all the results in the previous sections hold for the Sturm–Liouville operator ℓ. This shows that the class of Sturm–Liouville operators for which one can construct a positivity-preserving convolution structure includes irregular operators which are simultaneously degenerate (in the sense that the associated hyperbolic equation is parabolic at the initial line) and singular (in the sense that the first-order coefficient is unbounded near the left endpoint).

Acknowledgments

The first and third authors were partially supported by CMUP, which is financed by national funds through FCT – Fundação para a Ciência e a Tecnologia, I.P., under the project with reference UIDB/00144/2020. The first author was also supported by the grant PD/BD/135281/2017, under the FCT PhD Programme UC—UP MATH PhD Program. The second author was partially supported by the project CEMAPRE/REM – UIDB/05069/2020 financed by FCT through national funds.

References

[1] H. Bauer, *Probability Theory*, Walter De Gruyter, Berlin (1996).

[2] H. Bauer, *Measure and Integration Theory*, Walter De Gruyter, Berlin (2001).

[3] N. Ben Salem, Convolution semigroups and central limit theorem associated with a dual convolution structure, *J. Theor. Probab.* **7**, no. 2, pp. 417–436 (1994).

[4] C. Berg, G. Forst, *Potential Theory on Locally Compact Abelian Groups*, Springer, Berlin (1975).

[5] A. V. Bitsadze, *Equations of the Mixed Type*, Pergamon Press, Oxford, (1964).

[6] W. R. Bloom, H. Heyer, *Harmonic Analysis of Probability Measures on Hypergroups*, Walter de Gruyter, Berlin (1994).

[7] V. I. Bogachev, *Measure Theory*, Vol. II, Springer, Berlin (2007).

[8] A. N. Borodin, P. Salminen, *Handbook of Brownian Motion: Facts and Formulae*, Springer, Basel (2002).

[9] B. Böttcher, R. Schilling, J. Wang, *Lévy Matters III – Lévy-Type processes: Construction, approximation and sample path properties*, Lecture Notes in Mathematics vol. 2099, Springer, Berlin (2014).

[10] R. W. Carroll, *Transmutation Theory and Applications*, North-Holland, Amsterdam (1985).

[11] H. Chebli, Opérateus de translation généralisée et semi-groupes de convolution, in *Théorie du Potentiel et Analyse Harmonique* (J. Faraut, editor), Springer, Berlin, pp. 35–59 (1974).

[12] H. Chebli, Sturm-Liouville Hypergroups, in *Applications of Hypergroups and Related Measure Algebras: A Joint Summer Research Conference on Applications of Hypergroups and Related Measure Algebras*, July 31 to August 6, 1993, Seattle, WA, American Mathematical Society, Providence, RI, pp. 71–88 (1995).

[13] F. M. Cholewinski, A Hankel convolution complex inversion theory, *Mem. Am. Math. Soc.* **58**, 67 pp. (1965).

[14] D. L. Cohn, *Measure Theory*, Second Edition, Birkhäuser, New York (2013).

[15] W. C. Connett, C. Markett, A. L. Schwartz, Convolution and hypergroup structures associated with a class of Sturm-Liouville systems, *Trans. Am. Math. Soc.* **332**, no. 1, pp. 365–390 (1992).

[16] R. Courant, *Methods of Mathematical Physics – Vol. II: Partial Differential Equations*, Wiley, New York (1962).

[17] J. Delsarte, Sur une extension de la formule de Taylor, *J. Math. Pures Appl.* **17**, pp. 213–231 (1938).

[18] N. Dunford, J. T. Schwartz, *Linear Operators – Part II: Spectral Theory*, Wiley, New York (1963).

[19] J. Eckhardt, G. Teschl, Sturm-Liouville operators with measure-valued coefficients, *J. Anal. Math.* **120**, pp. 151–224 (2013).

[20] A. Erdélyi, W. Magnus, F. Oberhettinger, F. G. Tricomi, *Higher Transcendental Functions – Vol. II*, McGraw–Hill, New York (1953).

[21] S. N. Ethier, T. G. Kurtz, *Markov Processes: Characterization and Convergence*, Wiley, New York (1986).

[22] M. Flensted-Jensen, T. H. Koornwinder, The convolution structure for Jacobi function expansions, *Ark. Mat.* **11**, pp. 245–262 (1973).

[23] F. Fruchtl, Sturm-Liouville hypergroups and asymptotics, *Monatsh. Math.* **186**, no. 1, pp. 11–36 (2018).

[24] M. Fukushima, On general boundary conditions for one-dimensional diffusions with symmetry, *J. Math. Soc. Japan* **66**, no. 1, pp. 289–316 (2014).

[25] I. I. Hirschman, Variation diminishing Hankel transforms, *J. Analyse Math.* **8**, pp. 307–336 (1960).

[26] K. Itô, *Essentials of Stochastic Processes*, American Mathematical Society, Providence, RI (2006).

[27] R. I. Jewett, Spaces with an abstract convolution of measures, *Adv. Math.* **18**, no. 1, pp. 1–101 (1975).

[28] I. S. Kac, The existence of spectral functions of generalized second order differential systems with boundary conditions at the singular end, *Am. Math. Soc. Transl.* (2) **62**, pp. 204–262 (1967).

[29] V. A. Kakichev, On the convolution for integral transforms, *Izv Vyssh Uchebn Zaved Mat.* **2**, pp. 53–62 (1967) (in Russian).

[30] J. F. C. Kingman, Random walks with spherical symmetry, *Acta Math.* **109**, pp. 11–53 (1963).

[31] A. Klenke, *Probability Theory: A Comprehensive Course*, Second Edition, Springer, London (2014).

[32] T. H. Koornwinder, Jacobi functions and analysis on noncompact semisimple Lie groups, in *Special Functions: Group Theoretical Aspects and Applications* (R. A. Askey, T. H. Koornwinder, W. Schempp, editors), Reidel, Dordrecht, pp. 1–85 (1984).

[33] B. M. Levitan, Die Verallgemeinerung der Operation der Verschiebung im Zusammenhang mit fastperiodischen Funktionen, *Mat. Sb.* **7**, no. 49, pp. 449–478 (1940).

[34] B. M. Levitan, On a class of solutions of the Kolmogorov-Smoluchowski equation, *Vestnik Leningrad. Univ.* **15**, no. 7, pp. 81–115 (1960) (in Russian).

[35] V. Linetsky, The spectral decomposition of the option value, *Int. J. Theor. Appl. Finance* **7**, no. 3, pp. 337–384 (2004).

[36] J. V. Linnik, I. V. Ostrovskiĭ, *Decomposition of Random Variables and Vectors*, American Mathematical Society, Providence, RI (1977).

[37] G. L. Litvinov, Hypergroups and hypergroup algebras, *J. Soviet Math.* **38**, no. 2, pp. 1734–1761 (1987).

[38] N. K. Mamadaliev, On representation of a solution to a modified Cauchy problem, *Sib. Math. J.* **41**, no. 5, pp. 889–899 (2000).

[39] P. Mandl, *Analytical Treatment of One-Dimensional Markov Processes*, Springer, Berlin (1968).

[40] H. McKean, Elementary solutions for certain parabolic partial differential equations, *Trans. Am. Math. Soc.* **82**, 519-548 (1956).

[41] M. A. Naimark, *Linear Differential Operators. Part II: Linear Differential Operators in Hilbert Space*, Frederick Ungar Publishing Co., New York, 1968.

[42] F. W. J. Olver, D. W. Lozier, R. F. Boisvert, C. W. Clark (Eds.), *NIST Handbook of Mathematical Functions*, Cambridge University Press, Cambridge, 2010.

[43] E. V. Radkevich, Equations with nonnegative characteristic form II, *J. Math. Sci.* **158**, pp. 453–604 (2009).

[44] C. Rentzsch, M. Voit, Lévy processes on commutative hypergroups, in *Probability on Algebraic Structures: AMS Special Session on Probability on Algebraic Structures*, March 12–13, 1999, Gainesville, FL, American Mathematical Society, Providence, RI, pp. 83–105 (2000).

[45] M. Rösler, Convolution algebras which are not necessarily positivity-preserving, in *Applications of Hypergroups and Related Measure Algebras: A Joint Summer Research Conference on Applications of Hypergroups and Related Measure Algebras*, July 31 to August 6, 1993, Seattle, WA, American Mathematical Society, Providence, RI, pp. 71–88 (1995).

[46] R. Sousa, M. Guerra, S. Yakubovich, Lévy processes with respect to the index Whittaker convolution, Preprint, arXiv:1805.03051 (2018).

[47] R. Sousa, M. Guerra, S. Yakubovich, On the product formula and convolution associated with the index Whittaker transform, *J. Math. Anal. Appl.* **475**, no. 1, pp. 939–965 (2019).

[48] R. Sousa, M. Guerra, S. Yakubovich, Sturm-Liouville hypergroups without the compactness axiom, Preprint, arXiv:1901.11021 (2019).

[49] R. Sousa, S. Yakubovich, The spectral expansion approach to index transforms and connections with the theory of diffusion processes, *Commun. Pure Appl. Anal.* **17**, no. 6, pp. 2351–2378 (2018).

[50] H. M. Srivastava, Y. V. Vasil'ev, S. Yakubovich, A class of index transforms with Whittaker's function as the kernel, *Quart. J. Math. Oxford* **49**, no. 2, pp. 375–394 (1998).

[51] G. Teschl, *Mathematical Methods in Quantum Mechanics: With Applications to Schrödinger Operators*, Second Edition, American Mathematical Society, Providence, RI (2014).

[52] E. C. Titchmarsh, *Eigenfunction Expansions Associated with Second-Order Differential Equations*, Second Edition, Oxford University Press, Oxford (1962).

[53] K. Urbanik, Generalized convolutions, *Studia Math.* **23**, pp. 217–245 (1964).

[54] K. Urbanik, Analytical methods in probability theory, in *Transactions of the Tenth Prague Conference on Information Theory, Statistical Decision Functions, Random Processes, Vol. A*, Reidel, Dordrecht, pp. 151–163 (1988).

[55] N. Van Thu, S. Ogawa, M. Yamazato, A convolution approach to multivariate Bessel processes, in *Stochastic Processes and Applications to Mathematical Finance, Proceedings of the 6th Ritsumeikan International Symposium*, World Scientific, Singapore, pp. 233–244 (2007).

[56] V. E. Volkovich, Infinitely divisible distributions in algebras with stochastic convolution, *J. Soviet Math.* **40**, no. 4, pp. 459–467 (1988).

[57] V. E. Volkovich, On symmetric stochastic convolutions, *J. Theoret. Probab.* **5**, no. 3, pp. 417–430 (1992).

[58] G. N. Watson, *A Treatise on the Theory of Bessel Functions*, Second Edition, Cambridge University Press, Cambridge (1944).

[59] J. Weidmann, *Spectral Theory of Ordinary Differential Operators*, Springer, Berlin (1987).

[60] H. Weinberger, A maximum property of Cauchy's problem, *Ann. Math.* **64**, no. 2, pp. 505–513 (1956).

[61] S. Yakubovich, *Index Transforms*, World Scientific, Singapore (1996).

[62] S. Yakubovich, On the Plancherel theorem for the Olevskii transform, *Acta Math. Vietnam.* **31**, no. 3, pp. 249–260 (2006).

[63] H. Zeuner, Moment functions and laws of large numbers on hypergroups, *Math. Z.* **211**, no. 1, pp. 369–407 (1992).

7

Elements of Aomoto's Generalized Hypergeometric Functions and a Novel Perspective on Gauss' Hypergeometric Differential Equation

Yasuhiro Abe

Cereja Technology Co., Ltd.

CONTENTS

7.1 Introduction

Studies of physical phenomena such as scattering amplitudes (see, *e.g.*, [1–5]) and quantum Hall effects [6] in Grassmannian spaces $Gr(k+1, n+1)$ have been attentively carried out in recent years. This has revived an interest in a purely mathematical subject, *i.e.*, generalized hypergeometric functions on Grassmannian spaces, which were introduced and developed by Gelfand [7] and independently by Aomoto [8] many years ago. One of the main goals of this chapter is to present a clear and systematic review on these particular topics in mathematical physics. Particularly, we clarify integral representations of Aomoto's generalized hypergeometric functions in terms of twisted homology and cohomology. As a simplest example, we consider in detail Gauss' original hypergeometric functions in Aomoto's framework so as to familiarize ourselves to the concept of twisted homology and cohomology.

This chapter is organized as follows. In the next section we review some formal results of Aomoto's generalized hypergeometric functions on $Gr(k+1, n+1)$, based on Japanese textbooks [8,9]. We present a review in a pedagogical fashion since these results are not familiar enough to many physicists and mathematicians. In Section 7.3 we consider a particular case $Gr(2, n+1)$ and present its general formulation. In Section 7.4 we further study the case of $Gr(2,4)$ which reduces to Gauss' hypergeometric function. Imposing permutation invariance among branch points, we obtain new realizations of the hypergeometric differential equation in the form of a first- order Fuchsian differential equation. This is an original result first reported in [10], and this chapter is to be based mainly on some part of [10].

7.2 Elements of Aomoto's Generalized Hypergeometric Functions

7.2.1 Definition

Let Z be a $(k+1) \times (n+1)$ matrix

$$Z = \begin{pmatrix} z_{00} & z_{01} & z_{02} & \cdots & z_{0n} \\ z_{10} & z_{11} & z_{12} & \cdots & z_{1n} \\ \vdots & \vdots & \vdots & & \vdots \\ z_{k0} & z_{k1} & z_{k2} & \cdots & z_{kn} \end{pmatrix} \tag{7.1}$$

where $k < n$ and the matrix elements are complex, $z_{ij} \in \mathbb{C}$ $(0 \le i \le k;\ 0 \le j \le n)$. A function of Z, which we denote $F(Z)$, is defined as *a generalized hypergeometric function on Grassmannian space* $Gr(k+1, n+1)$ when it satisfies the following relations:

$$\sum_{j=0}^{n} z_{ij} \frac{\partial F}{\partial z_{pj}} = -\delta_{ip} F \quad (0 \le i, p \le k) \tag{7.2}$$

$$\sum_{i=0}^{k} z_{ij} \frac{\partial F}{\partial z_{ij}} = \alpha_j F \quad (0 \le j \le n) \tag{7.3}$$

$$\frac{\partial^2 F}{\partial z_{ip} \partial z_{jq}} = \frac{\partial^2 F}{\partial z_{iq} \partial z_{jp}} \quad (0 \le i, j \le k;\ 0 \le p, q \le n) \tag{7.4}$$

where the parameters α_j obey the noninteger conditions

$$\alpha_j \notin \mathbb{Z} \quad (0 \le j \le n) \tag{7.5}$$

$$\sum_{j=0}^{n} \alpha_j = -(k+1) \tag{7.6}$$

7.2.2 Integral Representation of $F(Z)$ and Twisted Cohomology

The essence of Aomoto's generalized hypergeometric function [8] is that, by use of the so-called twisted de Rham cohomology,[1] $F(Z)$ can be written in a form of an integral:

$$F(Z) = \int_{\Delta} \Phi\omega \tag{7.7}$$

where

$$\Phi = \prod_{j=0}^{n} l_j(\tau)^{\alpha_j} \tag{7.8}$$

$$l_j(\tau) = \tau_0 z_{0j} + \tau_1 z_{1j} + \cdots + \tau_k z_{kj} \quad (0 \leq j \leq n) \tag{7.9}$$

$$\omega = \sum_{i=0}^{k} (-1)^i \tau_i d\tau_0 \wedge d\tau_1 \wedge \ldots \wedge d\tau_{i-1} \wedge d\tau_{i+1} \wedge \ldots \wedge d\tau_k \tag{7.10}$$

The complex variables $\tau = (\tau_0, \tau_1, \cdots, \tau_k)$ are homogeneous coordinates of the complex projective space $\mathbb{CP}^k$, *i.e.*, $\mathbb{C}^{k+1} - \{0, 0, \cdots, 0\}$. The multivalued function Φ is then defined in a space

$$X = \mathbb{CP}^k - \bigcup_{j=0}^{n} \mathcal{H}_j \tag{7.11}$$

where

$$\mathcal{H}_j = \{\tau \in \mathbb{CP}^k ;\ l_j(\tau) = 0\} \tag{7.12}$$

We now consider the meaning of the integral path Δ. Since the integrand $\Phi\omega$ is a multivalued k-form, simple choice of Δ as a k-chain on X is not enough. *Upon the choice of Δ, we need to implicitly specify branches of Φ on Δ as well, otherwise we cannot properly define the integral.* In what follows we assume these implicit conditions.

Before considering further properties of Δ, we here notice that ω has an ambiguity in the evaluation of the integral (7.7). Suppose α is an arbitrary $(k-1)$-form defined in X. Then an integral over the exact k-form $d(\Phi\alpha)$ vanishes:

$$0 = \int_{\Delta} d(\Phi\alpha) = \int_{\Delta} \Phi \left(d\alpha + \frac{d\Phi}{\Phi} \wedge \alpha \right) = \int_{\Delta} \Phi \nabla \alpha \tag{7.13}$$

where ∇ can be interpreted as a covariant (exterior) derivative

$$\nabla = d + d\log\Phi\wedge = d + \sum_{j=0}^{n} \alpha_j \frac{dl_j}{l_j} \wedge \tag{7.14}$$

This means that $\omega' = \omega + \nabla\alpha$ is equivalent to ω in the definition of the integral (7.7). Namely, ω and ω' form an equivalent class, $\omega \sim \omega'$. This equivalent class is called the cohomology class.

To study this cohomology class, we consider the differential equation

$$\nabla f = df + \sum_{j=0}^{n} \alpha_j \frac{dl_j}{l_j} f = 0 \tag{7.15}$$

[1]The *twisted* de Rham cohomology is a version of the ordinary de Rham cohomology into which multivalued functions, such as Φ in (7.8), are incorporated. For mathematical rigor on this, see Section 7.2 in [8].

General solutions are locally determined by

$$f = \lambda \prod_{j=0}^{n} l_j(\tau)^{-\alpha_j} \qquad (\lambda \in \mathbb{C}^\times) \tag{7.16}$$

These local solutions are thus basically given by $1/\Phi$. The idea of locality is essential since even if $1/\Phi$ is multivalued within a local patch it can be treated as a single-valued function. Analytic continuation of these solutions forms a fundamental homotopy group of a closed path in X (or $1/X$ to be precise but it can be regarded as X by flipping the noninteger powers α_j in (7.8)). The representation of this fundamental group is called the *monodromy* representation. The monodromy representation determines the local system of the differential Equation (7.15). The general solution f or $1/\Phi$ gives a rank-1 local system in this sense[2]. We denote this rank-1 local system by $\mathcal{L}$. The above cohomology class is then defined as an element of the k-th cohomology group of X over $\mathcal{L}$, *i.e.*,

$$[\omega] \in H^k(X, \mathcal{L}) \tag{7.17}$$

This cohomology group $H^k(X, \mathcal{L})$ is also called *twisted* cohomology group.

7.2.3 Twisted Homology and Twisted Cycles

Having defined the cohomology group $H^k(X, \mathcal{L})$, we can now define the dual of it, *i.e.*, the k-th homology group $H_k(X, \mathcal{L}^\vee)$, known as the twisted homology group, where $\mathcal{L}^\vee$ is the rank-1 dual local system given by Φ. A differential equation corresponding to $\mathcal{L}^\vee$ can be written as

$$\nabla^\vee g = dg - \sum_{j=0}^{n} \alpha_j \frac{dl_j}{l_j} g = 0 \tag{7.18}$$

We can easily check that the general solutions are given by Φ:

$$g = \lambda \prod_{j=0}^{n} l_j(\tau)^{\alpha_j} = \lambda \Phi \qquad (\lambda \in \mathbb{C}^\times) \tag{7.19}$$

As before, an element of $H_k(X, \mathcal{L}^\vee)$ gives an equivalent class called a homology class.

In the following, we show that the integral path Δ forms an equivalent class and see that it coincides with the above homology class. Applying Stokes' theorem to (7.13), we find

$$0 = \int_\Delta \Phi \nabla \alpha = \int_{\partial \Delta} \Phi \alpha \tag{7.20}$$

where α is an arbitrary $(k-1)$-form as before. The boundary operator ∂ is in principle determined from Φ (with information on branches). Denoting $C_p(X, \mathcal{L}^\vee)$ as a p-dimensional chain group on X over $\mathcal{L}^\vee$, we can express the boundary operator as $\partial : C_p(X, \mathcal{L}^\vee) \longrightarrow C_{p-1}(X, \mathcal{L}^\vee)$. Since the relation (7.20) holds for an arbitrary α, we find that the k-chain Δ vanishes by the action of ∂:

$$\partial \Delta = 0 \tag{7.21}$$

The k-chain Δ satisfying the above is generically called the k-cycle. In the current framework it is also called the *twisted cycle*. Since the boundary operator satisfies $\partial^2 = 0$, the k-cycle has a redundancy in it. Namely, $\Delta' = \Delta + \partial C_{(+1)}$ also becomes the k-cycle where $C_{(+1)}$ is

[2]It is "rank-1" because each factor $l_j(\tau)$ in the local solutions (7.16) is first order in the elements of τ.

an arbitrary $(k+1)$-chain or an element of $C_{k+1}(X, \mathcal{L}^\vee)$. Thus Δ and Δ' form an equivalent class, $\Delta \sim \Delta'$, and this is exactly the homology class defined by $H_k(X, \mathcal{L}^\vee)$, *i.e.*,

$$[\Delta] \in H_k(X, \mathcal{L}^\vee) \tag{7.22}$$

To summarize, the generalized hypergeometric function (7.7) is determined by the following bilinear form

$$H_k(X, \mathcal{L}^\vee) \times H^k(X, \mathcal{L}) \longrightarrow \mathbb{C} \tag{7.23}$$

$$([\Delta], [\omega]) \longrightarrow \int_\Delta \Phi\omega \tag{7.24}$$

7.2.4 Differential Equations of $F(Z)$

The condition $l_j(\tau) = 0$ in (7.12) defines a hyperplane in $(k+1)$-dimensional spaces. To avoid redundancy in configuration of hyperplanes, we assume the set of hyperplanes is nondegenerate, that is, we consider the hyperplanes in *general position*. This can be realized by demanding that any $(k+1)$-dimensional minor determinants of the $(k+1) \times (n+1)$ matrix Z are nonzero. We then redefine X in (7.11) as

$$X = \{Z \in Mat_{k+1,n+1}(\mathbb{C}) |\ \text{any } (k+1)\text{-dim minor determinants of } Z \text{ are nonzero}\} \tag{7.25}$$

In what follows we implicitly demand this condition in Z. The configuration of $n+1$ hyperplanes in $\mathbb{CP}^k$ is determined by this matrix Z.

Apart from the concept of hyperplanes, we can also interpret that the above Z provides $n+1$ *distinct points* in $\mathbb{CP}^k$. Since a homogeneous coordinate of $\mathbb{CP}^k$ is given by $\mathbb{C}^{k+1} - \{0, 0, \ldots, 0\}$, we can consider each of the $n+1$ column vectors of Z as a point in $\mathbb{CP}^k$; the j-th column representing the j-th homogeneous coordinates of $\mathbb{CP}^k$ $(j = 0, 1, \ldots, n)$.

The scale transformation, under which the $\mathbb{CP}^k$ homogeneous coordinates are invariant, is realized by an action of $H_{n+1} = \{\text{diag}(h_0, h_1, \ldots h_n) | h_j \in \mathbb{C}^\times\}$ from right on Z. The general linear transformation of the homogeneous coordinates, on the other hand, can be realized by an action of $GL(k+1, \mathbb{C})$ from the left. These transformations are then given by

$$\text{Linear transformation:} \quad Z \to Z' = gZ \tag{7.26}$$

$$\text{Scale transformation:} \quad Z \to Z' = Zh \tag{7.27}$$

where $g \in GL(k+1, \mathbb{C})$ and $h \in H_{n+1}$. Under these transformations the integral $F(Z)$ in (7.7) behaves as

$$F(gZ) = (\text{det}g)^{-1} F(Z) \tag{7.28}$$

$$F(Zh) = F(Z) \prod_{j=0}^{n} h_j^{\alpha_j} \tag{7.29}$$

We now briefly show that the above relations lead to the defining equations of the generalized hypergeometric functions in (7.2) and (7.3), respectively. Let $\mathbf{1}_n$ be the n-dimensional identity matrix $\mathbf{1}_n = \text{diag}(1, 1, \ldots, 1)$, and $E_{ij}^{(n)}$ be an $n \times n$ matrix in which only the (i, j)-element is 1 and the others are zero. We consider g in a particular form of

$$g = \mathbf{1}_{k+1} + \epsilon E_{pi}^{(k+1)} \tag{7.30}$$

where ϵ is a parameter. Then gZ remains the same as Z except the p-th row which is replaced by $(z_{p0}+\epsilon z_{i0}, z_{p1}+\epsilon z_{i1}, \ldots, z_{pn}+\epsilon z_{in})$. Then the derivative of $F(gZ)$ with respect to ϵ is expressed as

$$\frac{\partial}{\partial \epsilon}F(gZ) = \sum_{j=0}^{n} z_{ij}\frac{\partial}{\partial z_{pj}}F(gZ) \tag{7.31}$$

On the other hand, using

$$\det g = \begin{cases} 1 & (i \neq p) \\ \epsilon & (i = p) \end{cases} \tag{7.32}$$

and (7.28), we find

$$\frac{\partial}{\partial \epsilon}F(gZ) = \begin{cases} 0 & (i \neq p) \\ -\frac{1}{\epsilon^2}F(Z) & (i = p) \end{cases} \tag{7.33}$$

Evaluating the derivative at $\epsilon = 0$ and $\epsilon = 1$ for $i \neq p$ and $i = p$, respectively, we then indeed find that (7.28) leads to the differential Equation (7.2).

Similarly, parametrizing h as

$$h = \mathrm{diag}(h_0, \ldots, h_{j-1}, (1+\epsilon)h_j, h_{j+1}, \ldots, h_n) \tag{7.34}$$

with $0 \leq j \leq n$, we find that Zh has only one ϵ-dependent column corresponding to the j-th column, $(z_{0j}(1+\epsilon)h_j, z_{1j}(1+\epsilon)h_j, \ldots, z_{kj}(1+\epsilon)h_j)^T$. The derivative of $F(Zh)$ with respect to ϵ is then expressed as

$$\frac{\partial}{\partial \epsilon}F(Zh) = \sum_{i=0}^{k} z_{ij}\frac{\partial}{\partial z_{ij}}F(Zh) = \sum_{i=0}^{k} z_{ij}\frac{\partial}{\partial z_{ij}}F(Z)(1+\epsilon)^{\alpha_j}\prod_{l=0}^{n} h_l^{\alpha_l} \tag{7.35}$$

where in the last step we use the relation from (7.29):

$$F(Zh) = F(Z)(1+\epsilon)^{\alpha_j}\prod_{l=0}^{n} h_l^{\alpha_l} \tag{7.36}$$

The same derivative can then be expressed as

$$\frac{\partial}{\partial \epsilon}F(Zh) = \alpha_j F(Z)(1+\epsilon)^{\alpha_j - 1}\prod_{l=0}^{k} h_l^{\alpha_l} \tag{7.37}$$

Setting $\epsilon = 0$, we can therefore derive the Equation (7.3).

The other Equation (7.4) for $F(Z)$ follows from the definition of Φ. From (7.8) and (7.9) we find that Φ satisfies

$$\frac{\partial \Phi}{\partial z_{ip}} = \frac{\alpha_i \tau_p}{l_i(\tau)}\Phi \tag{7.38}$$

This relation leads to

$$\frac{\partial^2 \Phi}{\partial z_{ip}\partial z_{jq}} = \frac{\alpha_i \alpha_j \tau_p \tau_q}{l_i(\tau) l_j(\tau)}\Phi = \frac{\partial^2 \Phi}{\partial z_{iq}\partial z_{jp}} \tag{7.39}$$

which automatically derives the Equation (7.4).

The integral $F(Z)$ in (7.7) therefore indeed satisfies the defining Equations (7.2)–(7.4) of the generalized hypergeometric functions on $Gr(k+1, n+1)$. *The Grassmannian space* $Gr(k+1, n+1)$ *is defined as a set of* $(k+1)$*-dimensional linear subspaces in* $(n+1)$*-dimensional complex vector space* $\mathbb{C}^{n+1}$. It is defined as

$$Gr(k+1, n+1) = \widetilde{Z}/GL(k+1, \mathbb{C}) \tag{7.40}$$

where $\widetilde{Z}$ is $(k+1) \times (n+1)$ complex matrices with $rank\widetilde{Z} = k+1$. Consider some matrix M and assume that there exists a nonzero r-dimensional minor determinant of M. Then the rank of M is in general defined by the largest number of such r's. Thus $\widetilde{Z}$ is not exactly same as Z defined in (7.25). $\widetilde{Z}$ is more relaxed since it allows some $(k+1)$-dimensional minor determinants vanish, that is, $Z \subseteq \widetilde{Z}$. In this sense $F(Z)$ is conventionally called the generalized hypergeometric functions on $Gr(k+1, n+1)$, and we follow this convention in the present chapter.

7.2.5 Nonprojected Formulation

In terms of the homogeneous coordinate $\tau = (\tau_0, \tau_1, \ldots, \tau_k)$ on $\mathbb{CP}^k$, coordinates on $\mathbb{C}^k$ can be parametrized as

$$t_1 = \frac{\tau_1}{\tau_0}, \; t_1 = \frac{\tau_1}{\tau_0}, \; \ldots, \; t_k = \frac{\tau_k}{\tau_0} \tag{7.41}$$

For simplicity, we now fix $(z_{00}, z_{10}, \ldots, z_{n0})^T$ at $(1, 0, \ldots, 0)^T$, *i.e.*,

$$Z = \begin{pmatrix} 1 & z_{01} & z_{02} & \cdots & z_{0n} \\ 0 & z_{11} & z_{12} & \cdots & z_{1n} \\ \vdots & \vdots & \vdots & & \vdots \\ 0 & z_{k1} & z_{k2} & \cdots & z_{kn} \end{pmatrix} \tag{7.42}$$

Then the integrand of $F(Z)$ can be expressed as

$$\begin{aligned} \Phi\omega &= \tau_0^{\alpha_0} \prod_{j=1}^{n} (\tau_0 z_{0j} + \tau_1 z_{1j} + \cdots + \tau_k z_{kj})^{\alpha_j} \\ &\quad \times \sum_{i=0}^{k} (-1)^i \tau_i d\tau_0 \wedge d\tau_1 \wedge \cdots \wedge d\tau_{i-1} \wedge d\tau_{i+1} \wedge \cdots \wedge d\tau_k \\ &= \prod_{j=1}^{n} \left(z_{0j} + \frac{\tau_1}{\tau_0} z_{1j} + \cdots + \frac{\tau_k}{\tau_0} z_{kj} \right)^{\alpha_j} d\left(\frac{\tau_1}{\tau_0}\right) \wedge d\left(\frac{\tau_2}{\tau_0}\right) \wedge \cdots \wedge d\left(\frac{\tau_k}{\tau_0}\right) \\ &= \widetilde{\Phi}\widetilde{\omega} \end{aligned} \tag{7.43}$$

where we use (7.6) and define $\widetilde{\Phi}$, $\widetilde{\omega}$ by

$$\widetilde{\Phi} = \prod_{j=1}^{n} \widetilde{l}_j(t)^{\alpha_j} \tag{7.44}$$

$$\widetilde{l}_j(t) = z_{0j} + t_1 z_{1j} + t_2 z_{2j} + \cdots + t_k z_{kj} \qquad (1 \le j \le n) \tag{7.45}$$

$$\widetilde{\omega} = dt_1 \wedge dt_2 \wedge \ldots \wedge dt_k \tag{7.46}$$

The exponents α_j $(j = 1, 2, \ldots, n)$ are also imposed to the noninteger conditions $\alpha_j \notin \mathbb{Z}$ and $\alpha_1 + \alpha_2 + \cdots + \alpha_n \notin \mathbb{Z}$. The multivalued function $\widetilde{\Phi}$ is now defined in the following space

$$\widetilde{X} = \mathbb{C}^k - \bigcup_{j=1}^{n} \widetilde{\mathcal{H}}_j \tag{7.47}$$

where

$$\widetilde{\mathcal{H}}_j = \{t \in \mathbb{C}^k \, ; \, \widetilde{l}_j(t) = 0\} \tag{7.48}$$

These are nonprojected versions of (7.11) and (7.12).

As before, from $\widetilde{\Phi}$ we can define rank-1 local systems $\widetilde{\mathcal{L}}$, $\widetilde{\mathcal{L}}^\vee$ on $\widetilde{X}$, which lead to the k-th homology and cohomology groups, $H_k(\widetilde{X}, \widetilde{\mathcal{L}}^\vee)$ and $H^k(\widetilde{X}, \widetilde{\mathcal{L}})$. Then the integral over $\widetilde{\Phi}\widetilde{\omega}$ is defined as

$$F(Z) = \int_{\widetilde{\Delta}} \widetilde{\Phi}\widetilde{\omega} \tag{7.49}$$

where $[\widetilde{\Delta}] = H_k(\widetilde{X}, \widetilde{\mathcal{L}}^\vee)$ and $[\widetilde{\omega}] = H^k(\widetilde{X}, \widetilde{\mathcal{L}})$.

In regard to the cohomology group $H^k(\widetilde{X}, \widetilde{\mathcal{L}})$, Aomoto shows the following theorem[3]:

1. The dimension of $H^k(\widetilde{X}, \widetilde{\mathcal{L}})$ is given by $\begin{pmatrix} n-1 \\ k \end{pmatrix}$.
2. The basis of $H^k(\widetilde{X}, \widetilde{\mathcal{L}})$ can be formed by $d\log \widetilde{l}_{j_1} \wedge d\log \widetilde{l}_{j_1} \wedge \ldots \wedge d\log \widetilde{l}_{j_k}$, where $1 \le j_1 < j_2 < \ldots < j_k \le n-1$.

Correspondingly, the homology group $H_k(\widetilde{X}, \widetilde{\mathcal{L}}^\vee)$ has dimension $\begin{pmatrix} n-1 \\ k \end{pmatrix}$ and its basis can be formed finite regions bounded by $\widetilde{\mathcal{H}}_j$. In terms of $\widetilde{l}_j$'s the basis of $H^k(\widetilde{X}, \widetilde{\mathcal{L}})$ can also be chosen as [9]

$$\varphi_{j_1 j_2 \cdots j_k} = d\log \frac{\widetilde{l}_{j_1+1}}{\widetilde{l}_{j_1}} \wedge d\log \frac{\widetilde{l}_{j_2+1}}{\widetilde{l}_{j_2}} \wedge \cdots \wedge d\log \frac{\widetilde{l}_{j_k+1}}{\widetilde{l}_{j_k}} \tag{7.50}$$

where $1 \le j_1 < j_2 < \ldots < j_k \le n-1$.

7.3 Generalized Hypergeometric Functions on $Gr(2, n+1)$

In this section we consider a particular case of $Gr(2, n+1)$. The corresponding configuration space is simply given by $n+1$ distinct points in $\mathbb{CP}^1$. This can be represented by a $2\times(n+1)$ matrix Z any of whose 2-dimensional minor determinants are nonzero. Allowing the freedom of the coordinate transformations $GL(2, \mathbb{C})$ from the left and the scale transformations $H_2 = \mathrm{diag}(h_0, h_1)$ from the right, we can uniquely parametrize Z as

$$Z = \begin{pmatrix} 1 & 0 & 1 & 1 & \cdots & 1 \\ 0 & 1 & -1 & -z_3 & \cdots & -z_n \end{pmatrix} \tag{7.51}$$

where $z_i \neq 0, 1, z_j$ $(i \neq j,\ 3 \le i, j \le n)$. Thus we can regard Z as

$$Z \simeq \{(z_3, z_4, \ldots, z_n) \in \mathbb{C}^{n-2} \,|\, z_i \neq 0, 1, z_j \ (i \neq j)\} \tag{7.52}$$

The three other points (z_0, z_1, z_2) can be fixed at $\{0, 1, \infty\}$. This agrees with the fact that the $GL(2, \mathbb{C})$ invariance fixes three points out of the $(n+1)$ distinct points in $\mathbb{CP}^1$.

In application of the previous section, we can carry out a systematic formulation of the generalized hypergeometric functions on $Gr(2, n+1)$ as follows. We begin with a multivalued function of the form

$$\Phi = 1^{\alpha_0} t^{\alpha_1} (1-t)^{\alpha_2} (1-z_3 t)^{\alpha_3} \ldots (1-z_n t)^{\alpha_n} = \prod_{j=1}^{n} l_j(t)^{\alpha_j} \tag{7.53}$$

where

$$l_0(t) = 1\,,\ l_1(t) = t\,,\ l_2(t) = 1-t\,,\ l_j(t) = 1 - z_j t \ \ (3 \le j \le n) \tag{7.54}$$

[3] Theorem 9.6.2 in [8]

As in (7.5) and (7.6), the exponents obey the noninteger conditions

$$\alpha_j \notin \mathbb{Z} \ \ (0 \leq j \leq n), \quad \sum_{j=0}^{n} \alpha_j = -2 \tag{7.55}$$

As considered before, the latter condition applies to the expressions (7.46)–(7.46), that is, when $F(Z)$ is expressed as $F(Z) = \int_\Delta \Phi dt$. The defining space of Φ is given by

$$X = \mathbb{CP}^1 - \{0, 1, 1/z_3, \ldots, 1/z_n, \infty\} \tag{7.56}$$

From Φ we can determine a rank-1 local system $\mathcal{L}$ on X and its dual local system $\mathcal{L}^\vee$. Applying the result in (7.50), the basis of the cohomology group $H^1(X, \mathcal{L})$ is then given by

$$d \log \frac{l_{j+1}}{l_j} \qquad (0 \leq j \leq n-1) \tag{7.57}$$

In the present case the basis of the homology group $H_1(X, \mathcal{L}^\vee)$ can be specified by a set of paths connecting the branch points. For example, we can choose these by

$$\Delta_{\infty 0}\,,\ \Delta_{01}\,,\ \Delta_{1\frac{1}{z_3}}\,,\ \Delta_{\frac{1}{z_3}\frac{1}{z_4}}\,,\ \ldots\,,\ \Delta_{\frac{1}{z_{n-1}}\frac{1}{z_n}} \tag{7.58}$$

where Δ_{pq} denotes a path on $\mathbb{CP}^1$ connecting branch points p and q. To summarize, for an element $\Delta \in H_1(X, \mathcal{L}^\vee)$ associated with Φ of (7.53), we can define a set of generalized hypergeometric functions on $Gr(2, n+1)$ as

$$f_j(Z) = \int_\Delta \Phi \, d \log \frac{l_{j+1}}{l_j} \tag{7.59}$$

where $0 \leq j \leq n-1$. In the next section we consider the case of $n = 3$, the simplest case where only one variable exists, which corresponds to Gauss' hypergeometric function.

7.4 Reduction to Gauss' Hypergeometric Function

7.4.1 Basics of Gauss' Hypergeometric Function

We first review the basics of Gauss' hypergeometric function. In power series, it is defined as

$$F(a, b, c; z) = \sum_{n=1}^{\infty} \frac{(a)_n (b)_n}{(c)_n \, n!} z^n \tag{7.60}$$

where $|z| < 1$, $c \notin \mathbb{Z}_{\leq 0}$ and

$$(a)_n = \begin{cases} 1 & (n = 1) \\ a(a+1)(a+2) \cdots (a+n-1) & (n \geq 1) \end{cases} \tag{7.61}$$

$F(a, b, c; z)$ satisfies the hypergeometric differential equation

$$\left[\frac{d^2}{dz^2} + \left(\frac{c}{z} + \frac{a+b+1-c}{z-1} \right) \frac{d}{dz} + \frac{ab}{z(z-1)} \right] F(a, b, c; z) = 0 \tag{7.62}$$

Euler's integral formula for $F(a, b, c; z)$ is written as

$$F(a,b,c;z) = \frac{\Gamma(c)}{\Gamma(a)\Gamma(c-a)} \int_0^1 t^{a-1}(1-t)^{c-a-1}(1-zt)^{-b}\,dt \tag{7.63}$$

where $|z| < 1$ and $0 < \Re(a) < \Re(c)$[4]. $\Gamma(a)$'s denote the Gamma functions

$$\Gamma(a) = \int_0^\infty e^{-t}t^{a-1}dt \qquad (\Re(a) > 0) \tag{7.64}$$

The second-order differential Equation (7.62) has regular singularities at $z = 0, 1, \infty$. Two independent solutions around each singular point are expressed as

$$z = 0 : \quad \begin{cases} f_1(z) = F(a,b,c;z) \\ f_2(z) = z^{1-c}F(a-c+1, b-c+1, 2-c; z) \end{cases} \tag{7.65}$$

$$z = 1 : \quad \begin{cases} f_3(z) = F(a,b,a+b-c+1; 1-z) \\ f_4(z) = (1-z)^{c-a-b}F(c-a, c-a, c-a-b+1; 1-z) \end{cases} \tag{7.66}$$

$$z = \infty : \quad \begin{cases} f_5(z) = z^{-a}F(a, a-c+1, a-b+1; 1/z) \\ f_6(z) = z^{-b}F(b-c+1, b, b-a+1; 1z) \end{cases} \tag{7.67}$$

where we assume $c \notin \mathbb{Z}$, $a+b-c \notin \mathbb{Z}$ and $a-b \notin \mathbb{Z}$ at $z = 0$, $z = 1$, and $z = \infty$, respectively.

7.4.2 Reduction to Gauss' Hypergeometric Function 1: From Defining Equations

From (7.63) we find the relevant 2×4 matrix in the form of

$$Z = \begin{pmatrix} 1 & 0 & 1 & 1 \\ 0 & 1 & -1 & -z \end{pmatrix} \tag{7.68}$$

The set of Equations (7.2)–(7.4) then reduces as follows:

$$(\partial_{00} + \partial_{02} + \partial_{03})F = -F \tag{7.69}$$
$$(\partial_{11} - \partial_{12} + z\partial_z)F = -F \tag{7.70}$$
$$(\partial_{10} + \partial_{12} - \partial_z)F = 0 \tag{7.71}$$
$$(\partial_{01} + \partial_{02} - \partial_{03})F = 0 \tag{7.72}$$
$$\partial_{00}F = \alpha_0 F \tag{7.73}$$
$$\partial_{11}F = \alpha_1 F \tag{7.74}$$
$$(\partial_{02} - \partial_{12})F = \alpha_2 F \tag{7.75}$$
$$(\partial_{03} + z\partial_z)F = \alpha_3 F \tag{7.76}$$
$$-\partial_z\partial_{02}F = \partial_{12}\partial_{03}F \tag{7.77}$$

where $\partial_{ij} = \frac{\partial}{\partial z_{ij}}$ and $\partial_{13} = -\frac{\partial}{\partial z} = -\partial_z$. The last relation (7.77) arises from (7.4); we here write down the one that is nontrivial and involves ∂_z. Since the sum of (7.69) and (7.70) equals to the sum of (7.73)–(7.76), we can easily find $\alpha_0 + \cdots + \alpha_3 = -2$ in accord with (7.55). The second-order Equation (7.77) is then expressed as

$$-\partial_z(\alpha_1 + \alpha_2 + 1 + z\partial_z)F = (\alpha_1 + 1 + z\partial_z)(\alpha_3 - z\partial_z)F \tag{7.78}$$

[4]This condition can be relaxed to $a \notin \mathbb{Z}$, $c - a \notin \mathbb{Z}$ by use of the well-known Pochhammer contour in the integral (7.63).

This can also be written as

$$\left[z(1-z)\partial_z^2 + (c-(a+b+1)z)\,\partial_z - ab\right] F = 0 \tag{7.79}$$

where

$$\begin{aligned} a &= \alpha_1 + 1 \\ b &= -\alpha_3 \\ c &= \alpha_1 + \alpha_2 + 2 \end{aligned} \tag{7.80}$$

We can easily check that (7.79) identifies with the hypergeometric differential Equation (7.62).

As seen in (7.51), there exist multiple complex variables for $n > 3$. In these cases reduction of the defining Equations (7.2)–(7.4) can be carried out in principle but, unfortunately, is not as straightforward as the case of $n = 3$.

7.4.3 Reduction to Gauss' Hypergeometric Function 2: Use of Twisted Cohomology

The hypergeometric Equation (7.62) is a second-order differential equation. Setting $f_1 = F$, $f_2 = \frac{z}{b}\frac{d}{dz}F$, we can express (7.62) in the form of a first-order Fuchsian differential Equation [8]:

$$\frac{d}{dz}\begin{pmatrix} f_1 \\ f_2 \end{pmatrix} = \left(\frac{A_0}{z} + \frac{A_1}{z-1}\right)\begin{pmatrix} f_1 \\ f_2 \end{pmatrix} \tag{7.81}$$

where

$$A_0 = \begin{pmatrix} 0 & b \\ 0 & 1-c \end{pmatrix}, \quad A_1 = \begin{pmatrix} 0 & 0 \\ -a & c-a-b-1 \end{pmatrix} \tag{7.82}$$

Using the results (7.53)–(7.59), we now obtain other first-order representations of the hypergeometric differential equation.

Let us start with a *nonprojected* multivalued function

$$\Phi = t^a(1-t)^{c-a}(1-zt)^{-b} \tag{7.83}$$

where

$$a\,,\, c-a\,,\, -b \notin \mathbb{Z} \tag{7.84}$$

Φ is defined on $X = \mathbb{CP}^1 - \{0, 1, 1/z, \infty\}$. From these we can determine a rank-1 local system $\mathcal{L}$ and its dual $\mathcal{L}^\vee$ on X. Then, using (7.57), we can obtain a basis of the cohomology group $H^1(X, \mathcal{L})$ given by the following set

$$\varphi_{\infty 0} = \frac{dt}{t} \tag{7.85}$$

$$\varphi_{01} = \frac{dt}{t(1-t)} \tag{7.86}$$

$$\varphi_{1\frac{1}{z}} = \frac{(z-1)dt}{(1-t)(1-zt)} \tag{7.87}$$

Similarly, from (7.58) a basis of the homology group $H_1(X, \mathcal{L}^\vee)$ is given by

$$\left\{\Delta_{\infty 0}\,,\, \Delta_{01}\,,\, \Delta_{1\frac{1}{z}}\right\} \tag{7.88}$$

In terms of these we can express Gauss' hypergeometric function as

$$f_{01}(Z) = \int_{\Delta_{01}} \Phi\varphi_{01} = \int_0^1 t^{a-1}(1-t)^{c-a-1}(1-zt)^{-b}dt \tag{7.89}$$

The derivative of $f_{01}(Z) = f_{01}(z)$ with respect to z is written as

$$\frac{d}{dz}f_{01}(z) = \frac{d}{dz}\int_{\Delta_{01}} \Phi\varphi_{01} = \int_{\Delta_{01}} \Phi\,\nabla_z\varphi_{01} \tag{7.90}$$

where

$$\nabla_z = \partial_z + \partial_z \log\Phi = \partial_z + \frac{bt}{1-zt} \tag{7.91}$$

Thus the derivative comes down to the computation of $\nabla_z\varphi_{01}$; notice that the choice of a twisted cycle Δ is irrelevant as far as the derivative itself is concerned. In order to make sense of (7.90) we should require $\nabla_z\varphi_{01} \in H^1(X, \mathcal{L})$, that is, it should be represented by linear combinations of (7.85)–(7.87). There is a caveat here, however. We know that an element of $H^1(X, \mathcal{L})$ forms an equivalent class as discussed earlier; see (7.13) and (7.14). In the present case ($k = 1$), α in (7.13) is a 0-form or a constant. So we can demand

$$d\log\Phi = a\frac{dt}{t} - (c-a)\frac{dt}{1-t} + b\frac{z\,dt}{1-zt} \equiv 0 \tag{7.92}$$

in the computation of $\nabla_z\varphi_{01}$. This means that the number of base elements can be reduced from 3 to 2. Namely, any elements of $H^1(X, \mathcal{L})$ can be expressed by a combination of an arbitrary pair in (7.85)–(7.87) under the condition (7.92). This explains the numbering discrepancies between (7.50) and (7.57) and agrees with the general result in the previous section that the dimension of the cohomology group is given by $\binom{n-1}{k} = \binom{2}{1} = 2.$

Choosing the pair of $(\varphi_{01}, \varphi_{\infty 0})$, we find

$$\begin{aligned}
\nabla_z\varphi_{\infty 0} &= \frac{bdt}{1-zt} \\
&\equiv \frac{1}{z}\left(-a\frac{dt}{t} + (c-a)\frac{dt}{1-t}\right) \\
&= \frac{c-a}{z}\varphi_{01} - \frac{c}{z}\varphi_{\infty 0}
\end{aligned} \tag{7.93}$$

$$\begin{aligned}
\nabla_z\varphi_{01} &= \nabla_z\left(\varphi_{\infty 0} + \frac{dt}{1-t}\right) \\
&= \nabla_z\varphi_{\infty 0} + \frac{b}{1-z}\left(\frac{dt}{1-t} - \frac{dt}{1-zt}\right) \\
&= \frac{z}{z-1}\nabla_z\varphi_{\infty 0} - \frac{b}{z-1}(\varphi_{01} - \varphi_{\infty 0}) \\
&\equiv \frac{c-a-b}{z-1}\varphi_{01} + \frac{b-c}{z-1}\varphi_{\infty 0}
\end{aligned} \tag{7.94}$$

where notation $\equiv$ means the use of condition (7.92). Using (7.90), we obtain a first-order differential equation

$$\frac{d}{dz}\begin{pmatrix} f_{01} \\ f_{\infty 0} \end{pmatrix} = \left(\frac{A_0^{(\infty 0)}}{z} + \frac{A_1^{(\infty 0)}}{z-1}\right)\begin{pmatrix} f_{01} \\ f_{\infty 0} \end{pmatrix} \tag{7.95}$$

where

$$A_0^{(\infty 0)} = \begin{pmatrix} 0 & 0 \\ c-a & -c \end{pmatrix}, \quad A_1^{(\infty 0)} = \begin{pmatrix} c-a-b & b-c \\ 0 & 0 \end{pmatrix} \tag{7.96}$$

Solving for f_{01}, we can easily confirm that (7.95) leads to Gauss' hypergeometric differential Equation (7.62).

Similarly, for the choice of $(\varphi_{01}, \varphi_{1\frac{1}{z}})$ we find

$$\nabla_z \varphi_{01} = \frac{b}{z-1}\varphi_{1\frac{1}{z}} \tag{7.97}$$

$$\begin{aligned} \nabla_z \varphi_{1\frac{1}{z}} &\equiv \nabla_z \left(-\frac{a}{b}\frac{z-1}{z}\varphi_{01} + \frac{c-a}{b}\frac{z-1}{z}\frac{dt}{(1-t)^2} \right) \\ &\equiv -\frac{a}{z}\varphi_{01} + \left(-\frac{c+1}{z} + \frac{c-a-b+1}{z-1} \right)\varphi_{1\frac{1}{z}} \end{aligned} \tag{7.98}$$

Notice that φ_{01} and $\varphi_{1\frac{1}{z}}$ have the same factor $(1-t)^{-1}$. This factor can be absorbed in the definition of Φ in (7.83). Thus, in applying the derivative formula (7.90), we should replace c by $c-1$. This leads to another first-order differential equation

$$\frac{d}{dz}\begin{pmatrix} f_{01} \\ f_{1\frac{1}{z}} \end{pmatrix} = \left(\frac{A_0^{(1\frac{1}{z})}}{z} + \frac{A_1^{(1\frac{1}{z})}}{z-1} \right)\begin{pmatrix} f_{01} \\ f_{1\frac{1}{z}} \end{pmatrix} \tag{7.99}$$

where

$$A_0^{(1\frac{1}{z})} = \begin{pmatrix} 0 & 0 \\ -a & -c \end{pmatrix}, \quad A_1^{(1\frac{1}{z})} = \begin{pmatrix} 0 & b \\ 0 & c-a-b \end{pmatrix} \tag{7.100}$$

Solving for f_{01}, we can also check that (7.99) becomes Gauss' hypergeometric differential Equation (7.62).

The representations (7.82) and (7.96) are obtained by Aomoto–Kita [8] and Haraoka [9], respectively. The last one (7.100) is not known in the literature as far as the author notices. Along the lines of the above derivation, we can also obtain the Aomoto–Kita representation (7.82) as follows. We introduce a new one-form

$$\widetilde{\varphi}_{1\frac{1}{z}} = \frac{z}{z-1}\varphi_{1\frac{1}{z}} = \frac{z\,dt}{(1-t)(1-zt)} \tag{7.101}$$

The corresponding hypergeometric function is given by $\widetilde{f}_{1\frac{1}{z}} = \int_{\Delta_{01}} \Phi \widetilde{\varphi}_{1\frac{1}{z}}$. From (7.97) we can easily see $\nabla_z \varphi_{01} = \frac{b}{z}\widetilde{\varphi}_{1\frac{1}{z}}$. This is consistent with the condition $f_1 = F$, $f_2 = \frac{z}{b}\frac{d}{dz}F$ in (7.81). Since z is defined as $z \neq 0, 1$, $\frac{b}{z}\widetilde{\varphi}_{1\frac{1}{z}}$ and $\frac{b}{z-1}\varphi_{1\frac{1}{z}}$ are equally well-defined one-forms. We can then choose the pair $(\varphi_{01}, \widetilde{\varphi}_{1\frac{1}{z}})$ as a possible basis of the cohomology group. The derivatives $\nabla_z \varphi_{01}$, $\nabla_z \widetilde{\varphi}_{1\frac{1}{z}}$ are calculated as

$$\nabla_z \varphi_{01} = \frac{b}{z}\widetilde{\varphi}_{1\frac{1}{z}} \tag{7.102}$$

$$\begin{aligned} \nabla_z \widetilde{\varphi}_{1\frac{1}{z}} &\equiv \nabla_z \left(-\frac{a}{b}\varphi_{01} + \frac{c-a}{b}\frac{dt}{(1-t)^2} \right) \\ &\equiv -\frac{a}{z-1}\varphi_{01} + \left(\frac{-c}{z} + \frac{c-a-b}{z-1} \right)\widetilde{\varphi}_{1\frac{1}{z}} \end{aligned} \tag{7.103}$$

where we use the relations

$$\frac{t\,dt}{(1-zt)(1-t)} = \frac{1}{z-1}\left(\frac{dt}{1-zt} - \frac{dt}{1-t} \right) \tag{7.104}$$

$$\frac{dt}{(1-t)^2} \equiv \frac{1}{c-a}\left(a\varphi_{01} + b\widetilde{\varphi}_{1\frac{1}{z}} \right) \tag{7.105}$$

As before, φ_{01} and $\widetilde{\varphi}_{1\frac{1}{z}}$ have the same factor $(1-t)^{-1}$. Thus, replacing c by $c-1$, we obtain a first-order differential equation

$$\frac{d}{dz}\begin{pmatrix} f_{01} \\ \widetilde{f}_{1\frac{1}{z}} \end{pmatrix} = \left(\frac{\widetilde{A}_0^{(1\frac{1}{z})}}{z} + \frac{\widetilde{A}_1^{(1\frac{1}{z})}}{z-1}\right)\begin{pmatrix} f_{01} \\ \widetilde{f}_{1\frac{1}{z}} \end{pmatrix} \tag{7.106}$$

where $\widetilde{f}_{1\frac{1}{z}} = \int_{\Delta_{01}} \Phi\widetilde{\varphi}_{1\frac{1}{z}}$ and

$$\widetilde{A}_0^{(1\frac{1}{z})} = \begin{pmatrix} 0 & b \\ 0 & 1-c \end{pmatrix}, \quad \widetilde{A}_1^{(1\frac{1}{z})} = \begin{pmatrix} 0 & 0 \\ -a & c-a-b-1 \end{pmatrix} \tag{7.107}$$

We therefore reproduce the Aomoto–Kita representation (7.81), (7.82) by a systematic construction of first-order representations of the hypergeometric differential equation.

Lastly, we note that $\varphi_{\infty 0} = \dfrac{dt}{t}$ and $\varphi_{01} = \dfrac{dt}{t(1-t)}$ have the same factor t^{-1} but we cannot absorb this factor into Φ. This is because we can not obtain dt as a base element of $H^1(X, \mathcal{L})$ which is generically given in the form of $d\log\dfrac{l_{j+1}}{l_j}$ as discussed in (7.57).

7.4.4 Reduction to Gauss' Hypergeometric Function 3: Permutation Invariance

The choice of twisted cycles or Δ's is irrelevant in the above derivations of the first-order Fuchsian differential equations. The hypergeometric function is therefore satisfied by a more general integral form, rather than (7.89), *i.e.*,

$$f_{01}^{(\Delta_{pq})}(z) = \int_{\Delta_{pq}} \Phi\varphi_{01} = \int_p^q t^{a-1}(1-t)^{c-a-1}(1-zt)^{-b}dt \tag{7.108}$$

where (p, q) represents an arbitrary pair among the four branch points $p, q \in \{0, 1, 1/z, \infty\}$. This means that we can impose *permutation invariance* on the branch points. Δ_{pq} is then given by the following set of twisted cycles:

$$\Delta_{pq} = \left\{\Delta_{\infty 0}\,, \Delta_{01}\,, \Delta_{1\frac{1}{z}}\,, \Delta_{1\infty}\,, \Delta_{\frac{1}{z}\infty}\,, \Delta_{0\frac{1}{z}}\right\} \tag{7.109}$$

so that the number of elements becomes $\begin{pmatrix} 4 \\ 2 \end{pmatrix} = 6$. Correspondingly, the base elements of the cohomology group also include

$$\varphi_{1\infty} = \frac{dt}{1-t} \tag{7.110}$$

$$\varphi_{\frac{1}{z}\infty} = \frac{z\,dt}{1-zt} \tag{7.111}$$

$$\varphi_{0\frac{1}{z}} = \frac{dt}{t(1-zt)} \tag{7.112}$$

besides (7.85)–(7.87). It is known that $f_{01}^{(\Delta_{pq})}(z)$ are related to the local solutions $f_i(z)$ $(i = 1, 2, \ldots, 6)$ in (7.65)–(7.67) by

$$f_{01}^{(\Delta_{01})}(z) = B(a, c-a)f_1(z) \tag{7.113}$$

$$f_{01}^{(\Delta_{\frac{1}{z}\infty})}(z) = e^{i\pi(a+b-c+1)}B(b-c+1, 1-b)f_2(z) \tag{7.114}$$

$$f_{01}^{(\Delta_{\infty 0})}(z) = e^{i\pi(1-a)} B(a, b-c+1) f_3(z) \tag{7.115}$$

$$f_{01}^{(\Delta_{1\frac{1}{z}})}(z) = e^{i\pi(a-c+1)} B(c-a, 1-b) f_4(z) \tag{7.116}$$

$$f_{01}^{(\Delta_{1\frac{1}{z}})}(z) = B(a, 1-b) f_5(z) \tag{7.117}$$

$$f_{01}^{(\Delta_{1\infty})}(z) = e^{-i\pi(a+b-c+1)} B(b-c+1, c-a) f_6(z) \tag{7.118}$$

where $B(a,b)$ is the beta function

$$B(a,b) = \frac{\Gamma(a)\Gamma(b)}{\Gamma(a+b)} = \int_0^1 t^{a-1}(1-t)^{b-1} dt \qquad (\Re(a) > 0\,,\, \Re(b) > 0) \tag{7.119}$$

(For derivations and details of these relations, see [9].)

The relevant configuration space represented by Z is given by $Gr(2,4)/\mathcal{S}_4$ where $\mathcal{S}_4$ denotes the rank-4 symmetry group. The permutation invariance can also be confirmed by deriving another set of the first-order differential equations with the choice of φ_{01} and one of (7.110)–(7.112). This is what we will present in the following.

For the choice of $(\varphi_{01}, \varphi_{1\infty})$ we find

$$\begin{aligned} \nabla_z \varphi_{1\infty} &= \frac{bt}{1-zt} \frac{dt}{(1-t)} \\ &\equiv -\frac{a}{z(z-1)}\varphi_{01} + \left(\frac{c}{z(z-1)} - \frac{b}{z-1}\right)\varphi_{1\infty} \end{aligned} \tag{7.120}$$

$$\begin{aligned} \nabla_z \varphi_{01} &= \nabla_z \frac{dt}{t} + \nabla_z \varphi_{1\infty} \\ &\equiv -\frac{a}{z-1}\varphi_{01} + \frac{c-b}{z-1}\varphi_{1\infty} \end{aligned} \tag{7.121}$$

The corresponding differential equation is then expressed as

$$\frac{d}{dz}\begin{pmatrix} f_{01} \\ f_{1\infty} \end{pmatrix} = \left(\frac{A_0^{(1\infty)}}{z} + \frac{A_1^{(1\infty)}}{z-1}\right)\begin{pmatrix} f_{01} \\ f_{1\infty} \end{pmatrix} \tag{7.122}$$

where

$$A_0^{(1\infty)} = \begin{pmatrix} 0 & 0 \\ a & -c \end{pmatrix}, \quad A_1^{(1\infty)} = \begin{pmatrix} -a & c-b \\ -a & c-b \end{pmatrix} \tag{7.123}$$

Solving for f_{01}, we can check that (7.122) indeed becomes Gauss' hypergeometric differential Equation (7.62).

Similarly, for $(\varphi_{01}, \varphi_{\frac{1}{z}\infty})$ we find

$$\begin{aligned} \nabla_z \varphi_{01} &= \frac{b\,dt}{(1-zt)(1-t)} \\ &\equiv -\frac{1}{z-1}\frac{ab}{c}\varphi_{01} - \frac{1}{z-1}\frac{b}{c}(b-c)\varphi_{\frac{1}{z}\infty} \end{aligned} \tag{7.124}$$

$$\begin{aligned} \nabla_z \varphi_{\frac{1}{z}\infty} &\equiv \nabla_z\left(-\frac{a}{b}\varphi_{01} + \frac{c}{b}\frac{dt}{1-t}\right) \\ &\equiv \frac{1}{z-1}\frac{a}{c}(a-c)\varphi_{01} + \left(\frac{1}{z-1}\frac{1}{c}(b-c)(a-c) - \frac{c}{z}\right)\varphi_{\frac{1}{z}\infty} \end{aligned} \tag{7.125}$$

The first-order differential equation is then expressed as

$$\frac{d}{dz}\begin{pmatrix} f_{01} \\ f_{\frac{1}{z}\infty} \end{pmatrix} = \left(\frac{A_0^{(\frac{1}{z}\infty)}}{z} + \frac{A_1^{(\frac{1}{z}\infty)}}{z-1}\right)\begin{pmatrix} f_{01} \\ f_{\frac{1}{z}\infty} \end{pmatrix} \tag{7.126}$$

where

$$A_0^{(\frac{1}{z}\infty)} = \begin{pmatrix} 0 & 0 \\ 0 & -c \end{pmatrix}, \quad A_1^{(\frac{1}{z}\infty)} = \begin{pmatrix} -\frac{ab}{c} & -\frac{b}{c}(b-c) \\ \frac{a}{c}(a-c) & \frac{1}{c}(b-c)(a-c) \end{pmatrix} \tag{7.127}$$

We can check that (7.126) reduces to the hypergeometric differential equation for f_{01}.

Lastly, for $(\varphi_{01}, \varphi_{0\frac{1}{z}})$ we find

$$\begin{aligned} \nabla_z \varphi_{01} &= \frac{b\,dt}{(1-zt)(1-t)} \\ &= -\frac{b}{z-1}\left(\varphi_{01} - \varphi_{0\frac{1}{z}}\right) \end{aligned} \tag{7.128}$$

$$\begin{aligned} \nabla_z \varphi_{0\frac{1}{z}} &\equiv \frac{b\,dt}{(1-zt)^2} \\ &\equiv -\frac{c-a}{z(z-1)}\varphi_{01} + \frac{c-az}{z(z-1)}\varphi_{0\frac{1}{z}} \end{aligned} \tag{7.129}$$

The corresponding differential equation becomes

$$\frac{d}{dz}\begin{pmatrix} f_{01} \\ f_{0\frac{1}{z}} \end{pmatrix} = \left(\frac{A_0^{(0\frac{1}{z})}}{z} + \frac{A_1^{(0\frac{1}{z})}}{z-1}\right)\begin{pmatrix} f_{01} \\ f_{0\frac{1}{z}} \end{pmatrix} \tag{7.130}$$

where

$$A_0^{(0\frac{1}{z})} = \begin{pmatrix} 0 & 0 \\ c-a & -c \end{pmatrix}, \quad A_1^{(0\frac{1}{z})} = \begin{pmatrix} -b & b \\ -c+a & c-a \end{pmatrix} \tag{7.131}$$

We can check that (7.130) reduces to the hypergeometric differential equation for f_{01} as well.

As in the case of (7.101), it is tempting to think of $\widetilde{\varphi}_{\frac{1}{z}\infty} = \frac{z-1}{z}\varphi_{\frac{1}{z}\infty} = \frac{(z-1)\,dt}{1-zt}$. But, with φ_{01} and $\widetilde{\varphi}_{\frac{1}{z}\infty}$, it is not feasible to obtain a first-order differential equation in the form of (7.126) which leads to the hypergeometric differential equation. This is because, if expanded in φ_{01} and $\widetilde{\varphi}_{\frac{1}{z}\infty}$, the z-dependence of the derivatives $\nabla_z\varphi_{01}$ and $\nabla_z\widetilde{\varphi}_{\frac{1}{z}\infty}$ cannot be written in terms of $\frac{1}{z}$ or $\frac{1}{z-1}$.

7.4.5 Summary

In this section we carry out a systematic derivation of first-order representations of the hypergeometric differential equation by the use of twisted cohomology as the simplest reduction of Aomoto's generalized hypergeometric function. The first-order equations are generically expressed as

$$\frac{d}{dz}\begin{pmatrix} f_{01} \\ f_{pq} \end{pmatrix} = \left(\frac{A_0^{(pq)}}{z} + \frac{A_1^{(pq)}}{z-1}\right)\begin{pmatrix} f_{01} \\ f_{pq} \end{pmatrix} = A_{01}^{(pq)}\begin{pmatrix} f_{01} \\ f_{pq} \end{pmatrix} \tag{7.132}$$

where (pq) denotes a pair of four branch points $\{0, 1, 1/z, \infty\}$ in $\Phi = t^a(1-t)^{c-a}(1-zt)^{-b}$. A list of the (2×2) matrices $A_{01}^{(pq)}$ obtained in this section is given by the following:

$$A_{01}^{(\infty 0)} = \begin{pmatrix} \dfrac{c-a-b}{z-1} & \dfrac{b-c}{z-1} \\ \dfrac{c-a}{z} & -\dfrac{c}{z} \end{pmatrix} \tag{7.133}$$

$$A_{01}^{(1\frac{1}{z})} = \begin{pmatrix} 0 & \dfrac{b}{z-1} \\ -\dfrac{a}{z} & -\dfrac{c}{z} + \dfrac{c-a-b}{z-1} \end{pmatrix} \tag{7.134}$$

$$\widetilde{A}_{01}^{(1\frac{1}{z})} = \begin{pmatrix} 0 & \dfrac{b}{z} \\ -\dfrac{a}{z-1} & -\dfrac{c-1}{z} + \dfrac{c-a-b-1}{z-1} \end{pmatrix} \tag{7.135}$$

$$A_{01}^{(1\infty)} = \begin{pmatrix} -\dfrac{a}{z-1} & \dfrac{c-b}{z-1} \\ \dfrac{a}{z} - \dfrac{a}{z-1} & -\dfrac{c}{z} + \dfrac{c-b}{z-1} \end{pmatrix} \tag{7.136}$$

$$A_{01}^{(\frac{1}{z}\infty)} = \begin{pmatrix} -\dfrac{1}{z-1}\dfrac{ab}{c} & -\dfrac{1}{z-1}\dfrac{b}{c}(b-c) \\ \dfrac{1}{z-1}\dfrac{a}{c}(a-c) & -\dfrac{c}{z} + \dfrac{1}{z-1}\dfrac{1}{c}(b-c)(a-c) \end{pmatrix} \tag{7.137}$$

$$A_{01}^{(0\frac{1}{z})} = \begin{pmatrix} -\dfrac{b}{z-1} & \dfrac{b}{z-1} \\ \dfrac{c-a}{z} - \dfrac{c-a}{z-1} & -\dfrac{c-1}{z} + \dfrac{c-a}{z-1} \end{pmatrix} \tag{7.138}$$

where we include the Aomoto–Kita representation $\widetilde{A}_{01}^{(1\frac{1}{z})}$. As far as the author notices, these expressions except (7.133, 7.135) are new for the description of the hypergeometric differential equation. A common feature among these matrices is that the determinant is identical:

$$\det A_{01}^{(pq)} = \frac{ab}{z(z-1)} \tag{7.139}$$

In terms of the first-order differential Equation (7.132), this means that the action of the derivative on the basis $\begin{pmatrix} f_{01} \\ f_{pq} \end{pmatrix}$ of the cohomology group $H^1(X, \mathcal{L})$ can be represented by a generator of the $SL(2,\mathbb{C})$ algebra. In other words, a change of the bases is governed by the $SL(2,\mathbb{C})$ symmetry. The $SL(2,\mathbb{C})$ invariance corresponds to the global conformal symmetry for holomorphic functions on $\mathbb{CP}^1$. In the present case we start from the holomorphic multivalued function Φ in (7.83) which is defined on $X = \mathbb{CP}^1 - \{0, 1, 1/z, \infty\}$. The result (7.139) is thus natural in concept but nontrivial in practice, because the equivalence condition $d\log\Phi \equiv 0$ in (7.92) is implicitly embedded into the expressions (7.133)–(7.138).

To conclude this chapter, we first review the definition of Aomoto's generalized hypergeometric functions on $Gr(k+1, n+1)$, interpreting their integral representations in terms of twisted homology and cohomology. We then consider reduction of the general $Gr(k+1, n+1)$ case to particular $Gr(2, n+1)$ cases. The case of $Gr(2,4)$ leads to Gauss' hypergeometric functions. We carry out a thorough study of this case in the this section. Much of the this chapter, by nature, deals with reviews of existed literature. But the results in (7.132)–(7.139) are new as far as the author notices.

References

[1] N. Arkani-Hamed, J. L. Bourjaily, F. Cachazo, A. B. Goncharov, A. Postnikov and J. Trnka, "Scattering amplitudes and the positive Grassmannian," arXiv:1212.5605 [hep-th].

[2] N. Arkani-Hamed and J. Trnka, "The amplituhedron," *JHEP* **1410**, 30 (2014).

[3] N. Arkani-Hamed and J. Trnka, "Into the amplituhedron," *JHEP* **1412**, 182 (2014).

[4] L. Ferro, T. Łukowski and M. Staudacher, "$\mathcal{N} = 4$ scattering amplitudes and the deformed Gramannian," *Nucl. Phys. B* **889**, 192 (2014).

[5] L. Ferro, T. Lukowski, A. Orta and M. Parisi, "Towards the amplituhedron volume," *JHEP* **1603**, 014 (2016). doi:10.1007/JHEP03(2016)014.

[6] F. Balli, A. Behtash, S. Kürkçüoğlu and G. Ünal, "Quantum Hall effect on the Grassmannians $\mathbf{Gr}_2(\mathbb{C}^N)$," *Phys. Rev. D* **89**, no. 10, 105031 (2014).

[7] I. M. Gelfand, General theory of hypergeometric functions, *Dokl. Akad. Nauk. SSSR* **288** (1986); English translation, *Soviet Math. Dokl.* **33** 573–577 (1986) .

[8] K. Aomoto and M. Kita, *Theory of Hypergeometric Functions* (in Japanese), Springer-Verlag, Tokyo (1994); English translation (by K. Iohara), Springer, Berlin (2011).

[9] Y. Haraoka, *Hypergeometric Functions* (in Japanese), Asakura-shoten, Tokyo (2002).

[10] Y. Abe, "A note on generalized hypergeometric functions, KZ solutions, and gluon amplitudes," *Nucl. Phys. B* **907**, 107 (2016).

8

Around Boundary Functions of the Right Half-Plane and the Unit Disc

F.-H. Li

Henan Suda Electric Vehicle Technology Co., Ltd.

Shigeru Kanemitsu

Kyushu Institute of Technology

CONTENTS

8.1 Prerequisites

References [17, Chapter 6] and [18, Chapter 6] were devoted to the basics of control theory when compared with number theory, especially in the light of the deep theory of zeta-functions including number-theoretic mean averages, whose view is rather unique. Prior to this, there were two trends of applications of mathematics to control theory, i.e. application of functional analysis, cf. e.g. [7,8], and that of algebraic geometry [4,5].

There we remarked that there is great similarity in control theory and number theory in their treatment of the signals in time domain (t) and frequency domain (ω) or an expanded frequency domain ($s = \sigma + j\omega$) which is conducted by the Laplace transform in the case of control theory while in the theory of zeta-functions, this role is played by the Mellin transform, both of which are special cases of the Fourier transforms and convert the signals in time domain to those in the right half-plane (expanded frequency domain), cf. e.g. [12, Chapter 7]. This remark is quite natural from our higher standpoint that *all phenomena are avatars of the niryana (core) and that all are waves and thus Fourier analysis comes in at the fundamental stage.* On comparing [9]., in the study of linear systems, in place of a state space equation in the time domain, one considers the transfer function (which is the ratio of the Laplace transforms of input and output signals) in the frequency domain. By the visualization principle, in its state space representation, it is to be embedded in the state equation. It is true in general that analysis of signals in frequency domain is far easier than that in time domain, and it is the governing procedure to study the signals in frequency domain and pull back the results to those in the time domain. This idea applies to many problems in life, taking averages and drawing out a general tendency which one applies to the analysis of general phenomena including anomalies or singularities.

Here the variables are connected by

$$s = \sigma + j\omega = -i\tau, \qquad w = e^{i\tau}, \quad \operatorname{Im}\tau > 0. \tag{8.1}$$

By the second transformation, the upper half-plane is mapped into the inside of the unit circle. This may be replaced by the transformation

$$w = e^{i\psi_0}\frac{z - z_0}{z - \bar{z_0}}, \qquad z.z_0 \in \mathcal{H} \tag{8.2}$$

and ψ_0 is any real constant [12, Section 3.3.1, pp. 73–78] and [12, section 6.3.4, pp. 206–209].

Definition 8.1. The domain where the complex frequency variable $s = \sigma + j\omega$ lies is the right half-plane, which we denote

$$\mathcal{RHP} = \{z \in \mathbb{C} | \sigma = \operatorname{Re} s > 0\} \tag{8.3}$$

once and for all. We call the ratio of output/input of the signal form/into a system (in frequency domain), the **transfer function**, which is in general just a rational function, and indeed, in the form of a linear fraction. We call a pole as transfer function **unstable** if it lies in $\mathcal{RHP}$. One of the aims of control theory is to cancel the unstable poles of transfer functions to make them stable. Let H^∞ denote the space of all functions analytic in the right half-plane $\mathcal{RHP}$ with H^∞ norm $\|f\|_\infty = \operatorname{ess\,sup}|f(j\omega)| < \infty$. A Hardy space H^p is defined by replacing the infinity norm by p-norm. BH^∞ indicates the subspace of H^∞ whose elements are bounded, i.e. have the H^∞ norm < 1, i.e. $|f(j\omega)| < 1$ for all ω ([15, p. 42]).

In the case of the Riemann zeta-function $\zeta(s)$, canceling the simple pole at $s = 1$ is simple enough. In order to keep the functional equation valid, we multiply by $s(1-s)$ to introduce the ξ-function

$$\xi(x) := s(1-s)_1^{-\frac{\pi}{2}s}\Gamma\left(\frac{s}{2}\right)\zeta(s) = \xi(1-s). \tag{8.4}$$

However, in the case of control theory, an unstable pole is to be canceled by a J-lossless conjugator.

Recall that a matrix $\Sigma = \Sigma(s)$ is J**-lossless** means that the identity (8.12) holds true, which in turn means that the system is power-preserving. [15, Definition 5.1, p. 82]. Let

$$\Sigma = \begin{pmatrix} \Sigma_{11} & \Sigma_{12} \\ \Sigma_{21} & \Sigma_{22} \end{pmatrix} \tag{8.5}$$

be a system (matrix) and

$$\begin{pmatrix} \boldsymbol{a}_1 \\ \boldsymbol{a}_2 \end{pmatrix} = \Sigma \begin{pmatrix} \boldsymbol{b}_1 \\ \boldsymbol{b}_2 \end{pmatrix} = \begin{pmatrix} \Sigma_{11} & \Sigma_{12} \\ \Sigma_{21} & \Sigma_{22} \end{pmatrix} \begin{pmatrix} \boldsymbol{b}_1 \\ \boldsymbol{b}_2 \end{pmatrix} \tag{8.6}$$

with two kinds of inputs $\boldsymbol{a}_1, \boldsymbol{a}_2$ and outputs $\boldsymbol{b}_1, \boldsymbol{b}_2$. The system $Sigma(s)$ is said to be **lossless** if it is stable and power-preserving, i.e.

$$|\boldsymbol{a}_1(s)|^2 + |\boldsymbol{a}_2(s)|^2 = |\boldsymbol{b}_1(s)|^2 + |\boldsymbol{b}_2(s)|^2 \tag{8.7}$$

on the imaginary axis, i.e. for all $s = j\omega$, $\omega \in \mathbb{R}$. This may be stated as [15, (4.48), p. 82].

$$\Sigma(-s)\Sigma(s) = I, \quad \forall s = j\omega, \quad \omega \in \mathbb{R}. \tag{8.8}$$

First the chain scattering representation of Σ is given by [15, (4.2), p. 68]. This depends on the regularity of the matrix Σ_{21}, in which the restriction has been removed in [30]. The chain scattering representation has been used in classical circuit analysis, and Kimura [15] is devoted to this subject, and one of the merits is that cascade connection can be expressed simply as the product of two chain scattering matrices. Another feature is the **homographic transformation** [15, (4.82), p. 93], [17, p. 165]. This reduces to an ordinary linear fractional transformation (8.80), and in the general case, it amounts to Sigel modular forms.

On a controller S, the chain scattering matrix

$$\mathrm{CHAIN}(\Sigma) = \Theta = \begin{pmatrix} \Theta_{11} & \Theta_{12} \\ \Theta_{21} & \Theta_{22} \end{pmatrix}. \tag{8.9}$$

acts as

$$\Theta S = (\Theta_{11}S + \Theta_{12})(\Theta_{21}S + \Theta_{22})^{-1} =: \mathrm{HM}(\Theta, S). \tag{8.10}$$

Then we have the concatenation rule (cf. [16, pp. 16–17], [17, p. 42])

$$\mathrm{HM}(\Theta_2, \mathrm{HM}(\Theta_1, S)) = \mathrm{HM}(\Theta_2\Theta_1, S). \tag{8.11}$$

(8.8) reads for

$$\Theta^*(j\omega)J\Theta(j\omega) = J, \tag{8.12}$$

where J is the signature matrix

$$J = \begin{pmatrix} I & O \\ O & -I \end{pmatrix}. \tag{8.13}$$

Given a transfer function $G(s)$, we say that the J-lossless matrix $\Theta = \Theta(s)$ is a ***J*-lossless stabilizing conjugator** of if it stabilizes G in the sense of [15, Definition 5.1, p. 108], i.e. if $\Theta(s)$ is a rational function with minimum degree which cancels all the unstable poles of $G(s)$:

(i) $G(s)\Theta(s)$ is stable;

(ii) $\deg \Theta(s)$ is eqaul to the number of unstable poles of $G(s)$ (which rules out the case that the zeros of $G(s)$ are canceled out by the poles of $\Theta(s)$).

8.2 $\mathcal{RHP}$ vs. $\mathcal{D}$

In this section, we describe the difference between features of number theory and control theory in the unit disc and in the half-plane.

8.2.1 Laurent Expansion to q-Expansion

Recall rudiments of Fourier expansion [3, pp. 159–160], [13], [20]. Given a function f on $[-T, T]$, we consider its **periodic extension** $\bar{f}$ coinciding f on $[-T, T]$ and often denote it by the same symbol f: $\bar{f} = f$ and speak of f as a **periodic function of period** $2T$. It suffices to consider f over any interval of length $2T$: $[0, 2T]$, say. If f is piecewise smooth and continuous on $[-T, T]$ (or more generally of L^p), then (its periodic extension) f can be expanded into a **Fourier series**:

$$f(t) = \sum_{n=-\infty}^{\infty} a_n \, e^{i\lambda_n t}, \quad \lambda_n = \frac{2\pi}{2T} n, \tag{8.14}$$

where the nth **Fourier coefficient** a_n is given by

$$a_n = \frac{1}{2T} \int_{-T}^{T} f(t) \, e^{-i\lambda_n t} \, dt. \tag{8.15}$$

Compare [3, Theorem 2.36]. In view of (8.14), the period is often taken to be 1 or 2π.

Corollary 8.1. *If $f(z)$ is analytic on the unit circle $C : |z| = 1$, consider the composite function*

$$\tilde{f}(\zeta) = f(e^{i\zeta}); \quad \tilde{f}(\zeta + 2\pi) = \tilde{f}(\zeta). \tag{8.16}$$

Then we have the **Fourier expansion as a complex series** *for $\tilde{f}(\zeta)$:*

$$f(e^{i\theta}) = \sum_{n=-\infty}^{\infty} a_n e^{in\zeta}, \tag{8.17}$$

where a_n coincides with the n-th Fourier coefficient given by (8.15)

Proof. Since $f(z)$ is analytic in an annulus

$$D : \rho < |z| < r, \quad 0 < \rho < 1 < r \tag{8.18}$$

containing the unit circle, it is expanded into a Laurent series

$$f(z) = \sum_{n=-\infty}^{\infty} a_n z^n, \tag{8.19}$$

where $(\rho < r_1 < r)$

$$a_n = \frac{1}{2\pi i} \int_{|z|=r_1} \frac{f(z)}{z^{n+1}} \, dz. \tag{8.20}$$

Writing

$$z = e^{i\zeta}, \quad \zeta = \xi + i\eta, \tag{8.21}$$

(8.19) leads to (8.17). The n-th Laurent coefficient becomes (8.17) on substituting (8.21) into (8.20).

Remark 8.1. *Under the correspondence* (8.21), *the domain* (8.18) *is mapped onto a horizontal strip*

$$\xi \in \mathbb{R}, \quad -\log r < \eta < \log \frac{1}{\rho}. \tag{8.22}$$

On the unit circle ($\zeta = \theta \in \mathbb{R}$), (8.17) may be viewed as an ordinary Fourier series (8.14) for the complex-valued periodic functions $f(\theta)$ of period $2T = 2\pi$. The Fourier series converges to $f(\theta)$ for almost all θ if $f \in L^p$, say. Therefore we may view (8.17) as a complex extension of the Fourier expansion of f periodic of period 2π. By considering the real and imaginary parts, we may infer the same results for real-valued periodic functions.

It seems rather customary in literature to choose $\zeta = 2\pi z$ and consider the composite function $\tilde{f}(z) = f(e^{2\pi i z})$ and speak of periodic functions of period 1 or functions defined on the (one-dimensional) torus $\mathbb{T} = \mathbb{T}^1 = \mathbb{R}/\mathbb{Z} \simeq [0, 1]$.

Example 8.1. *If $f(z)$ is analytic in $\mathcal{H}$, where $\mathcal{H}$ is the upper half-plane ($\operatorname{Im} z > 0$), then we still want to have the Fourier expansion* (8.17) *for the periodic functions $\tilde{f}(z)$ of period 2π in* (8.16). *If $\tilde{f}(z)$ is analytic in $\bar{\mathcal{H}}$, with a bar indicating its closure, then it is analytic in a domain containing the real axis ($\operatorname{Im} z = 0$), i.e. of the form* (8.22), *and we have the Fourier expansion.*

But what we would like to study are analytic functions in $\mathcal{H}$ and the real axis is the boundary corresponding to the circle of convergence of a power series. Therefore, the real axis is a rather delicate stuff and in general we can expect limit values on the boundary at best. To circumvent this, we note that under the correspondence (8.21) *with a slight change of notation*

$$q = e^{2\pi i z}, \quad z \in \mathcal{H}, \tag{8.23}$$

the upper half-plane is mapped into the inside of the unit circle with the origin removed:

$$\mathcal{H} \leftrightarrow 0 < |q| < 1 \tag{8.24}$$

and the **point at infinity** ∞ *corresponds to the origin. Compare Lemma 8.6 and the passage preceding it. We make a convention that we understand the behavior of $f(z)$ at ∞ through the behavior of $\tilde{f}(z)$ at 0. The symbol q is traditional in the theory for more than 100 years. The variable in $\mathcal{H}$ is often taken as τ and then $q = e^{2\pi i\tau}$ with $\operatorname{Im}\tau > 0$.*

Now we have an annulus (8.24), *and if $\tilde{f}(z)$ is meromorphic at the origin, then we may consider the Laurent expansion of $\tilde{f}(z) = f(q)$ at $q = 0$*

$$\tilde{f}(z) = f(q) = \sum_{n=-\infty}^{\infty} a_n q^n = \sum_{n=-\infty}^{\infty} a_n e^{2\pi i n z}, \tag{8.25}$$

where a_{-n} are 0 for some $M > 0$ onwards. We may speak of (8.25) *as a Fourier expansion of $\tilde{f}(z)$ which is a substitute for the "periodicity"*

$$\tilde{f}(z+1) = \tilde{f}(z). \tag{8.26}$$

Note that this not an ordinary periodicity but the coincidence of values in strips of width 1, *or $\mathcal{H}/\langle T\rangle$, where $Tz = z + 1$ is the translation.*

Definition 8.2. If $f(z)$ is a meromorphic function in $\mathcal{H}$ having the Fourier expansion (8.25) and satisfying the relation

$$f\left(\frac{-1}{z}\right) = z^{2k} f(z), \tag{8.27}$$

then f is called a (**holomorphic**) modular form of weight $2k \in \mathbb{N}$ if $f(z)$ is analytic at ∞, i.e. the series (8.25) starts from $n = 0$. If $a_0 = 0$, then a modular form is called a cusp form.

The holomoprhic modular forms are associated with the full modular group $\mathrm{SL}_2(\mathbb{Z})$ while in Section 8.3.1 we encounter $\mathrm{SL}_2(\mathbb{C})$. Non-holomorphic (or non-analytic) automorphic forms have been studied since the time of Maass wave forms, cf. [10].

8.2.2 $\mathcal{RHP} \to \mathcal{D}$

As a rather unique transition from the right half-plane $\mathcal{RHP}$ to the unit disc $\mathcal{D} = \{z \in \mathbb{C} | |z| < 1\}$, we give a new proof of the celebrated Abel continuity theorem based on the convergence of generalized Dirichlet series and conformality of an analytic function.

The traditional proof depends on the partial summation without integration which we also reproduce here. Our new proof depends on the partial summation for a general sequence and originates from Stieltjes integrals. This shows the power of a proper generalization and problem setting.

Following the tradition, we use $s = \sigma + it$ for the complex variable. Let $\{\lambda_n\} \subset \mathbb{R}$ be an increasing sequence for which we may suppose $\lambda_1 > 0$. For a sequence $\{a_n\} \subset \mathbb{C}$, the series

$$f(s) = \sum_{n=1}^{\infty} a_n e^{-\lambda_n s} \tag{8.28}$$

convergent in some half-plane is called a generalized Dirichlet series. 1. If $\lambda_n = \log n$ with log denoting the principal value, $f(s) = \sum_{n=1}^{\infty} \frac{a_n}{n^s}$ is (an ordinary) Dirichlet series.
2. If $\lambda_n = n$ and $e^{-s} = w$, $f(z) = f(z) = \sum_{n=1}^{\infty} a_n w^n$ is the power series.

Partial summation is thought of as a discrete version of integration by parts and is stated in its simplest form as

Corollary 8.2. (Partial summation)
Let $A_n = \sum_{k=1}^{n} a_k$ be the nth partial sum of $\{a_n\}$. Then

$$\sum_{k\le n} a_k b_k = B_n b_n - \sum_{k=1}^{n-1} A_k(b_{k+1} - b_k). \tag{8.29}$$

In [11] this is treated as a special case of integration by parts for Stieltjes integral.

Lemma 8.1. [21, Satz 1.4, p. 371] *Let $\{\lambda_n\} \subset \mathbb{R}$ be an increasing sequence with $\lambda_1 > 0$. Then for any complex sequence $\{a_n\}$ and $x > 0$ we write*

$$A(x) = \sum_{\lambda_1 \le \lambda_n \le x} a_n \tag{8.30}$$

with empty sum being 0. Suppose $g(t)$ is of class C^1 on $[\lambda_1, \infty)$, then prove the partial summation formula

$$\sum_{\lambda_1 \le \lambda_k \le x} a_k g(\lambda_k) = A(x)g(x) - \int_{\lambda_1}^{x} A(t)g'(u)\, du. \tag{8.31}$$

Proof.

$$\begin{aligned} A(x)g(x) - \sum_{\lambda_k \le x} a_k g(\lambda_k) &= \sum_{\lambda_k \le x} a_k(g(x) - g(\lambda_k)) \qquad (8.32)\\ = \sum_{\lambda_k \le x} \int_{\lambda_k}^{x} a_k g'(u)\, du &= \int_{\lambda_1}^{x} \sum_{\lambda_k \le u} a_k g'(u)\, du, \end{aligned}$$

which is the last term of the right-hand side of (8.31) with minus sign. We note the last step in (8.32), which is usually regarded as a coincidence of the exchange of the sum and the integral, but what we really made is the change of repeated (Stieltjes) integrals. Compare [28].

Lemma 8.2. *Let $\{\lambda_n\} \subset \mathbb{R}$ be an increasing sequence with $\lambda_1 > 0$. Then for any complex sequence $\{a_n\}$ and $x > 0$ we write*

$$A_m(x) = \sum_{\lambda_{m+1} \le \lambda_n \le x} a_n \tag{8.33}$$

with the empty sum being 0. *Suppose* $g(t)$ *is of class* C^1 *on* $[\lambda_1, \infty)$. *Then we have a variant of the partial summation formula*

$$\sum_{\lambda_{m+1}\le\lambda_k\le x} a_k g(\lambda_k) = A_m(x)g(x) - \int_{\lambda_{m+1}}^{x} A(u)g'(u)\,du. \tag{8.34}$$

This follows from Lemma 8.1 by subtracting the sum $A(\lambda_m)$ in (8.30).

In literature [6, pp. 2–4], [25, pp. 65–66], this lemma is stated and used in the form of the sum as follows.

Lemma 8.3. *For* $m \in \mathbb{N}$ *we write*

$$A_m(x) = \sum_{m+1\le n\le x} a_n. \tag{8.35}$$

Then

$$\sum_{m\le k\le x} a_k g(k) = A(m,x)g([x]) - \sum_{k=1}^{n-1} A(m, x-1)\Delta a_k. \tag{8.36}$$

Lemma 8.4. *If* $\sigma \ne 0$, *then*

$$|\Delta e^{-\lambda_n s}| \le \frac{|s|}{\sigma}\Delta e^{-\lambda_n \sigma}. \tag{8.37}$$

A common proof of Theorem 8.1 depends on Lemma 8.3 and 8.4 and can be found in many standard books, cf. [6], [25]. Our proof depending on Lemma 8.2 is very smooth.

Theorem 8.1. *If the series* (8.28) *is convergent for* $s = s_0 = \sigma_0 + it_0$, *then* $f(s)$ *is uniformly convergent in the right half-plane* $\sigma > \sigma_0$ *in the wide sense and represents an analytic function there. More precisely, let* D *be an angular domain*

$$\sigma - \sigma_0 \ge 0, \quad \arg(s - s_0) \le \delta \tag{8.38}$$

with $0 < \delta < \frac{\pi}{2}$. *Then* $f(s)$ *is uniformly convergent on* D *in the wide sense.*

Proof. We interpret the series (8.28) as

$$f(s) = \sum_{\lambda_1\le\lambda_n} a_n e^{-s\lambda_n}. \tag{8.39}$$

Then $f(s_0) = \sum_{\lambda_1\le\lambda_n} a_n e^{-\lambda_n s_0}$ is convergent. Let $S_0(x) = \sum_{\lambda_k\le x} a_n e^{-\lambda_n s_0}$ be the $[z]$th partial sum. Then the modified Cauchy condition holds true: For every $\varepsilon > 0$, there exists an N such that $n > m > N$ implies

$$|S_0(\lambda_n) - S_0(\lambda_m)| < \varepsilon.$$

We apply Lemma 8.2 with $A_m(x) = S_0(x) - S_0(\lambda_m)$ and $g(u) = e^{-(s-s_0)u}$ to $S(\lambda_n) - S(\lambda_m)$ with

$$S(x) = \sum_{\lambda_1\le\lambda_k\le x} a_k e^{-s_0 k} e^{-(s-s_0)k}.$$

Then

$$S(\lambda_n) - S(\lambda_m) = A_m(\lambda_n)e^{-(s-s_0)\lambda_n} + (s - s_0)\int_{\lambda_{m+1}}^{\lambda_n} A_m(u)e^{-(s-s_0)u}\,du.$$

It follows that

$$|S(\lambda_n) - S(\lambda_m)| = \varepsilon e^{-(\sigma-\sigma_0)\lambda_n} + |s - s_0|\varepsilon \int_{\lambda_{m+1}}^{\lambda_n} e^{-(\sigma-\sigma_0)u}\, du \leq \frac{|s - s_0|}{\sigma - \sigma_0}\varepsilon e^{-(\sigma-\sigma_0)\lambda_{m+1}}. \tag{8.40}$$

Since

$$\frac{|s - s_0|}{\sigma - \sigma_0} \leq \frac{1}{\cos\delta},$$

we see that the Cauchy condition is satisfied, so that $\{S(n)\}$ is uniformly convergent in the angular domain in the wide sense.

Corollary 8.3. (Counterpart of the Abel continuity theorem) *$f(s)$ approaches to $f(s_0)$ as $s \to s_0$ in the angular domain* (8.38).

This follows from the fact that the sum of uniformly convergent series of continuous functions is continuous.

Corollary 8.4. (Abel continuity theorem) *Suppose a power series $f(z) = \sum_{n=1}^{\infty} a_n z^n$ converges at the point z_0 on its circle of convergence. Draw two chords (inside the circle) that start from z_0 and form an angle δ with the tangent at z_0 of the circle ($0 < \delta < \frac{\pi}{2}$). Let Δ be the (closure of) intersection of this angular subdomain and the disc of convergence. Then as $f(z)$ approaches $f(s_0)$ as $z \to z_0$ in the angular domain inside Δ.*

Proof. For the special case 2, Corollary 8.3 implies the convergence. The corresponding domain under the analytic map $e^{-s} = z$ is the disc with radius $|e^{-s}| = e^{-\sigma}$. To make this in conformity with the radius of convergence, we need to magnify it. The angle is preserved since this is a conformal mapping.

8.2.3 Fourier Series as an Intrinsic Property of the Monolog

Proposition 8.1. *Let $\bar{B}_1(x) = x - [x] - \frac{1}{2}$ be the first periodic Bernoulli polynomial, where $[x]$ indicates the greatest integer function. For $x \notin \mathbb{Z}$ and $1 < N \in \mathbb{N}$, the approximation*

$$\bar{B}_1(x) = -\sum_{n=1}^{N} \frac{\sin 2\pi n x}{\pi n} + O\left(\min\left(1, \frac{1}{N\|x\|}\right)\right) \tag{8.41}$$

holds true, where $\|x\| = \min\{1 - \{x\}, \{x\}\}$ is the distance from x to the nearest integer with $\{x\}$ $x - [x]$ the integral part of x. Hence the Fourier expansion

$$\bar{B}_1(x) = -\frac{1}{\pi}\sum_{n=1}^{\infty} \frac{\sin 2\pi n x}{n} = -\frac{1}{2\pi i}\sum_{0\neq n=-\infty}^{\infty} \frac{e^{2\pi i n x}}{n} \tag{8.42}$$

follows which is uniformly convergent in any interval not containing an integer.

Proof. First we check oddness

$$\bar{B}_1(1 - x) = -\bar{B}_1(x). \tag{8.43}$$

This follows from $[1 - x] = [-[x] + 1 - \{x\}] = -[x]$. By periodicity and oddness, it suffices to consider the case $0 < x \leq \frac{1}{2}$. Note that for $m \neq 0$

$$\int_{\frac{1}{2}}^{x} e^{2\pi i m t}\, dt = \frac{1}{2\pi i m}\left(e^{2\pi i m x} - e^{2\pi i m}\right).$$

Summing this over $m = -N, \ldots, N, m \neq 0$, we deduce that

$$\int_{\frac{1}{2}}^{x} \sum_{\substack{m=-N \\ m\neq 0}}^{N} e^{2\pi imt}\, dt = \sum_{\substack{m=-N \\ m\neq 0}}^{N} \frac{1}{2\pi im} e^{2\pi imx}. \tag{8.44}$$

We use the elementary formula

$$\sum_{k=-N}^{N} e^{2\pi ikt} = \frac{\sin(2N+1)\pi t}{\sin \pi t}. \tag{8.45}$$

Adding $x - \frac{1}{2}$ to both sides of (8.44) and using this formula, we derive that

$$\sum_{\substack{m=-N \\ m\neq 0}}^{N} \frac{1}{2\pi im} e^{2\pi imx}\, dt + x - \frac{1}{2} = R(x) \tag{8.46}$$

where $R(x) = \int_{\frac{1}{2}}^{x} \frac{\sin(2N+1)\pi t}{\sin \pi t}\, dt$. Since $\frac{1}{\sin \pi t}$ is decreasing on $[x, \frac{1}{2}]$, the second mean value theorem for integrals applies to give

$$|R(x)| = \left| \frac{1}{\sin \pi x} \int_{\frac{1}{2}}^{\xi} \sin(2N+1)\pi t\, dt \right| \leq \frac{2}{x} \frac{2}{(2N+1)\pi}, \tag{8.47}$$

where we have used Jordan's inequality $\sin |x| \geq \frac{2}{\pi}|x|$ in the last step. Taking the range of x into account, we conclude that (8.47) leads to the estimate in (8.41) and so (8.46) proves the assertion.

We note that the argument in existing literature as well as the above proof suggests that the Fourier series is in the sense of symmetric limit $\lim_{N\to\infty} \sum_{-N}^{N}$, However, we have revealed [27] that this is just the difference of the Lerch zeta-function which we now introduce.

Lemma 8.5. (Generalization of Dirichlet's test) *Suppose that the partial sums of $a_n(x)$ are bounded uniformly in x, $x \in \mathrm{I} = [a, b]$, $\lim_{n\to\infty} b_n(s) = 0$ uniformly in $\sigma > 0$ and $|b_n(s) - b_{n+1}(s)| \leq c_n$ and $\sum_{n=1}^{\infty} c_n < \infty$. Then the series $\sum_{n=1}^{\infty} a_n(x) b_n(s)$ is uniformly convergent in $\sigma > 0$ and $x \in \mathrm{I}$.*

Proof can be given using Corollary 8.2.

Example 8.2. *The polylogarithm function of order s or the Lerch zeta-function*

$$l_s(z) = \sum_{n=1}^{\infty} \frac{e^{2\pi inz}}{n^s} \tag{8.48}$$

is absolutely convergent for $\sigma > 1$, $z \in \mathbb{R}$ or $\mathrm{Im}\, z > 0$, $s \in \mathbb{C}$. For $s = 1$, the series on the right of (8.48) is uniformly convergent in an interval not containing an integer and defines the polylogarithm function of order 1, which is indeed, the monologarithm function:

$$l_1(x) = \sum_{n=1}^{\infty} \frac{e^{2\pi inx}}{n} = -\log\left(1 - e^{2\pi ix}\right). \tag{8.49}$$

For fixed $s \in \mathbb{C}$, the Lerch zeta-function is a one-valued analytic function on the z-plane with slit along negative imaginary axis and its translations by integers, i.e.

$$\mathbb{C}_1 = \mathbb{C} - \{n + iy | n \in \mathbb{Z}, y \leq 0\}. \tag{8.50}$$

To prove absolute convergence of (8.48) for $\sigma > 1$, $zx \in \mathbb{R}$ we use for $0 < x < 1$

$$A_n(x) = \sum_{k=1}^{n} e^{2\pi ikx} = e^{2\pi ix} \frac{1 - e^{2\pi inx}}{1 - e^{2\pi ix}} \tag{8.51}$$

instead of (8.45). The denominator of the right-hand side of (8.51) is $|1 - e^{2\pi ix}| = 2 \sin \pi x = O(\|x\|)$. Since $b_n(s) = n^{-s}$ satisfies

$$|b_n(s) - b_{n+1}(s)| = O(n^{-\sigma-1}),$$

we appeal to the convergence of $\zeta(\sigma + 1)$ for $\sigma > 0$ and we may apply Lemma 8.5. To prove the last half of the example, standard Schwarz lemma is not enough and we need the following theorem

Theorem 8.2. (Generalization of Schwarz reflection principle) *Suppose a domain D has a line segment l as part of its boundary and that $w = f(z)$ is analytic in D and continuous on l such that z describes l, w describes a segment λ as part of the boundary of the domain $w(D)$. For any point $z \in D$ let ζ denote the reflection with respect to l and let ω be the reflection of w with respect to λ. Then $\omega = \omega(\zeta)$ is an analytic continuation of w.*

8.2.4 Boundary Functions of Certain Lambert Series

For details of this section, we refer to [18, Chapter 6]. Let $\{a_n\} \subset \mathbb{C}$ be such that

$$\limsup |a_n|^{1/n} \leq 1, \tag{8.52}$$

and *a fortiori* such that the power series $\sum_{n=1}^{\infty} a_n z^n$ is absolutely convergent in $|z| < 1$. Then the **Lambert series**

$$f(z) = \sum_{n=1}^{\infty} a_n \frac{z^n}{1 - z^n} \tag{8.53}$$

is absolutely convergent in $|z| < 1$ and represents an analytic function, and moreover, the power series of this function can be obtained by formal rearrangement of (8.53), i.e.

$$f(z) = \sum_{n=1}^{\infty} b_n z^n \quad (|z| < 1), \tag{8.54}$$

where

$$b_n = \sum_{d|n} a_d, \tag{8.55}$$

where d runs over all divisors of n.

Wintner's paper [29], elucidating Riemann's procedure in Fragment [23], [22] seems to be the first instance of considering the behavior of boundary function, see the passage after (8.57). Riemann divides the Lambert series by z and integrates the result from 0 to $z = re^{i\theta}$, $r = |z| < 1$, and then puts $r = 1$ to obtain his results, i.e. "Fourier expansions", in a formal way. Wintner's legitimation reads as follows. Slightly modifying Riemann's procedure by the trivial factor $e^{i\theta}$, which makes the function $F = F(r, \theta)$ a function in z, one obtains

$$F(re^{i\theta}) = \int_0^r \frac{f(re^{i\theta})}{r} dr = \sum_{n=0}^{\infty} \frac{b_n}{n} (re^{i\theta})^n,$$

or

$$F(z) = \int_0^r \frac{f(re^{i\theta})}{r} dr = \sum_{n=0}^{\infty} c_n z^n, \tag{8.56}$$

with

$$c_n = \frac{1}{n} b_n = \frac{1}{n} \sum_{d|n} a_n. \tag{8.57}$$

Then Wintner *asks if $F(z)$ tends to a measurable boundary function $F(e^{i\theta})$ as $r \to 1$ (within the Stolz path), and if so, then whether or not the boundary function is of class L^p, so that Riemann's formal trigonometric series actually is the Fourier series of the boundary function.*

Their use has been made from several ℓ^2-type conditions. We consider conditions

$$\sum_{n=-\infty}^{\infty} |n|^{\varepsilon} |c_n|^{\delta} < \infty, \tag{8.58}$$

which is referred to as the Riesz–Fischer condition for $\varepsilon = 0$, $\delta = 2$, the L^p-condition ($p > 2$) for $\varepsilon = 0$, $\delta = \frac{p}{p-1}$, and the Hausdorff–Paley extension of Riesz–Fischer theorem for $\varepsilon > 0$, $\delta = 2$ ($p \geq 2$). The last intermediate condition trigonometric series $\sum_{n=-\infty}^{\infty} c_n e^{in\theta}$ is convergent a.e. (almost everywhere) and represents the values of the function a.e.

Theorem 8.3. (Wintner) *Suppose*

$$a_n = O(n^{\lambda - \delta}) \tag{8.59}$$

for some $\delta > 0$ and a fixed $0 < \lambda \leq \frac{1}{2}$. Then the boundary function $F(e^{i\theta})$ exists and is measurable such that

$$F(re^{i\theta}) \to F(e^{i\theta}) \quad \text{a.e., as } r \to 1 \tag{8.60}$$

along the Stoltz path. If $\lambda < \frac{1}{2}$, then $F(e^{i\theta})$ is of class $L^{1/\lambda}$ and if in (8.59), the exponent can be taken arbitrarily small, then it is of class L^{∞}.

Proof. We may show that

$$c_n = O(n^{\lambda - 1 - \delta}) \tag{8.61}$$

for some $\delta > 0$. Hence if $\lambda < \frac{1}{2}$, then the L^p-condition in (8.58) is satisfied and if $\lambda = \frac{1}{2}$, then the series for $F(re^{i\theta})$ is Cauchy in L^2, and so there exists a function $F(e^{i\theta})$ of class L^2 such that

$$F(e^{i\theta}) \sim \sum_{n=-\infty}^{\infty} c_n e^{in\theta}. \tag{8.62}$$

This together with the Hausdorff–Paley condition in (8.58) implies that

$$F(e^{i\theta}) = \sum_{n=-\infty}^{\infty} c_n e^{in\theta} \quad \text{a.e.} \tag{8.63}$$

Hence (8.60) follows by the Abel continuity theorem, Corollary 8.4.

Example 8.3. *In the case* $a_{2n} = 0, a_{2n+1} = 4(-1)^{n+1}$, *we obtain the Lambert series*

$$f(z) = \frac{2K}{\pi} = 4\sum_{n=1}^{\infty} \frac{(-1)^{n+1} z^{2n+1}}{1 - z^{2n+1}} + 1, \tag{8.64}$$

which in the notation of (8.54) *reads*

$$f(z) = \frac{2K}{\pi} = 4 \sum_{\ell,m,n=0}^{\infty} d_{4,1}(n) z^{2^{\ell}(4m-1)^2 n} + 1, \tag{8.65}$$

with $d_{4,1}(n)$ *denoting the number of divisors of* n *of the form* $4k+1$. *Hence*

$$F(z) = \sum_{n=0}^{\infty} \frac{d_{4,1}(n)}{n} z^n, \quad |z| < 1. \tag{8.66}$$

Hence by Theorem 8.3, the boundary function $F(e^{i\theta})$ *exists and*

$$F(e^{i\theta}) = \sum_{n=1}^{\infty} \frac{d_{4,1}(n)}{n} e^{in\theta} \quad a.e. \tag{8.67}$$

In the case $a_n = 1$, *we obtain the Lambert series considered by Lambert himself.*

8.2.5 Control Theory in the Unit Disc

This section is a counterpart of control theory in contrast with number theory in the unit disc, which is in contrast with function theory in the complex plane and that in the unit ball [24]. The following table is a sequel to Table 8.1.

As is suggested by the title of the book of Rudin [24], "function theory in the unit ball" and that in the $\mathcal{RHR}$ can go parallel. This corresponds to the relation between the Lambert series and zeta-function developed in Chapter 4. This section describes some relevance of control theory in the unit disc developed in [8, pp. 120–123] to the number theory in the unit disc.

Definition 8.3. A complex function $f(s)$ (defined on a symmetric domain w.r.t.the real axis) satisfying the condition

$$f(s) = \overline{f(\bar{s})} \tag{8.68}$$

for all s is called **real** (on the real axis).

We denote by $\mathcal{A}_{\mathcal{RHP}}$ the (real) vector space of all functions $f(s)$ which are analytic in the $\mathcal{RHP}$ and which are real, bounded, and continuous in the closed $\mathcal{RHP}$. By $\mathcal{A}$ we denote

TABLE 8.1
Domains and Group Structure

Domain	Positive Real Axis	Right-Half Plane	Upper Half-Plane
Name	Time domain	Expanded frequency domain	Lobachevski plane
Variable	$t > 0$	$\sigma = \operatorname{Re} s > 0$	$\operatorname{Im} \tau > 0$
Mapping	$t \leftrightarrow \frac{1}{t}$	$s \leftrightarrow 1 - s$	$\tau \leftrightarrow -\frac{1}{\tau}$
Group	$\mathbb{R}_+^{\times}$	Symmetry: $s \leftrightarrow 1 - s$	$\Gamma \subset \mathrm{SL}_2(\mathbb{R})$
Object	Lambert series	Modular form	Zeta-function

the disc counterpart of $\mathcal{A}_{\mathcal{RHP}}$, i.e. the vector space of functions which are analytic in the $\mathcal{D}$ and which are real, bounded and continuous in the closed $\bar{\mathcal{D}}$.

Let $H^\infty(\mathcal{D})$ be the algebra of analytic functions on the unit disc $\mathcal{D} \subset \mathbb{C}$. A function in $H^\infty\mathcal{D}$ is called **inner** if it has a radial limit a.e. on the unit circle. Compare e.g. [2].

Let

$$f(a, z) = \frac{\bar{a}}{|a|} \frac{a - z}{1 - \bar{a}z} \tag{8.69}$$

and let $\{z_n\} \subset D$ satisfy the Blaschke condition

$$\sum_n (1 - |z_n|) < \infty. \tag{8.70}$$

Then the **Blaschke product**

$$B(z) = e^{ic} z^m \prod f(z_n, z) \tag{8.71}$$

is an inner function.

8.3 Robust Controller

8.3.1 GNP

The **generalized Nevanlinna–Pick problem** (GNP) states that given $(\alpha_i, \beta_i) \in \mathbb{C}$, $1 \le j \le \ell$ and additionally $\beta_0, \beta_{\ell+1} \in \mathbb{C}$ such that

$$\operatorname{Re} \alpha_i > 0, \quad |\beta_i| < 1, \quad 1 \le i \le \ell, \tag{8.72}$$

one is supposed to find a function $u \in \mathrm{BH}^\infty$ satisfying the interpolation conditions

$$u(\alpha_i) = \beta_i, \quad 1 \le i \le \ell, \quad u(0) = \beta_0, \; u(\infty) = \beta_{\ell+1}. \tag{8.73}$$

Compare Definition 8.1. The original Nevanlinna–Pick problem—NP problem is concerned with interpolation without additional condition $\beta_0, \beta_{\ell+1} \in \mathbb{C}$. In its classical version, the NP theorem reads

Theorem 8.4. *The solvability of the NP problem is equivalent to the positive definiteness of the Pick matrix.*

The Hermitian matrix P defined by

$$P = \begin{pmatrix} \frac{1-\beta_1\bar{\beta}_1}{\alpha_1+\bar{\alpha}_1} & \cdots & \frac{1-\beta_1\bar{\beta}_\ell}{\alpha_1+\bar{\alpha}_\ell} \\ \cdots & \cdots & \cdots \\ \frac{1-\beta_\ell\bar{\beta}_1}{\alpha_\ell+\bar{\alpha}_1} & \cdots & \frac{1-\beta_\ell\bar{\beta}_\ell}{\alpha_\ell+\bar{\alpha}_\ell} \end{pmatrix} \tag{8.74}$$

is called the Pick matrix. Let

$$B_i(s) = \frac{\alpha_i - s}{\bar{\alpha}_i + s}, \; 1 \le i \le \ell, \tag{8.75}$$

which is a BH^∞-function satisfying $|B_i(j\omega)| = 1$ for all $\omega \in \mathbb{R}$. Let $\beta_{i,k+1}$ denote the Fenyves array defined by

$$\beta_{i,1} = \beta_i, \; 1 \le i \le \ell, \tag{8.76}$$

$$\beta_{i,k+1} = \frac{(\alpha_i + \bar{\alpha}_i)(\beta_{i,k} - \beta_{k,k})}{(\alpha_i - \alpha_k)(1 - \bar{\beta}_{k,k}\beta_{i,k})}, \; 1 \le k \le i - 1 \le \ell - 1.$$

Let

$$\rho_k = \beta_{k,k}, \; 1 \le k \le \ell. \tag{8.77}$$

It is well known that P is positive definite if and only if $|\beta_{i,h}| < 1$. Define

$$\Theta_k(s) = \frac{1}{\sqrt{1 - |\rho_k|^2}} \begin{pmatrix} -B_k(s) & \rho_k \\ -\bar{\rho}_k B_k(s) & 1 \end{pmatrix}, \; 1 \le k \le \ell. \tag{8.78}$$

Remark 8.2. *The additional interpolation conditions in* (8.73) *need some explanation. The value at infinity pertains to the properness of the controller* $c(s)$ *expressed in (A1) in Theorem 8.6 in terms of* $u(s)$ *by* (8.102) *and* $\beta_{\ell+1} = 0$ *will suffice.*

On the other hand, Lemma 8.8 implies that all the unstable poles of $p_0(s)$ *are to be canceled out by the zeros of* $q(s)$*, including the simple pole at* $s = 0$*, which imposes the other interpolation condition on* $u(s)$*, i.e.*

$$|u(0)| = |\beta_0|, \tag{8.79}$$

which is the essential addition to the original NP problem.

We recall the definition of the homographic transformation (8.10). In our present case, the controller is a scalar $S = s \in \mathbb{C}$ and (8.10) amounts to the linear fractional transformation

$$\mathrm{HM}(\Theta, s) = \Theta s = \frac{t_{11}s + t_{12}}{t_{21}s + t_{22}}, \tag{8.80}$$

where we put

$$\Theta = \begin{pmatrix} t_{11} & t_{12} \\ t_{21} & t_{22} \end{pmatrix} \tag{8.81}$$

which may be thought of as $\in \mathrm{SL}_2(\mathbb{C})$. Since the origin and the point at infinity are symmetric points (relative to the fractional transformations Θ, cf. [1]), the values at those points may be conveniently used to determine Θ.

Lemma 8.6. *Suppose*

$$u_1 = \Theta u_{\ell+1} = \frac{t_{11}u_{\ell+1} + t_{12}}{t_{21}u_{\ell+1} + t_{22}}. \tag{8.82}$$

and that the inverse transformation of (8.82) *is given by*

$$u_{\ell+1}(s) = f_1(s)\frac{u_1(s) + f_3(s)}{1 + f_2(s)u_1(s)}, \tag{8.83}$$

where

$$f_1(s) = \frac{t_{22}(s)}{t_{11}(s)}, \; f_2(s) = -\frac{t_{21}(s)}{t_{11}(s)}, \; f_3(s) = -\frac{t_{12}(s)}{t_{22}(s)}. \tag{8.84}$$

Let

$$\delta_0(= u_{\ell+1}(0)) = f_1(0)\frac{u_1(0) + f_3(0)}{1 + f_2(0)u_1(0)} \tag{8.85}$$

$$\delta_\infty(= u_{\ell+1}(\infty)) = f_1(\infty)\frac{u_1(\infty) + f_3(\infty)}{1 + f_2(\infty)u_1(\infty)}.$$

and suppose

$$|\delta_0| < 1, \quad |\delta_{\ell+1}| < 1. \tag{8.86}$$

Then the functions $u_{\ell+1} \in \mathrm{BH}^\infty$ *satisfying* (8.95) *are given by*

$$u_{\ell+1}(s) = \frac{cu_{\ell+1}(\infty)s + du_{\ell+1}(0)}{cs+d}. \tag{8.87}$$

Proof. By (8.86) we may choose the values of $u_{\ell+1}$ at $0, \infty$ as in (8.85) (cf. (8.95)) to obtain

$$|u_{\ell+1}(0)| < 1, \quad |u_{\ell+1}(\infty)| < 1, \tag{8.88}$$

whence we only need to determine the form of the linear transformation $u_{\ell+1}$. This is a well-known procedure (e.g. [1]) and we readily conclude (8.87).

The following theorem is a slightly modified version of [15, Theorem 5.8, p. 115] in conjunction with [14] and is crucial for the proof of Theorem 8.6.

Theorem 8.5. *The GNP is solvable if and only if the pair of matrices*

$$A = \begin{pmatrix} \alpha_1 & 0 & \cdots & 0 \\ 0 & \alpha_2 & \cdots & 0 \\ & & \cdots & \\ 0 & 0 & \cdots & \alpha_\ell \end{pmatrix}, \quad B = \begin{pmatrix} 1 & -\beta_1 \\ 1 & -\beta_2 \\ & \cdots \\ 1 & -\beta_\ell \end{pmatrix} \tag{8.89}$$

has a J-lossless stabilizing conjugator Θ *for the signature matrix*

$$J = \begin{pmatrix} I & O \\ O & -I \end{pmatrix}, \tag{8.90}$$

and

$$|\beta_0| < 1, \quad |\beta_\ell| < 1. \tag{8.91}$$

If this is the case, then all the parametric solutions u_1 *are obtained by the procedure*

$$u_k = \mathrm{HM}(\Theta_k, u_{k+1}), \quad 1 \le k \le \ell \tag{8.92}$$

starting from any $u_{\ell+1} \in \mathrm{BH}^\infty$ *satisfying Condition* (8.96), *where* Θ_k *is defined by*

$$\Theta_k(s) = \frac{1}{\sqrt{1-|\rho_k|^2}} \begin{pmatrix} -B_k(s) & \rho_k \\ -\bar{\rho}_k B_k(s) & 1 \end{pmatrix}, \ 1 \le k \le \ell. \tag{8.93}$$

in place of (8.78).

That is, if Θ *is defined by*

$$\Theta = \prod_{k=1}^{\ell} \Theta_k, \tag{8.94}$$

then $\mathrm{HM}(\Theta, u_{\ell+1})$ *solves the NP problem with the additional interpolation conditions enunciated in* (8.73)

$$u_1(0) = \beta_0, \quad u_1(\infty) = \beta_{\ell+1}. \tag{8.95}$$

We refer to [18, Chapter 6] for a detailed proof. We remark that in proving necessity, by Lemma 8.6, we may choose a suitable $u_{\ell+1}(s) \in \mathrm{BH}^\infty$ such that

$$u_{\ell+1}(0) = \delta_0, \quad u_{\ell+1}(\infty) = \delta_\infty \tag{8.96}$$

as in (8.85), then $u_1(s)$ satisfies (8.95) since (8.82) and (8.83) are equivalent.

8.3.2 Robust Stabilizer

Definition 8.4. A minimal phase factor $f_m(s)$ of $f(s)$ is a function having no zero in $\operatorname{Re} s \geq 0$ such that $|f_m(j\omega)| = |f(j\omega)|$.

Let r be the uncertainty (band) function defined by

$$r(s) = \frac{\rho(s)}{s}, \tag{8.97}$$

where $\rho(s) = \rho_m(s)$ is the minimal phase function.

Let a rational transfer matrix $p_0(s)$ be the nominal model of the plant dynamics with a simple pole at $s = 0$. This occurs when a plant is imprecisely given and the true plant lies within some uncertainty range of $p_0(s)$ as given below.

A transfer function $p(s)$ is said to belong to the class $C(p_0, r)$ if

(i) $p(s)$ has the same number of unstable poles as $p_0(s)$.

(ii) $|p(j\omega) - p_0(j\omega)| \leq |r(j\omega)|, \quad |r(j\omega)| > 0$ for $\forall \omega \in \mathbb{R}$.

This is a candidate for the true transfer function. Compare e.g. [26].

Definition 8.5. Consider a unity feedback system Σ consisting of the input r, output y, the controller (compensator) C (also often denoted K), and the plant P, with intermediate input to and output from C denoted by e and u, respectively. The class $C(p_0, r)$ is said to be robustly stabilizable if there exists a controller C such that the unity feedback system Σ is stable for each $p \in C(p_0, r)$. Such a controller is called a **robust stabilizer**.

The main result of [3, Section 1.12], [18, Chapter 6] is the following theorem.

Theorem 8.6. ([14, Theorem 2]) *Suppose*

(A1) *All the unstable poles (those lying in $\mathcal{RHP}$) of $p_0(s)$ are simple and $p_0(s)$ has a simple pole at $s = 0$*

(A2) *The relative degree of the denominator of $r(s)$ is at most* 1.

Then the class of plants are robustly stabilizable if and only if the Pick matrix P in (8.74) *is positive definite and*

$$|\beta_0| = |r'_m(0)/\tilde{p}_0(0)| < 1, \tag{8.98}$$

where $\tilde{p}_0$ is defined by (8.100) *and is a nominal plant $p_0(s)$ for all the poles being eliminated.*

In the proof, the Blascke product plays an important role. Suppose $\alpha_1, \cdots, \alpha_\ell$ are the unstable poles of $p_0(s)$. Let $B(s)$ be the product of all $B_i(s)$'s in (8.75)—the Blaschke product (for which see e.g. [19])

$$B(s) = \prod_{i=1}^{\ell} B_i(s), \tag{8.99}$$

which is a BH^∞-function satisfying $|B(j\omega)| = 1$ and cancels all unstable poles of $p_0(s)$. We modify $p_0(s)$ by eliminating all the poles by multiplying the BH^∞-function ([14, (35]):

$$\tilde{p}_0(s) := sB(s)p_0(s) \tag{8.100}$$

is a stable function. We refer to [18, Chapter 6] for details of proof, and we just state necessary lemmas.

[31] introduced the function

$$q(s) = c(s)(1 + p_0(s)c(s))^{-1} \tag{8.101}$$

with nominal function $p_0(s)$ introduced in Definition 8.4, which can be solved for the stabilizer $c(s)$:

$$c(s) = q(s)(1 - p_0(s)q(s))^{-1} \tag{8.102}$$

Lemma 8.7. *The controller $c(s)$ is a robust stabilizer for the class $C(p_0, r)$ if and only if Σ is stable for $p = p_0$ and*

$$|r(j\omega)q(j\omega)| < 1 \tag{8.103}$$

for all $\omega \in \mathbb{R}$.

Lemma 8.8. *The unity feedback system Σ is stable for $p = p_0$ if and only if*

(i) *$q(s)$ is stable*

(ii) *$1 - p_0(s)q(s)$ has the zeros at the unstable poles of $p_0(s)$ up to multiplicity. This implies that all the unstable poles of $p_0(s)$ are canceled out by the zeros of $q(s)$.*

References

[1] Ahlfors, L., *Complex Analysis*, McGraw-Hill, New York (1996).

[2] Beurling, A., On two problems concerning linear transformations in Hilbert space, *Acta Math.* **81** (1949) 239–255.

[3] Chakraborty, K., Kanemitsu, S., and Kuzumaki, T., *A Quick Introduction to Complex Analysis*, World Scientific Publishing, Singapore (2016).

[4] Falb, P., *Methods of Algebraic Geometry in Control Theory: Part I Scalar Linear Systems and Affine Algebraic Geometry*, Birkhäuser, Basel (1990).

[5] Falb, P., *Methods of Algebraic Geometry in Control Theory: Part II Multivariable Linear Systems and Projective Algebraic Geometry*, Birkhäuser, Basel (2000).

[6] Hardy, G. H. and Riesz, M., *The General Theory of Dirichlet's Series*, Cambridge University Press, Cambridge (1915); reprint, Hafner, New York (1972).

[7] Helton, J. W., Non-Euclidean functional analysis and electronics, *Bull. Am. Math. Soc.* **7** (1982) 1–64.

[8] Helton, J. W. and Merino, O., *Classical Control Using H^∞ Methods, Theory, Optimization and Design*, SIAM, Philadelphia, PA (1998).

[9] Kamata, Y., Ma, T.-L., Sun, Y., and Kanemitsu, S., Electric current as a continuous flow, Chapter 10 of this volume.

[10] Kanemitsu, S., Agarwal, P. G., and Li, H.-L., Note on the Kuznetsov sum formula, *Basic Sci. J. Text. Univ.* **30**, No. 3 (2017) 293–301.

[11] Kanemitsu, S., Kuzumaki, T., and Liu, J. -Y., Analysis and betond, to appear.

[12] Kanemitsu, S. and Tsukada, H., *Contributons to the Theory of Zeta-Functions: The Modular Relation Supremacy*, World Scientific Publishing, Singapore (2015).

[13] Katznelson, Y., *Introduction to Harmonic Analysis*, 3rd ed., Cambridge University Press, Cambridge (2004).

[14] Kimura, H., Robust stabilizability for a class of transfer functions, *IEEE Trans. Autom. Control* **27** (1984) 788–793.

[15] Kimura, H., *Chain Scattering Approach to H^∞-Control*, Birkhäuser, Boston, Basel, Berlin (1997).

[16] Kuroš, A. G., *The Theory of Groups*, Vol. II, AMS-Chelsea, Providence, RI (1974).

[17] Li, F. -H., Wang, N. -L., and Kanemitsu, S., *Number Theory and Its Applications*, World Scientific Publishing, London, Singapore, New Jersey, November 28 (2012).

[18] Li, H. -L., Li, F. -H., Wang, N. -L., and Kanemitsu, S., *Number Theory and Its Applications II*, 285 p., World Scientific Publishing, London, Singapore, New Jersey, November 30 (2017).

[19] Marshall, D. E. and Stray, A., Interpolating Blaschke products, *Pac. J. Math.* **173** (1992) 759–769.

[20] Papoulis, A., *The Fourier Integral and Its Applications*, McGraw-Hill, New York (1962).

[21] Prachar, K., *Primzahlverteilung*, Springer Verlage, Berlin (1957).

[22] Riemann, B., Fragmente über Grenzfälle der ellipitischen Modulfunctionen, Posthumous fragment, 455–465 (1953).

[23] Riemann, B., *Collected Works of Bernhard Riemann*, ed. by H. Weber, 2nd ed. Dover, New York (1953).

[24] Rudin, W., *Function Theory in the Unit Ball of $\mathbb{C}^n$*, Springer Verlage, Berlin, Heidelberg (2008).

[25] Serre, J. -P., *A Course in Arithmetic*, Springer Verlage, New York, Berlin, Heidelberg (1973).

[26] Vidyasagar, M. and Kimura, H., Robust controllers for uncertain linear multivariable systems, *Automatica* **22** (1983) 585–601.

[27] Wang, X.-H., Mehta, J., and Kanemitsu, S., The boundary Lerch zeta-function and short character sums a la Y. Yamamoto, Kyushu J. Math (2019).

[28] Widder, D. W., *The Laplace Transform*, Princeton University Press, Princeton, NJ (1946).

[29] Wintner, A., On Riemann's fragment concerning elliptic modular functions, *Am. J. Math.* **63** (1941) 628–634.

[30] Xin, X. and Kimura, H., Singular (J, J')-lossless factorization for strictly proper functions, *Int. J. Control* **59**, No. 6 (1994), 1383–1400.

[31] Zames, G. and Francis, B. A., Feedback minimax sensitivity, and optimal robustness *IEEE Trans. Autom. Control* **28** (1992) 759–769.

9

The Stankovich Integral Transform and Its Applications

Arsen Pskhu
Institute of Applied Mathematics and Automation, Kabardino-Balkar Scientific Center Russian Academy of Sciences (KBSC RAS)

CONTENTS

9.1 Introduction

In [1], B. Stakovich introduced an integral transform given by

$$G(x) = \int_0^\infty \phi\left(-\nu, 0; -\frac{y}{x^\nu}\right) \frac{g(y)}{\nu y}\, dy \qquad (0 < \nu < 1), \tag{9.1}$$

where $\phi(-\nu, 0; \cdot)$ stands for the Wright function [2,3], which is defined by

$$\phi(\delta, \mu; z) = \sum_{k=0}^{\infty} \frac{z^k}{k!\Gamma(\mu + \delta k)} \qquad (\delta > -1).$$

Paper [1] was one of a series of works devoted to the generalization of the Hankel transform [4,5]. A study of the transform (9.1) was continued in [6–8].

As it became clear, transforms of this kind play a special role in fractional calculus due to its various effective applications, in particular, in the theory of fractional differential equations and in the theory of special functions [9–12].

This paper overviews two adjoint integral transforms involving the Wright function as the kernel, discusses their basic properties, especially emphasizing those related to operators of fractional integration and differentiation, and describe their application to solve evolutionary equations of fractional order and to study special functions arising in fractional calculus.

The paper has the following structure. In Section 9.2 we give the definitions of the Stankovich transforms. Section 9.3 presents the properties of the transforms, including images of some special functions. Some applications are indicated in Section 9.4.

9.2 Transforms Definition

Let

$$0 < \alpha, \beta < 1, \qquad \text{and} \qquad \mu, \nu \in \mathbb{R}.$$

The main object we study here is the pair of integral transforms

$$A^{\alpha,\mu}u(x) \equiv \left(A^{\alpha,\mu}u\right)(x) = \int_0^{\infty} u(t)\, x^{\mu-1}\phi\left(-\alpha, \mu; -\frac{t}{x^{\alpha}}\right) dt, \tag{9.2}$$

and

$$B^{\beta,\nu}v(x) \equiv \left(B^{\beta,\nu}v\right)(x) = \int_0^{\infty} v(t)\, t^{\nu-1}\phi\left(-\beta, \nu; -\frac{x}{t^{\beta}}\right) dt. \tag{9.3}$$

We write A^{α} and B^{β} instead of $A^{\alpha,0}$ and instead of $B^{\beta,0}$, respectively, for short.

It is assumed that

$$u(t)\phi\left(-\alpha, \mu; -\frac{t}{x^{\alpha}}\right) \in L\left(\mathbb{R}_+\right) \quad \text{and} \quad v(t)t^{\nu-1}\phi\left(-\beta, \nu; -\frac{x}{t^{\beta}}\right) \in L\left(\mathbb{R}_+\right) \tag{9.4}$$

for any positive x.

In what follows, we denote by $\mathcal{A}^{\alpha,\mu}$ the set of locally integrable functions (i.e. $u(x) \in L(a, b)$ for all $a > b > 0$) such that

$$\lim_{x \to 0} u(x)x^{\varepsilon} < \infty$$

where $\varepsilon < 1$ for $(-\mu) \notin \mathbb{N} \cup \{0\}$, and $\varepsilon < 2$ otherwise; and

$$\lim_{x\to\infty} u(x)\exp(-\rho x^{\delta}) < \infty \quad \text{for some} \quad \delta < \frac{1}{1-\alpha} \quad \text{and} \quad \rho > 0.$$

Likewise, $\mathcal{B}^{\beta,\nu}$ denotes the set of functions $v(x) \in L(a, b)$ $(a > b > 0)$ satisfying

$$\lim_{x\to 0} v(x)\exp(-\rho x^{-\gamma}) < \infty, \qquad \gamma < \frac{\beta}{1-\beta}, \quad \rho > 0;$$

and

$$\lim_{x\to\infty} v(x)x^{\varepsilon} < \infty,$$

where $\varepsilon > \nu$ for $(-\nu) \notin \mathbb{N}\cup\{0\}$, and $\varepsilon < \nu - \beta$ otherwise;

From the asymptotic expansion of the Wright function (see [3]), an inequality follows:

$$|\phi(-\delta,\, \varepsilon;\, z)| \le C\, \exp\left(-\nu\, |z|^{\frac{1}{1-\delta}}\right), \qquad C = C(\delta, \varepsilon, \nu), \tag{9.5}$$

where $\delta \in (0,1)$, $\varepsilon \in \mathbb{R}$, and

$$\nu < (1-\delta)\delta^{\frac{\delta}{1-\delta}} \cos\frac{\pi - |\arg z|}{1-\delta}, \quad \pi \ge |\arg z| > \frac{1+\delta}{2}\pi.$$

Note that the inclusions

$$u(x) \in \mathcal{A}^{\alpha,\mu} \qquad \text{and} \qquad v(x) \in \mathcal{B}^{\beta,\nu}, \tag{9.6}$$

and (9.5) provide (9.4). Therefore, each of the inclusions (9.6) ensures the convergence of the corresponding integrals in (9.2) and (9.3).

9.3 Properties

9.3.1 Adjointness Property

Lemma 9.1. *Let*

$$u(t)v(x)x^{\mu-1}\phi\left(-\alpha, \mu; -\frac{t}{x^{\alpha}}\right) \in L\left(\mathbb{R}^2_+\right). \tag{9.7}$$

Then

$$\int_0^{\infty} v(x)\, A^{\alpha,\mu}u(x)\, dx = \int_0^{\infty} u(x)\, B^{\alpha,\mu}v(x)\, dx \tag{9.8}$$

Proof. The equality (9.8) is a corollary of inclusion (9.7) and Fubini's theorem. □

In particular, by virtue of $B^{\beta}1 = 1$ (see (9.13) below), formula (9.8) implies that

$$\int_0^{\infty} A^{\alpha} f(x)\, dx = \int_0^{\infty} f(x)\, dx.$$

9.3.2 Transforms of Power Functions

Lemma 9.2. *Let ν be such that $x^{\nu-1} \in \mathcal{A}^{\alpha,\mu}$. Then*

$$A^{\alpha,\mu}x^{\nu-1} = x^{\alpha\nu+\mu-1}\frac{\Gamma(\nu)}{\Gamma(\alpha\nu+\mu)}, \tag{9.9}$$

and

$$A^{\alpha}x^{-1} = \alpha x^{-1}. \tag{9.10}$$

Proof. Substituting sx^{α} for t in (9.2), we get

$$A^{\alpha,\mu}x^{\nu-1} = x^{\alpha\nu+\mu-1}\int_0^{\infty} s^{\nu-1}\phi\,(-\alpha, \mu; -s)\, ds.$$

Now, using formulas [1,13]

$$\int_0^\infty s^{\nu-1}\phi\,(-\alpha,\mu;-s)\,ds = \frac{\Gamma(\nu)}{\Gamma(\alpha\nu+\mu)} \quad \text{and} \quad \int_0^\infty \frac{1}{s}\phi\,(-\alpha,0;-s)\,ds = \alpha \tag{9.11}$$

yields (9.9) and (9.10) □

Lemma 9.3. *Let δ be such that $x^\delta \in \mathcal{B}^{\beta,\nu}$. Then*

$$B^{\beta,\nu}x^\delta = \frac{\Gamma(-(\nu+\delta)/\beta)}{\beta\Gamma(-\delta)}x^{(\nu+\delta)/\beta}, \tag{9.12}$$

and

$$B^\beta x^\delta = \frac{\Gamma(1-\delta/\beta)}{\Gamma(1-\delta)}x^{\delta/\beta}. \tag{9.13}$$

Proof. After a change of variable in the integral in (9.3) we get

$$B^{\beta,\nu}x^\delta = \frac{1}{\beta}x^{\frac{\nu+\delta}{\beta}}\int_0^\infty s^{-\frac{\nu+\delta}{\beta}-1}\phi\,(-\beta,\nu;-s)\,ds.$$

Combining this with (9.11) leads to (9.12). Using (9.12) and the formula $z\Gamma(z) = \Gamma(z+1)$ we obtain (9.13) □

9.3.3 Convolution Property

In what follows, $(u*v)(x)$ denotes the Laplace convolution of the functions $u(x)$ and $v(x)$, i.e.

$$(u*v)(x) = \int_0^x u(t)v(x-t)\,dt.$$

Lemma 9.4. *Let $u(x) \in \mathcal{A}^{\alpha,\mu} \cap L(0,a)$, $v(x) \in \mathcal{A}^{\alpha,\nu} \cap L(0,a)$, $(u*v)(x) \in \mathcal{A}^{\alpha,\mu+\nu}$, $A^{\alpha,\mu}u \in L(0,a)$, and $A^{\alpha,\nu}v \in L(0,a)$ for any positive a. Then*

$$A^{\alpha,\mu+\nu}(u*v)(x) = (A^{\alpha,\mu}u * A^{\alpha,\nu}v)(x). \tag{9.14}$$

Proof. By the definition (9.2), we obtain

$$\begin{aligned}
A^{\alpha,\mu+\nu}(u*v)(x) &= x^{\mu+\nu-1}\int_0^\infty \phi\,(-\alpha,\mu+\nu;-tx^{-\alpha})\int_0^t u(s)v(t-s)\,ds\,dt \\
&= x^{\mu+\nu-1}\int_0^\infty u(s)\int_s^\infty \phi\,(-\alpha,\mu+\nu;-tx^{-\alpha})\,v(t-s)\,dt\,ds \\
&= x^{\mu+\nu-1}\int_0^\infty\int_0^\infty u(s)v(t)\phi\left(-\alpha,\mu+\nu;-\frac{t+s}{x^\alpha}\right)dt\,ds.
\end{aligned}$$

Taking into account the formula

$$x^{\mu+\nu-1}\phi\left(-\alpha,\mu+\nu;-\frac{t+s}{x^\alpha}\right)$$

$$= \int_0^x \xi^{\mu-1}\phi\left(-\alpha,\mu;-\frac{t}{\xi^\alpha}\right)(x-\xi)^{\nu-1}\phi\left(-\alpha,\nu;-\frac{s}{(x-\xi)^\alpha}\right)d\xi$$

we get (9.14). □

9.3.4 Composition Rules

Lemma 9.5. *1) If* $u(x) \in \mathcal{A}^{\beta,\nu} \cap \mathcal{A}^{\alpha\beta,\alpha\nu+\mu}$ *and* $\left(A^{\beta,\nu}u\right)(x) \in \mathcal{A}^{\alpha,\mu}$*, then*

$$A^{\alpha,\mu}A^{\beta,\nu}u(x) = A^{\alpha\beta,\alpha\nu+\mu}u(x). \tag{9.15}$$

2) If $v(x) \in \mathcal{B}^{\beta,\nu} \cap \mathcal{B}^{\alpha\beta,\beta\mu+\nu}$ *and* $\left(B^{\beta,\nu}v\right)(x) \in \mathcal{B}^{\alpha,\mu}$*, then*

$$B^{\alpha,\mu}B^{\beta,\nu}v(x) = B^{\alpha\beta,\beta\mu+\nu}v(x). \tag{9.16}$$

Proof. The formulas (9.15) and (9.16) are consequences of the identities

$$A^{\alpha,\mu}\left[x^{\nu-1}\phi\left(-\beta,\nu;-\frac{t}{x^{\beta}}\right)\right] = x^{\alpha\nu+\mu-1}\phi\left(-\alpha\beta,\alpha\nu+\mu;-\frac{t}{x^{\alpha\beta}}\right),$$

and

$$B^{\alpha,\mu}\left[t^{\nu-1}\phi\left(-\beta,\nu;-\frac{x}{t^{\beta}}\right)\right] = t^{\beta\mu+\nu-1}\phi\left(-\alpha\beta,\beta\mu+\nu;-\frac{x}{t^{\alpha\beta}}\right),$$

which are given below (see (9.37) and (9.38)). □

9.3.5 Laplace Transform

In what follows, $\mathcal{L}\{f(x);p\}$ denotes the Laplace transform of the function $f(x)$,

$$\mathcal{L}\{f(x);p\} = \int_0^{\infty} e^{-xp}f(x)\,dx. \tag{9.17}$$

Lemma 9.6. *Let* $u(x) \in L(0,a)$, $\left(A^{\alpha,\mu}u\right)(x) \in L(0,a)$ *for any positive* a*;*

$$\lim_{x\to\infty} u(x)e^{-\sigma x} < \infty, \qquad \lim_{x\to\infty} e^{-\sigma x}A^{\alpha,\mu}u(x) < \infty \quad \text{for some } \sigma;$$

and $U(p) = \mathcal{L}\{u(x);p\}$*. Then*

$$\mathcal{L}\{\left(A^{\alpha,\mu}u\right)(x);p\} = p^{-\mu}U(p^{\alpha}). \tag{9.18}$$

Proof. Using the formula

$$\mathcal{L}\left\{x^{\mu-1}\phi\left(-\alpha,\mu;-cx^{-\alpha}\right);p\right\} = p^{-\mu}\exp\left(-cp^{\alpha}\right) \qquad (c>0)$$

we get

$$\mathcal{L}\{\left(A^{\alpha,\mu}u\right)(x);p\} = p^{-\mu}\int_0^{\infty} u(t)\exp\left(-tp^{\alpha}\right)dt.$$

This gives (9.18). □

9.3.6 Mellin Transform

From now on, $\mathcal{M}\{g(x);s\}$ denotes the Mellin transform of the function $g(x)$, i.e.

$$\mathcal{M}\{g(x);s\} = \int_0^{\infty} x^{s-1}g(x)\,dx$$

Lemma 9.7. *1) Let* $u(x) \in \mathcal{A}^{\alpha,\mu}$*,*

$$x^{s-1}u(x) \in L(0,\infty) \qquad \text{and} \qquad x^{s-1}A^{\alpha,\mu}u(x) \in L(0,\infty)$$

when $\sigma_0 < s < \sigma_1$ for some σ_0 and σ_1; and $u^*(s) = \mathcal{M}\{u(x); s\}$. Then

$$\mathcal{M}\{A^{\alpha,\mu}u(x); s\} = u^*(1 + (s + \mu - 1)/\alpha)\frac{\Gamma((1 - s - \mu)/\alpha)}{\alpha\Gamma(1 - s)}, \tag{9.19}$$

2) Let $v(x) \in \mathcal{B}^{\beta,\nu}$,

$$x^{s-1}v(x) \in L(0, \infty) \qquad \text{and} \qquad x^{s-1}B^{\beta,\nu}v(x) \in L(0, \infty)$$

when $\sigma_0 < s < \sigma_1$ for some σ_0 and σ_1; and $v^*(s) = \mathcal{M}\{v(x); s\}$. Then

$$\mathcal{M}\left\{B^{\beta,\nu}v(x); s\right\} = v^*(s\beta + \nu)\frac{\Gamma(s)}{\Gamma(s\beta + \nu)}. \tag{9.20}$$

Proof. It follows from (9.8) that

$$\mathcal{M}\{A^{\alpha,\mu}u(x); s\} = \int_0^\infty u(x)B^{\alpha,\mu}x^{s-1}ds.$$

Using (9.12) gives (9.19).

Similarly, (9.8) implies

$$\mathcal{M}\left\{B^{\beta,\nu}v(x); s\right\} = \int_0^\infty v(x)A^{\beta,\nu}x^{s-1}ds.$$

So, (9.9) yields (9.20). □

9.3.7 Fractional Differentiation and Integration

Here and subsequently (see [14–16]), if $\delta < 0$, D_{0x}^δ and W_x^δ stand for the Riemann–Liouville and Weyl fractional integrals, respectively, i.e.

$$D_{0x}^\delta g(x) = \int_0^x \frac{(x - t)^{-\delta-1}}{\Gamma(-\delta)}g(t)\,dt \quad \text{and} \quad W_x^\delta g(x) = \int_x^\infty \frac{(t - x)^{-\delta-1}}{\Gamma(-\delta)}g(t)\,dt.$$

For $\delta > 0$, D_{0x}^δ, ∂_{0x}^δ, and W_x^δ denote the Riemann–Liouville, Caputo, and Weyl fractional derivatives, respectively,

$$D_{0x}^\delta g(x) = \frac{d^n}{dx^n}D_{0x}^{\delta-n}g(x), \qquad \partial_{0x}^\delta g(x) = D_{0x}^{\delta-n}\frac{d^n}{dx^n}g(x),$$

and

$$W_x^\delta g(x) = (-1)^n W_x^{\delta-n}\frac{d^n}{dx^n}g(x)$$

where n is chosen such that $\delta \in (n - 1, n]$ and $n \in \mathbb{N}$. Moreover, it is assumed that

$$D_{0x}^0 g(x) = g(x), \quad \partial_{0x}^0 g(x) = g(x), \quad \text{and} \quad W_x^0 g(x) = g(x).$$

Lemma 9.8. *1) Let $\delta > 0$ and $\delta \in (n - 1, n]$ for some $n \in \mathbb{N}$, $u(x) \in \mathcal{A}^{\alpha,\mu} \cap \mathcal{A}^{\alpha,\mu-\alpha\delta}$, $D_{0x}^{\delta-n}u(x) \in C^n[0, \infty)$, $D_{0x}^{\delta-k}u(x) \in \mathcal{A}^{\alpha,\mu-\alpha k}$ for $k = 0, 1, ..n$, and let $A^{\alpha,\mu}u(x) \in L(0, a)$ for some $a > 0$ if $\alpha\delta \notin \mathbb{N}$. Then*

$$D_{0x}^{\alpha\delta}A^{\alpha,\mu}u(x) = A^{\alpha,\mu}\left[D_{0x}^\delta u(x)\right] + \sum_{k=0}^{n-1}\left[D_{0x}^{\delta-k-1}u(x)\right]_{x=0}\frac{x^{\mu-\alpha k-1}}{\Gamma(\mu - \alpha k)}. \tag{9.21}$$

2) Let $\delta < 0$, $u(x) \in \mathcal{A}^{\alpha,\mu} \cap \mathcal{A}^{\alpha,\mu-\alpha\delta}$, $D_{0x}^\delta u(x) \in \mathcal{A}^{\alpha,\mu}$, and $A^{\alpha,\mu}u(x) \in L(0, a)$ for some $a > 0$. Then

$$D_{0x}^{\alpha\delta}A^{\alpha,\mu}u(x) = A^{\alpha,\mu}\left[D_{0x}^\delta u(x)\right]. \tag{9.22}$$

Proof. We put $n = 0$ if $\delta < 0$. Using the formulas of differentiation for the Wright function [17,18]

$$D^{\varepsilon}_{0x}\left[x^{\mu-1}\phi\left(-\alpha,\mu;-\frac{c}{x^{\alpha}}\right)\right] = x^{\mu-\varepsilon-1}\phi\left(-\alpha,\mu-\varepsilon;-\frac{c}{x^{\alpha}}\right), \tag{9.23}$$

$$W^{\varepsilon}_{t}\phi\left(-\alpha,\mu;-ct\right) = c^{\varepsilon}\phi\left(-\alpha,\mu-\alpha\varepsilon;-ct\right) \qquad (c>0,\ \varepsilon\in\mathbb{R}), \tag{9.24}$$

and [2,3]

$$\frac{d}{dz}\phi\left(-\alpha,\mu;z\right) = \phi\left(-\alpha,\mu-\alpha;z\right), \tag{9.25}$$

we get

$$\begin{aligned} D^{\alpha\delta}_{0x}A^{\alpha,\mu}u(x) &= x^{\mu-\alpha\delta-1}\int_0^{\infty} u(t)\,\phi\left(-\alpha,\mu-\alpha\delta;-tx^{-\alpha}\right)dt \\ &= x^{\mu-1}\int_0^{\infty} u(t)W^{\delta}_{t}\phi\left(-\alpha,\mu;-tx^{-\alpha}\right)dt \\ &= (-1)^n x^{\mu-1}\int_0^{\infty} D^{\delta-n}_{0t}u(t)\cdot\frac{d^n}{dt^n}\phi\left(-\alpha,\mu;-tx^{-\alpha}\right)dt. \end{aligned}$$

If $\delta < 0$ (i.e. $n = 0$), this proves (9.22). In the case $\delta > 0$, integrating by parts leads to (9.21). □

Lemma 9.9. *Let $\delta \in \mathbb{R}$ and n be the smallest non-negative integer not exceeding $\beta\delta$, i.e. $n \in \mathbb{N}$ and $n-1 < \beta\delta \le n$ if $\delta > 0$, and $n = 0$ if $\delta \le 0$, and let $\frac{d^n}{dx^n}B^{\beta,\nu}v(x) \in L(a,\infty)$ for some a if $\delta \notin \mathbb{N}$. If $v(x) \in \mathcal{B}^{\beta,\nu}$, $v^{(n-k)}(x) \in \mathcal{B}^{\beta,\nu-\beta\delta-k}$ $(k = 0,1,...,n)$, and $W^{\beta\delta}_{x}v(x) \in \mathcal{B}^{\beta,\nu}$, then*

$$W^{\delta}_{x}B^{\beta,\nu}v(x) = B^{\beta,\nu}W^{\beta\delta}_{x}v(x). \tag{9.26}$$

Proof. By (9.23), (9.24), and (9.25) we obtain

$$W^{\delta}_{x}B^{\beta,\nu}v(x) = \int_0^{\infty} v(t)\,t^{\nu-\beta\delta-1}\phi\left(-\beta,\nu-\beta\delta;-xt^{-\beta}\right)dt$$

$$= \int_0^{\infty} v(t)\,D^{\beta\delta}_{0t}t^{\nu-1}\phi\left(-\beta,\nu;-xt^{-\beta}\right)dt.$$

Integrating by parts yields

$$W^{\delta}_{x}B^{\beta,\nu}v(x) = (-1)^n\int_0^{\infty}\frac{d^n}{dt^n}v(t)\cdot D^{\beta\delta-n}_{0t}t^{\nu-1}\phi\left(-\beta,\nu;-xt^{-\beta}\right)dt.$$

This proves (9.26). □

Remark 9.1. *A special case of* (9.21) *is the equality*

$$D^{\alpha}_{0x}A^{\alpha,\mu}u(x) = A^{\alpha,\mu}u'(x) + u(0)\frac{x^{\mu-1}}{\Gamma(\mu)}, \tag{9.27}$$

which occurs when $\delta = 1$. If $\mu = 0$ in (9.27)*, we get*

$$D^{\alpha}_{0x}A^{\alpha}u(x) = A^{\alpha}u'(x). \tag{9.28}$$

Moreover, (9.27) *with $\mu = 1-\alpha$ implies*

$$\partial^{\alpha}_{0x}A^{\alpha,1-\alpha}v(x) = A^{\alpha,1-\alpha}v'(x). \tag{9.29}$$

This follows from the formulas

$$\partial_{0x}^{\alpha} g(x) = D_{0x}^{\alpha} g(x) - g(0) \frac{x^{-\alpha}}{\Gamma(1-\alpha)}$$

and (see (9.33) *below)*

$$\lim_{x \to 0} A^{\alpha, 1-\alpha} u(x) = u(0).$$

Likewise, putting $\delta = 1$ *in* (9.26) *leads to*

$$\frac{d}{dx} B^{\beta,\nu} v(x) = -B^{\beta,\nu} W_x^{\beta} v(x). \tag{9.30}$$

9.3.8 Limit Behavior

Lemma 9.10. *Let* $\varepsilon = (\delta - \mu)/\alpha$ *and* $\varepsilon \leq 1$. *If* $u(x) \in \mathcal{A}^{\alpha,\mu} \cap \mathcal{A}^{\alpha,\mu-\delta+1}$, $D_{0x}^{\varepsilon-1} u(x) \in \mathcal{A}^{\alpha,1-\alpha}$, *and if*

$$\lim_{x \to 0} D_{0x}^{\varepsilon-1} u(x) = u_0 < \infty,$$

then

$$\lim_{x \to 0} D_{0x}^{\delta-1} A^{\alpha,\mu} u(x) = u_0. \tag{9.31}$$

Proof. By (9.23) and (9.24) we have

$$\begin{aligned} D_{0x}^{\delta-1} A^{\alpha,\mu} u(x) &= x^{\mu-\delta} \int_0^{\infty} u(t)\, \phi\left(-\alpha, \mu - \delta + 1; -t x^{-\alpha}\right) dt \\ &= x^{-\alpha} \int_0^{\infty} u(t)\, W_t^{\varepsilon-1} \phi\left(-\alpha, 1-\alpha; -t x^{-\alpha}\right) dt \\ &= x^{-\alpha} \int_0^{\infty} D_{0t}^{\varepsilon-1} u(t) \cdot \phi\left(-\alpha, 1-\alpha; -t x^{-\alpha}\right) dt. \end{aligned}$$

Making a change of variable $t = s x^{\alpha}$ and letting $x \to 0$, by (9.9) we get (9.31). □

Lemma 9.11. *Let* $v(x) \in \mathcal{B}^{\beta,\nu}$, $\nu \geq 0$ *and*

$$\lim_{x \to 0} W_x^{-\nu} v(x) = v_0 < \infty.$$

Then

$$\lim_{x \to 0} B^{\beta,\nu} v(x) = v_0. \tag{9.32}$$

Proof. If $\nu > 0$ we get

$$\lim_{x \to 0} B^{\beta,\nu} v(x) = \frac{1}{\Gamma(\nu)} \int_0^{\infty} v(t)\, t^{\nu-1} dt.$$

This proves (9.32) for positive ν. For $\nu = 0$, by (9.13) we have

$$B^{\beta} v(x) = \int_0^{\infty} [v(t) - v(0)]\, t^{\nu-1} \phi\left(-\beta, 0; -x t^{-\beta}\right) dt + v(0)$$

and, for an arbitrary positive ε,

$$\left(\int_0^{\varepsilon} + \int_{\varepsilon}^{\infty} \right) [v(t) - v(0)] \frac{1}{t} \phi\left(-\beta, 0; -x t^{-\beta}\right) dt = I_1 + I_2.$$

It is easy to see that $\lim_{x \to 0} I_2 = 0$. Let $\omega(\varepsilon)$ stands for $\sup_{t \in (0,\varepsilon)} |v(t) - v(0)|$. We obtain

$$|I_1| \leq \text{const} \cdot \omega(\varepsilon).$$

Due to arbitrary choice of ε, we get $\lim_{x \to 0} I_1 = 0$, which completes the proof.

Remark 9.2. *Taking $\delta = \alpha + \mu$ in (9.31) gives*

$$\lim_{x\to 0} D_{0x}^{\alpha+\mu-1} A^{\alpha,\mu} u(x) = u(0).$$

In particular, this yields

$$\lim_{x\to 0} D_{0x}^{\alpha-1} A^{\alpha} u(x) = u(0), \qquad \text{and} \qquad \lim_{x\to 0} \mathrm{A}^{\alpha,1-\alpha}\mathrm{u(x)} = \mathrm{u(0)}. \tag{9.33}$$

Similarly, (9.32) *gives*

$$\lim_{x\to 0} B^{\beta} v(x) = v(0).$$

9.3.9 Transforms of Some Special Functions

In this subsection, we assume that $\lambda \in \mathbb{C}$, $c > 0$, $\mu \in \mathbb{R}$, and $\nu \in \mathbb{R}$.

9.3.9.1 Mittag-Leffler Function

In what follows, $E_{\xi,\eta}(z)$ stands for the Mittag-Leffler function, which is defined by

$$E_{\xi,\eta}(z) = \sum_{k=0}^{\infty} \frac{z^k}{\Gamma(\xi k + \eta)}.$$

By (9.9) we get

$$A^{\alpha,\mu}\left[x^{\eta-1} E_{\xi,\eta}\left(\lambda x^{\xi}\right)\right] = x^{\alpha\eta+\mu-1} E_{\alpha\xi,\alpha\eta+\mu}\left(\lambda x^{\xi}\right) \quad (\eta > 0). \tag{9.34}$$

9.3.9.2 Exponential, Trigonometric, and Hyperbolic Functions

By virtue of equalities

$$\exp z = E_{1,1}(z), \qquad \sin z = z E_{2,2}\left(-z^2\right), \qquad \cos z = E_{2,1}\left(-z^2\right),$$

$$\sinh z = z E_{2,2}\left(z^2\right), \qquad \text{and} \qquad \cosh z = E_{2,1}\left(z^2\right),$$

special cases of (9.34) are

$$A^{\alpha,\mu} \exp(\lambda x) = x^{\alpha+\mu-1} E_{\alpha,\alpha+\mu}\left(\lambda x^{\alpha}\right), \tag{9.35}$$

$$\begin{aligned} A^{\alpha,\mu} \sin(\lambda x) &= \lambda x^{2\alpha+\mu-1} E_{2\alpha,2\alpha+\mu}\left(-\lambda^2 x^{2\alpha}\right), \\ A^{\alpha,\mu} \cos(\lambda x) &= x^{\alpha+\mu-1} E_{2\alpha,\alpha+\mu}\left(-\lambda^2 x^{2\alpha}\right), \\ A^{\alpha,\mu} \sinh(\lambda x) &= \lambda x^{2\alpha+\mu-1} E_{2\alpha,2\alpha+\mu}\left(\lambda^2 x^{2\alpha}\right); \\ A^{\alpha,\mu} \cosh(\lambda x) &= x^{\alpha+\mu-1} E_{2\alpha,\alpha+\mu}\left(\lambda^2 x^{2\alpha}\right). \end{aligned} \tag{9.36}$$

Also, by (9.20) we get

$$B^{\beta,\nu} \exp(-cx) = c^{-\nu} \exp(-c^{\beta} x).$$

9.3.9.3 Wright Functions

From (9.9), (9.12), and (9.18) it follows that

$$A^{\alpha,\mu}\left[x^{\eta-1}\phi\left(\omega,\eta;\lambda x^{\omega}\right)\right]=x^{\alpha\eta+\mu-1}\phi\left(\alpha\omega,\alpha\eta+\mu;\lambda x^{\alpha\omega}\right) \qquad (\omega>0),$$

$$A^{\alpha,\mu}\left[x^{\eta-1}\phi\left(-\xi,\eta;-cx^{-\xi}\right)\right]=x^{\alpha\eta+\mu-1}\phi\left(-\alpha\xi,\alpha\eta+\mu;-cx^{-\alpha\xi}\right) \tag{9.37}$$

$$B^{\beta,\nu}\phi\left(-\xi,\eta;-cx\right)=c^{-\nu}\phi\left(-\beta\xi,\xi\nu+\eta;-c^{\beta}x\right), \tag{9.38}$$

and

$$A^{\alpha,\mu}\left[x^{\delta-1}e^{\delta,\varepsilon}_{\xi,\eta}\left(-cx^{\xi}\right)\right]=x^{\alpha\delta+\mu-1}e^{\alpha\delta+\mu,\varepsilon}_{\alpha\xi,\eta}\left(-cx^{\alpha\xi}\right) \quad (0<\xi<1,\ \delta>0). \tag{9.39}$$

Here and subsequently, $e^{\delta,\varepsilon}_{\xi,\eta}(z)$ denotes the Wright-type function [17]

$$e^{\delta,\varepsilon}_{\xi,\eta}(z)=\sum_{k=0}^{\infty}\frac{z^k}{\Gamma(\xi k+\delta)\Gamma(-\eta k+\varepsilon)}.$$

9.3.9.4 Nu-Function

Let $\nu(x,\delta)$ be defined by

$$\nu(x,\delta)=\int_0^{\infty}\frac{x^{\delta+t}\,dt}{\Gamma(\delta+t+1)}.$$

By (9.9), it can be shown that

$$A^{\alpha,\mu}\nu(x,\delta)=\frac{1}{\alpha}\nu(x,\alpha\delta+\alpha+\mu-1) \qquad (\delta>-1). \tag{9.40}$$

A consequence of (9.40) is that the function $f(x)=\nu(x,\delta)$ with $\delta=\frac{\mu}{1-\alpha}-1$ is a solution of the equation $A^{\alpha,\mu}f(x)=\frac{1}{\alpha}f(x)$.

9.4 Application

In this section, we give some examples of application for the transforms under consideration.

9.4.1 Integral Representation of the Wright Function

By (see [1])

$$\frac{1}{\sqrt{\pi}}\exp\left(-\frac{z^2}{4}\right)=\phi\left(-\frac{1}{2},\frac{1}{2};-z\right),$$

the equality (see, e.g. [19])

$$\int_0^{\infty}\exp\left(-ys^2\right)\cos(xs)\,ds=\frac{1}{2}\left(\frac{\pi}{y}\right)^{1/2}\exp\left(-\frac{x^2}{4y}\right),$$

can be rewritten as

$$y^{-\frac{1}{2}}\phi\left(-\frac{1}{2},\frac{1}{2};-\frac{x}{y^{1/2}}\right)=\frac{2}{\pi}\int_0^{\infty}\exp\left(-ys^2\right)\cos(xs)\,ds.$$

By (9.35) and (9.37), applying to both parts of this equality the transform $A^{2\eta,\varepsilon-\eta}$ (with respect to y) and a simple computation give

$$\phi(-\eta, \varepsilon; -x) = \frac{2}{\pi} \int_0^\infty E_{2\eta,\eta+\varepsilon}\left(-s^2\right) \cos(xs)\, ds. \tag{9.41}$$

The formula (9.41) is proved under the assumption $0 < \eta < 1/2$. However, it holds for $\eta \in (0, 1)$, as is easy to check.

Moreover, combining (9.36), (9.39), and (9.41) yields

$$e_{\xi,\eta}^{\delta,\varepsilon}(-x) = \frac{2}{\pi} \int_0^\infty E_{2\xi,\xi}\left(-x^2 s^2\right) E_{2\eta,\eta+\varepsilon}\left(-s^2\right) ds.$$

9.4.2 Evaluation of Improper Integrals

As an example of using (9.8), we give a simple proof of the formula

$$\int_0^\infty t^{\delta-1} E_{\xi,\eta}\left(-\lambda t^\varepsilon\right) dt = \lambda^{-\frac{\delta}{\varepsilon}} \frac{\Gamma\left(\frac{\delta}{\varepsilon}\right)\Gamma\left(1-\frac{\delta}{\varepsilon}\right)}{\varepsilon\Gamma\left(\eta - \frac{\xi\delta}{\varepsilon}\right)}, \tag{9.42}$$

where $|\arg\lambda| < \frac{\pi}{2}$, $\xi \in (0, 2)$, $\varepsilon > 0$, $\delta \in (0, \varepsilon)$ if $\eta \neq \xi$, and $\delta \in (0, 2\varepsilon)$ if $\eta = \xi$.

Without loss of generality we can assume $\varepsilon = \xi$. Indeed, this will be the case after a change of variable by $t = y^{\xi/\varepsilon}$.

Consider the integral

$$I = \int_0^\infty y^{\delta-1} E_{\xi,\eta}\left(-\lambda y^\xi\right) dy.$$

Let first $\xi \in (0, 1)$. By (9.8), (9.12), and (9.34) we have

$$I = \int_0^\infty y^{\delta-\eta} A^{\xi,\eta-\xi} e^{-\lambda y} dy = \int_0^\infty e^{-\lambda y} B^{\xi,\eta-\xi} y^{\delta-\eta} dy$$

$$= \frac{\Gamma\left(1-\frac{\delta}{\xi}\right)}{\xi\Gamma(\eta-\delta)} \int_0^\infty e^{-\lambda y} y^{\frac{\delta}{\xi}-1} dy = \lambda^{-\frac{\delta}{\xi}} \frac{\Gamma\left(\frac{\delta}{\xi}\right)\Gamma\left(1-\frac{\delta}{\xi}\right)}{\xi\Gamma(\eta-\delta)}. \tag{9.43}$$

This gives (9.42) for $\xi < 1$.

Now let $\xi \geq 1$. Choose a positive number ω such that $\omega\xi < 1$. By (9.8) and (9.12) we can write

$$\begin{aligned} I &= \frac{\Gamma(1-\omega(\delta-\eta))}{\Gamma(1-\delta+\eta)} \int_0^\infty y^{\eta-1} E_{\xi,\eta}\left(-\lambda y^\xi\right) B^\omega y^{\omega(\delta-\eta)} dy \\ &= \frac{\Gamma(1-\omega(\delta-\eta))}{\Gamma(1-\delta+\eta)} \int_0^\infty y^{\omega\delta-1} E_{\omega\xi,\omega\eta}\left(-\lambda y^{\omega\xi}\right) dy. \end{aligned}$$

That is, we come to the already considered case, because $\omega\xi < 1$. Applying (9.43) completes the proof of (9.42).

9.4.3 Fractional Differential Equations

We briefly sketch a method for solving fractional differential equations using the transformations (9.2) and (9.3), and then, as an example, construct a fundamental solution for a higher order parabolic equation with time fractional derivative.

9.4.3.1 Equations with Riemann–Liouville Derivatives

Let $\sigma \in (n-1, n]$, $n \in \mathbb{N}$. Consider the initial value problem

$$(D_{0x}^{\sigma} - L)\, u(x) = f(x), \qquad \lim_{x\to 0} D_{0x}^{\sigma-k} u(x) = a_k \quad (k = 1, ..., n). \tag{9.44}$$

Here, L denotes a linear operator independent of x (i.e. $u(x)$ as well as L may depend on variables other than x, but we omit them).

The solution of (9.44) can be found in the form

$$u(x) = \sum_{j=0}^{n} u_j(x),$$

where $u_j(x)$ are solutions of the following problems

$$(D_{0x}^{\sigma} - L)\, u_0(x) = f(x), \qquad \lim_{x\to 0} D_{0x}^{\sigma-k} u_0(x) = 0 \quad (k = 1, ..., n), \tag{9.45}$$

and

$$(D_{0x}^{\sigma} - L)\, u_j(x) = 0, \qquad \lim_{x\to 0} D_{0x}^{\sigma-k} u_j(x) = \begin{cases} a_j & \text{if} \quad k = j, \\ 0 & \text{else.} \end{cases} \tag{9.46}$$

Together with the problems (9.45) and (9.46), we consider the corresponding integer-order problems

$$\left(\frac{d^n}{dx^n} - L\right) v_0(x) = g(x), \qquad v_0^{(n-k)}(0) = 0 \quad (k = 1, ..., n),$$

and

$$\left(\frac{d^n}{dx^n} - L\right) v_j(x) = 0, \qquad v_j^{(n-k)}(0) = \begin{cases} a_j & \text{if} \quad k = j, \\ 0 & \text{else.} \end{cases}$$

By (9.14), (9.21), and (9.31), we can state that, under some relevant assumptions, the functions $u_j(x)$ can be given by

$$u_0(x) = A^{\omega} v_0(x) \quad \text{and} \quad u_j(x) = A^{\omega,(1-j)(1-\omega)} v_j(x) \quad (j = 1, ..., n),$$

where $\omega = \frac{\sigma}{n}$, and $f(x) = A^{\omega} g(x)$.

9.4.3.2 Equations with Caputo and Weyl Derivatives

Here, for simplicity we assume $\sigma \in (0, 1)$. It follows from (9.14), (9.29), and (9.33) that the function

$$u(x) = A^{\sigma,1-\sigma} v(x)$$

is a solution of the problem

$$(\partial_{0x}^{\sigma} - L)\, u(x) = f(x), \qquad u(0) = a,$$

if $v(x)$ satisfies

$$\left(\frac{d}{dx} - L\right) v(x) = g(x) \qquad v(0) = a.$$

Here, L is the same as above, and $f(x) = A^{\sigma,1-\sigma} g(x)$.

Likewise, (9.30) yields that solutions of the equation

$$(W_x^{\sigma} - L)\, u(x) = f(x),$$

should be sought among solutions of

$$v(x) = B^{\sigma,\nu}u(x),$$

where $v(x)$ is a solution of

$$\left(-\frac{d}{dx} - L\right) v(x) = B^{\sigma,\nu} f(x).$$

9.4.3.3 Fundamental Solution for a Higher-Order Parabolic Equation

Let $\sigma \in (0,1)$, $t > 0$, $(x,y) \in \mathbb{R}^2$, and $m \in \mathbb{N}$. Consider the equation

$$\left[\frac{\partial^\sigma}{\partial t^\sigma} + (-\Delta_{x,y})^m\right] u(x,y,t) = f(x,y,t), \tag{9.47}$$

where $\Delta_{x,y}$ denotes the Laplace operator with respect to x and y,

$$\Delta_{x,y} = \frac{\partial^2}{\partial x^2} + \frac{\partial^2}{\partial y^2},$$

and

$$\frac{\partial^\sigma}{\partial t^\sigma} = D^\sigma_{0t} \qquad \text{or} \qquad \frac{\partial^\sigma}{\partial t^\sigma} = \partial^\sigma_{0t}.$$

The integer-order equation corresponding to (9.47) is

$$\left[\frac{\partial}{\partial t} + (-\Delta_{x,y})^m\right] v(x,y,t) = g(x,y,t),$$

and its fundamental solution can be written in the form

$$F(x,y,t) = \frac{1}{4\pi^2}\int_{\mathbb{R}^2} \exp\left[-t(\xi^2+\eta^2)^m + i(x\xi + y\eta)\right] d\xi\, d\eta.$$

A simple computation gives

$$F(x,y,t) = \frac{1}{4\pi^2}\int_{\mathbb{R}^2} \exp\left[-t(\rho^2+\theta^2)^m + i\theta r\right] d\rho\, d\theta$$

$$= \frac{1}{4\pi^2}\int_0^\infty s\exp\left(-ts^{2m}\right)\int_{-\pi}^{\pi} \exp\left(i\,s\,r\sin\varphi\right) d\varphi\, ds \qquad (r = \sqrt{x^2+y^2}).$$

Finally, we get

$$F(x,y,t) = \frac{1}{2\pi}\int_0^\infty s\exp\left(-ts^{2m}\right) J_0\left(s\sqrt{x^2+y^2}\right) ds,$$

where $J_0(z)$ stands for the Bessel function,

$$J_0\left(z\right) = \frac{1}{2\pi}\int_{-\pi}^{\pi} \exp\left(i\,z\sin\varphi\right) d\varphi.$$

Let $F_{\sigma,\mu}(x,y,t)$ be defined by

$$F_{\sigma,\mu}(x,y,t) = \frac{t^{\mu-1}}{2\pi}\int_0^\infty sE_{\sigma,\mu}\left(-t^\sigma s^{2m}\right) J_0\left(s\sqrt{x^2+y^2}\right) ds.$$

That is, $F_{\sigma,\mu}(x,y,t) = A^{\sigma,\mu-\sigma}F(x,y,t)$ with respect to t (see (9.35)).

By (9.14), (9.28), (9.29), and (9.33), we can assert the function $F_{\sigma,\sigma}(x, y, t)$ is the fundamental solution for (9.47). Moreover, if the fractional derivative in (9.47) is given by the Riemann–Liouville sense, i.e. $\frac{\partial^\sigma}{\partial t^\sigma} = D^\sigma_{0t}$, then a solution of (9.47) satisfying

$$\lim_{t \to 0} D^{\sigma-1}_{0t} u(x, y, t) = \tau(x, y),$$

can be written in the form

$$u(x, y, t) = \int_{\mathbb{R}^2} \tau(\xi, \eta)\, F_{\sigma,\sigma}(x - \xi, y - \eta, t)\, d\xi\, d\eta + U(x, y, t),$$

where

$$U(x, y, t) = \int_0^y \int_{\mathbb{R}^2} f(\xi, \eta, s)\, F_{\sigma,\sigma}(x - \xi, y - \eta, t - s)\, d\xi\, d\eta\, ds.$$

Similarly, the function

$$u(x, y, t) = \int_{\mathbb{R}^2} \tau(\xi, \eta)\, F_{\sigma,1}(x - \xi, y - \eta, t)\, d\xi\, d\eta + U(x, y, t)$$

is a solution of (9.47) with $\frac{\partial^\sigma}{\partial t^\sigma} = \partial^\sigma_{0t}$, and satisfies the condition $u(x, y, 0) = \tau(x, y)$.

Acknowledgments

The reported study was funded by RFBR according to the research project no 18–51–45005.

References

[1] B. Stanković, O jednoj klasi singularnih integralnih jednačina (on a class of singular integral equations), *Zbornik Radova SAN*, **43**(4) (1955), 81–130 (Serbian).

[2] E. M. Wright, On the coefficients of power series having exponential singularities, *J. London Math. Soc.*, **8**(29) (1933), 71–79.

[3] E. M. Wright, The generalized Bessel function of order greater than one, *Q. J. Math. Oxford Ser.*, **11** (1940), 36–48.

[4] R. P. Agarwal, Sur une généralisation de la transformation de Hankel, *Ann. soc. sci. Bruxelles.*, **64**(1) (1950), 164–168.

[5] B. Stanković, Inversion et invariantes de la transformation généralisée de Hankel, *Publ. Inst. Math. Beograde*, **8** (1955), 37–52.

[6] B. Stanković, Inversion d'une transformation intégrale, *Publ. Inst. math. Acad. serbe sci.*, **10** (1956), 85–88.

[7] B. Stanković, Abbildung gewisser Operationen durch die zweidimensionale Laplace-transformation, *Publ. Inst. math. Acad. serbe sci.*, **10** (1957), 1–8.

[8] B. Stanković, On the function of E. M. Wright, *Publ. Inst. math. Nouvélle seria*, **10**(24) (1970), 113–124.

[9] E. Bazhlekova, Subordination principle for fractional evolution equations, *Fract. Calc. Appl. Anal.*, **3**(3) (2000), 213–230.

[10] E. G. Bazhlekova, *Fractional Evolution Equations in Banach Spaces*, Eindhoven: Technische Universiteit Eindhoven (2001).

[11] A. V. Pskhu, On the real zeros of functions of Mittag-Leffler type, *Math. Notes*, **77**(3–4) (2005), 546–552.

[12] A. V. Pskhu, The fundamental solution of a diffusion-wave equation of fractional order, *Izv. Math.*, **73**(2) (2009), 351–392.

[13] M. M. Dzhrbashyan and R. A. Bagiyan, On integral representations and measures associated with functions of Mittag-Leffler type, *Sov. Math. Dokl.*, **16** (1975), 1072–1076.

[14] A. M. Nakhushev, On the theory of fractional calculus, *Differ. Equ.*, **24**(2) (1988), 239–247.

[15] K. S. Miller and B. Ross, *An Introduction to the Fractional Calculus and Fractional Differential Equations*, Hoboken, NJ: Wiley (1993).

[16] A. A. Kilbas, H. M. Srivastava and J. J. Trujillo, *Theory and Applications of Fractional Differential Equations*, North-Holland Mathematics Studies, **204**, Amsterdam: Elsevier (2006).

[17] A. V. Pskhu, Solution of a boundary value problem for a fractional partial differential equation, *Differ. Equ.*, **39**(8) (2003), 1150–1158.

[18] A. V. Pskhu, On stabilization of solutions of the Cauchy problem for fractional diffusion-wave equation, *Itogi Nauki i Tekhniki. Ser. Sovrem. Mat. Pril. Temat. Obz.*, **149** (2018), VINITI, Moscow, 84–94 (Russian).

[19] M. Lavrentiev et B. Chabat, *Méthodes de la théorie des fonctions d'une variable complexe*, Moscow: Mir (1972).

10

Electric Current as a Continuous Flow

Yusuke Kamata, Tingli Ma, Yong Sun, and Shigeru Kanemitsu
Kyushu Institute of Technology

CONTENTS

10.1 Introduction

As in many other disciplines, there exists a set of DEs that fixes the setting of electromagnetism. These are Maxwell equations governing the relation among electric and magnetic fields. The subject electromagnetism is thought of as a hard one particularly because of Maxwell equations, and indeed in many textbooks on physics (e.g. [1]), the treatment is not transparent and would force the learners to overcome undue pain. One of the reasons for this difficulty may lie in the fact that Maxwell equations do not explicitly exhibit current to be a continuous flow. In this paper we shall prove that electric current is a *continuous flow*. The continuous-part is shown by the equation of continuity as with fluid flows. This part is in common with fluid dynamics [2,3] in which a vectorial version of the general Stokes theorem is used to deduce the equation of continuity of fluid most naturally, reproduced in Section 10.5. We show the flow-part by reducing the Maxwell equations to two partial DEs and one *wave equation* satisfied by the shift which appears as a auxiliary term (Theorem 10.6) by slightly modifying the argument in [4, pp. 367–370].

Since both fluid and electric current are flows, it is natural to apply differential forms rather than treating the 2- and 3-dimensional flows separately. In this paper we shall give a rather transparent treatment by an efficient use of differential forms (cf. e.g. [5]) which has been applied successfully in reference [2] in fluid dynamics. As is known [6, pp. 150–162], the use of 4-dimensional expression for electromagnetism in special relativity theory leads to the most concise form of Maxwell equations. We shall give a slight modification of [7, pp. 452–454].

After formulating Maxwell equations in 4-dimensional form in Section 10.1 we provide a theorem that transforms them into wave equations in Section 10.2. This exhibits the

TABLE 10.1
Corresponding Equations

Ours	(10.2)	(10.3)	(10.4)	(10.5)	(10.14)	(10.15)
[3]	(17.47a)	(17.47d)	(17.47b)	(17.47c)	(17.15a)	(17.15b)
This Paper	(10.16)	(10.17)	(10.27)	(10.29)	(10.30)	
[3]	(17.48a)	(17.48d)	(17.61)	(17.71b)	(17.71a)	

similarity of flows and waves and gives an evidence that current is a wave and hence a flow (Table 10.1).

Let $\boldsymbol{E}$ and $\boldsymbol{H}$ denote the **electric field** and the **magnetic field**, respectively. Additionally, let

$$\boldsymbol{D} = 4\pi \boldsymbol{E}, \quad \boldsymbol{B} = \mu \boldsymbol{H} \tag{10.1}$$

denote the **electric induction field** and the **magnetic induction field**, respectively, where μ is the permeability of free space. In the case of formal treatment of symbols, we ignore the coefficients and think of them as involved in the definitions of quantities. Classical electromagnetism is concerned with the relationship among current, electrons, and electric and magnetic fields.

Thus classical electromagnetism is governed by **Maxwell equations** which are as follows. For notation, cf. Section 10.5.

Faraday's induction law:

$$\operatorname{curl}(\boldsymbol{E}) = -\frac{1}{c}\frac{\partial \boldsymbol{B}}{\partial t} = -\frac{\mu}{c}\frac{\partial \boldsymbol{H}}{\partial t}, \tag{10.2}$$

Absence of magnetic charge:

$$\operatorname{div}(\boldsymbol{B}) = 0, \tag{10.3}$$

Generalized Oersted's law:

$$\operatorname{curl}(\boldsymbol{H}) = 4\pi \boldsymbol{i} + \frac{1}{c}\frac{\partial \boldsymbol{D}}{\partial t}, \tag{10.4}$$

and

$$\operatorname{div}(\boldsymbol{D}) = K\nabla \cdot \boldsymbol{E} = K\rho, \tag{10.5}$$

where the scalar function ρ describes the distribution of charge and the vector function

$$\boldsymbol{j} = {}^t(j_1, j_2, j_3) \tag{10.6}$$

is the distribution of current.

In special relativity theory, the 4-dimensional expression is known for electromagnetism. We introduce the 4-dimensional electric current

$$\boldsymbol{J} = {}^t(\boldsymbol{j}, \rho) = {}^t(j_1(\boldsymbol{x}, t), j_2(\boldsymbol{x}, t), j_3(\boldsymbol{x}, t), \rho(\boldsymbol{x}, t)), \quad \boldsymbol{x} = {}^t(x, y, z). \tag{10.7}$$

Then the **equation of continuity** of current reads

$$\operatorname{div} \boldsymbol{j} = -\frac{1}{c}\frac{\partial \rho}{\partial t}. \tag{10.8}$$

Indeed, we have from (10.4) and (10.32)

$$0 = c \operatorname{div} \operatorname{curl} \boldsymbol{H} = 4\pi c \operatorname{div} \boldsymbol{j} + \operatorname{div} \frac{\partial \boldsymbol{D}}{\partial t} \tag{10.9}$$

and from (10.5) by differentiation:

$$\frac{\partial}{\partial t}\operatorname{div}\boldsymbol{D} = 4\pi\frac{\partial\rho}{\partial t}. \tag{10.10}$$

Hence noting that $\operatorname{div}\frac{\partial}{\partial t} = \frac{\partial}{\partial t}\operatorname{div}$, we replace the last term in (10.9) by (10.10) to obtain

$$0 = c\operatorname{div}\boldsymbol{j} + \frac{\partial\rho}{\partial t}, \tag{10.11}$$

proving (10.8). Equation (10.11) is the ordinary **equation of continuity**. Compare Example 10.1 for a proof.

Theorem 10.1. (i) *The equation of continuity* (10.8) *may be expressed as*

$$\mathrm{d}\overset{3}{\mathcal{J}} = 0, \tag{10.12}$$

where the 3*-form* $\mathcal{J}$ *is defined by*

$$\overset{3}{\mathcal{J}} = j_1\mathrm{d}y\wedge\mathrm{d}z\wedge\mathrm{d}t + j_2\mathrm{d}z\wedge\mathrm{d}t\wedge\mathrm{d}x + j_3\mathrm{d}t\wedge\mathrm{d}x\wedge\mathrm{d}y + \rho\mathrm{d}x\wedge\mathrm{d}y\wedge\mathrm{d}z. \tag{10.13}$$

(ii) *Define*

$$\overset{2}{\mathcal{E}} = c(E_1\mathrm{d}x + E_2\mathrm{d}y + E_3\mathrm{d}z)\wedge\mathrm{d}t + B_1\mathrm{d}y\wedge\mathrm{d}z + B_2\mathrm{d}z\wedge\mathrm{d}x + B_3\mathrm{d}x\wedge\mathrm{d}y \tag{10.14}$$

and

$$\overset{2}{\mathcal{H}} = -c(H_1\mathrm{d}x + H_2\mathrm{d}y + H_3\mathrm{d}z)\mathrm{d}t + D_1\mathrm{d}y\wedge\mathrm{d}z + D_2\mathrm{d}z\wedge\mathrm{d}x + D_3\mathrm{d}x\wedge\mathrm{d}y. \tag{10.15}$$

Then the equations (10.2) *and* (10.3) *may be expressed as*

$$\mathrm{d}\overset{2}{\mathcal{E}} = 0, \tag{10.16}$$

while (10.4) *and* (10.5) *may be by*

$$\mathrm{d}\overset{2}{\mathcal{H}} + 4\pi\overset{3}{\mathcal{J}} = 0. \tag{10.17}$$

Proof. We apply the rule (10.51). (i)

$$\begin{aligned}\mathrm{d}j_1\mathrm{d}y\wedge\mathrm{d}z\wedge\mathrm{d}t &= \frac{\partial j_1}{\partial x}\mathrm{d}x\wedge\mathrm{d}y\wedge\mathrm{d}z\wedge\mathrm{d}t + \frac{\partial j_1}{\partial y}\mathrm{d}y\wedge\mathrm{d}y\wedge\mathrm{d}z\wedge\mathrm{d}t + \cdots \\ &= \frac{\partial j_1}{\partial x}\mathrm{d}x\wedge\mathrm{d}y\wedge\mathrm{d}z\wedge\mathrm{d}t\end{aligned} \tag{10.18}$$

in view of $\mathrm{d}y\wedge\mathrm{d}y = 0, \cdots$. Hence

$$\mathrm{d}\overset{3}{\mathcal{J}} = \operatorname{div}(\boldsymbol{J})\mathrm{d}x\wedge\mathrm{d}y\wedge\mathrm{d}z\wedge\mathrm{d}t, \tag{10.19}$$

whence (10.12).
(ii) We have

$$\begin{aligned}\mathrm{d}\overset{2}{\mathcal{E}} = &\left(\operatorname{curl}(\boldsymbol{E}) + \frac{\partial\boldsymbol{B}}{\partial t}\right)\cdot{}^t(\mathrm{d}y\wedge\mathrm{d}z\wedge\mathrm{d}t, \mathrm{d}z\wedge\mathrm{d}x\wedge\mathrm{d}t, \mathrm{d}x\wedge\mathrm{d}y\wedge\mathrm{d}t) \\ &+ \operatorname{div}(\boldsymbol{B})\mathrm{d}x\wedge\mathrm{d}y\wedge\mathrm{d}z.\end{aligned} \tag{10.20}$$

Hence (10.16) implies the coefficients are 0, which amounts to (10.2) and (10.3).

Similar to (10.20) we have

$$\mathrm{d}\overset{2}{\mathcal{H}} = -c\mathrm{curl}(\boldsymbol{H}) \cdot {}^t(\mathrm{d}y \wedge \mathrm{d}z \wedge \mathrm{d}t, \mathrm{d}z \wedge \mathrm{d}x \wedge \mathrm{d}t, \mathrm{d}x \wedge \mathrm{d}y \wedge \mathrm{d}t) + \mathrm{div}(\boldsymbol{D})\mathrm{d}x \wedge \mathrm{d}y \wedge \mathrm{d}z. \tag{10.21}$$

Since (10.4) reads

$$c\mathrm{curl}(\boldsymbol{H}) = 4\pi\left(c\boldsymbol{j} + \frac{\partial \boldsymbol{E}}{\partial t}\right), \tag{10.22}$$

the left-hand side of (10.17) reads

$$\left(-c\mathrm{curl}(\boldsymbol{H}) + 4\pi\left(c\boldsymbol{j} + \frac{\partial \boldsymbol{E}}{\partial t}\right)\right) \cdot {}^t(\mathrm{d}y \wedge \mathrm{d}z \wedge \mathrm{d}t, \mathrm{d}z \wedge \mathrm{d}x \wedge \mathrm{d}t, \mathrm{d}x \wedge \mathrm{d}y \wedge \mathrm{d}t) + (\mathrm{div}(\boldsymbol{D}) - 4\pi\rho)\,\mathrm{d}x \wedge \mathrm{d}y \wedge \mathrm{d}z. \tag{10.23}$$

Hence (10.17) implies (10.4) and (10.5), completing the proof. □

Theorem 10.2. (Poincaré's lemma) *If the domain Ω is smoothly contractible, then if $\mathrm{d}\overset{k}{\omega} = 0$, there is a $(k-1)$-form $\overset{k-1}{\omega}$ such that*

$$\mathrm{d}\overset{k-1}{\omega} = \overset{k}{\omega}. \tag{10.24}$$

The solution of (10.24) *is unique up to* $\mathrm{d}\overset{k-2}{\omega}$.

Using the notion of cohomology, the condition in the theorem may be stated as

$$H^k(\Omega) = 0.$$

We apply Poincaré's lemma, Theorem 10.2 with $k = 2$ to the following sequence

$$C : \to \vartheta^0(\Omega) \xrightarrow{\mathrm{d}} \vartheta^1(\Omega) \xrightarrow{\mathrm{d}} \vartheta^2(\Omega) \xrightarrow{\mathrm{d}} \vartheta^3(\Omega). \tag{10.25}$$

Then the following theorem is an immediate consequence.

Theorem 10.3. *Suppose that* (10.16) *holds on a differentiably contractable set, say on $\mathbb{R}^4$. Then there exists a 1-form*

$$\overset{1}{\omega} = A_1\mathrm{d}x + A_2\mathrm{d}y + A_3\mathrm{d}z + A_0\mathrm{d}t \tag{10.26}$$

such that

$$\overset{2}{\mathcal{E}} = \mathrm{d}\overset{1}{\omega}. \tag{10.27}$$

In vector analysis notation,

$$f \longrightarrow \mathrm{grad}(f) = \nabla f \longrightarrow \mathrm{curl}(\boldsymbol{f}) = \nabla \times \boldsymbol{f} \longrightarrow \mathrm{div}(\boldsymbol{f}) = \nabla \cdot \boldsymbol{f} \tag{10.28}$$

is the complex corresponding to (10.25). In the terminology of vector analysis, Theorem 10.3 means

Theorem 10.4. *There exists a vector potential $\boldsymbol{V} = {}^t(V_1, V_2.V_3)$ such that*

$$\boldsymbol{H} = \mathrm{curl}\,\boldsymbol{V} \tag{10.29}$$

and

$$\boldsymbol{E} = \nabla A_0 - \frac{\partial}{\partial t}\boldsymbol{V}. \tag{10.30}$$

Note that Theorem 10.7, (ii) entails

Lemma 10.1. (i)

$$\nabla \times (\nabla f) = \mathrm{curl}(\mathrm{grad}\, f) = \boldsymbol{o}, \tag{10.31}$$

and

(ii)

$$\nabla \cdot (\nabla \times \boldsymbol{A}) = \mathrm{div}(\mathrm{curl}\, \boldsymbol{A}) = 0. \tag{10.32}$$

In this setting, Poincaré's lemma, Theorem 10.2, reads for vector fields

Theorem 10.5. (Poincaré's lemma for space) *If the domain Ω is smoothly contractible (to a point) and $\boldsymbol{A}$ is a vector field of class C^1 on Ω, then*

(i) $\mathrm{curl}\, \boldsymbol{A} = 0 \Longleftrightarrow$ *there exists a function $f \in C^1(\Omega)$, unique up to a constant, such that $\boldsymbol{A} = \mathrm{grad}\, f = \nabla f$.*

(ii) $\mathrm{div}\, \boldsymbol{A} = 0 \Longleftrightarrow$ *there exists a vector field $\boldsymbol{V}$ of class C^1, unique up to ∇g, where g is an arbitrary scalar function, such that $\boldsymbol{A} = \mathrm{curl}\, \boldsymbol{V}$.*

Remark 10.1. *An ordinary procedure for deriving* (10.29) *and* (10.30) *is as follows.*

Applying Theorem 10.5, (ii) to (10.35)*, we find a vector potential $\boldsymbol{V}$ of class C^1 satisfying* (10.29)*. Substituting this in* (10.34)*, we find*

$$\boldsymbol{o} = \nabla \times \boldsymbol{E} + \frac{\partial}{\partial t} \mathrm{curl}\, \boldsymbol{V} = \nabla \times \left(\boldsymbol{E} + \frac{\partial}{\partial t} \boldsymbol{V} \right),$$

whence, by Theorem 10.5, (i), there exists a function A_0, which we now write $f \in C^1(\Omega)$, unique up to a constant satisfying (10.29)*:*

$$\boldsymbol{E} = \nabla f - \frac{\partial}{\partial t} \boldsymbol{V}. \tag{10.33}$$

10.2 Maxwell Equations as Wave Equations

Now we shall treat the following slightly modified form (with $c = 1$) for simplicity.

$$\mathrm{curl}(\boldsymbol{E}) = \nabla \times \boldsymbol{E} = -\frac{\partial \boldsymbol{H}}{\partial t}, \tag{10.34}$$

$$\mathrm{div}(\boldsymbol{H}) = \nabla \cdot \boldsymbol{H} = 0, \tag{10.35}$$

$$\mathrm{curl}(\boldsymbol{H}) = \nabla \times \boldsymbol{H} = 4\pi \left(\boldsymbol{J} + \frac{\partial \boldsymbol{E}}{\partial t} \right), \tag{10.36}$$

$$\mathrm{div}(\boldsymbol{E}) = \nabla \cdot \boldsymbol{E} = \rho. \tag{10.37}$$

We shall prove the following theorem based on Theorem 10.3.

Theorem 10.6. *Suppose the domains of all relevant vector fields are smoothly contractible. Then the solutions of the Maxwell equations amount to those of* (10.41) *and* (10.42) *plus the shift ψ which are solutions of the inhomogeneous wave equation*

$$\nabla^2 \psi - 4\pi \frac{\partial^2 \psi}{\partial t^2} = -h. \tag{10.38}$$

In coordinate form this reads

$$\frac{\partial^2\psi}{\delta x^2}+\frac{\partial^2\psi}{\partial y^2}+\frac{\partial^2\psi}{\partial z^2}-4\pi\frac{\partial^2\psi}{\partial t^2}+h=0, \tag{10.39}$$

where h is a constant given by

$$-h=4\pi\left(\nabla\cdot(\boldsymbol{V}+\text{grad}g)-\frac{\partial f}{\partial t}\right). \tag{10.40}$$

Here

$$\nabla^2 f-4\pi\frac{\partial^2 f}{\partial t^2}=\rho \tag{10.41}$$

and

$$\nabla^2\boldsymbol{V}-\frac{\partial^2\boldsymbol{V}}{\partial t^2}=-\boldsymbol{J}, \tag{10.42}$$

under the constraint

$$\frac{1}{4\pi}\nabla\cdot\boldsymbol{V}-\frac{\partial f}{\partial t}=0. \tag{10.43}$$

In addition to the formulas in Lemma 10.1, which are useful in vector analysis, we have

$$\text{curl}(\text{curl}(\boldsymbol{V}))=\nabla\times(\nabla\times\boldsymbol{V})=\nabla(\nabla\cdot\boldsymbol{V})-\nabla^2\boldsymbol{V}=\text{grad}\,\text{div}(\boldsymbol{V})-\nabla^2\boldsymbol{V}. \tag{10.44}$$

Proof. Now substituting (10.33), (10.37) reads

$$\rho=\nabla\cdot(\nabla f-\frac{\partial}{\partial t}\boldsymbol{V})=\nabla^2 f-\frac{\partial}{\partial t}(\nabla\cdot\boldsymbol{V}) \tag{10.45}$$

and (10.36) reads on substituting (10.29)

$$\boldsymbol{J}=\frac{1}{4\pi}\nabla\times(\nabla\times\boldsymbol{V})-\frac{\partial}{\partial t}\left(\nabla f-\frac{\partial\boldsymbol{V}}{\partial t}\right),$$

which becomes by (10.44)

$$\boldsymbol{J}=\frac{1}{4\pi}\nabla(\nabla\cdot\boldsymbol{V})-\nabla^2\boldsymbol{V}-\nabla\frac{\partial f}{\partial t}+\frac{\partial^2\boldsymbol{V}}{\partial t^2}, \tag{10.46}$$

and which in turn becomes

$$\nabla^2\boldsymbol{V}-\frac{\partial^2\boldsymbol{V}}{\partial t^2}=-\boldsymbol{J}+\nabla\left(\frac{1}{4\pi}\nabla\cdot\boldsymbol{V}-\frac{\partial f}{\partial t}\right). \tag{10.47}$$

Under (10.43), Eqn. (10.47) takes a particularly simple form, and (10.45) and (10.47) become (10.41) and (10.42), respectively.

Finally, we prove the following: Suppose $\boldsymbol{V}$ is shifted as in Theorem 10.5, (ii): $\boldsymbol{V}+\text{grad}g$. Then the shift $f+\frac{\partial g}{\partial t}$ also satisfies (10.33) if f does so.

$$\nabla\left(f+\frac{\partial g}{\partial t}\right)-\frac{\partial}{\partial t}(\boldsymbol{V}+\text{grad}g)=\nabla f-\frac{\partial}{\partial t}\boldsymbol{V}+\nabla\frac{\partial}{\partial t}g-\frac{\partial}{\partial t}\nabla g=\boldsymbol{E}.$$

We need to choose g so as to satisfy (10.43), i.e.

$$\frac{1}{4\pi}\nabla\cdot(\boldsymbol{V}+\text{grad}g)-\frac{\partial}{\partial t}(f+\frac{\partial g}{\partial t})=0.$$

or

$$\nabla^2 g-4\pi\frac{\partial^g}{\partial t^2}=4\pi\left(\nabla\cdot(\boldsymbol{V}+\text{grad}g)-\frac{\partial f}{\partial t}\right). \tag{10.48}$$

which is of the form (10.38)

Here f and $\boldsymbol{V}$ are particular solutions of (10.41) and (10.42), respectively. Once the shift function g is obtained, then all the solutions of the respective equations are determined. Hence Theorem 10.6 holds true.

10.3 A Titbit about Differential Forms

Definition 10.1. Let $\Omega \subset \mathbb{R}^n$ be an open region. Then with coefficients $a_I = a_I(\boldsymbol{x})$ of class C^∞, we define the (differential) k-**form** $\omega = \overset{k}{\omega}$ on Ω by

$$\omega = \overset{k}{\omega}(\boldsymbol{x}) = \sum_I a_I(\boldsymbol{x}) \mathrm{d}\boldsymbol{x}_I \tag{10.49}$$
$$= \sum_{i_1=1}^{n} \cdots \sum_{i_k=1}^{n} a_{i_1,\cdots,i_k}(\boldsymbol{x}) \mathrm{d}x_{i_1} \wedge \cdots \wedge \mathrm{d}x_{i_k},$$

where I denotes the k-ple index $I = \{i_1, \cdots, i_k\}$.

Let $\overset{k}{\vartheta}(\Omega)$ be the vector space of all infinitely many times differentiable k-forms on Ω.

The wedge product is computed according to the change-of-order formula:

$$\overset{k}{\omega} \wedge \overset{l}{\omega} = (-1)^{kl} \overset{l}{\omega} \wedge \overset{k}{\omega}. \tag{10.50}$$

This implies in particular that

$$\mathrm{d}\boldsymbol{x}_i \wedge \mathrm{d}\boldsymbol{x}_j = (\delta_{ij} - 1)\mathrm{d}\boldsymbol{x}_j \wedge \mathrm{d}\boldsymbol{x}_i. \tag{10.51}$$

where δ_{ij} is the Kronecker delta.

Definition 10.2. The (inner) **differential operator** (or derivation)

$$\mathrm{d} : \overset{k}{\vartheta}(\Omega) \to \overset{k+1}{\vartheta}(\Omega) \tag{10.52}$$

is defined for $\overset{k}{\omega} = \sum_I \overset{k}{\omega}_I = \sum_I a_I(\boldsymbol{x})\mathrm{d}\boldsymbol{x}_I$ *by*

$$\mathrm{d}\overset{k}{\omega} = \sum_I \mathrm{d}\overset{k}{\omega}_I = \sum_I \mathrm{d}a_I \wedge \mathrm{d}\boldsymbol{x}_I \tag{10.53}$$

and

$$\mathrm{d}\overset{k}{\omega}_I = \mathrm{d}a_I \wedge \mathrm{d}\boldsymbol{x}_I = \sum_{i=1}^{n} \frac{\partial a_I(x)}{\partial x_i} \mathrm{d}x_i \wedge \mathrm{d}\boldsymbol{x}_I \tag{10.54}$$

Theorem 10.7. (i)

$$\mathrm{d}(\overset{k}{\omega} \wedge \overset{l}{\omega}) = \mathrm{d}\overset{k}{\omega} \wedge \overset{l}{\omega} + (-1)^k \overset{k}{\omega} \wedge \mathrm{d}\overset{l}{\omega}. \tag{10.55}$$

(ii) *For every* $\omega \in \vartheta(\Omega)$

$$\mathrm{d}^2\omega = \mathrm{d}(\mathrm{d}\omega) = 0. \tag{10.56}$$

Remark 10.2. *If we denote the derivation* d *from* $\vartheta^{k-1}(\Omega) \to \vartheta^k(\Omega)$ *by* δ^{k-1}, *i.e. we consider the complex* $C = \{\vartheta^k(\Omega), \delta^k\}$,

$$C : \cdots \to \vartheta^{k-1}(\Omega) \xrightarrow{\delta^{k-1}} \vartheta^k(\Omega) \xrightarrow{\delta^k} \vartheta^{k+1}(\Omega) \to \cdots, \tag{10.57}$$

with

$$\delta^k \circ \delta^{k-1} = 0, \tag{10.58}$$

then

$$B^k(\Omega) = \mathrm{Im}\, \delta^{k-1}, \quad Z^k(\Omega) = \mathrm{Ker}\, \delta^k, \tag{10.59}$$

where Ker δ^k *is, by definition, the inverse image* $\left(\delta^k\right)^{-1}(\{0\})$ *of the identity element* 0.

Lemma 10.2. *$Z^k(\Omega)$ and $B^k(\Omega)$ form vector spaces and $B^k(\Omega) \subset Z^k(\Omega)$.*

Proof. The second assertion follows from (10.58) which is (10.56). □

Definition 10.3. The factor group

$$H^k(\Omega) = Z^k(\Omega)/B^k(\Omega) \tag{10.60}$$

is called the k**th cohomology group** of Ω.

Definition 10.4. A topological space X is called **simply connected** if any continuous closed curve lying in X can be contracted to one point without passing outside of X. For Poincar'e's lemma to hold, we need to impose a more stringent condition to the effect that the open set $\Omega \subset \mathbb{R}^n$ can be **differentiably contracted to one point** $\boldsymbol{x}_0$, say, i.e. there exists a continuos C^1 class mapping $h = h(\boldsymbol{x}, t) : \Omega \times [0, 1] \to \Omega$ such that for every $\boldsymbol{x} \in \Omega$,

$$h(\boldsymbol{x}, 0) = \boldsymbol{x}_0, \quad h(\boldsymbol{x}, 1) = \boldsymbol{x}. \tag{10.61}$$

We refer to such a mapping h a homotopy of Ω to $\boldsymbol{x}_0$ and we refer to Ω as **smoothly contractible** in what follows. A region $\Omega \subset \mathbb{R}^n$ is called **starlike** with respect to a point $\boldsymbol{x}_0$ if for every point *boldsymbol* $x \in \Omega$ the line segment joining them is contained in Ω. If $\boldsymbol{x}_0$ can run all the points, i.e. for any two points, the line segment joining them belongs to Ω, then it is called **convex**.

10.4 Algebraic Introduction to Cohomology

Definition 10.5. Let I be an index set which is a commutative semigroup with identity, e.g. $\mathbb{N} \cup \{0\}$ and R be a ring with additive unity 0. The direct sum $M = \sum_{k \in I} M_k$ of R-modules (hereafter modules) M_k is called a **graded module of type** I. Every element of M_k is called a homogenous element of degree k. Any submodule of M of the form $\sum_{k \in I} N_k$ is called a **graded submodule**, where N_k are submodules of M_k. From this the graded quotient module $\sum_{k \in I} M_k / \sum_{k \in I} N_k$ is constructed. When we speak of a cohomology group, we usually mean the graded module $H^*(C) = \sum_{k \in I} H^k(\Omega)$, where each component is the kth cohomology group in (10.60).

Graded modules with multiple index can be defined similarly: If $M = \sum_{k,l \in I} M_{k,l}$, then putting $M_m = \sum_{k+l=m \in I} M_{k,l}$, we have $M = \sum_{m \in I} M_m$ If R is a commutative ring and $A = \sum A_k$ is a graded module in which multiplication is defined and A forms an algebra over R satisfying $A_k A_l \subset A_{k+l}$, then we call A a **graded algebra**. Any graded submodule of A which forms an algebra [ideal] is called a **graded subalgebra**[**homogeneous ideal**].

Let M be a module over a commutative ring R. From the k-ple tensor product $T^k(M) = M \otimes \cdots \otimes M$ and $T^0(M) = R$ we form the graded module $T(M) = \sum_k T^k(M)$ which becomes a graded algebra by interpreting the natural isomorphism as multiplication:

$$T^k(M) \otimes T^l(M) \simeq T^{k+l}(M). \tag{10.62}$$

We call the graded algebra $T(M)$ the **tensor algebra** of M.

Theorem 10.8. *Let M [A] be an R-module [algebra]. Any linear map $f : M \to A$ can be uniquely extended to a module [algebra] homomorphism $T(f) : T(M) \to A$*

On M_k it is given by $T(f)(x_1 \otimes \cdots \otimes x_k) = f(x_1) \cdots f(x_k)$.

$T(M)$ is a universal tensor algebra whose specialization gives rise to various important classes of algebras. The alternating algebra $\Lambda(M)$ is one of them which we now introduce.

k-ple linear map $f : M^k \to N$ is called **alternating** if it satisfies one of the equivalent conditions (A_1) $f(x_1, \cdots, x_k) = 0$ if $x_i = x_j$ occurs for some $i \neq j$;

(A_1) $f(x_1, \cdots, x_k)$ changes sign if $x_i = x_j$ $i \neq j$ are changed with each other. Compare (10.51).

In $T^k(M)$ let $\mathfrak{a}_k$ denote the set of all elements $x_1 \otimes \cdots \otimes x_k$ such that $x_i = x_j$ for some $i \neq j$. Then the quotient $\Lambda^k(M) := T^k(M)/\mathfrak{a}_k$ is the totality of all classes $x_1 \otimes \cdots \otimes x_k$ mod $\mathfrak{a}$, which may be thought of as elements $x_1 \otimes \cdots \otimes x_k$ which are 0 if $x_i = x_j$ occurs for some $i \neq j$, i.e. the differential forms. Thus naturally we denote the image of the canonical projection $\pi_{\mathfrak{a}}$ by $x_1 \wedge \cdots \wedge x_k$:

$$\pi_{\mathfrak{D}} : T^k(M) \to \Lambda^k(M); \quad \pi_{\mathfrak{a}}(x_1 \otimes \cdots \otimes x_k) = x_1 \wedge \cdots \wedge x_k.$$

Proposition 10.1. *If M is a free module with basis $\{u_m\}$, then $T^k(M)$ is also a free module with basis of the form $\{u_{i_1} \otimes \cdots \otimes u_{i_k}\}$.*

This explains the situation in Definition 10.1 by choosing $T^k(M) = \overset{k}{\vartheta}(\Omega)$ k-differential forms with $M = \oplus_{m=1}^{n} R\mathrm{d}x_m$, and R is the algebra of all infinitely differentiable functions defined on an open domain $\Omega \in \mathbb{R}$.

Formula (10.50) holds true:

$$\overset{k}{\omega} \wedge \overset{l}{\omega} = (-1)^{kl} \overset{l}{\omega} \wedge \overset{k}{\omega}, \overset{k}{\omega} \in T^k(M), \overset{l}{\omega} \in T^k(M). \tag{10.63}$$

With $\mathfrak{a}_0 = 0$, the direct sum $\mathfrak{a} = \sum \mathfrak{a}_k$ forms a homogeneous ideal of $T(M)$ and the quotient $\Lambda(M) = T(M)//\mathfrak{a}$ is called the **alternating algebra** of M. This consists of coordinates whose entries are k-forms with distinct basis elements arranged in $k = 0, 1, \ldots$. In view of (10.62), the product of two basis elements is again the wedge product, so that we use the wedge product symbol for the product in $\Lambda(M)$.

Proposition 10.2. *We have*

$$\Lambda(M_1 \oplus M_2) \simeq \Lambda(M_1) \otimes \Lambda(M_2)$$

10.5 Vectorial Stokes Theorem

This Appendix is for deducing the equation of continuity (10.11) for the flow (10.6) most naturally from the vector Stokes theorem. We recall basic notation.

Definition 10.6. Let $X \subset \mathbb{R}$ be a domain with sufficiently smooth boundary. For a vector-valued function $\boldsymbol{f} = \begin{pmatrix} P \\ Q \\ R \end{pmatrix} \in C^1(X)$ (i.e. $P, Q, R \in C^1(X)$), we define its **divergence** $\operatorname{div} \boldsymbol{f}$ and **curl** $\operatorname{curl} \boldsymbol{f}$ (also called rotation $\operatorname{rot} \boldsymbol{f}$) by

$$\operatorname{div} \boldsymbol{f} = \frac{\partial P}{\partial x} + \frac{\partial Q}{\partial y} + \frac{\partial R}{\partial z} = \nabla \cdot \boldsymbol{f} \tag{10.64}$$

and

$$\operatorname{curl} \boldsymbol{f} = {}^t\left(\begin{vmatrix} \frac{\partial}{\partial y} & Q \\ \frac{\partial}{\partial z} & R \end{vmatrix}, - \begin{vmatrix} \frac{\partial}{\partial x} & P \\ \frac{\partial}{\partial z} & R \end{vmatrix}, \begin{vmatrix} \frac{\partial}{\partial x} & P \\ \frac{\partial}{\partial y} & Q \end{vmatrix} \right) \tag{10.65}$$

$$= {}^t\left(\frac{\partial R}{\partial y} - \frac{\partial Q}{\partial z}, -\left(\frac{\partial R}{\partial x} - \frac{\partial P}{\partial z}\right), \frac{\partial Q}{\partial x} - \frac{\partial P}{\partial y}\right)$$
$$= \nabla \times \boldsymbol{f}$$

Definition 10.7. We extend the Definition (10.64) of the divergence to a matrix. Let

$$P = (\boldsymbol{p}_1, \cdots, \boldsymbol{p}_n), \tag{10.66}$$

where $\boldsymbol{p}_j, 1 \le j \le n$ are of C^1 class as in Definition 10.6. Then

$$\nabla \cdot P = \nabla \cdot (\boldsymbol{p}_1, \cdots, \boldsymbol{p}_n) = (\nabla \cdot \boldsymbol{p}_1, \cdots, \nabla \cdot \boldsymbol{p}_n). \tag{10.67}$$

Hereafter, a manifold is a differential manifold with a sufficiently smooth boundary.

Theorem 10.9. (General Stokes' theorem) *Let $\mathcal{M}$ be a $(k+1)$-dimensional manifold, $\partial\mathcal{M}$ its boundary, and let ω be a differential form of degree k in n variables. Then the identity*

$$\underbrace{\int \cdots \int}_{k+1}{}_{\mathcal{M}} \, \mathrm{d}\omega = \underbrace{\int \cdots \int}_{k}{}_{\partial\mathcal{M}} \, \omega \tag{10.68}$$

holds true; in a more concise form it reads

$$\mathrm{d}\omega(\mathcal{M}) = \omega(\partial\mathcal{M}). \tag{10.69}$$

Furthermore, suppose $\omega = \boldsymbol{f} \cdot \mathrm{d}A$ with "area element" $A = A(\partial\mathcal{M})$, we have $\mathrm{d}\omega = \nabla \cdot \boldsymbol{f}\mathrm{d}V$ with "volume element" $V = V(\mathcal{M})$. Then (10.68) *reads*

$$\underbrace{\int \cdots \int}_{k+1}{}_{\mathcal{M}} \, \nabla \cdot \boldsymbol{f}\mathrm{d}V = \underbrace{\int \cdots \int}_{k}{}_{\partial\mathcal{M}} \, \boldsymbol{f} \cdot \mathrm{d}A. \tag{10.70}$$

In vectorial form, (10.70) *reads*

$$\underbrace{\int \cdots \int}_{k+1}{}_{\mathcal{M}} \, \nabla \cdot \boldsymbol{P}\mathrm{d}V = \underbrace{\int \cdots \int}_{k}{}_{\partial\mathcal{M}} \, \boldsymbol{P} \cdot \mathrm{d}A, \tag{10.71}$$

where $\boldsymbol{P} = (\boldsymbol{p}_1, \ldots, \boldsymbol{p}_n)$ is an $n \times n$-matrix under the inner product rule

$$\boldsymbol{P} \cdot \mathrm{d}A = (\boldsymbol{p}_1 \cdot \mathrm{d}A, \cdots, \boldsymbol{p}_n \cdot \mathrm{d}A) \tag{10.72}$$

Corollary 10.1. (Unicity theorem) *Suppose $n = k+1$ and that we are given the identity*

$$\underbrace{\int \cdots \int}_{k+1}{}_{\mathcal{M}} \, \nabla \cdot \boldsymbol{P}\mathrm{d}V = \underbrace{\int \cdots \int}_{k+1}{}_{\mathcal{M}} \, \nabla \cdot \boldsymbol{Q}\mathrm{d}V \tag{10.73}$$

for any bounded domain $V \subset \mathcal{M}$. Then

$$\boldsymbol{P} = \boldsymbol{Q} \tag{10.74}$$

throughout $\mathcal{M}$. Furthermore, the Condition (10.73) *may be replaced by*

$$\underbrace{\int \cdots \int}_{k+1}{}_{\mathcal{M}} \, \nabla \cdot \boldsymbol{P}\mathrm{d}V = \underbrace{\int \cdots \int}_{k}{}_{\partial\mathcal{M}} \, \boldsymbol{Q} \cdot \mathrm{d}A. \tag{10.75}$$

Proof. Equation (10.73) amounts to

$$\underbrace{\int\cdots\int}_{k+1}{}_{\mathcal{M}}\ \nabla\cdot(\boldsymbol{P}-\boldsymbol{Q})\mathrm{d}V=0 \tag{10.76}$$

for any bounded domain V. Since the integrand is continuous, if it is non-zero at a point $\boldsymbol{z}_0$, say $c>0$. Then in a neighborhood V_0 of the point it is positive and so the integral over V_0 is positive, a contradiction.

By (10.71), the second term on the right of (10.75) is $\underbrace{\int\cdots\int}_{k+1}{}_{\mathcal{M}}\ \nabla\cdot\boldsymbol{Q}\mathrm{d}V$ leading to (10.73).

Given (10.75), the general Stokes theorem implies (10.73), and we still have the validity of the corollary, completing the proof. □

Remark 10.3. *The condition $n=k+1$ indicates the saturation of the integrals in* (10.73), *i.e. they are scalar integrals including the case* (10.71) *of matrices. In view of* (10.67) *the latter case amounts to*

$$\left(\underbrace{\int\cdots\int}_{k+1}{}_{\mathcal{M}}\nabla\cdot\boldsymbol{p}_1,\cdots,\underbrace{\int\cdots\int}_{k+1}{}_{\mathcal{M}}\nabla\cdot\boldsymbol{p}_n\right)=\left(\underbrace{\int\cdots\int}_{k+1}{}_{\mathcal{M}}\nabla\cdot\boldsymbol{q}_1,\cdots,\underbrace{\int\cdots\int}_{k+1}{}_{\mathcal{M}}\nabla\cdot\boldsymbol{q}_n\right), \tag{10.77}$$

which implies coincidence of components, and so the unicity still holds.

To describe the equation of continuity we need some preparations.

Definition 10.8. A smooth (n-dimensional) curve $\boldsymbol{x}=\boldsymbol{x}(t)$ is a C^1 class map $\boldsymbol{x}:[a,b]\to\mathbb{R}^n$ in the form $\boldsymbol{x}=\boldsymbol{x}(t)={}^t(x_1,\ldots,x_n)$ and the tangent vector is $\dot{\boldsymbol{x}}(t)={}^t(\dot{x_1},\ldots,\dot{x_n})$.

Definition 10.9. A (smooth) surface $S\subset\mathbb{R}^3$ is a (smooth) mapping from $D=\overset{2}{\mathbb{I}}\subset\mathbb{R}^2\to\mathbb{R}^3$. We restrict ourselves to the parametric representation with a 2-dimensional interval D.

$$\boldsymbol{\phi}:D\to S\subset\mathbb{R}^3;\quad {}^t(x,y,z)=\boldsymbol{x}=\boldsymbol{\phi}(\boldsymbol{t}),\boldsymbol{t}={}^t(u,v) \tag{10.78}$$

in component form

$$\begin{cases}x=x(u,v)\\ y=y(u,v)\\ z=z(u,v),\end{cases}\qquad {}^t(u,v)\in\overset{2}{\mathbb{I}}. \tag{10.79}$$

We assume throughout that the functions are C^1 and so the gradient of $\boldsymbol{\phi}$ is

$$\nabla\boldsymbol{\phi}=\begin{pmatrix}\nabla x\\ \nabla y\\ \nabla z\end{pmatrix}=\begin{pmatrix}\frac{\partial x}{\partial u}&\frac{\partial x}{\partial v}\\ \frac{\partial y}{\partial u}&\frac{\partial y}{\partial v}\\ \frac{\partial z}{\partial u}&\frac{\partial z}{\partial v}\end{pmatrix}=(\boldsymbol{x}_u,\boldsymbol{x}_v). \tag{10.80}$$

We remark that $\boldsymbol{x}_u(u_0,v_0),\boldsymbol{x}_v(u_0,v_0)$ are tangent vectors of the curves $\boldsymbol{x}(u,v_0)$ and $\boldsymbol{x}(u_0,v)$ on S, respectively, at the point $\boldsymbol{x}_0=\boldsymbol{x}_0(u_0,v_0)$ on the surface. Hence the infinitesimal parallelogram formed by them has the area element

$$\mathrm{d}A=|\boldsymbol{x}_u\times\boldsymbol{x}_v|\mathrm{d}u\mathrm{d}v. \tag{10.81}$$

Hence for a continuous function defined on S, its integral over S is to be defined as

$$\iiint_S f = \iint_D f(\boldsymbol{\phi}(\boldsymbol{t}))|\boldsymbol{x}_u \times \boldsymbol{x}_v| \, \mathrm{d}u\mathrm{d}v, \tag{10.82}$$

where

$$|\boldsymbol{x}_u \times \boldsymbol{x}_v| = \sqrt{\left(\frac{\partial(y,z)}{\partial(u,v)}\right)^2 + \left(\frac{\partial(x,z)}{\partial(u,v)}\right)^2 + \left(\frac{\partial(x,y)}{\partial(u,v)}\right)^2}. \tag{10.83}$$

The unit vector

$$\boldsymbol{n} = \frac{1}{|\boldsymbol{x}_u \times \boldsymbol{x}_v|}\boldsymbol{x}_u \times \boldsymbol{x}_v = |\boldsymbol{x}_u \times \boldsymbol{x}_v|^t \left(\frac{\partial(y,z)}{\partial(u,v)}, -\frac{\partial(x,z)}{\partial(u,v)}, \frac{\partial(x,y)}{\partial(u,v)}\right) \tag{10.84}$$

is the (outward) **normal vector** of S at $\boldsymbol{x} = \boldsymbol{x}(u, v)$ when the tangent vectors exist.

More precisely, one would need to argue as follows. By a division of D into subdomains D_{ij} gives rise to that of S into subsurfaces S_{ij}. Choosing an arbitrary point $p_{ij} \in S_{ij}$ and forming the Riemann sum

$$\sum f(p_{ij})A(S_{ij}),$$

where A means the area. Then letting the division smaller, prove the existence of the limit just as in the case of the ordinary Riemann integral. Equation (10.82) allows to compute the integral over a surface as an ordinary double integral.

We write

$$f_{\boldsymbol{n}} = \boldsymbol{f} \cdot \boldsymbol{n} \tag{10.85}$$

for the **component of $\boldsymbol{f}$ in the direction of $\boldsymbol{n}$**. Hence $f_{\boldsymbol{n}}$ *expresses the ratio of mass flowing out at* P.

Lemma 10.3. *Let* $\boldsymbol{f} = {}^t(P, Q, R) \in C(\Omega)$ *be a continuous vector field in* $\Omega \subset \mathbb{R}^3$,$\omega = P\mathrm{d}y\mathrm{d}z + Q\mathrm{d}z\mathrm{d}x + R\mathrm{d}x\mathrm{d}y$ *a differential form of degree 2,* $S \subset \Omega$ *a smooth surface given by* $S : \boldsymbol{x} = \boldsymbol{x}(u, v)$, $(u, v) \in \mathbb{R}^2$. *Then we have*

$$\iint_S (P\mathrm{d}y\mathrm{d}z + Q\mathrm{d}z\mathrm{d}x + R\mathrm{d}x\mathrm{d}y) = \iint_S f_{\boldsymbol{n}} \, \mathrm{d}A, \tag{10.86}$$

where the right-hand side is the surface integral defined by

$$\iint_S f_{\boldsymbol{n}} \, \mathrm{d}A = \iint_D f_{\boldsymbol{n}}(\boldsymbol{x}(u,v))|\boldsymbol{x}_u \times \boldsymbol{x}_v| \, \mathrm{d}u\mathrm{d}v. \tag{10.87}$$

Proof. By (10.84) and (10.85), we have

$$f_{\boldsymbol{n}}|\boldsymbol{x}_u \times \boldsymbol{x}_v| = \boldsymbol{f} \cdot (\boldsymbol{x}_u \times \boldsymbol{x}_v). \tag{10.88}$$

Substituting this, we have

$$\iint_S f_{\boldsymbol{n}} \, \mathrm{d}A = \iint_D \left(P\frac{\partial(y,z)}{\partial(u,v)} - Q\frac{\partial(z,x)}{\partial(u,v)} + R\frac{\partial(x,y)}{\partial(u,v)}\right) \mathrm{d}u\mathrm{d}v, \tag{10.89}$$

which amounts to the assertion. □

Therefore its integral over all ∂X

$$Q = Q(\partial X) := \iint_{\partial X} f_{\boldsymbol{n}} \tag{10.90}$$

expresses the **totality of masses flowing out of X in a unit time interval.**

Example 10.1 (Equation of continuity) *Let $\boldsymbol{v}$ be the vector field of velocity of the fluid flowing in a domain $\Omega \subset \mathbb{R}^3$, and let $X \subset \Omega$ be a bounded closed domain with its boundary ∂X forming a surface. Let ρ denote the continuous density (distribution) function of the fluid in Ω. Then $\boldsymbol{j} = \rho\boldsymbol{v}$ is the vector field describing the flow of mass distribution of the fluid. Suppose there is no source or sink in X. Then since $Q(\partial X)$ indicates the rate of decrease of all masses in X, we have*

$$\iint_{\partial X} j_{\boldsymbol{n}} = -\frac{\mathrm{d}}{\mathrm{d}t}\iiint_X \rho. \tag{10.91}$$

By Lemma 10.3, the left-hand side is $\iiint_X \operatorname{div} \boldsymbol{f}$:

$$\iint_{\partial X} \boldsymbol{j} \cdot \mathrm{d}A = \iiint_X \operatorname{div} \boldsymbol{j}. \tag{10.92}$$

Since $\frac{\mathrm{d}\rho}{\mathrm{d}t}$ is continuous, we may change the integration and differentiation on the right of (10.91), *and we obtain*

$$\iiint_X \operatorname{div} \boldsymbol{j} = -\iiint_X \frac{\partial \rho}{\partial t}. \tag{10.93}$$

Hence, by Corollary 10.1 we conclude (10.11).

The next example is not directly relevant to electromagnetism but is the fundamental result in fluid dynamics, which together with another result of Stokes leads to the fundamental Navier-Stokes equation, a counterpart of Maxwell equations. This reveals that a flow also obeys a variant of Newton's equation of motion.

Example 10.2 (Cauchy equation of motion) *In this example we show that Newton' second law of motion entails the Cauchy equation of motion* (10.98).

We consider a flow in a domain $\Omega \subset \mathbb{R}^3$ and let $V \subset \Omega$ be a bounded closed domain with its boundary surface $A = \partial V$. Let ρ denote the continuous density (distribution) function of the fluid in Ω. Then ***Newton' second law of motion*** *reads*

$$M\boldsymbol{a} = \boldsymbol{F}, \tag{10.94}$$

where $M = \int_V \rho \mathrm{d}V$ indicates the mass and $\boldsymbol{F}$ is the sum of all external forces acting on the fluid volume, and it consists of body force $\boldsymbol{F}_b$ and surface force $\boldsymbol{F}_s$. The body forces act on the center of mass of the fluid volume, and we assume it to be the gravitational force

$$\boldsymbol{F}_b = \int_V \rho\boldsymbol{g}\,\mathrm{d}V \tag{10.95}$$

and the surface force is given by stress dyadic

$$\boldsymbol{F}_s = \int_A \boldsymbol{P} \cdot \mathrm{d}A = \int_V \nabla \cdot \boldsymbol{P}\,\mathrm{d}V. \tag{10.96}$$

Hence

$$\int_V (\boldsymbol{g} - \boldsymbol{a})\rho \mathrm{d}V + \int_V \nabla \cdot \boldsymbol{P}\,\mathrm{d}V = 0. \tag{10.97}$$

Corollary 10.1 implies the **Cauchy equation of motion**

$$(\boldsymbol{g} - \boldsymbol{a})\rho + \nabla \cdot P = \boldsymbol{o}. \tag{10.98}$$

We state well-known special cases given in Table 10.2 of the general Stokes theorem in [2].

TABLE 10.2
Special Cases of General Stokes' Theorem

Name	Degree	Number of var.
General Stokes' theorem	k	n
Green's theorem	1	2
Gauss divergence theorem	2	3
Stokes' theorem	1	3

Acknowledgments

We would like to express our hearty thanks to Professor M. Mrugala for pointing out the reference [6] which contains some relevant contents to the present paper.

Dedication

We dedicate to Professor Wenpeng Zhang with friendship and respect on the occasion of his 60th birthday.

References

[1] S. Koide, *Physics*, 6th ed., Shokabo, Tokyo (1971).

[2] S. Kanemitsu, F. Li, Y. Kamata, and K. Maruyama, Mathematical foundations of fluid dynamics, *Basic Sci. J. Text. Univ.*, **29** No. 3 (2016), 275–290.

[3] B. Lal, *Mathematical Theory of Electromagnetism*, Asia Publishing House, London (1965).

[4] C. G. Buck, *Advanced Calculus*, McGraw-Hill, New York (1956).

[5] H. Flanders, *Differential Forms with Applications to the Physical Sciences*, Academic Press, New York, London (1963).

[6] R. S. Ingarden and A. Jamiołkowski, *Classical Electrodynamics*, Elesevier, Oxford (1989).

[7] K. Maurin, *Analiza*, Part I, PWN, Warszawa (1973).

11

On New Integral Inequalities Involving Generalized Fractional Integral Operators

M. Emin Özdemir

Bursa Uludağ University

Ahmet Ocak Akdemir

Ağrı İbrahim Çeçen University

Erhan Set

Ordu University

Alper Ekinci

Ağrı İbrahim Çeçen University
Bandırma 17 Eylül University

CONTENTS

11.1 Introduction

Inequality theory plays an important role in almost all areas of mathematics. The foundations of the inequality theories were laid in the 18th and 19th centuries by mathematicians such as K. F. Gauss (1777–1855), A. L. Cauchy (1789–1857), and P. L. Chebyshev (1821–1894). In later years, the importance of inequalities has increased, and inequality has attracted the attention of famous mathematicians such as H. Poincare (1854–1912), A. M. Lyapunav (1857–1918), O. Hölder (1859–1937), and J. Hadamard (1865–1963). Although this field played a role in all branches of mathematics, the development of its subjects as a branch of mathematics was realized by G. H. Hardy, J. E. Littlewood, and G. Polya in 1934 by the study of "Inequalities". These theoretical foundations developed by other mathematicians have led to the discovery of many new inequalities and interesting applications of mathematics in various fields such as mathematical analysis, applied mathematics, and probability theory. Studies on inequalities are based on discovering new inequalities and strengthening classical approaches. Modern inequality theory continues as an area of mathematics based on deep foundations without losing its importance for centuries. Inequality theory continues to be a field that is constantly studied and still plays an active role in research and is fascinating.

In these studies some classical methods of classical analysis have been used, but in recent years, the usage of fractional integral operators in this field has led to the beginning of a new perspective to the inequality theory.

Fractional analysis can be considered as an expansion of classical analysis. In the fractional analysis the presence of more than one derivative definition gives the opportunity to obtain the best solution of the problem.

Fractional analysis has been studied by many scientists, and they have expressed the fractional derivative and integral in different ways with different notations. However, although there are transitions between these expressions, their definitions and definitions differ in terms of their physical interpretation.

Recently, a new definition of fractional derivative, which is considered to be the natural extension of the classical derivative, was given by Khalil et al. This new definition draws attention with its variability to classical derivative. This new fractional derivative definition, called conformable fractional derivative, has attracted a great deal of attention due to these properties, and in a short time, many studies related to this new definition have been made.

This new definition introduced by Khalil et al. [1] has a limit form similar to that of the classical type. They also described the Rolle's theorem and the mean value theorem for conformable fractional derivatives. Conformable fractional derivative analysis was developed by Abdeljawad [2]. In his work, Abdeljawad presented the concepts of fractional derivatives, the fractional chain rule and Gronwall's inequality, and left- and right-sided conformable fractional derivatives. He also gave the concept of coherent fractional sequence derivative, two kinds of fractional partial integration formulas for $0 < \alpha \leq 1$, coherent fractional sequence expansion, fractional Taylor inequality, and finally fractional Laplace transformation.

Some fractional integral operators generalize some other fractional integrals in special cases, as in the following integral operator. Recently, Jarad et al. have defined a new fractional integral operator. Also, they gave some properties and relations between some other fractional integral operators, such as Riemann–Liouville fractional integral, Hadamard fractional integrals, and generalized fractional integral operators, with this operator.

In the present chapter, we shall provide a collection of fractional integral operators, which generalizes the classical integrals. We shall recall some historical remarks, concepts, and definitions to refresh our memories about some of the remarkable milestones in the theory of fractional calculus. Later, we will give some important integral inequalities involving the generalized fractional integral operators. Finally, we will call attention to some special cases on inequalities of Hermite–Hadamard type involving generalized fractional integral operators.

11.2 Concepts of Fractional Integral Operators

We will give the following well-known definitions in the fractional calculus.

Definition 11.1. [3] Let $f \in L_1[a,b]$. The Riemann–Liouville integrals $J^{\alpha}_{a+}f$ and $J^{\alpha}_{b-}f$ of order $\alpha > 0$ are defined by

$$J^{\alpha}_{a+}f(t) = \frac{1}{\Gamma(\alpha)} \int_a^t (t-x)^{\alpha-1} f(x)dx, \quad t > a \tag{11.1}$$

and

$$J^{\alpha}_{b-}f(t) = \frac{1}{\Gamma(\alpha)} \int_t^b (x-t)^{\alpha-1} f(x)dx, \quad t < b \tag{11.2}$$

Respectively, where $\Gamma(\alpha) = \int_0^{\infty} e^{-t} t^{\alpha-1} dt$.

Here $J_{a+}^{0}f(t) = J_{b-}^{0}f(t) = f(t)$. In the case of $\alpha = 1$, the fractional integral reduces to classical integral.

In Ref. [2], Abdeljawad gave the following definitions of Right-Left conformable fractional integrals:

Definition 11.2. [2] Let $\alpha \in (n, n+1]$, $n = 0, 1, 2, \dots$ and set $\beta = \alpha - n$. Then the left conformable fractional integral of any order $\alpha > 0$ is defined by

$$\left(I_{\alpha}^{a} f\right)(t) = \frac{1}{n!}\int_{a}^{t}(t-x)^{n}(x-a)^{\beta-1}f(x)dx.$$

Definition 11.3. Analogously, the right conformable fractional integral of any order $\alpha > 0$ is defined by

$$\left({}^{b}I_{\alpha} f\right)(t) = \frac{1}{n!}\int_{t}^{b}(x-t)^{n}(b-x)^{\beta-1}f(x)dx.$$

Notice that if $\alpha = n+1$ then $\beta = \alpha - n = n+1-n = 1$; hence, $\left(I_{n+1}^{a} f\right)(t) = \left(J_{a+}^{n+1} f\right)(t)$ and $\left({}^{b}I_{n+1} f\right)(t) = \left(J_{b-}^{n+1} f\right)(t)$.

In Ref. [1], Khalil et al. gave the following definition:

Definition 11.4. [1] (Conformable fractional integral) Let $\alpha \in (0,1]$, $0 \le \kappa_1 < \kappa_2$. A function $h\colon [\kappa_1, \kappa_2] \to \mathbb{R}$ is α-fractional integrable on $[\kappa_1, \kappa_2]$ if the integral

$$\int_{\kappa_1}^{\kappa_2} h(x)d_{\alpha}x = \int_{\kappa_1}^{\kappa_2} h(x)x^{\alpha-1}dx$$

exists and is finite. All α*-fractional integrable functions on* $[\kappa_1, \kappa_2]$ *is indicated by* $L_{\alpha}([\kappa_1, \kappa_2])$.

Remark 11.1. [1].

$$I_{\alpha}^{\kappa_1}(h_1)(s) = I_{1}^{\kappa_1}\left(s^{\alpha-1}h_1\right) = \int_{\kappa_1}^{s}\frac{h_1(x)}{x^{1-\alpha}}dx$$

where the integral is the usual Riemann improper integral, and $\alpha \in (0,1]$.

In Ref. [4], Jarad et al. has defined a new fractional integral operator. Also, they have proved some properties and relations between the some other fractional integral operators.

Definition 11.5. [4] Let $\beta \in \mathbb{C}$, $\mathrm{Re}(\beta) > 0$, then the left- and right-sided fractional conformable integral operators has defined, respectively, as follows:

$${}_{a}^{\beta}\mathfrak{J}^{\alpha} f(x) = \frac{1}{\Gamma(\alpha)}\int_{a}^{x}\left(\frac{(x-a)^{\alpha} - (t-a)^{\alpha}}{\alpha}\right)^{\beta-1}\frac{f(t)}{(t-a)^{1-\alpha}}dt$$

and

$${}^{\beta}\mathfrak{J}_{b}^{\alpha} f(x) = \frac{1}{\Gamma(\beta)}\int_{x}^{b}\left(\frac{(b-x)^{\alpha} - (b-t)^{\alpha}}{\alpha}\right)^{\beta-1}\frac{f(t)}{(b-t)^{1-\alpha}}dt.$$

The results presented here, being general, can be reduced to yield many relatively simple inequalities and identities for functions associated with certain fractional integral operators. For example, the case $\alpha = 1$ in the obtained results is found to yield the same results involving Riemann–Liouville fractional integrals, given before, in literatures. Further, getting more knowledge, see the paper given in [4].

Katugampola (see Refs. [5] and [6]) considered the following iterative process in 2011:

$$\int_a^x t_1^\rho dt_1 \int_a^{t_1} t_2^\rho dt_2 \ldots \int_a^{t_{n-1}} t_n^\rho f(t_n)\, dt_n = \frac{(\rho+1)^{1-n}}{(n-1)!}\int_a^x \left(t^{\rho+1} - \tau^{\rho+1}\right)^{n-1} \tau^\rho f(\tau)\, d\tau,$$

for $n \in \mathbb{N}$. This generates the Katugampola's concept of fractional integral, defined in [5] and also in [6].

Definition 11.6. [5] Let $f \in [a, b]$, the left-sided Katugampola fractional integral ${}^\rho I_{a+}^\alpha f$ of order $\alpha \in \mathbb{C}$, $\text{Re}(\alpha) > 0$ *is defined by*

$${}^\rho I_{a+}^\alpha f(x) = \frac{\rho^{1-\alpha}}{\Gamma(\alpha)}\int_a^x \frac{t^{\rho-1}}{(x^\rho - t^\rho)^{1-a}} f(t)\, dt, \;\; x > a,$$

the right-sided Katugampola fractional integral ${}^\rho I_{b-}^\alpha f$ of order $\alpha \in \mathbb{C}$, $Re(\alpha) > 0$ is defined by

$${}^\rho I_{b-}^\alpha f(x) = \frac{\rho^{1-\alpha}}{\Gamma(\alpha)}\int_x^b \frac{t^{\rho-1}}{(t^\rho - x^\rho)^{1-a}} f(t)\, dt, \;\; x < b.$$

Katugampola's operators are generalizations of A. Erdélyi and H. Kober operators introduced in 1940 (see Refs. [7] and [8]), as well. Other similar approaches on moving iterative integrals and derivatives into fractional framework in connection with theoretic and practical applications are in the mathematical literature of the last decade.

Remark 11.2. *If* $\rho = 1$ *then the Katugampola fractional integrals become Riemann–Liouville fractional integrals.*

In Ref. [8], Raina introduced a class of functions defined formally by

$$\mathcal{F}_{\rho,\lambda}^\sigma(x) = \mathcal{F}_{\rho,\lambda}^{\sigma(0),\sigma(1)} \ldots (x) = \sum_0^\infty \frac{\sigma(k)}{\Gamma(\rho k + \lambda)} x^k (\rho, \lambda > 0; |x| < \mathbb{R}) \tag{11.3}$$

where the coefficients $\sigma(k)(k \in \mathbb{N} = \mathbb{N} \cup \{0\}$ is a bounded sequence of positive real numbers and $\mathbb{R}$ is the set of real numbers. With the help of (11.3), Raina [9] defined the following left-sided and right-sided fractional integral operators, respectively, as follows:

$$\mathcal{J}_{\rho,\lambda,a^+;w}^\sigma \varphi(x) = \int_a^x (x-t)^{\lambda-1} \mathcal{F}_{\rho,\lambda}^\sigma [w(x-t)^\rho]\, \varphi(t)\, dt \;\; (x > a > 0) \tag{11.4}$$

and

$$\mathcal{J}_{\rho,\lambda,b^-;w}^\sigma \varphi(x) = \int_x^b (t-x)^{\lambda-1} \mathcal{F}_{\rho,\lambda}^\sigma [w(t-x)^\rho]\, \varphi(t)\, dt \;\; (0 < x < b) \tag{11.5}$$

where $\lambda, \rho > 0$, $w \in \mathbb{R}$, and $\varphi(t)$ is such that the integral on the right side exits.

It is easy to verify that $\mathcal{J}_{\rho,\lambda,a^+;w}^\sigma \varphi(x)$ and $\mathcal{J}_{\rho,\lambda,b^-;w}^\sigma \varphi(x)$ are bounded integral operators on $L(a, b)$, if

$$\mathfrak{M} = \mathcal{F}_{\rho,\lambda+1}^\sigma [w(b-a)^\rho] < \infty \tag{11.6}$$

In fact, for $\varphi \in L(a, b)$, we have

$$\left\| \mathcal{J}_{\rho,\lambda,a^+;w}^\sigma \varphi(x) \right\|_1 \le \mathfrak{M}(b-a)^\lambda \|\varphi\|_1 \tag{11.7}$$

and

$$\left\| \mathcal{J}_{\rho,\lambda,b^-;w}^\sigma \varphi(x) \right\|_1 \le \mathfrak{M}(b-a)^\lambda \|\varphi\|_1 \tag{11.8}$$

where

$$||\varphi||_p \colon = \left(\int_a^b |\varphi(t)|^p dt \right)^{\frac{1}{p}}.$$

Here, many useful fractional integral operators can be obtained by specializing the coefficient $\sigma(k)$. For instance the classical Riemann–Liouville fractional integrals J_{a+}^{α} and J_{b-}^{α} of order α follow easily by setting $\lambda = \alpha$, $\sigma(0) = 1$, and $w = 0$ in (11.4) and (11.5).

In Ref. [10], authors defined the generalized fractional integral operators as follows:

Definition 11.7. Let $f \in L[a, b]$. The Generalized Fractional integrals ${}_{a^+\varphi}I_{\varphi}f$ and ${}_{b^-\varphi}I_{\varphi}f$ with $a \geq 0$ are defined by

$${}_{a^+\varphi}I_{\varphi}f(x) = \int_a^x \frac{\varphi(x-t)}{x-t} f(t)\, dt, \quad x > a \tag{11.9}$$

and

$${}_{b^-\varphi}I_{\varphi}f(x) = \int_x^b \frac{\varphi(t-x)}{t-x} f(t)\, dt, \quad x < b, \tag{11.10}$$

where $\varphi\colon\ [0, \infty) \to [0, \infty)$ *a function which satisfies* $\int_0^1 \frac{\varphi(t)}{t} dt < \infty$.

These important special cases of the integral operators (11.9) and (11.10) are mentioned below.

1. If we take $\varphi(t) = t$, the operators (11.9) and (11.10) reduce to the Riemann integral as follows:
$$I_{a^+}f(x) = \int_a^x f(t)dt, \quad x > a,$$
and
$$I_{b^-}f(x) = \int_x^b f(t)dt, \quad x < b.$$

2. If we take $\varphi(t) = \frac{t^{\alpha}}{\Gamma(\alpha)}$, the operators (11.9) and (11.10) reduce to the Riemann–Liouville integral as (11.1) and (11.2).

3. If we take $\varphi(t) = \frac{1}{k\Gamma_k(\alpha)} t^{\frac{\alpha}{k}}$, the operators (11.9) and (11.10) reduce to the k-Riemann–Liouville integral as follows:
$$I_{a^+,k}^{\alpha}f(x) = \frac{1}{k\Gamma_k(\alpha)} \int_a^x (x-t)^{\frac{\alpha}{k}-1} f(t)dt, \quad x > a,$$
and
$$I_{b^-,k}^{\alpha}f(x) = \frac{1}{k\Gamma_k(\alpha)} \int_x^b (t-x)^{\frac{\alpha}{k}-1} f(t)dt, \quad x < b.$$
where
$$\Gamma_k(\alpha) = \int_0^{\infty} t^{\alpha-1} e^{-\frac{t^k}{k}} dt, \ \mathrm{R}(\alpha) > 0$$
and
$$\Gamma_k(\alpha) = k^{\frac{\alpha}{k}-1} \Gamma\left(\frac{\alpha}{k}\right), \ \mathrm{R}(\alpha) > 0, k > 0$$
are given by Mubeen and Habibullah in [11].

11.3 Integral Inequalities via Fractional Integral Operators

During the last decades, several researches have studied about integral inequalities by using different kinds of fractional integral operators. In this section, we will give some fundamental results to put forth historical improvement of this field.

The following integral identiy was proved by Set in 2012.

Lemma 11.1. [12] *Let $f\colon [a,b] \to \mathbb{R}$ be a differentiable mapping on (a,b) with $a < b$. If $f' \in L[a,b]$, then for all $x \in [a,b]$ and $\alpha > 0$, we have*

$$\left(\frac{(x-a)^{\alpha}+(b-x)^{\alpha}}{b-a}\right) f(x) - \frac{\Gamma(\alpha+1)}{b-a}\left[J_{x^-}^{\alpha} f(a) + J_{x^+}^{\alpha} f(b)\right]$$
$$= \frac{(x-a)^{\alpha+1}}{b-a}\int_0^1 t^{\alpha} f'\left(tx+(1-t)a\right)dt - \frac{(b-x)^{\alpha+1}}{b-a}\int_0^1 t^{\alpha} f'\left(tx+(1-t)b\right)dt$$

where $\Gamma(\alpha) = \int_0^{\infty} e^{-1}u^{\alpha-1}du$.

The author has established Ostrowski-type integral inequalities for s-convex functions via Riemann–Liouville fractional integrals.

Theorem 11.1. [12] *Let $f\colon [a,b] \subseteq (0,\infty) \to \mathbb{R}$, be a differentiable mapping on (a,b) with $a < b$ such that $f' \in L[a,b]$. If $|f'|$ is s-convex in the second sense on $[a,b]$ for some fixed $s \in (0,1]$ and $|f'(x)| \leq M, x \in a,b]$, then the following inequality for fractional integrals with $\alpha > 0$ holds:*

$$\left|\left(\frac{(x-a)^{\alpha}+(b-x)^{\alpha}}{b-a}\right) f(x) - \frac{\Gamma(\alpha+1)}{(b-a)}\left[J_{x^-}^{\alpha} f(a) + J_{x^+}^{\alpha} f(b)\right]\right|$$
$$\leq \frac{M}{b-a}\left(1+\frac{\Gamma(\alpha+1)\Gamma(s+1)}{\Gamma(\alpha+s+1)}\right)\left[\frac{(x-a)^{\alpha+1}+(b-x)^{\alpha+1}}{\alpha+s+1}\right]$$

where Γ is Euler gamma function.

Theorem 11.2. [12] *Let $f\colon [a,b] \subseteq (0,\infty) \to \mathbb{R}$, be a differentiable mapping on (a,b) with $a < b$ such that $f' \in L[a,b]$. If $|f'|^q$ is s-convex in the second sense on $[a,b]$ for some fixed $s \in (0,1], p, q > 1$ and $|f'(x)| \leq M, x \in a,b]$, then the following inequality for fractional integrals holds:*

$$\left|\left(\frac{(x-a)^{\alpha}+(b-x)^{\alpha}}{b-a}\right) f(x) - \frac{\Gamma(\alpha+1)}{(b-a)}\left[J_{x^-}^{\alpha} f(a) + J_{x^+}^{\alpha} f(b)\right]\right|$$
$$\leq \frac{M}{(1+p\alpha)^{\frac{1}{p}}}\left(\frac{2}{s+1}\right)^{\frac{1}{q}}\left[\frac{(x-a)^{\alpha+1}+(b-x)^{\alpha+1}}{b-a}\right]$$

where $\dfrac{1}{p}+\dfrac{1}{q}=1, \alpha > 0$, and Γ is the Euler gamma function.

Theorem 11.3. [12] *Let $f\colon [a,b] \subseteq (0,\infty) \to \mathbb{R}$, be a differentiable mapping on (a,b) with $a < b$ such that $f' \in L[a,b]$. If $|f'|^q$ is s-convex in the second sense on $[a,b]$ for some fixed $s \in (0,1], q \geq 1$, and $|f'(x)| \leq M, x \in a,b]$, then the following inequality for fractional integrals holds:*

$$\left|\left(\frac{(x-a)^{\alpha}+(b-x)^{\alpha}}{b-a}\right) f(x) - \frac{\Gamma(\alpha+1)}{(b-a)}\left[J_{x^-}^{\alpha} f(a) + J_{x^+}^{\alpha} f(b)\right]\right| \leq M\left(\frac{1}{1+\alpha}\right)^{1-\frac{1}{q}}$$
$$\times \left(\frac{1}{\alpha+s+1}\right)^{\frac{1}{q}}\left(1+\frac{\Gamma(\alpha+1)\Gamma(s+1)}{\Gamma(\alpha+s+1)}\right)^{\frac{1}{q}}\left[\frac{(x-a)^{\alpha+1}+(b-x)^{\alpha+1}}{b-a}\right]$$

where $\alpha > 0$ and Γ is the Euler gamma function.

In the following theorems, authors have given the Hadamard-type inequalities for product of different kinds of convex functions via different kinds of fractional integral operators such as Riemann–Liouville, Katugampola, and conformable fractional integral operators.

Some Hermite–Hadamard type inequalities for products of two different functions are proposed by Chen and Wu in [13] as follows:

Theorem 11.4. *Let $f, g\colon [a,b] \to \mathbb{R}$, $a, b \in [0,\infty)$, $a < b$ be functions such that and $g, fg \in L[a,b]$. If f is convex and nonnegative and g is s-convex on $[a,b]$ for some fixed $s \in [0,1]$, then the following inequality for fractional integrals holds:*

$$\frac{\Gamma(\alpha)}{(b-a)^\alpha}\left[J_{a+}^{\alpha} f(b)g(b) + J_{b-}^{\alpha} f(a)g(a)\right] \leq \left(\frac{1}{\alpha+s+1} + B(\alpha, s+2)\right) M(a,b)$$
$$+ \left(B(\alpha+1, s+1) + \frac{1}{(\alpha+s)(\alpha+s+1)}\right) N(a,b),$$

where $M(a,b) = f(a)g(a) + f(b)g(b)$, $N(a,b) = f(a)g(b) + f(b)g(a)$.

Theorem 11.5. *Let $f, g\colon [a,b] \to \mathbb{R}$, $a, b \in [0,\infty)$, $a < b$ be functions such that $f, g, fg \in L[a,b]$. If f is s_1-convex and g is s_2-convex function on $[a,b]$ for some fixed $s_1, s_2 \in [0,1]$, then the following inequality for fractional integrals holds:*

$$\frac{\Gamma(\alpha)}{(b-a)^\alpha}\left[J_{a+}^{\alpha} f(b)g(b) + J_{b-}^{\alpha} f(a)g(a)\right] \leq \left(\frac{1}{\alpha+s_1+s_2} + B(\alpha, s_1+s_2+1)\right) M(a,b)$$
$$+ \left[B\left(\alpha+s_1, s_2+1\right) + B\left(\alpha+s_2, s_1+1\right)\right] N(a,b),$$

where $M(a,b) = f(a)g(a) + f(b)g(b)$, $N(a,b) = f(a)g(b) + f(b)g(a)$.

Theorem 11.6. *Let $f, g\colon [a,b] \to \mathbb{R}$, $a, b \in [0,\infty)$, $a < b$ be functions such that $fg \in L[a,b]$. If f is convex and nonnegative on $[a,b] g$ is s_2-convex function on $[a,b]$ for some fixed $s \in [0,1]$, then*

$$2^s f\left(\frac{a+b}{2}\right) g\left(\frac{a+b}{2}\right) \leq \frac{\Gamma(\alpha+1)}{2(b-a)^\alpha}\left[J_{a+}^{\alpha} f(b)g(b) + J_{b-}^{\alpha} f(a)g(a)\right]$$
$$+ \frac{1}{2}M(a,b)\left(B(\alpha+1, s+1)\frac{1}{(\alpha+s)(\alpha+s+1)}\right) + \frac{1}{2}N(a,b)\left(B(\alpha, s+2) + \frac{1}{\alpha+s+1}\right),$$

where $M(a,b) = f(a)g(a) + f(b)g(b)$, $N(a,b) = f(a)g(b) + f(b)g(a)$.

In Ref. [14], authors have proved the following inequalities for different kinds of convexity via conformable fractional integrals:

Theorem 11.7. *Let $f, g\colon [a,b] \to \mathbb{R}$, be functions with $0 \leq a < b$ and $f, g, fg \in L_1[a,b]$. If f is convex and nonnegative and g is s-convex on $[a,b]$ for some fixed $s \in [0,1]$, then one has the following inequality for conformable fractional integrals:*

$$\frac{1}{(b-a)^\alpha}\left[I_\alpha^a f(b)g(b) +^b I_\alpha f(a)g(a)\right]$$
$$\leq \frac{M(a,b)}{n!}\left[B(n+s+2, \alpha-n) + B(n+1, \alpha-n+s+1)\right]$$
$$+ \frac{N(a,b)}{n!}\left[B(n+2, \alpha-n+s) + B(s+n+1, \alpha-n+1)\right] \tag{11.11}$$

with $\alpha \in (n, n+1]$. ($M(a,b) = f(a)g(a) + f(b)g(b)$, $N(a,b) = f(a)g(b) + f(b)g(a)$).

Theorem 11.8. *Suppose that $f, g\colon [a,b] \to \mathbb{R}$ be functions with $0 \le a < b$ and $f, g, fg \in L_1[a,b]$. If f is s_1-convex and g is s_2-convex function on $[a,b]$ for some fixed $s_1, s_2 \in [0,1]$, then one has the following inequality for conformable fractional integrals:*

$$\begin{aligned}
&\frac{1}{(b-a)^{\alpha}}\left[I_{\alpha}^{a} f(b) g(b)+{ }^{b} I_{\alpha} f(a) g(a)\right] \\
&\quad \le \frac{1}{n!} M(a,b)\left[B\left(s_1+s_2+n+1, \alpha-n\right)+B\left(n+1, s_1+s_2+\alpha-n\right)\right] \\
&\quad +\frac{1}{n!} N(a,b)\left[B\left(n+s_1+1, \alpha-n+s_2\right)+B\left(n+s_2+1, \alpha-n+s_1\right)\right]. \qquad (11.12)
\end{aligned}$$

where $\alpha \in (n, n+1]$ with $M(a,b)$ and $N(a,b)$ as in Theorem 11.7.

Theorem 11.9. *Let $f, g\colon [a,b] \to \mathbb{R}$, be functions with $0 \le a < b$ and $f, g, fg \in L_1[a,b]$. If f is convex and g is s-convex on $[a,b]$ for some fixed $s \in [0,1]$, then one has the following inequality for conformable fractional integrals:*

$$\begin{aligned}
&2^{s} B(n+1, \alpha-n) f\left(\frac{a+b}{2}\right) g\left(\frac{a+b}{2}\right) \le \frac{\Gamma(n+1)}{2(b-a)^{\alpha}}\left[I_{\alpha}^{a} f(b) g(b)+{ }^{b} I_{\alpha} f(a) g(a)\right] \\
&\quad +\frac{1}{2} M(a,b)[B(n+2, \alpha-n+s)+B(s+n+1, \alpha-n+1)] \\
&\quad +\frac{1}{2} N(a,b)[B(n+1, \alpha-n+s+1)+B(n+s+2, \alpha-n)] \qquad (11.13)
\end{aligned}$$

$\alpha \in (n, n+1]$ and $M(a,b)$ and $N(a,b)$ as in Theorem 11.7.

In Ref. [15], Akdemir et al. have established some new integral inequalities for different kinds of convexity but now for new conformable fractional integrals:

Theorem 11.10. *Let $f, g\colon [a,b] \to \mathbb{R}$, be functions with $0 \le a < b$ and $f, g, fg \in L_1[a,b]$. If f is convex and nonnegative and g is s-convex on $[a,b]$ for some fixed $s \in [0,1]$, then one has the following inequality for new conformable fractional integrals:*

$$\begin{aligned}
&\left(\frac{1}{b-a}\right)^{\alpha\beta} \Gamma(\beta)\left[{}_{a}^{\beta}\mathcal{J}_{\alpha} fg(b)+{}^{\beta}\mathcal{J}_{b}^{\alpha} fg(a)\right] \\
&\quad \le\left[\beta_1(s+2, \alpha)-\beta_1(s+2, \alpha\beta)+\frac{1}{\alpha+s+1}-\frac{1}{\alpha\beta+s+1}\right] M(a,b) \\
&\quad +\left[\beta_1(2, \alpha+s)-\beta_1(2, \alpha\beta+s)+\beta_1(s+1, \alpha+1)-\beta_1(s+1, \alpha\beta+1)\right] N(a,b) \quad (11.14)
\end{aligned}$$

where $\alpha, \beta > 0$ and β_1 is Euler Beta function, $M(a,b)$ and $N(a,b)$ as in Theorem 11.7.

Theorem 11.11. *Suppose that $f, g\colon [a,b] \to \mathbb{R}$ be functions with $0 \le a < b$ and $f, g, fg \in L_1[a,b]$. If f is s_1-convex and g is s_2-convex function on $[a,b]$ for some fixed $s_1, s_2 \in [0,1]$, then one has the following inequality for new conformable fractional integrals:*

$$\begin{aligned}
&\alpha^{\beta-1}\left(\frac{1}{b-a}\right)^{\alpha\beta} \Gamma(\beta)\left[{}_{a}^{\beta}\mathcal{J}_{\alpha} fg(b)+{}^{\beta}\mathcal{J}_{b}^{\alpha} fg(a)\right] \\
&\quad \le\left[\beta_1\left(s_1+s_2+1, \alpha\right)-\beta_1\left(s_1+s_2+1, \alpha\beta\right)+\frac{1}{\alpha+s_1+s_2}-\frac{1}{\alpha\beta+s_1+s_2}\right] M(a,b) \\
&\quad +\left[\beta_1\left(s_1+1, \alpha+s_2\right)-\beta_1\left(s_1+1, \alpha\beta+s_2\right)+\beta_1\left(s_2+1, \alpha+s_1\right)\right. \\
&\quad \left.-\beta_1\left(s_2+1, \alpha\beta+s_1\right)\right] N(a,b) \qquad (11.15)
\end{aligned}$$

where $\alpha, \beta > 0$ and β_1 is the Euler Beta function with $M(a,b)$ and $N(a,b)$ as in Theorem 11.7.

Theorem 11.12. *Let $f, g\colon [a,b] \to \mathbb{R}$, be functions with $0 \le a < b$ and $f, g, fg \in L_1[a,b]$. If f is convex and g is s-convex on $[a,b]$ for some fixed $s \in [0,1]$, then one has the following inequality for new conformable fractional integrals:*

$$\begin{aligned}\frac{2^{s+1}}{\beta\alpha^{\beta}} fg\left(\frac{a+b}{2}\right) &\le \frac{2^{s+1}}{(b-a)^{\alpha\beta}}\Gamma(\beta)\left[{}_{a}^{\beta}\mathcal{J}_{\alpha} fg(b) + {}^{\beta}\mathcal{J}_{b}^{\alpha} fg(a)\right]\\ &\le [\beta_1(2,s+1) - \beta_1(2,\alpha\beta-\alpha+s+1) + \beta_1(s+1,2) - \beta_1(s+1,\alpha\beta-\alpha+2)]\,M(a,b)\\ &\quad + [\beta_1(s+2,s+2) - \beta_1(s+2,\alpha\beta-\alpha+s+2)]\,N(a,b)\end{aligned}$$

where $\alpha, \beta > 0$ and β_1 is the Euler Beta function with $M(a,b)$ and $N(a,b)$ as in Theorem 11.7.

In Ref. [16], Çelik et al. have proved Hadamard-type integral inequalities via Katugampola fractional integrals:

Theorem 11.13. *Let $f, g\colon [a^{\rho}, b^{\rho}] \to \mathbb{R}_0^{+}$, be functions with $0 \le a < b$ and $f, g, fg \in X_c^p(a^{\rho}, b^{\rho})$. If f is convex and g is s-convex on $[a^{\rho}, b^{\rho}]$ for some fixed $s \in 0,1]$, then one has the following inequality for Katugampola fractional integrals:*

$$\begin{aligned}&\frac{\Gamma(\alpha)}{\rho^{1-\alpha}(b^{\rho}-a^{\rho})^{\alpha}}\left[{}^{\rho}\mathcal{J}_{a+}^{\alpha}[fg\circ h](b) + {}^{\rho}\mathcal{I}_{b-}^{\alpha}[fg\circ h](a)\right]\\ &\qquad \le M(a^{\rho}, b^{\rho})\frac{1}{\rho}\left[\frac{1}{\alpha+s+1} + B(\alpha, s+2)\right]\\ &\qquad + N(a^{\rho}, b^{\rho})\frac{1}{\rho}\left[B(\alpha+1, s+1) + \frac{1}{(\alpha+s)(\alpha+s+1)}\right],\end{aligned} \tag{11.16}$$

where $\alpha, \rho \in \mathbb{R}^{+}$, with $h(u) = u^{\rho}$,

$$M(a^{\rho}, b^{\rho}) = f(a^{\rho})g(a^{\rho})f(b^{\rho})g(b^{\rho}) \text{ and } N(a^{\rho}, b^{\rho}) = f(a^{\rho})g(b^{\rho}) + f(b^{\rho})g(a^{\rho}).$$

Theorem 11.14. *Suppose that $f, g\colon [a^{\rho}, b^{\rho}] \to \mathbb{R}_0^{+}$, be functions with $0 \le a < b$ and $f, g, fg \in X_c^p(a^{\rho}, b^{\rho})$. If f is s_1-convex and g is s_2-convex function on $[a^{\rho}, b^{\rho}]$ for some fixed $s_1, s_2 \in 0,1]$, then one has the following inequality for Katugampola fractional integrals:*

$$\begin{aligned}&\frac{\Gamma(\alpha)}{\rho^{1-\alpha}(b^{\rho}-a^{\rho})^{\alpha}}\left[{}^{\rho}\mathcal{J}_{a+}^{\alpha}[fg\circ h](b) + {}^{\rho}\mathcal{J}_{b-}^{\alpha}[fg\circ h](a)\right]\\ &\qquad \le M(a^{\rho}, b^{\rho})\frac{1}{\rho}\left[\frac{1}{\alpha+s_1+s_2} + B(\alpha, s_1+s_2+1)\right]\\ &\qquad + N(a^{\rho}, b^{\rho})\frac{1}{\rho}\left[B(\alpha+s_1, s_2+1) + B(\alpha+s_2, s_1+1))\right]\end{aligned} \tag{11.17}$$

where $\alpha, \rho \in \mathbb{R}^{+}$, and $M(a^{\rho}, b^{\rho})$ and $N(a^{\rho}, b^{\rho})$ are the same as given in Theorem 11.13 with $h(u) = u^{\rho}$.

Theorem 11.15. *Let $f, g\colon [a^{\rho}, b^{\rho}] \to \mathbb{R}_0^{+}$, be functions with $0 \le a < b$ and $f, g, fg \in X_c^p(a^{\rho}, b^{\rho})$. If f is convex and g is s-convex on $[a^{\rho}, b^{\rho}]$ for some fixed $s \in 0,1]$, then one has the following inequality for Katugampola fractional integrals:*

$$\begin{aligned}\frac{2^{s}}{\rho} f\left(\frac{a^{\rho}+b^{\rho}}{2}\right) g\left(\frac{a^{\rho}+b^{\rho}}{2}\right) &\le \frac{\Gamma(\alpha+1)}{2\rho^{1-\alpha}(b^{\rho}-a^{\rho})^{\alpha}}\left[{}^{\rho}\mathcal{J}_{a+}^{\alpha}[fg\circ h](b) + {}^{\rho}\mathcal{J}_{b-}^{\alpha}[fg\circ h](a)\right]\\ &\quad + \frac{1}{2}M(a^{\rho}, b^{\rho})\left[\frac{1}{\rho}B(\alpha+1, s+1) + \frac{1}{(\rho\alpha+s)(\rho\alpha+s+1)}\right]\\ &\quad + \frac{1}{2}N(a^{\rho}, b^{\rho})\left[\frac{1}{\rho}B(\alpha, s+2) + \frac{1}{\rho\alpha+\rho s+\rho}\right]\end{aligned} \tag{11.18}$$

where $\alpha, \rho \in \mathbb{R}^+$, and $M(a^\rho, b^\rho)$ and $N(a^\rho, b^\rho)$ are the same as given in Theorem 11.13 with $h(u) = u^\rho$.

In Ref. [17], Set et al. have proved Hadamard-type integral inequalities via generalized fractional integrals:

Theorem 11.16. *Let $f, g\colon [a,b] \to \mathbb{R}_0^+$ $(a < b)$ with $a, b \in (0,\infty)$ be functions such that $f,\ g \in L[a,b]$. Also, let f be convex and g be s-convex on $[a,b]$ for some fixed $s \in (0,1]$. Then*

$$\begin{aligned}\frac{1}{(b-a)^\alpha}&\left[\left(\mathcal{J}^\sigma_{\rho,\lambda,a^+;w}\right)(fg(b)) + \left(\mathcal{J}^\sigma_{\rho,\lambda,b^-;w}\right)(fg(a))\right]\\ &\leq M(a,b)\,\mathcal{F}^{\sigma_1}_{\rho,\alpha}\left[w(b-a)^\rho\right] + N(a,b)\mathcal{F}^{\sigma_2}_{\rho,\alpha}\left[w(b-a)^\rho\right]\end{aligned} \tag{11.19}$$

where $\alpha, \rho, w \in \mathbb{R}^+$, $M(a,b)$ and $N(a,b)$ are the same as in Theorem 11.7,

$$\sigma_1(k)\colon = \sigma(k)\left(\frac{1}{\alpha+\rho k+s+1} + B(\alpha+\rho k, s+2)\right)$$

and

$$\sigma_2(k)\colon = \sigma(k)\left(B(\alpha+\rho k+1, s+1) + \frac{1}{(\alpha+\rho k+s)(\alpha+\rho k+s+1)}\right).$$

Theorem 11.17. *Let $f, g\colon [a,b] \to \mathbb{R}_0^+$ $(a < b)$ with $a, b \in (0,\infty)$ be functions such that $f,\ g,\ fg \in L[a,b]$. Also, let f and g be s_1-convex and s_2-convex on $[a,b]$ for some fixed $s_1,\ s_2 \in (0,1]$, respectively. Then*

$$\begin{aligned}\frac{1}{(b-a)^\alpha}&\left[\left(\mathcal{J}^\sigma_{\rho,\lambda,a^+;w}\right)(fg(b)) + \left(\mathcal{J}^\sigma_{\rho,\lambda,b^-;w}\right)(fg(a))\right]\\ &\leq M(a,b)\,\mathcal{F}^{\sigma_3}_{\rho,\alpha}\left[w(b-a)^\rho\right] + N(a,b)\mathcal{F}^{\sigma_4}_{\rho,\alpha}\left[w(b-a)^\rho\right]\end{aligned} \tag{11.20}$$

where $\alpha, \rho, w \in \mathbb{R}^+$, and $M(a,b)$ and $N(a,b)$ are the same as in Theorem 11.7,

$$\begin{aligned}\sigma_3(k)\colon &= \sigma(k)\left\{\frac{1}{\alpha+\rho k+s_1+s_2} + B(\alpha+\rho k, s_1+s_2+1)\right\},\\ \sigma_4(k)\colon &= \sigma(k)\left\{B(\alpha+\rho k+s_1, s_2+1) + B(\alpha+\rho k+s_2, s_1+1)\right\}.\end{aligned}$$

Theorem 11.18. *Let $f, g\colon [a,b] \to \mathbb{R}_0^+$ $(a < b)$ with $a, b \in (0,\infty)$ be functions such that $fg \in L[a,b]$. Also, let f be convex and g be s-convex on $[a,b]$ for some fixed $s \in (0,1]$. Then*

$$\begin{aligned}&2^{s+1} f\left(\frac{a+b}{2}\right) g\left(\frac{a+b}{2}\right) \mathcal{F}^{\sigma_7}_{\rho,\alpha}\left[w(b-a)^\rho\right]\\ &\leq \frac{1}{(b-a)^\alpha}\left[\left(\mathcal{J}^\sigma_{\rho,\lambda,a^+;w}\right)(fg(b)) + \left(\mathcal{J}^\sigma_{\rho,\lambda,b^-;w}\right)(fg(a))\right]\\ &\leq M(a,b)\,\mathcal{F}^{\sigma_2}_{\rho,\alpha}\left[w(b-a)^\rho\right] + N(a,b)\mathcal{F}^{\sigma_1}_{\rho,\alpha}\left[w(b-a)^\rho\right]\end{aligned} \tag{11.21}$$

where $\alpha, \rho, w \in \mathbb{R}^+$, $M(a,b)$ and $N(a,b)$ are the same as in Theorem 11.7, and σ_1 and σ_2 are the same as given in Theorem 11.16:

$$\sigma_7(k)\colon = \frac{\sigma(k)}{\alpha+\rho k}.$$

11.4 New Results via Generalized Fractional Integral Operators

In this section, we will obtain some new Hadamard-type integral inequalities for product of two convex functions via the generalized fractional operators. We will also give some special cases of our main theorems.

Theorem 11.19. *Let $f, g\colon [a,b] \to \mathbb{R}^+$ be two functions with $0 \le a < x < b$ and $f, g, fg \in L[a,b]$. If f and g are convex, then one has the following inequalities for generalized fractional operators;*

$$\begin{aligned}\int_a^x \frac{\varphi(x-s)}{x-s} fg(s)\,ds = {}_{a^+}\mathrm{J}_\varphi fg(s) &\le f(x)g(x)\int_a^x \frac{\varphi(x-s)}{x-s}\left(\frac{s-a}{x-a}\right)^2 ds \\ &+ f(a)g(b)\int_a^x \frac{\varphi(x-s)}{x-s}\left(\frac{x-s}{x-a}\right)^2 ds \\ &+ [f(x)g(a)+f(a)g(x)]\int_a^x \frac{\varphi(x-s)}{x-s}\frac{(x-s)(s-a)}{(x-a)^2} ds \end{aligned} \tag{11.22}$$

and

$$\begin{aligned}\int_x^b \frac{\varphi(s-x)}{(s-x)} fg(s)\,ds = {}_{b-}\mathcal{J}_\varphi fg(s) &\le f(x)g(x)\int_x^b \frac{\varphi(s-x)}{(s-x)}\left(\frac{b-s}{b-x}\right)^2 ds \\ &+ f(a)g(b)\int_x^b \frac{\varphi(s-x)}{(s-x)}\left(\frac{s-x}{b-x}\right)^2 ds \\ &+ [f(x)g(b)+f(b)g(x)]\int_x^b \frac{\varphi(s-x)}{(s-x)}\frac{(b-s)(s-x)}{(b-x)^2} ds \end{aligned} \tag{11.23}$$

Proof. Since f and g are convex functions, we can write

$$f(tx+(1-t)a) \le tf(x)+(1-t)f(a) \tag{11.24}$$

and

$$g(tx+(1-t)a) \le tg(x)+(1-t)g(a) \tag{11.25}$$

If we multiply (11.24) and (11.25), we get

$$\begin{aligned}f(tx+(1-t)a)g(tx+(1-t)a) &\le t^2 f(x)g(x)+(1-t)^2 f(a)g(a)\\ &+ t(1-t)f(x)g(a)+t(1-t)f(a)g(x)\end{aligned} \tag{11.26}$$

By multiplying both sides of (11.26) by $\dfrac{\varphi(x-a)(1-t)}{(x-a)(1-t)}$ and integrating with respect to t over $[0,1]$, we have

$$\begin{aligned}\int_0^1 \frac{\varphi(x-a)(1-t)}{(x-a)(1-t)} fg(tx+(1-t)a)\,dt &\le f(x)g(x)\int_0^1 \frac{\varphi(x-a)(1-t)}{(x-a)(1-t)} t^2 dt \\ &+ f(a)g(b)\int_0^1 \frac{\varphi(x-a)(1-t)}{(x-a)(1-t)}(1-t)^2 dt \\ &+ [f(x)g(a)+f(a)g(x)]\int_0^1 \frac{\varphi(x-a)(1-t)}{(x-a)(1-t)} t(1-t)dt.\end{aligned} \tag{11.27}$$

If we use the change of the variables $s = tx + (1-t)a$, we obtain

$$\int_a^x \frac{\varphi(x-s)}{(x-s)} fg(s)\,ds = {}_{a+}\mathcal{J}_{\varphi} fg(s)$$

$$\leq f(x)g(x) \int_a^x \frac{\varphi(x-s)}{(x-s)} \left(\frac{s-a}{x-a}\right)^2 ds + f(a)g(b) \int_a^x \frac{\varphi(x-s)}{(x-s)} \left(\frac{x-s}{x-a}\right)^2 ds$$

$$+ [f(x)g(a) + f(a)g(x)] \int_a^x \frac{\varphi(x-s)}{(x-s)} \frac{(x-s)(s-a)}{(x-a)^2} ds$$

By a similar argument, we can easily get $\left(\frac{\varphi(b-x)t}{(b-x)t}\right)$

$$\int_x^b \frac{\varphi(s-x)}{(s-x)} fg(s)\,ds = {}_{b-}\mathcal{J}_{\varphi} fg(s)$$

$$\leq f(x)g(x) \int_x^b \frac{\varphi(s-x)}{(s-x)} \left(\frac{b-s}{b-x}\right)^2 ds + f(a)g(b) \int_x^b \frac{\varphi(s-x)}{(s-x)} \left(\frac{s-x}{b-x}\right)^2 ds$$

$$+ [f(x)g(b) + f(b)g(x)] \int_x^b \frac{\varphi(s-x)}{(s-x)} \frac{(b-s)(s-x)}{(b-x)^2} ds.$$

which completes the proof.

Corollary 11.1. *If we choose,* $\varphi(t) = \frac{t^\alpha}{\Gamma(\alpha)}$, *the inequalities (11.22) and (11.23) reduce to*

$$\mathcal{J}^{\alpha}_{a+} fg(s) \leq \frac{f(x)g(x)}{\Gamma(\alpha)(x-a)^2} \frac{2(x-a)^{\alpha+2}}{\alpha(\alpha+1)(\alpha+2)} + \frac{f(x)g(x)}{\Gamma(\alpha)(x-a)^2} \frac{a^{\alpha+2} - x^{\alpha+2}}{(\alpha+2)}$$

$$+ \frac{f(x)g(a) + f(a)g(x)}{\Gamma(\alpha)(x-a)^2} \frac{(x-a)^{\alpha+2}}{(\alpha+1)(\alpha+2)}$$

and

$$\mathcal{J}^{\alpha}_{b-} fg(s) \leq \frac{f(x)g(x)}{\Gamma(\alpha)} \frac{(b-x)^\alpha}{\alpha} + \frac{f(b)g(b)}{\Gamma(\alpha)(b-x)^2} \frac{(b-2x)^{\alpha+2} - (-x)^{\alpha+2}}{(\alpha+2)}$$

$$+ \frac{f(x)g(b) + f(b)g(x)}{\Gamma(\alpha)(b-x)^2} \frac{(b-x)^{\alpha+2}}{(\alpha+1)(\alpha+2)}.$$

Corollary 11.2. *If we choose* $\varphi(t) = t(x-t)^{\alpha-1}$, *the inequalities reduce to*

$$\mathcal{J}^{\alpha}_{a+} fg(s) \leq \frac{f(x)g(x)}{(x-a)^2} \left[\frac{a^2 x^\alpha}{\alpha} - \frac{2ax^{\alpha+1}}{\alpha+1} + \frac{x^{\alpha+2}}{\alpha+2} - \frac{2a^{\alpha+2}}{\alpha(\alpha+1)(\alpha+2)}\right]$$

$$+ \frac{f(a)g(a)}{(x-a)^2} \left[\frac{2x^{\alpha+2}}{\alpha(\alpha+1)(\alpha+2)} - \frac{a^{\alpha+2}}{\alpha+2} + \frac{2xa^{\alpha+1}}{\alpha+1} - \frac{a^\alpha x^2}{\alpha}\right] + \frac{f(a)g(x) + f(x)g(a)}{(x-a)^2}$$

$$\times \left[\frac{-a^{\alpha+2}}{(\alpha+1)(\alpha+2)} + \frac{x^{\alpha+2}}{(\alpha+1)(\alpha+2)} - \frac{xa^{\alpha+1}}{\alpha(\alpha+1)} - \frac{ax^{\alpha+1}}{(\alpha)(\alpha+1)}\right]$$

and

$$\mathcal{J}_{b-}^{\alpha} fg(s) \leq \frac{f(x)g(x)}{(b-x)^2}\left[\frac{2b^{\alpha+2}}{\alpha(\alpha+1)(\alpha+2)} + \frac{2bx^{\alpha+1}}{\alpha+1} - \frac{x^{\alpha+2}}{\alpha+2} - \frac{b^2x^{\alpha}}{\alpha}\right]$$
$$+\frac{f(b)g(b)}{(b-x)^2}\left[\frac{b^{\alpha}x^2}{\alpha} - \frac{2xb^{\alpha+1}}{\alpha+1} + \frac{b^{\alpha+2}}{\alpha+2} - \frac{2x^{\alpha+2}}{(\alpha)(\alpha+1)(\alpha+2)}\right] + \frac{f(b)g(x)+f(x)g(b)}{(b-x)^2}$$
$$\times\left[\frac{b^{\alpha+2}}{(\alpha+1)(\alpha+2)} - \frac{x^{\alpha+2}}{(\alpha+1)(\alpha+2)} - \frac{xb^{\alpha+1}}{\alpha(\alpha+1)} + \frac{b^{\alpha+1}}{\alpha(\alpha+1)}\right]$$

Theorem 11.20. *Let $f, g\colon [a,b] \to \mathbb{R}^+$ be two functions with $0 \leq a < x < b$ and $f, g, fg \in L[a,b]$. If f and g are convex, then one has the following inequalities for generalized fractional operators;*

$$\frac{g(a)}{x-a}\int_a^x \frac{\varphi(x-s)}{x-s}\left(\frac{s-a}{x-a}\right) f(s)\,ds + \frac{f(x)}{x-a}\int_a^x \frac{\varphi(x-s)}{x-s}\left(\frac{s-a}{x-a}\right) g(s)\,ds$$
$$+\frac{f(a)}{x-a}\int_a^x \frac{\varphi(x-s)}{x-s}\left(\frac{s-a}{x-a}\right) g(s)\,ds + \frac{g(x)}{x-a}\int_a^x \frac{\varphi(x-s)}{x-s}\left(\frac{s-a}{x-a}\right) f(s)\,ds$$
$$\leq \frac{1}{x-a}\int_a^x \frac{\varphi(x-s)}{x-s} fg(s)\,ds + \frac{1}{x-a}\int_a^x \frac{\varphi(x-s)}{x-s}\left(\frac{s-a}{x-a}\right)^2 f(x)g(x)\,ds$$
$$+\frac{1}{x-a}\int_a^x \frac{\varphi(x-s)}{x-s}\left(\frac{s-a}{x-a}\right)^2 f(a)g(a)\,ds$$
$$+\frac{[f(x)g(a)+f(a)g(x)]}{x-a}\int_a^x \frac{\varphi(x-s)}{x-s}\frac{(x-s)(s-a)}{(x-a)^2}ds$$
$$+\frac{g(x)}{x-a}\int_a^x \frac{\varphi(x-s)}{x-s}(s-a)f(s)\,ds + \frac{g(a)}{x-a}\int_a^x \varphi(x-s)f(s)\,ds$$
$$+\frac{f(x)}{x-a}\int_a^x \frac{\varphi(x-s)}{x-s}(s-a)g(s)\,ds + \frac{f(a)}{x-a}\int_a^x \varphi(x-s)g(s)\,ds \leq J_{a+}^{\varphi} fg(s)$$
$$+\frac{f(x)g(x)}{(x-a)^2}\int_a^x \frac{\varphi(x-s)}{x-s}(s-a)^2\,ds + \frac{f(a)g(a)}{(x-a)^2}\int_a^x \varphi(x-s)(x-s)\,ds$$
$$+\frac{[f(x)g(a)+f(a)g(x)]}{(x-a)^2}\int_a^x \varphi(x-s)(s-a)ds \tag{11.28}$$

Proof. From the definition of convexity, we can write

$$f(tx+(1-t)a) \leq tf(x)+(1-t)f(a)$$

and

$$g(tx+(1-t)a) \leq tg(x)+(1-t)g(a).$$

By using the elementary inequality, $e \leq f$ and $p \leq r$, then $er+fp \leq ep+fr$ for $e, f, p, r \in \mathbb{R}$, we get

$$[tf(tx+(1-t)a)g(x)+tg(tx+(1-t)a)f(x)+(1-t)g(tx+(1-t)a)f(a)$$
$$+(1-t)f(tx+(1-t)a)g(a)] \leq f(tx+(1-t)a)\,g(tx+(1-t)a)+t^2f(x)g(x)$$
$$+(1-t)^2f(a)g(a)+t(1-t)\left[f(x)g(a)+f(a)g(x)\right].$$

By multiplying both sides of the above inequality by $\dfrac{\varphi(x-a)(1-t)}{(x-a)(1-t)}$ and integrating with respect to t over [0,1], we have

$$\begin{aligned}
&\int_0^1 \frac{\varphi((x-a)(1-t))}{(x-a)(1-t)} tf(tx+(1-t)a)g(x)dt \\
&\quad + \int_0^1 \frac{\varphi((x-a)(1-t))}{(x-a)(1-t)} (1-t)f(tx+(1-t)a)g(a)dt \\
&\quad + \int_0^1 \frac{\varphi((x-a)(1-t))}{(x-a)(1-t)} tg(tx+(1-t)a)f(x)dt \\
&\quad + \int_0^1 \frac{\varphi((x-a)(1-t))}{(x-a)(1-t)} (1-t)g(tx+(1-t)a)f(a)dt \\
&\leq \int_0^1 \frac{\varphi((x-a)(1-t))}{(x-a)(1-t)} fg(tx+(1-t)a)dt + \int_0^1 \frac{\varphi((x-a)(1-t))}{(x-a)(1-t)} t^2 f(x)g(x)dt \\
&\quad + \int_0^1 \frac{\varphi((x-a)(1-t))}{(x-a)(1-t)} (1-t)^2 f(a)g(a)dt \\
&\quad + \int_0^1 \frac{\varphi((x-a)(1-t))}{(x-a)(1-t)} t(1-t)[f(x)g(a)+f(a)g(x)]dt.
\end{aligned}$$

By tracking into account the change of the variables, $s = tx + (1-t)a$, we obtain

$$\begin{aligned}
&\frac{g(a)}{x-a}\int_a^x \frac{\varphi(x-s)}{x-s}\left(\frac{s-a}{x-a}\right) f(s)\,ds + \frac{f(x)}{x-a}\int_a^x \frac{\varphi(x-s)}{x-s}\left(\frac{s-a}{x-a}\right) g(s)\,ds \\
&\quad + \frac{f(a)}{x-a}\int_a^x \frac{\varphi(x-s)}{x-s}\left(\frac{s-a}{x-a}\right) g(s)\,ds + \frac{g(x)}{x-a}\int_a^x \frac{\varphi(x-s)}{x-s}\left(\frac{s-a}{x-a}\right) f(s)\,ds \\
&\leq \frac{1}{x-a}\int_a^x \frac{\varphi(x-s)}{x-s} fg(s)\,ds + \frac{1}{x-a}\int_a^x \frac{\varphi(x-s)}{x-s}\left(\frac{s-a}{x-a}\right)^2 f(x)\,g(x)\,ds \\
&\quad + \frac{1}{x-a}\int_a^x \frac{\varphi(x-s)}{x-s}\left(\frac{s-a}{x-a}\right)^2 f(a)\,g(a)\,ds \\
&\quad + \frac{[f(x)\,g(a)+f(a)\,g(x)]}{x-a}\int_a^x \frac{\varphi(x-s)}{x-s}\frac{(x-s)(s-a)}{(x-a)^2}ds \\
&\quad + \frac{g(x)}{x-a}\int_a^x \frac{\varphi(x-s)}{x-s}(s-a)\,f(s)\,ds + \frac{g(a)}{x-a}\int_a^x \varphi(x-s)\,f(s)\,ds \\
&\quad + \frac{f(x)}{x-a}\int_a^x \frac{\varphi(x-s)}{x-s}(s-a)\,g(s)\,ds + \frac{f(a)}{x-a}\int_a^x \varphi(x-s)\,g(s)\,ds \\
&\leq J_{a+}^{\varphi} fg(s) + \frac{f(x)\,g(x)}{(x-a)^2}\int_a^x \frac{\varphi(x-s)}{x-s}(s-a)^2\,ds \\
&\quad + \frac{f(a)\,g(a)}{(x-a)^2}\int_a^x \varphi(x-s)\,(x-s)\,ds + \frac{[f(x)\,g(a)+f(a)\,g(x)]}{(x-a)^2}\int_a^x \varphi(x-s)\,(s-a)ds
\end{aligned}$$

By a similar argument, we can obtain

$$\begin{aligned}
&\frac{g(x)}{b-x}\int_x^b \frac{\varphi(s-x)}{s-x}(b-s)f(s)ds + \frac{g(b)}{b-x}\int_x^b \varphi(s-x)f(s)ds \\
&\quad + \frac{f(x)}{b-x}\int_x^b \frac{\varphi(s-x)}{s-x}(b-s)g(s)ds + \frac{f(b)}{b-x}\int_x^b \varphi(s-x)g(s)ds \leq_{b-} \mathrm{J}^{\varphi} fg\,(s)
\end{aligned}$$

$$+\frac{f(x)g(x)}{(b-x)^2}\int_x^b \frac{\varphi(s-x)}{(s-x)}(b-s)^2 ds+\frac{f(b)g(b)}{(b-x)^2}\int_x^b \varphi(s-x)(s-x)ds$$
$$+\frac{f(x)g(b)+f(b)g(x)}{(b-x)^2}\int_x^b \varphi(s-x)(b-s)ds.$$

This completes the proof.

Corollary 11.3. *If we choose* $\varphi(t)=\frac{t^{\alpha}}{\Gamma(\alpha)}$*, the inequalities (11.28) reduce to*

$$\frac{g(x)}{\Gamma(\alpha)(x-a)}\int_a^x (x-s)^{\alpha-1}(s-a)f(s)ds+\frac{g(a)}{\Gamma(\alpha)(x-a)}\int_a^x (x-s)^{\alpha}f(s)ds$$
$$+\frac{f(x)}{\Gamma(\alpha)(x-a)}\int_a^x (x-s)^{\alpha-1}(s-a)g(s)ds+\frac{f(a)}{\Gamma(\alpha)(x-a)}\int_a^x (x-s)^{\alpha}g(s)ds$$
$$\leq \int_a^x (x-s)^{\alpha-1}fg(s)ds+\frac{f(x)g(x)}{\Gamma(\alpha)(x-a)^2}\int_a^x (x-s)^{\alpha-1}(s-a)^2ds$$
$$+\frac{f(a)g(a)}{\Gamma(\alpha)(x-a)^2}\int_a^x (x-s)^{\alpha+1}ds+\frac{f(x)g(a)+f(a)g(x)}{\Gamma(\alpha)(x-a)^2}\int_a^x (x-s)^{\alpha}(s-a)ds.$$

By computing the above integrals, we get

$$\frac{g(x)}{\Gamma(\alpha)(x-a)}\int_a^x (x-s)^{\alpha-1}(s-a)f(s)ds+\frac{g(a)}{\Gamma(\alpha)(x-a)}\int_a^x (x-s)^{\alpha}f(s)ds$$
$$+\frac{f(x)}{\Gamma(\alpha)(x-a)}\int_a^x (x-s)^{\alpha-1}(s-a)g(s)ds+\frac{f(a)}{\Gamma(\alpha)(x-a)}\int_a^x (x-s)^{\alpha}g(s)ds$$
$$\leq J_{a+}^{\alpha}fg(s)+\frac{f(x)g(x)}{\Gamma(\alpha)(x-a)^2}\left[\frac{2(x-a)^{\alpha+2}}{\alpha(\alpha+1)(\alpha+2)}\right]+\frac{f(a)g(a)}{\Gamma(\alpha)(x-a)^2}\left[\frac{a^{\alpha+2}-x^{\alpha+2}}{(\alpha+2)}\right]$$
$$+\frac{f(x)g(a)+f(a)g(x)}{\Gamma(\alpha)(x-a)^2}\left[\frac{(x-a)^{\alpha+2}}{(\alpha+1)(\alpha+2)}\right].$$

Corollary 11.4. *If we choose* $\phi(t)=t(x-t)^{\alpha-1}$*, the following inequality holds:*

$$\frac{g(x)}{x-a}\int_a^x s^{\alpha-1}(s-a)f(s)ds+\frac{g(a)}{x-a}\int_a^x (x-s)S^{\alpha-1}f(s)ds$$
$$+\frac{f(x)}{x-a}\int_a^x s^{\alpha-1}(s-a)g(s)ds+\frac{f(a)}{x-a}\int_a^x (x-s)s^{\alpha-1}g(s)ds$$
$$\leq J_{a+}^{\alpha}fg(s)+\frac{f(x)g(x)}{(x-a)^2}\left[\frac{a^2x^{\alpha}}{\alpha}-\frac{2ax^{\alpha+1}}{\alpha+1}+\frac{x^{\alpha+2}}{\alpha+2}-\frac{2a^{\alpha+2}}{\alpha(\alpha+1)(\alpha+2)}\right]$$
$$+\frac{f(a)g(a)}{(x-a)^2}\left[\frac{2x^{\alpha+2}}{\alpha(\alpha+1)(\alpha+2)}-\frac{a^{\alpha+2}}{\alpha+2}+\frac{2xa^{\alpha+1}}{\alpha+1}-\frac{a^{\alpha}x^2}{\alpha}\right]$$

References

[1] R. Khalil, M. Al Horani, A. Yousef and M. Sababheh, A new definition of fractional derivative. *Journal of Computational and Applied Mathematics*, 264 (2014), 65–70.

[2] T. Abdeljawad, On conformable fractional calculus. *Journal of Computational and Applied Mathematics*, 279 (2015), 57–66.

[3] A.A. Kilbas, H.M. Sirivastava and J.J. Trujillo, Theory and applications of fractional differential equations, In: *Nort-Holland Mathematics Studies*, vol. 204, Elsevier Science B. V., Amsterdam, 2006, pp. 154.

[4] F. Jarad, E. Uğurlu, T. Abdeljawad and D. Baleanu, On a new class of fractional operators, *Advances in Difference Equations*, 2017(1) (2017), Article No. 247. doi: 10.1186/s13662-017-1306-z.

[5] U.N. Katugampola, New approach to a generalized fractional integrals, *Applied Mathematics and Computation*, 218(4) (2011), 860–865.

[6] U.N. Katugampola, New approach to a generalized fractional derivatives, *Bulletin of Mathematical Analysis and Applications*, 6(4) (2014), 1–15.

[7] A. Erdélyi, On fractional integration and its application to the theory of Hankel transforms, *The Quarterly Journal of Mathematics*, Oxford, Second Series, 11 (1940), 293–303.

[8] H. Kober, On fractional integrals and derivatives, *The Quarterly Journal of Mathematics* (Oxford Series), 11(1) (1940), 193–211.

[9] R.K. Raina, On generalized Wright's hypergeometric functions and fractional calculus operators, *East Asian Mathematical Journal*, 21(2) (2005), 191–203.

[10] F. Ertuğral and M.Z. Sarıkaya, Some Trapezoid type inequalities for generalized fractional operator, https://www.researchgate.net/publication/331277449.

[11] S. Mubeen and G.M. Habibullah, k-Fractional integrals and application, *International Journal of Contemporary Mathematical Sciences*, 7(2) (2012), 89–94.

[12] E. Set, New inequalities of Ostrowski type for mappings whose derivatives are s-convex in the second sense via fractional integrals, *Computers and Mathematics with Applications*, 63 (2012), 1147–1154.

[13] F. Chen and S. Wu, Several complementary inequalities to inequalities of Hermite-Hadamard type for s-convex functions, *Journal of Nonlinear Science and Applications*, 9 (2016), 705–716.

[14] E. Set, A.O. Akdemir and B. Çelik, Some Hermite-Hadamard type ınequalities for products of two different convex functions via conformable fractional ıntegrals, *X. Statistical Days*, Giresun-Turkey, 11–15 November 2016.

[15] A.O. Akdemir, E. Set and A. Ekinci, On new conformable fractional integral inequalities for product of different kinds of convexity, *TWMS Journal of Applied and Engineering Mathematics*, 9(1), Special Issue (2019), 142–150.

[16] B. Çelik, E. Set and A.O. Akdemir, On generalizations of Hermite-Hadamard type ınequalities for products of different kinds of convex functions via Katugampola Fractional integrals, *Proceedings of the International Conference on Mathematical Studies and Applications 2018 Karamanoglu Mehmetbey University*, Karaman, Turkey, 4–6 October 2018.

[17] E. Set, J. Choi and B. Çelik, New Hermite-Hadamard type inequalities for product of different convex functions involving certain fractional integral operators, *Journal of Mathematics and Computer Science*, 18 (2018), 29–36.

12

A Note on Fox's H Function in the Light of Braaksma's Results

Dmitrii B. Karp

Holon Institute of Technology

CONTENTS

12.1 Introduction

Fox's H function is a general function defined by the Mellin–Barnes integral (12.1). It was introduced by Fox [4] in the context of symmetrical Fourier kernels and has been studied by a number of authors thereafter [1,9,11,14]. Fox's H function has a number of important applications, most notably in statistics [2,5,11] and fractional calculus [10,11]. Braaksma's mammoth manuscript [1] remains among the deepest investigations on the topic. In this note we will be only interested in the case when the parameter μ defined in (12.2) equals zero (Braaksma's work contains a careful study of both cases $\mu > 0$ and $\mu = 0$, while the case $\mu < 0$ reduces to $\mu > 0$ by a simple change of variable). When $\mu = 0$ the integral (12.1) defining the H function only converges if $|z| < \beta^{-1}$, where β is given in (12.2). It also converges for $|z| > \beta^{-1}$ but over a different integration contour so that the two functions obtained in this way are not, generally speaking, analytic continuations of each other. Braaksma constructed analytic continuations of the H function to an infinite set of sectors forming a partition of the Riemann surface of the logarithm. The points of intersection of the circle $|z| = \beta^{-1}$ with the rays bounding these sectors are singular points of H. He further derived analytic continuation from each such sector to the domain $|z| > \beta^{-1}$ on this Riemann surface.

Under certain restrictions to be specified below the positive semiaxis $\arg(z) = 0$ forms the boundary between two such sectors, so that the point $z = \beta^{-1}$ is a singular point of Fox's H function. We studied the behavior of a particular case $H^{q,0}_{p,q}$ of Fox's H function in the neighborhood of this point in our recent papers [6,7] (as Braaksma's parameter μ is called Δ in our papers following [9], the case $\mu = 0$ was given the denomination "delta-neutral"). In particular, we found a convergent inverse factorial series for the integrand of the delta-neutral $H^{q,0}_{p,q}$ function and utilized it to derive a power series expansion of this function in a neighborhood of the singular point $z = \beta^{-1}$. We also deduced recursive and determinantal formulas for the coefficients of this expansion. However, our work contained an error caused by an incorrect citing and use of Nørlund's results in [6, Theorem 2]. The consequence of this error was the omission of the condition that the scaling factors α_i, β_j in (12.1) should all be strictly greater than 1/6 (and not only positive as claimed) in [6, Theorem 5] and [7, Theorem 1].

The purpose of this note is to correct, simplify, refine, and generalize our results presented in [6,7]. We give a simpler proof, refine the convergence abscissa of the inverse factorial series and present another expansion that works for arbitrary positive scaling factors α_i, β_j. This is done in Section 12.3. In order to obtain these results we needed a slight generalization of a particular case of [1, Theorem 2], which we establish in Section 12.2. As some parts of Braaksma's work do not seem to be well-understood we also rewrite some of his ideas here.

12.2 Braaksma Revisited

As we will need to modify some parts of Braaksma's proof, for the reader's convenience we adhere to the notation of [1]. Define

$$H_{p,q}^{m,n}\left(z\left|\begin{matrix}(\alpha_1,a_1),\ldots(\alpha_p,a_p)\\(\beta_1,b_1),\ldots(\beta_q,b_q)\end{matrix}\right.\right)=H(z)=\frac{1}{2\pi i}\int_C h(s)z^s ds, \tag{12.1}$$

where

$$z^s=\exp\{s(\log|z|+i\arg z)\},$$

and $\arg z$ may take any real value. Here

$$h(s)=\frac{\prod_{j=1}^{n}\Gamma(1-a_j+\alpha_j s)\prod_{j=1}^{m}\Gamma(b_j-\beta_j s)}{\prod_{j=m+1}^{q}\Gamma(1-b_j+\beta_j s)\prod_{j=n+1}^{p}\Gamma(a_j-\alpha_j s)}$$
$$=\underbrace{\frac{\prod_{j=1}^{p}\Gamma(1-a_j+\alpha_j s)}{\prod_{j=1}^{q}\Gamma(1-b_j+\beta_j s)}}_{h_0(s)}\times\underbrace{\pi^{m+n-p}\frac{\prod_{j=n+1}^{p}\sin[\pi(a_j-\alpha_j s)]}{\prod_{j=1}^{m}\sin[\pi(b_j-\beta_j s)]}}_{h_1(s)}.$$

The contour C is the right loop separating the poles $s=(b_j+\nu)/\beta_j$, $j=1,\ldots,m$, $\nu=0,1,\ldots$ (lying to the right of C, hence inside the contour) from the poles $s=(a_j-1-\nu)/\alpha_j$, $j=1,\ldots,n$, $\nu=0,1,\ldots$ (lying to the left of C, hence outside the contour). Denote

$$\mu=\sum_{j=1}^{q}\beta_j-\sum_{j=1}^{p}\alpha_j,\qquad \beta=\prod_{j=1}^{p}\alpha_j^{\alpha_j}\prod_{j=1}^{q}\beta_j^{-\beta_j}, \tag{12.2}$$

and

$$\delta_0=\left(\sum_{j=1}^{m}\beta_j-\sum_{j=n+1}^{p}\alpha_j\right)\pi. \tag{12.3}$$

Next, by [1, (4.26),(4.27), p. 270]

$$Q(z)=\sum \text{residues of } h(s)z^s \text{ in the points } s=(a_j-1-\nu)/\alpha_j, j=1,\ldots,n,\nu=0,1,\ldots \tag{12.4}$$

$$P(z)=\sum \text{residues of } h_0(s)z^s \text{ in the points } s=(a_j-1-\nu)/\alpha_j, j=1,\ldots,p,\nu=0,1,\ldots \tag{12.5}$$

In [1, Lemma 4, p. 263] Braaksma uses $\sin(z) = e^{iz}(1 - e^{-2iz})/(2i)$ to expand:

$$h_1(s) = (2\pi i)^{m+n-p} \frac{\prod_{j=n+1}^{p} e^{i\pi(a_j - \alpha_j s)} \left(1 - e^{-2i\pi(a_j - \alpha_j s)}\right)}{\prod_{j=1}^{m} e^{i\pi(b_j - \beta_j s)} \left(1 - e^{-2i\pi(b_j - \beta_j s)}\right)}$$

$$= \underbrace{(2\pi i)^{m+n-p} e^{i\pi\{\sum_{j=n+1}^{p} a_j - \sum_{j=1}^{m} b_j\}}}_{c_0} \underbrace{e^{i\pi s\{\sum_{j=1}^{m} \beta_j - \sum_{j=n+1}^{p} \alpha_j\}}}_{=e^{i\gamma_0 s}} \frac{\prod_{j=n+1}^{p} \left(1 - e^{2i\pi(\alpha_j s - a_j)}\right)}{\prod_{j=1}^{m} \left(1 - e^{2i\pi(\beta_j s - b_j)}\right)}.$$

For sufficiently large positive $\Im(s)$ we will have $\Re(2i\pi(\beta_j s - b_j)) < 0$ so that $|e^{2i\pi(\beta_j s - b_j)}| < 1$, and we can expand $\left(1 - e^{2i\pi(\beta_j s - b_j)}\right)^{-1}$ in geometric series:

$$\frac{1}{1 - e^{2i\pi(\beta_j s - b_j)}} = \sum_{\nu=0}^{\infty} e^{2i\pi\nu(\beta_j s - b_j)} = \sum_{\nu=0}^{\infty} \underbrace{e^{-2i\pi\nu b_j}}_{B_{j,\nu}} e^{2i\pi\nu\beta_j s}.$$

By expanding we get

$$\prod_{j=n+1}^{p} \left(1 - e^{2i\pi(\alpha_j s - a_j)}\right) = 1 - A_{n+1} e^{2i\pi\alpha_{n+1} s} - A_{n+2} e^{2i\pi\alpha_{n+2} s} - \ldots - A_p e^{2i\pi\alpha_p s}$$
$$+ \ldots + A_{n+1} A_{n+2} \cdots A_p e^{2i\pi(\alpha_{n+1} + \alpha_{n+2} + \cdots + \alpha_p) s},$$

where $A_k = e^{-2i\pi a_k}$. Multiplying all these expressions yields

$$h_1(s) = c_0 e^{i\gamma_0 s} \left(1 + \hat{c}_1 e^{i\gamma_1 s} + \hat{c}_2 e^{i\gamma_2 s} + \cdots\right)$$

with $0 < \gamma_1 < \gamma_2 < \cdots$, where

$$\gamma_0 = \pi \left\{\sum_{j=1}^{m} \beta_j - \sum_{j=n+1}^{p} \alpha_j\right\} \tag{12.6}$$

and

$$\gamma_1 = 2\pi \min(\alpha_{n+1}, \alpha_{n+2}, \ldots, \alpha_p, \beta_1, \ldots, \beta_m). \tag{12.7}$$

If $\Im(s)$ is negative and of large absolute value, we can use $\sin(z) = -e^{-iz}(1 - e^{2iz})/(2i)$ to similarly expand

$$h_1(s) = d_0 e^{-i\gamma_0 s} \left(1 + \hat{d}_1 e^{-i\gamma_1 s} + \hat{d}_2 e^{-i\gamma_2 s} + \cdots\right),$$

where

$$d_0 = (-2\pi i)^{m+n-p} e^{i\pi\{\sum_{j=1}^{m} b_j - \sum_{j=n+1}^{p} a_j\}} = (2\pi)^{2(m+n-p)}/c_0,$$

and $d_j = 1/c_j$. In [1, Definition I, p. 265] Braaksma takes the (unordered) multiset

$$\{\ldots, -\gamma_2, -\gamma_1, -\gamma_0, \gamma_0, \gamma_1, \gamma_2, \ldots\},$$

sorts it in ascending order, removes the repeated elements, and calls the new set

$$\{\ldots, \delta_{-2}, \delta_{-1}, \delta_0, \delta_1, \delta_2, \ldots\} \text{ with } \delta_0 = \gamma_0.$$

We next assume $\gamma_0 = \delta_0 = 0$. Then $\delta_j = -\delta_{-j}$ and $\delta_j > 0$ for $j = 1, 2, \ldots$, and, according to [1, Definition II, p. 265], we see that $\kappa = 0$ (by definition κ is the index such that $\delta_\kappa = -\delta_0$). Further, if an integer $r > 0$ we will have case a) in [1, Definition II, p. 265] and $C_r = c_r$, $D_r = 0$; if $r < 0$ we have case b) and $C_r = 0$, $D_r = -d_r$; if $r = 0$, then we have case c) and $C_0 = c_0$, $D_0 = -d_0 = -(2\pi)^{2(m+n-p)}/c_0$. Note that the above formulas imply that (this is a slight modification of [1, (4.10), p. 266])

$$h_1(s) = \sum_{j=0}^{\infty} c_j e^{i\delta_j s}, \quad c_j = c_0 \hat{c}_j \text{ for } j = 1, 2, \ldots \tag{12.8}$$

for $\Im(s) > 0$, and

$$h_1(s) = \sum_{j=0}^{\infty} d_j e^{-i\delta_j s} \quad c_j = d_0 \hat{d}_j \text{ for } j = 1, 2, \ldots \tag{12.9}$$

for $\Im(s) < 0$; each series obviously uniformly converges: first for $\Im(s) \geq \epsilon > 0$, while the second for $\Im(s) \leq -\epsilon < 0$. Hence, c_j, d_j can be viewed as generalized Fourier coefficients in the upper and lower half-planes, while δ_j are "generalized frequencies".

Having made these preparations we can formulate (a part of) [1, Theorem 3, p. 280] as follows.

Theorem 12.1. *Suppose $\mu = \gamma_0 = 0$ (see (12.2) and (12.6) for definitions). Then $H(z)$ can be continued analytically into the sector*

$$\Delta_- = \{z : -\gamma_1 < \arg(z) < 0\},$$

where γ_1 is defined in (12.7), by

$$H(z) = \frac{1}{2\pi i} \int_D h_0(s) \{h_1(s) - c_0\} z^s ds - V(z);$$

and into the sector

$$\Delta_+ = \{z : 0 < \arg(z) < \gamma_1\}$$

by

$$H(z) = \frac{1}{2\pi i} \int_D h_0(s) \{h_1(s) - d_0\} z^s ds - V(z);$$

the contour D starts at $s = -i\infty + \sigma$ and terminates at $s = i\infty + \sigma$, σ is arbitrary real number, leaves the points $s = (a_j - \nu - 1)/\alpha_j$, $j = 1, \ldots, p$, $\nu = 0, 1, \ldots$ (the poles of $h_0(s)$) on the left and the points $s = (b_j + \nu)/\beta_j$, $j = 1, \ldots, m$, $\nu = 0, 1, \ldots$, which do not coincide with the points $(a_j - \nu - 1)/\alpha_j$, $j = 1, \ldots, p$, $\nu = 0, 1, \ldots$, (the poles of $h(s)$ which are not poles of $h_0(s)$) on the right. The function $V(z)$ is the sum of residues of $h(s)z^s$ in the points which are poles of both $h(s)$ and $h_0(s)$. This function is equal to zero if

$$\frac{b_j + \nu}{\beta_j} \neq \frac{a_h - 1 - \lambda}{\alpha_h}$$

for all $j = 1, \ldots, m$, $h = n + 1, \ldots, p$, $\nu, \lambda = 0, 1, 2, \ldots$

Moreover, the function $H(z)$ can be continued analytically from the sector Δ_- to the domain $|z| > \beta^{-1}$ by the formula

$$H(z) = Q(z) - c_0 P(z),$$

and from the sector Δ_+ to the domain $|z| > \beta^{-1}$ by the formula

$$H(z) = Q(z) - d_0 P(z).$$

Here Q and P are defined in (12.4) and (12.5), respectively. Both series (12.4) and (12.5) converge for $|z| > \beta^{-1}$.

Our next goal is to find conditions that would guarantee that the analytic continuations to the sectors Δ_- and Δ_+ are analytic in the "large" sector $\Delta_- \cup [0,\infty) \cup \Delta_+$. First, we have to require $c_0 = d_0$. Writing the definitions of c_0 and d_0 and performing some simple calculations we arrive at the condition

$$\sum_{j=n+1}^{p} a_j - \sum_{j=1}^{m} b_j = \frac{p-m-n}{2} + \eta,$$

where η is any integer. In this case we have

$$c_0 = d_0 = (2\pi)^{m+n-p} e^{i\pi(m+n-p)/2} e^{i\pi(p-m-n)/2} e^{i\pi\eta} = (-1)^\eta (2\pi)^{m+n-p}.$$

In particular, if $n = 0$, $m = q$, and $\eta = 1$ we obtain

$$\sum_{j=1}^{p} a_j - \sum_{j=1}^{q} b_j = \frac{p-q}{2} + 1,$$

which corresponds to the case considered in [7]. Now in this particular situation ($\mu = \delta_0 = 0$, $c_0 = d_0$) we obtain a strengthened version of [1, Lemma 4a, p. 265]:

Lemma 12.1. *Suppose $\mu = \delta_0 = 0$, $c_0 = d_0$. Then there exists a positive constant $K = K(\epsilon)$, independent of s, such that for any $\epsilon > 0$*

$$|(h_1(s) - c_0)e^{-i\delta_1 s}| \leq K \text{ if } \Im(s) \geq \epsilon$$

and

$$|(h_1(s) - c_0)e^{i\delta_1 s}| \leq K \text{ if } \Im(s) \leq -\epsilon.$$

For the proof note that $c_0 = d_0$ and use (12.8) and (12.9) to get straightforward term-by-term estimates, similarly to the proof of [1, Lemma 4a, p. 265]. Note, however, that we get a larger difference $2\delta_1$ between the exponents of the multipliers $e^{-i\delta_1 s}$ and $e^{i\delta_1 s}$ in the first and second bounds, than that in [1, Lemma 4a, p. 265]. This becomes possible by virtue of the condition $c_0 = d_0$.

We now formulate an extension of a particular case of [1, Theorem 2, p. 280] for the values of parameters $\mu = \delta_0 = 0$ and $c_0 = d_0 = (-1)^\eta (2\pi)^{m+n-p}$ ($\eta \in \mathbb{Z}$ is arbitrary).

Theorem 12.2. *Suppose $\mu = \delta_0 = 0$, where μ and δ_0 are defined in (12.2) and (12.3), respectively, and $c_0 = d_0$. Then $c_0 = d_0 = (-1)^\eta (2\pi)^{m+n-p}$, where*

$$\eta = \sum_{j=n+1}^{p} a_j - \sum_{j=1}^{m} b_j - \frac{p-m-n}{2} \in \mathbb{Z}.$$

Furthermore, the function $H^{m;n}_{p,q}(z)$ defined in (12.1) can be continued analytically to the sector

$$\Delta_{\gamma_1} = \{z : -\gamma_1 < \arg(z) < \gamma_1\}, \tag{12.10}$$

where γ_1 is defined in (12.7) by means of

$$H(z) = \frac{1}{2\pi i} \int_D h_0(s) \{h_1(s) - c_0\} z^s ds - V(z); \tag{12.11}$$

the contour D and the function $V(z)$ have been defined in Theorem 12.1. Moreover, the function $H(z)$ can be continued analytically from this sector to the domain $|z| > \beta^{-1}$ on the Riemann surface of the logarithm by the formula

$$H(z) = Q(z) - c_0 P(z),$$

Here Q and P are defined in (12.4) and (12.5), respectively. Both series (12.4) and (12.5) converge for $|z| > \beta^{-1}$.

Proof. We follow the proof of [1, Theorem 2, p. 280] for $r = 1$ and [1, (6.10), p. 280] taking the form:

$$f(s) = h_0(s)\beta^{-s}\{h_1(s) - c_0\}e^{-i\gamma_1 s},$$

so that, in view of the first bound in Lemma 12.1, we can apply [1, Lemma 6, p. 271] with $\beta z e^{i\gamma_1}$ playing the role of z. Then [1, Lemma 6, p. 271] applies for $\arg(z) > -\gamma_1$. Instead of formula [1, (6.11), p. 280] we take

$$f(s) = h_0(s)\beta^{-s}\{h_1(s) - c_0\}e^{i\gamma_1 s}.$$

In view of the second bound in Lemma 12.1, we can now apply [1, Lemma 6a, p. 277] with $\beta z e^{-i\gamma_1}$ playing the role of z, and hence [1, Lemma 6a, p. 277] applies for $\arg(z) < \gamma_1$. This leads to the validity of [1, (6.12), p. 280] for $-\gamma_1 < \arg(z) < \gamma_1$. The remaining part of the proof of [1, Theorem 2, p. 280] remains intact. □

For the function $H^{q,0}_{p,q}(z)$ ($m = q$, $n = 0$) we have $Q(z) = 0$ by (12.4), so that if $\mu = 0$ (and hence in this case $\delta_0 = \pi\mu = 0$) and

$$\eta = \sum_{j=1}^{p} a_j - \sum_{j=1}^{q} b_j + \frac{q-p}{2} \in \mathbb{Z}, \tag{12.12}$$

(note that $\eta = \alpha - 1/2$, where α is defined in [1, (3.24)]), then $H^{q,0}_{p,q}(z)$ is analytic in the sector

$$-\gamma_1 < \arg(z) < \gamma_1,$$

where γ_1 is defined in (12.7), its analytic continuation to this sector is given by (12.11) (with $c_0 = (-1)^{\eta}(2\pi)^{q-p}$) and from this sector to the domain $|z| > \beta^{-1}$ by

$$\begin{aligned}(-1)^{\eta+1}(2\pi)^{q-p} \sum &\text{residues of } h_0(s)z^s \text{ in the points } s \\ &= (a_j - 1 - \nu)/\alpha_j, j \in \{1,\ldots,p\},\ \nu \in \mathcal{N}_0,\end{aligned}$$

according to (12.5).

12.3 Expansion in the Neighborhood of the Singular Point

We now focus on the function $H^{q,0}_{p,q}(z)$ with $\mu = 0$. For further convenience we also apply the change of variable $s \to -s$ in the integral (12.1) to obtain for $|z| < \beta^{-1}$ the definition

$$H^{q,0}_{p,q}\left(z \,\middle|\, \begin{matrix}(\alpha_1, a_1), \ldots (\alpha_p, a_p) \\ (\beta_1, b_1), \ldots (\beta_q, b_q)\end{matrix}\right) = \frac{1}{2\pi i}\int_{C'} \frac{\prod_{j=1}^{q}\Gamma(b_j + \beta_j s)}{\prod_{j=1}^{p}\Gamma(a_j + \alpha_j s)} z^{-s} ds, \tag{12.13}$$

where the contour $C' = -C$, it is the left loop starting at $-\infty - ik$, terminating at $-\infty + ik$ and encompassing all the poles $s = (b_j - \nu)/\beta_j$, $j = 1, \ldots, q$, $\nu = 0, 1, \ldots$ (so that $k > \Im(b_j/\beta_j)$, $j = 1, \ldots, q$). To be consistent with [6,7] denote

$$\rho = \beta^{-1} > 0.$$

Assuming that $\eta = 1$, we see by Theorem 12.2 that the function $H^{q,0}_{p,q}(\rho t)$ is analytic in the domain $G := U_- \cup \Delta_{\gamma_1}$, where $U_- = \{t : |t| < 1\} \setminus (-1, 0]$ denotes the unit disk cut along the interval $(-1, 0]$ (we need a cut in view of the branch point at $t = 0$) and

$$\Delta_{\gamma_1} = \{t \in \mathbb{C} : -\gamma_1 < \arg(t) < \gamma_1\}, \quad \gamma_1 = 2\pi \min(\alpha_1, \ldots, \alpha_p, \beta_1, \ldots, \beta_q),$$

is the sector defined in (12.10) (note that change of variable $z \to \rho t$ does not alter this sector). Hence, for arbitrary complex σ we have the power series expansion

$$\phi(t) = t^{-\sigma} H_{p,q}^{q,0}(\rho t) = \sum_{n=0}^{\infty} V_n(\sigma)(1-t)^n \tag{12.14}$$

convergent in the disk $|1-t| < R$, where R is the distance from $t = 1$ to the boundary of the domain G. Elementary geometry shows that $R = 1$ if $\gamma_1 \geq \pi/3$ (in view of (12.7), this is equivalent to $\min(\alpha_1, \dots, \alpha_p, \beta_1, \dots, \beta_q) \geq 1/6$) and $R = 2\sin(\gamma_1/2)$ if $\gamma_1 < \pi/3$. Assume for a moment that $\gamma_1 \geq \pi/3$. Multiplying the above expansion by t^{z-1} and integrating term by term (to be justified below) we get

$$\int_0^1 t^{z+\sigma-1} t^{-\sigma} H_{p,q}^{q,0}(\rho t)dt = \sum_{n=0}^{\infty} V_n(\sigma) \int_0^1 t^{z+\sigma-1}(1-t)^n dt = \sum_{n=0}^{\infty} \frac{V_n(\sigma)n!}{(z+\sigma)_{n+1}}.$$

On the other hand, by [8, Theorem 6]

$$\int_0^1 t^{z-1} H_{p,q}^{q,0}(\rho t)dt = \rho^{-z} \int_0^{\rho} u^{z-1} H_{p,q}^{q,0}(u)du = \rho^{-z} \frac{\prod_{j=1}^q \Gamma(\beta_j z + b_j)}{\prod_{j=1}^p \Gamma(\alpha_j z + a_j)}$$

or

$$W(z) = \rho^{-z} \frac{\prod_{j=1}^q \Gamma(\beta_j z + b_j)}{\prod_{j=1}^p \Gamma(\alpha_j z + a_j)} = \sum_{n=0}^{\infty} \frac{V_n(\sigma)n!}{(z+\sigma)_{n+1}}. \tag{12.15}$$

This inverse factorial series was derived in [6, Theorem 1], where the coefficients were found explicitly. However, the condition $\gamma_1 \geq \pi/3$ was mistakenly omitted in this reference.

Here we give a more direct proof of this inverse factorial series expansion than that presented in [6], which will lead to a corrected proof of the expansion [7, Theorem 1] generalizing (12.14) to the case $\eta \neq 1$. Our proof will be based on a theorem due to Nørlund. To formulate this theorem we need to define the Hadamard order of a function $f = \sum_{n \geq 0} f_n x^n$ holomorphic inside the unit disk $|x| < 1$. According to [12, 23(1), p. 46] the Hadamard order of f on the circle $|x| = 1$ is defined by

$$\omega = \limsup_{n \to \infty} \frac{\log |n f_n|}{\log(n)}. \tag{12.16}$$

Nørlund proved the following theorem (the formulation below is a combination of [12, Theorem S87, p. 187] and [12, Theorem S88, p. 189]).

Theorem 12.3. *Suppose*

$$\phi(t) = \sum_{s=0}^{\infty} a_s (1-t)^s$$

is holomorphic for $|1-t| < 1$ and has a finite order on the circle $|1-t| = 1$. Then

$$F(w) = \int_0^1 t^{w-1} \phi(t) dt$$

can be expanded in the inverse factorial series

$$F(w) = \sum_{s=0}^{\infty} \frac{a_s s!}{(w)_{s+1}}$$

convergent in a half-plane $\Re(w) > \lambda$ (save the points $w = 0, -1, -2, \dots$) and divergent for $\Re(w) < \lambda$. If the series $\sum_{s=0}^{\infty} a_s$ diverges, the convergence abscissa $\lambda = h - 1$, where h is the order of $t^{-1}\phi(t)$ on the the circle $|1-t| = 1$; if $\sum_{s=0}^{\infty} a_s$ converges, then $\lambda = h' - 1$, where h' is the order of $[\phi(t) - \phi(0+)]/t$ on the circle $|1-t| = 1$.

Assume that $\gamma_1 > \pi/3$. Then, according to the above, for some $\varepsilon > 0$ and any $0 < \phi < \pi/2$ the function $H^{q,0}_{p,q}(\rho t)$ is holomorphic in the domain

$$\Delta(\phi,\varepsilon) = \{t : |1-t| \le 1+\varepsilon \ \ and \ \ -\pi+\phi \le \arg(t) \le \pi-\phi\},$$

which is larger than the disk $|1-t| < 1$. We want to apply Theorem 12.3 with $\phi(t) = t^{-\sigma}H^{q,0}_{p,q}(\rho t)$ and $w = z+\sigma$. Hence we need to determine the order of $\phi(t)t^{-1} = t^{-\sigma-1}H^{q,0}_{p,q}(\rho t)$ on the circle $|1-t| = 1$ and the order of $[\phi(t)-\phi(0+)]/t$ when $\phi(0+) \neq \infty$. First recall the shifting property [9, (2.1.5)]

$$t^{-\sigma}H^{q,0}_{p,q}\left(\rho t \,\middle|\, \begin{matrix}(\alpha_1,a_1),\dots(\alpha_p,a_p)\\(\beta_1,b_1),\dots(\beta_q,b_q)\end{matrix}\right) = \rho^{\sigma}H^{q,0}_{p,q}\left(\rho t \,\middle|\, \begin{matrix}(\alpha_1,a_1-\sigma\alpha_1),\dots(\alpha_p,a_p-\sigma\alpha_p)\\(\beta_1,b_1-\sigma\beta_1),\dots(\beta_q,b_q-\sigma\beta_q)\end{matrix}\right)$$

Next, suppose

$$\mathcal{P}_\sigma = \{(s,r) : s = \text{pole of } h_\sigma(s); r = \text{its multiplicity}\},$$

where s runs over all distinct poles of the integrand of $t^{-\sigma}H^{q,0}_{p,q}(\rho t)$ which is given by

$$h_\sigma(s) = \frac{\prod_{j=1}^{q}\Gamma(b_j-\sigma\beta_j+\beta_j s)}{\prod_{j=1}^{p}\Gamma(a_j-\sigma\alpha_j+\alpha_j s)},$$

and r stands for the multiplicity of the corresponding pole. We need the following definitions: denote by $\mathcal{P}'_\sigma$ the set $\mathcal{P}_\sigma$, where the elements of the form $(m,1)$, $m \in \mathbb{Z}_{\le 0}$, have been removed (in other words, simple poles at nonpositive integers are removed); and define

$$\hat{\beta}(\sigma) = \max\{\Re(s) : (s,r) \in \mathcal{P}'_\sigma\}. \tag{12.17}$$

Note that $\hat{\beta}(\sigma) = \max_{1\le j\le q}\Re(\sigma - b_j/\beta_j)$ except when this maximum is attained for $\sigma - b_j/\beta_j \in \mathbb{Z}_{\le 0}$ and the corresponding pole is simple. It is also convenient to write $\mathcal{P}_{\sigma,1}$ for the set of the first components of the elements of $\mathcal{P}_\sigma$ (i.e. the set of poles of $h_\sigma(s)$ without multiplicities). Similarly for $\mathcal{P}'_\sigma$.

Lemma 12.2. *Let $\phi(t)$ be defined in* (12.14).

If the series $\sum_{n\ge 0}V_n(\sigma)$ diverges, then the Hadamard order h of $t^{-1}\phi(t)$ on the circle $|1-t| = 1$ is given by $h = \hat{\beta}(\sigma)+1$.

If the series $\sum_{n\ge 0}V_n(\sigma)$ converges, then the Hadamard order h' of $[\phi(t)-\phi(0+)]/t$ on the circle $|1-t| = 1$ is given by the same expression $h' = \hat{\beta}(\sigma)+1$.

Proof. Denote

$$I_\sigma := -\mathcal{P}_{\sigma,1} \setminus \mathcal{P}'_{\sigma,1}.$$

By definition $I_\sigma \subset \mathbb{Z}_{\ge 0}$. The residue theorem applied to the definition of $t^{-\sigma}H^{q,0}_{p,q}(\rho t)$ gives as $t \to 0$ [9, Theorem 1.5]:

$$\begin{aligned}\phi(t) = t^{-\sigma}H^{q,0}_{p,q}(\rho t) &= \rho^\sigma \sum \operatorname*{res}_{s\in\mathcal{P}'_{\sigma,1}} h_\sigma(s)(\rho t)^{-s} + \rho^\sigma \sum \operatorname*{res}_{s\in\mathcal{P}_{\sigma,1}\setminus\mathcal{P}'_{\sigma,1}} h_\sigma(s)(\rho t)^{-s}\\ &= (\log t)^{r-1}\sum_{k=1}^{l} C_k t^{-\hat{b}_k}(1+o(1)) + \sum_{j\in I_\sigma\subset\mathbb{Z}_{\ge 0}} A_j t^j,\end{aligned} \tag{12.18}$$

where $\hat{b}_k = \sigma - b_{j_k}/\beta_{j_k}$, $k = 1,2,\dots,l$, are the elements of $\mathcal{P}'_{\sigma,1}$ with $\Re(\hat{b}_k) = \hat{\beta}(\sigma)$ and having maximal multiplicity (which we denoted by r) among all poles with real part $\hat{\beta}(\sigma)$. The precise values of the nonzero constants $C_1,\dots,C_l$ can be found in [9, Theorem 1.5], but they are immaterial for our purposes here.

To determine the order note first that if the series $\sum_{n\geq 0} V_n(\sigma)$ converges, then $\lim_{t\to 0}\phi(t) < \infty$ by Abel's theorem. Hence, by (12.18) this series diverges if $\hat{\beta}(\sigma) > 0$ or $\hat{\beta}(\sigma) = 0$ and $r > 1$ or even when $\hat{\beta}(\sigma) = 0$, $r = 1$ and $\Im(\hat{b}_k) \neq 0$ for some $k \in \{1,\ldots,l\}$ (because no finite limit $\lim_{t\to 0}\phi(t)$ exists in all these cases). As $\hat{b}_k \in \mathcal{P}'_{\sigma,1}$ the case $\hat{b}_k = 0$ is impossible by definition of $\mathcal{P}'_\sigma$. Hence, we conclude that $\sum_{n\geq 0} V_n(\sigma)$ diverges if $\hat{\beta}(\sigma) \geq 0$. To determine the Hadamard order of $t^{-1}\phi(t)$ in these situations we apply [3, Theorem VI.4] with

$$f(x) = (1-x)^{-\sigma-1} H^{q,0}_{p,q}(\rho(1-x))$$

and $\zeta = 1$. This gives the following asymptotic relation for the power series coefficients $V_n(\sigma+1)$

$$V_n(\sigma+1) = A_0 + \sum_{k=1}^{l} D_k n^{\hat{b}_k} (\log n)^{r-1}(1+o(1)) + o(1) \tag{12.19}$$

with some complex constants D_k and $o(1)$ being different in each term. By definition of the Hadamard order (12.16) the above asymptotic formula leads immediately to $h = \hat{\beta}(\sigma) + 1$.

Next, we assume that the series $\sum_{n\geq 0} V_n(\sigma)$ converges, so that $\lim_{t\to 0}\phi(t) < \infty$. By the above argument this is only possible if $\hat{\beta}(\sigma) < 0$. Then, by (12.18) we have

$$\lim_{t\to 0}\phi(t) = A_0$$

(A_0 may vanish). The last term in (12.18) for $t^{-1}(\phi(t) - A_0)$ takes the form

$$\sum_{j\in I_\sigma\setminus\{0\}\subset\mathbb{Z}_{>0}} A_j t^{j-1}$$

which is holomorphic around $t = 0$ and does not affect the order of $t^{-1}(\phi(t) - A_0)$ (see [12, Section 26]). Hence (12.19) with A_0 removed holds true again, and the order h' of $t^{-1}(\phi(t) - A_0)$ still equals $h' = \hat{\beta}(\sigma) + 1$. □

The coefficients $V_n(\sigma)$ have been computed in [6, Corollary 1] and [7, Theorem 1] via rearrangement of Poincaré asymptotics of $W(z)$ defined in (12.15) as $z \to \infty$ into inverse factorial series (this method dates back to Stirling). Putting these facts together we arrive at the following statements.

Theorem 12.4. *Suppose* $\gamma_1 = 2\pi\min(\alpha_1,\ldots,\alpha_p,\beta_1,\ldots,\beta_q) > \pi/3$, $\mu = 0$ *and* $\eta = 1$, *where* μ *and* η *are defined in* (12.2) *and* (12.12), *respectively. Then the function*

$$t^{-\sigma} H^{q,0}_{p,q}\left(\rho t \,\middle|\, \begin{matrix}(\alpha_1,a_1),\ldots(\alpha_p,a_p)\\(\beta_1,b_1),\ldots(\beta_q,b_q)\end{matrix}\right)$$

defined in (12.13) *with* $\rho = \beta^{-1}$ *and* β *given in* (12.2) *is analytic in the sector* $-\gamma_1 < \arg(z) < \gamma_1$ *by Theorem 12.2 and can be developed in convergent power series* (12.14) *with coefficients given by*

$$V_n(\sigma) = (2\pi)^{(q-p)/2}\prod\nolimits_{k=1}^{q}\beta_k^{b_k-1/2}\prod\nolimits_{j=1}^{p}\alpha_k^{1/2-a_j}\sum_{k=0}^{n}\frac{(-1)^k l_{n-k}}{k!(n-k)!}\mathcal{B}_k^{(n+1)}(1-\sigma), \tag{12.20}$$

where $\mathcal{B}_k^{(n+1)}(\cdot)$ *is Bernoulli-Nørlund polynomial [13, (1)]. The coefficients* l_r *satisfy the recurrence relation (with* $l_0 = 1$*)*

$$l_r = \frac{1}{r}\sum_{m=1}^{r} q_m l_{r-m}, \quad q_m = \frac{(-1)^{m+1}}{m+1}\left[\sum_{k=1}^{q}\frac{\mathcal{B}^{(1)}_{m+1}(b_k)}{\beta_k^m} - \sum_{j=1}^{p}\frac{\mathcal{B}^{(1)}_{m+1}(a_j)}{\alpha_j^m}\right]. \tag{12.21}$$

The following theorem is a corrected and refined version of [6, Theorem 5].

Theorem 12.5. *Under conditions of Theorem 12.4 the inverse factorial series* (12.15) *converges in the half plane* $\Re(z) > \lambda$ *(save the points* $w = 0, -1, -2, \ldots$*) and diverges in* $\Re(z) < \lambda$. *The convergence abscissa* $\lambda = \hat{\beta}(\sigma) - \Re(\sigma)$, *where* $\hat{\beta}(\sigma)$ *is defined in* (12.17).

Remark 12.1. *If* $\hat{\beta}(\sigma) = \max_{1\le j\le q} \Re(\sigma - b_j/\beta_j)$ *(which will be the case if this maximum is not attained for* $\sigma - b_j/\beta_j \in \mathbb{Z}_{\le 0}$*), then, clearly,*

$$\lambda = -\min_{1\le j\le q} \Re(b_j/\beta_j).$$

In the boundary case $\gamma_1 = \pi/3$ the function $\phi(t)$ has three singularities of the circle $|1-t| = 1$, and our calculation of order in Lemma 12.2 becomes inapplicable. Hence, the convergence abscissa will, generally speaking, be different from that given in Theorem 12.5.

To get rid of the restriction $\eta = 1$ we applied the following simple trick in [6, Corollary 1]. Given $W(z)$ from (12.15) with $\eta \neq 1$, consider

$$W_1(z) = W(z)\frac{\Gamma(z+\theta+\eta)}{\Gamma(z+\theta+1)} = \rho^{-z}\frac{\Gamma(z+\theta+\eta)\prod_{j=1}^{q}\Gamma(\beta_j z + b_j)}{\Gamma(z+\theta+1)\prod_{j=1}^{p}\Gamma(\alpha_j z + a_j)},$$

where θ is arbitrary real number. Obviously, the parameter η defined in (12.12) computed for $W_1(z)$ takes the value unity: $\eta_1 = 1$. Hence, we can apply (12.15) to $W_1(z)$. Taking $\sigma = \theta + \eta$ we obtain

$$W(z) = W_1(z)\frac{\Gamma(z+\theta+1)}{\Gamma(z+\theta+\eta)} = \sum_{n=0}^{\infty}\frac{h_n\Gamma(z+\theta+1)}{(z+\theta+\eta)_{n+1}\Gamma(z+\theta+\eta)} = \sum_{n=0}^{\infty}\frac{h_n\Gamma(z+\theta+1)}{\Gamma(z+\theta+\eta+n+1)}, \tag{12.22}$$

where the coefficients $h_n = V_n'(\theta+\eta)n!$ are obtained by applying formulas (12.20), (12.21) to the function $W_1(z)$. In other words, in these formulas we have to make the following replacements:

$$\begin{aligned} (\alpha_1,\ldots,\alpha_p) &\to (\alpha_1,\ldots,\alpha_p,1), & (a_1,\ldots,a_p) &\to (a_1,\ldots,a_p,\theta+1),\\ (\beta_1,\ldots,\beta_q) &\to (\beta_1,\ldots,\beta_q,1), & (b_1,\ldots,b_q) &\to (b_1,\ldots,b_q,\theta+\eta). \end{aligned} \tag{12.23}$$

The resulting expressions are given explicitly in [6, Corollary 1]. The convergence abscissa of the above expansion equals that for $W_1(z)$ and hence by Theorem 12.5 is computed as follows: remove nonpositive integers from the set of simple poles of $W_1(z)$ and denote by $\hat{\beta}_1(\theta+\eta)$ the maximum of the real parts of the remaining poles. The convergence abscissa then equals $\hat{\beta}_1(\theta+\eta) - \theta + \Re(\eta)$.

Given the convergent inverse factorial series (12.22) we can substitute the beta integral

$$\frac{\Gamma(z+\theta+1)\Gamma(\eta+n)}{\Gamma(z+\theta+\eta+n+1)} = \int_0^1 t^{z+\theta}(1-t)^{n+\eta-1}dt$$

and exchange the order of summation in integration. Then uniqueness of the inverse truncated Mellin transform immediately leads to

$$H^{q,0}_{p,q}\left(\rho t\,\middle|\, \begin{matrix}(\alpha_1,a_1),\ldots,(\alpha_p,a_p)\\(\beta_1,b_1),\ldots,(\beta_q,b_q)\end{matrix}\right) = t^{\theta+1}(1-t)^{\eta-1}\sum_{n=0}^{\infty}\frac{h_n}{\Gamma(\eta+n)}(1-t)^n. \tag{12.24}$$

These steps have been rigorously justified in [7, Theorem 1], where we also presented alternative formulas for the coefficients. Finally we remark that the radius of convergence of (12.24) is the same as for the expansion (12.14), where $\eta = 1$. Indeed, this series has the same radius of convergence as the series $\sum_{n\ge0} h_n(1-t)^n/n!$ which is an instance of (12.14) with substitutions (12.23). Altogether, this yields the following statement generalizing Theorem 12.4 to arbitrary values of parameters.

Theorem 12.6. *For arbitrary positive α_i, β_j and complex a_i, b_j, the series in* (12.24) *with coefficients $h_n = V_n'(\theta+\eta)n!$, where $V_n'(\theta+\eta)$ are given by* (12.20), (12.21) *with substitutions* (12.23), *has the radius of convergence $R = 1$ if*

$$\gamma_1 = 2\pi \min(\alpha_1, \ldots, \alpha_p, \beta_1, \ldots, \beta_q) \geq \pi/3$$

and $R = 2\sin(\gamma_1/2)$ if $\gamma_1 < \pi/3$.

In case $\gamma_1 \leq 1/3$ we can easily get an expansion convergent in the disk of radius 1, but at the price of having powers of $1 - t^{\omega}$ instead of powers of $1 - t$ with some $\omega > 0$. Indeed, the substitution $t = u^{1/\omega}$ with $\omega > 1$ extends the sector Δ_{γ_1} to the sector $\Delta_{\omega\gamma_1}$, where $\omega\gamma_1$ can be made greater than $\pi/3$ by choosing the appropriate ω. Hence, $u \to H^{q,0}_{p,q}(\rho u^{1/\omega})$ is analytic in $|1 - u| < 1$ for such ω. On the other hand, according to [9, (2.1.4)]

$$H^{q,0}_{p,q}\left(\rho u^{1/\omega} \,\middle|\, \begin{matrix} (\alpha_1, a_1), \ldots, (\alpha_p, a_p) \\ (\beta_1, b_1), \ldots, (\beta_q, b_q) \end{matrix}\right) = \omega H^{q,0}_{p,q}\left(\rho^{\omega} u \,\middle|\, \begin{matrix} (\omega\alpha_1, a_1), \ldots, (\omega\alpha_p, a_p) \\ (\omega\beta_1, b_1), \ldots, (\omega\beta_q, b_q) \end{matrix}\right).$$

Writing $\hat{\alpha}_i = \omega\alpha_i$ and $\hat{\beta}_j = \omega\beta_j$, formula (12.2) and the definition $\rho = \beta^{-1}$ imply that $\hat{\rho} = \rho^{\omega}$, in view of $\sum_{i=1}^{p} \alpha_i = \sum_{j=1}^{q} \beta_j$. Hence, we can apply Theorem 12.6 to the function on the right-hand side of the above relation. Returning to the original variable $t = u^{1/\omega}$, an extended version of expansion (12.24) then takes the form

$$H^{q,0}_{p,q}\left(\rho t \,\middle|\, \begin{matrix} (\alpha_1, a_1), \ldots, (\alpha_p, a_p) \\ (\beta_1, b_1), \ldots, (\beta_q, b_q) \end{matrix}\right) = \omega t^{\omega(\theta+1)} (1 - t^{\omega})^{\hat{\eta}-1} \sum_{n=0}^{\infty} \frac{\hat{h}_n}{\Gamma(\hat{\eta} + n)} (1 - t^{\omega})^n, \quad (12.25)$$

where, obviously, the coefficients are calculated as described in Theorem 12.6, but writing $\hat{\alpha}_i$, $\hat{\beta}_j$ instead of α_i, β_j. The series (12.25) converges in the domain $|1 - t^{\omega}| < 1$.

References

[1] B.L.J. Braaksma, Asymptotic expansions and analytic continuation for a class of Barnes integrals, *Composito Mathematica* 15(3) (1962–1964), 239–341.

[2] C.A. Coelho and B.C. Arnold, Instances of the Meijer *G* and Fox *H* functions and of the distribution of the product of independent beta random variables with finite representations, *AIP Conference Proceedings* 1479(1) (2012), 1133.

[3] Ph. Flajolet and R. Sedgewick, *Analytic Combinatorics*, Cambridge University Press, Cambridge (2009).

[4] C. Fox, The *G* and *H* functions as symmetrical Kernels, *Transactions of the American Mathematical Society* 98 (1961), 395–429.

[5] A.K. Gupta and J. Tang, On a general distribution for a class of likelihood ratio criteria, *Australian Journal of Statistics* 30(3) (1988), 359–366.

[6] D. Karp and E. Prilepkina, An inverse factorial series for a general gamma ratio and related properties of the Nørlund-Bernoulli polynomials, *Journal of Mathematical Sciences*, 234(5) (2018), 680–696.

[7] D. Karp and E. Prilepkina, Some new facts around the delta neutral *H* function of Fox, *Computational Methods and Function Theory*, 17(2) (2017) 343–367.

[8] D. Karp and E. Prilepkina, Completely monotonic gamma ratio and infinitely divisible H-function of Fox, *Computational Methods and Function Theory*, 16 (2016), 135–153.

[9] A.A. Kilbas and M. Saigo, *H-Transforms: Theory and Applications*, Analytical Methods and Special Functions, Volume 9, Chapman & Hall/CRC Press, Boca Raton, FL (2004).

[10] A.A. Kilbas, H.M. Srivastava, and J.J. Trujillo, *Theory and Applications of Fractional Differential Equations*, North-Holland Mathematics Studies, Volume 204, Elsevier, Amsterdam (2006).

[11] A.M. Mathai, R.K. Saxena, and H.J. Haubold, *The H-Function: Theory and Applications*, Springer, New York (2010).

[12] N.E. Nörlund, *Leçons sur les séries d'interpolation*, Gauthier-Villars, Paris (1926).

[13] N.E. Nørlund, Sur les Valeurs Asymptotiques des Nombres et des Polynômes de Bernoulli, *The Rendiconti del Circolo Matematico di Palermo* 10(1) (1961), 27–44.

[14] R.B. Paris and D. Kaminski, *Asymptotics and Mellin-Barnes Integrals*, Cambridge University Press, Cambridge (2001).

13

Categories and Zeta & Möbius Functions: Applications to Universal Fractional Operators

Philippe Riot
Zeta Innovation

Alain Le Méhauté
Materials Design Inc.
Kazan Federal University

Dmitrii Tayurskii
Kazan Federal University

CONTENTS

13.1 Riemann Zeta Function and Heuristic Approach of Riemann Hypothesis

This chapter is mainly addressed to physicists and engineers with a good knowledge of mathematics. It takes into account the current progresses in pure mathematics and the mental representations often close to emerging engineering issues. We hope that the analysis will thrust the practicians to look at what the mathematicians are currently doing. Let us give an example. A recent step in modeling physical exchanges in fractal geometries highlights the existence of homomorphism between Zeta Riemann function $\zeta(s)$ or Dirichlet series [1,2] and canonical transfer functions based on fractional derivatives [3]. These impedance $Z_\alpha(\eta)$ are termed Cole and Cole universal transfer functions [4]. These functions represent any kind of dynamics based on a local first - order test function, applied on self-similar distributions of sites of exchanges of energy (interface, surface, etc.), whose non-integer dimension is d_f with $1 \leq d_f \leq 2$. Zeta Riemann function $\zeta(s)$ with $s = \alpha(d_f) + i\theta$ appears [3,5,6], as a fibration along θ above $Z_\alpha(\eta)$. In this engineering approach of $\zeta(s)$, the occurrence of the non-trivial zeros of zeta function arises related to a metric transition over the dynamic processes locking the parameter of order, namely a Cantor dimension $\alpha = 1/d_f$, so that $d_f = 2$ (immanent Peano interface of exchange). The "experimental" confirmation of the Riemann Hypothesis emerges as a minoration of the entropic production and as a maximal scattering of the site of transfer. This analysis made several attempts for approaching the Riemann's Hypothesis from physical analogies of universal mathematical functions [3,8], including (i) quantum mechanics using stochastic matrix [9] or (ii) self-similar mechanical lattices [10]. The analysis of the algebra then at play in fractal context [3,11] can advantageously be developed from the theory of category [12–14], which, in dynamic automorphisms, distinguishes multiplication (product) namely cartesian projection $n \leftarrow n \times m \rightarrow m$ or partition and addition (co-product as opposite categories), namely $n \rightarrow n + m \leftarrow m$ or sum. In particular, this generalization allows to replace the standard and local exponential function (as the expression of morphism of the additive group $(\mathbb{R}, +)$ upon the multiplicative group $(\mathbb{R}, \times)$ to express the coarse graining) by a generalization over dual and subcategories [15]. In this context, the natural extension of the characteristics of *standard set theory* to *presheaf categorical approach* (namely object $X \in C$ is replaced by $\mathrm{Hom}_{C^{op}}(Y, X)$ of its links in C^{op}[16]) departs from the usual analytic analysis "stalemate" and, by using the categorical notion of limits, renews the approach of Riemann Hypothesis. For example, the automorphisms over $Z_\alpha(\eta)$ bases the universal function $\zeta(s)$. Likewise preliminary analysis [17,18] based on the major rule of arithmetic $n = \prod_{p \in \text{primes}} p_i^{ri}$ leads to consider $\zeta(s)$ as a $(\mathbb{N}, +/\times)$-measure of exponential operator in an infinite vectorial space $\Xi_\alpha(s)$ in which an infinite vector basis is squared according to the unit $-\alpha\log(p_i)$. This approach ensures the condition of self-similarity required by Bagchi [19] while the categorical duality s^{-n} *versus* n^s ensures the validity of Voronin's Theorem [20]. In addition this approach emphasizes the issues about the fonctorial adjunction $\Xi_\alpha(s) \rightleftharpoons \Xi_{1-\alpha}(1-s)$ in relation with the geometrical completion of the fractal dynamics $Z_\alpha(\eta) \rightleftharpoons \overleftarrow{Z}_{1-\alpha}(\tau)$, completion suggested at this stage, only by engineering data. Moreover if mathematically relevant, this completeness would be able to address the fundamental issue of the arrow of time [21,22].

13.1.1 Introduction to Zeta Riemann Function Based on Universal Transfer Function: $\mathbb{N}-$Measure in Prime $\Xi(s)-$Space

The new approach of Riemann Hypothesis [3,6,17,18] is based on Voronin and Bagchi's works [19,20]. This approach points out the role of self-similarity [12–14] in particular when category theory is considered [23–25]. This self-similarity can be reduced to the one of simple randomness when the self-similarity is shortened to the one of the set of integers $\mathbb{N} = \mathbb{N}{+}\mathbb{N} = \mathbb{N} \times \mathbb{N}$ [17,18]. As already proved, the structure of the infinite vectorial space $\Xi_\alpha(s)$, whose exponential measure is $\zeta(s)$, leads to the equation $\mathbb{N} = \mathbb{N} \times \mathbb{N}$, by using the fibered products: $\{n \leftarrow n{\times}\theta \rightarrow \theta\}_\alpha$. We are then able to account for the emergence of a $\zeta(s)-$manifold which, via a parametrization along $i\theta$, if $s = \alpha + i\theta(mod 2\pi)$, covers the complex plane in which $Z_\alpha(n)$ is expressed.[1]. The additive expression of zeta function $\zeta(s) = \sum\limits_{n\in\mathbb{N}} n^{-s}$ then refers to the additive total order monoid $(\mathbb{N}{+} \leq)$, while the multiplicative: $\zeta(s) = \prod_{p_i\in\wp}(1-p_i^{-s})^{-1}$ refers to Euler spectral decomposition of $\zeta(s)$ pointing out the role of the partial order monoid $(\mathbb{N}, \div)$. This division asserts that a morphism $p_i \rightarrow n$ exists if and only if $p_i \leq n$. The binding between the different types of order (in physics star series and parallel diagrams based triangle) can be imaged through the following analytical development: if we write $q = p_i^{-s}$, with $p_i \in \wp$ the set of primes, the analytic series $(1-q)^{-1} \sim 1{+}q{+}q^2{+}\cdots{+}q^r{+}\cdots$ appearing in the spectral expression of $\zeta(s)$, involves a covering by log-periodic polycycles which give birth to a manifold. The complex base of this manifold is the impedance $Z_\alpha(n)$ which defines the automorphisms over a set $\{n^{-\alpha}\}$ characterizing a self-similar dynamics[4] [5–7]. In a second step this base is extended along $i\theta$ giving birth to a new set $\{n^{-(\alpha\pm i\theta)}\}$ in which sections unveil the self-similar chaos previously hidden behind the Fourier Transform of the dynamics over $Z_\alpha(n)$ namely $i^\alpha n = (i\omega\tau)^\alpha$. In this dual approach, $\zeta(s)$ appears as a fuzzy equality operator between co-product over $n \in \mathbb{N}$, namely addition and spectral analysis (as inverse product based on p_i). Some endomorphisms can then find their origin in the self-similar structures of the dynamic "measure" induced by $\mathbb{N}$. This dual approach exemplifies and patterns the Alain Connes' remark, cited by K. Sabbagh [28] stating that "Riemann Hypothesis (RH) is probably the most basic problem of mathematics; it expresses a (singular) entanglement between addition and multiplication. RH reveals the presence of a conceptual gap that breaks the coherence of our current understanding ..." in particular by forgetting the intrinsic self-similarity of $\mathbb{N}$ (authors remark).

To understand the fuzzy print imposed by the zeta function, the analysis must be shifted into category theory by pointing out that, when $\zeta(s)$ is defined through an additive Dirichlet series $\zeta(s) = \sum\limits_{n\in\mathbb{N}} n^{-s}$, zeta function must be expressed by means of a convolution product [27] based on multiplicative monoidal structure $(\mathbb{N}, \times) \rightarrow \sum\limits_{n_i \times n_j}(\ n_i \times n_j)^{-s}$. Conversely the analytic function used by Voronin as a weighting function [19,20] $f(s)$ is expressed via a polynomial power series associated with an additive monoid, namely: $(\mathbb{N}, +) :\rightarrow \sum\limits_{n_i+n_j} s^{n_i}s^{n_j}$. In both cases, these equations must be considered in accordance with Voronin complex shift

[1]Based on distribution theory and convolution [27] the TEISI model [5,6] can be expressed through a universal transfer function [4] by means of a fractional differential equation $Z_\alpha(\omega\tau) \sim 1/1{+}(i\omega\tau)^\alpha$. The test function is a first-order transfer of energy, performed across a fractal interface with d_f fractal dimension and $\alpha = 1/d_f$. $\zeta(s)$ Riemann function is obtained as a result of the fibration of $Z_\alpha(\omega\tau)$ along $\theta \in \mathbb{C}$: $s = \alpha + i\theta$ with $i^2 = -1$. This extension points out the role of the phase $i^\alpha \rightarrow \varphi = \frac{\pi}{2}(1-\alpha)$. It can also be referred to as spectral dimension with $i^{2\alpha} \rightarrow \psi = \frac{\pi}{4}(2\alpha - 1)$

operator $|\zeta(s+it) - f(s)| \leq \varepsilon$. The zeta function $\zeta(s)$ being paired with Möbius function as dual (inverse in convolution) the additional morphism must be partly found upon its combinatorial properties. Starting from all of that, Bagchi hypothesis asserts that self-similarity as a result of the adjunctions $+ \rightleftharpoons \times$ and $n^{-s} \rightleftharpoons s^n$ and offers the opportunity to localize the zeta function over its non-trivial zeros by writing $\zeta_{1/2}(s) = \sum s^{n_i}$ according to Riemann Hypothesis. The role of the pair of commutations must be proved in the framework of enriched category theory [17,18]. The current approach will focus on (i) *the concept of fibered bi-category* (n, s), extended to $\zeta(s)$-manifold endomorphisms while the θ–fibration over $Z_\alpha(n)$ addresses the question of coupling between the $i\theta$ parameter and the Fourier parameter of the dynamics $i\omega\tau$. (ii) Except for $\alpha = 1/2$, *the absence of completeness of fractional operators* leads that the closure of fractional differential representation requires the design of a virtual non-causal dynamics category $\overleftarrow{Z}_{1-\alpha}(\tau)$ with the emergence of an arrow of "neotime". (iii) *The structure of the closure* will be based on the chain $\Xi_1(-) \rightarrow \Xi_\alpha(s) \rightarrow \Xi_\alpha(1-s)$ and therefore obviously related to the functional relationship $\zeta(s) \longleftrightarrow \zeta(1-s)$. This constraint implicitly involves the step $\Xi_{1-\alpha}(1-s) \rightarrow \Xi_1(-)$ for commutating the whole diagram. This brief introduction emphasizes the role of the relationship between mirror structures and categories. Let us analyze this issue more deeply.

13.2 Dual Structures in Category Theory

13.2.1 An Outlook about Category Theory

Due to its ontological power [29] and *far beyond the set theory*, categories theory has been considered by Francis Lawvere and Alexander Grothendieck as one of the main foundations of pure mathematics. Indeed this theory no longer considers solely objects (x) but also morphisms $f = \mathrm{Hom}(x, y)$, termed *arrows*, attached to these objects; $(x, \xrightarrow{f})$. This theory is based on the generalization of the notion of function. Thanks to Sauders Mac Lane and Samuel Eilenberg for the idea of studying the most general structures based on the algebras ruling the composition of morphisms. Converse to the usual approach and since the work of Noether, algebras are no more analyzed by exploring the properties from the inside but by outlining the operations, preserving the overall structure of the diagrams of arrows. In this framework the existence of arrows imposes the notion of reversible duality as a substitute of the usual notion of physical equilibrium. Indeed any mathematical object defined by means of arrows has certainly its *domain* of definition (the empennage of the arrow) but also its *co-domain* (its targets; its butt, aim, and meaning). The domain and the co-domain being able to have a kind of reversible similarity, so that such an "identity" —or quasi-identity based on differences—fulfills a specific role in the theory. Identity is partly defined through mirror reflections able to be "distorted" (it is the role of adjunction to do that). For instance the question of coarse graining is at the heart of the issue determined by ordering relationships (often reduced to inclusion). The arrow being a non-symmetrical operator (similar to physical relation in space or time or both) the modalities of compositions (namely the algebra) must be exactly defined: (i) as a simple 2D composition $(\xrightarrow{f}\xrightarrow{g}\approx\xrightarrow{g\circ f})$; (ii) using associaticity which introduces tetrahedral relations, namely 3D structures $(f \circ (g \circ h) = (f \circ g) \circ h)$. In this algebra the *identity-arrow* I_d appears as the neutral element. In this framework any indexation of successive steps builds a chain in accordance with an arithmetic related to the monoid $(\mathbb{N},+)$(implicit axiom of the infinity via a conoyau). The notion of functor emerges from these principles when *categories* itself become *objects* associated to arrows. A functor is an enriched notion of the concept of a function. A functor does not always exist. A functor, namely a morphism between morphisms, is a "transducer" ensuring the analogy between different categories and their diagrams. In particular if the functor hides any inclusion rules or division (ordered categories, exponentiation functor), then the functor will bring to light *approximations and coarse graining* (in usual analysis it is a convolution procedure using Fourier/Laplace transforms for instance). On the other hand, just like groups that can be nested in each other, if the ordering rules are conservative, self-similarity will naturally emerge from functor chains because a category of different categories is almost surely in other categories. Let us observe that the Russel's Paradox comes from the fact that, contrary to category theory, a set cannot carry any difference between domain and co-domain of relations. Conversely any intrinsic difference is associated with the definition of a functor F such that $\mathrm{Hom}(F(X),Y)$ concerns the morphisms of the sets category (among other characteristics, functor Hom can grant a status of a "distance"). One of the most important notions for thinking the categories C (or its diagrams) is the notion of disjoint sum when $\bigsqcup_{i\in\mathbb{N}} \tau_i$ is naturally equipped with arrow reflexivity. The disjoint union of a family of sets corresponds to their sum in category theory.

The reciprocal arrows and thus the "opposite" category C^{op}plays the role of a global unifier for the "stability" and universality of a category $(1_X \leq C^{op} \circ C)$ [12–14,23–25]. Then any diagram becomes a constructive self-consistent mathematical object, and the disjoint sum of arrows fixed on an object becomes its best approximation as a "quasi-set" of *presheaves.* In the above context Yoneda asserts that it is possible to define all objects living in local small categories C by the set of arrows attached to it. Thus the morphisms between objects become similar to the natural transformations between these functors without extra conditions. Yoneda's lemma is naturally related to the concept of limit, and co-limit is considered as a universal capability of *distinguo* between two objects of the category. Any reciprocal arrow is similar to a generalized inversion (just as an adjoint matrix in vectorial space is used to create an inverse operator). In particular for posets and ordered categories, any arrow must be divided, and the functorization then leads to a lattice where the algebra of subobjects responds to a Heyting algebra (if the symbol $\leq$ terms the arrow, the sum becomes a Galois connection). The associated concepts of lattice (with its fundamental notions of filter ideals and scales) then arise. In particular, in the case of indexation via integers the notion of scale can naturally involve the notion of self-similarity through a functorization implying indexation for the set $\mathbb{N}$ including its partition over prime numbers $\Im$. Then, *Riemann zeta function can be introduced as a measure in categories through the natural transform involving the well-known quadratic self-similarity of* $\mathbb{N}+\mathbb{N}=\mathbb{N}=\mathbb{N}\times\mathbb{N}$ [17,18]. The remarkable steadiness of a mathematical landscape attached to $\zeta(s)$ is updated here. Thus the theory of categories allows us to justify the universality of $\zeta(s)$ and thus on the universality of Cole and Cole transfer function $Z_\alpha(n)$ [4]; otherwise, it is easily related neither to the set theory nor to the standard complex analysis. Moreover, the logic is likely associated with the categories available as intuitionistic logics; thus categories can be implemented for facing fuzzy problems such as non-local distributions or distributors.

The concept of duality [26] is then enriched with diagrammatic properties (commutation of diagram) more subtle than the simple binary relationship (dialectic as logical duality). Every diagram has its dual (any vector space its linear forms; any Fourier space its temporal dynamics; any extensive variable its intensity in thermodynamics; in algebraic Geometry the torsion-extension pair; in mechanics the transformations of Hamilton and Lagrange via the Legendre transform if and only if the process involved is a differomorphism; in the logic of predicates the couple of statements $\exists$ and $\forall$ are a dual of each other; etc.). Any pair of dual categorical objects builds an involution via the cartesian square of internal functions. Categories build some kind of overall invariance-based internal *versus* external relationships. This is what Gilles Chatelet names "the offensive of the lateral" [26]. For instance, the lateral axis can unfold any element of a overall diagram with an upper limit over a set of "ideal" giving rise to a spectrum, which like a "ghost" moves with him, not only its image but also its virtualities (all possibles). The involution as monadic and modular functor plays a major role in category theory because this kind of functorization can be performed only once. Indeed $T^2 = T = A \circ B$ gives rise to the emergence of universal invariants that why the *T-monad* based on $(T, m, e)_X$ ($\mu_X : TTX \to TX$ and $e_X : X \to TX$) as well as the *Kleisli Algebra* [23–25] is important in self-similar like analysis. The duplication induced by involution sheds light to monadic processes which found the fractal structure construction step after step; hence, the implicit link between the categorical issues and the fractal geometries is especially analyzed by Tom Leinster [12,13]. If due to its arrows, a category is able to give rise to another one, the unfolded dualities mimic a physical action embedded in a space-time which is no longer homogeneous and symmetric. This space-time involves a temporal irreversibility emerging from fogginess, uncertainty, non-decidability, non-separability, incompleteness, and from existence of remnants after set-partition. In this framework, the aim of physics based on categories demands to go up inside the chain of dualities for reaching a point of universality (an extremal categorical mapping) that ensures the completeness of a "scientific" statement. These extremal duality is what Grothendieck has named Topoi either based on topology or on its generalization through new objects termed "sites".[2]

13.2.2 Lattices and Exponentiation: First Step to Applications in Physics

Lattice appears when two categories are combined in the framework of an algebraic structure. For instance when a functor is able to naturally adjoint an arrow and its subarrows: exponential operator, as binary relation which implement the link, emerges. This operator converges, i.e. it gives birth to a pair of extremal behaviors (upper and lower limits). Likely when a monadic structure that is equipped with an endofunctor (a functor mapping a category to itself), together with two natural transformations is concerned, it gives birth to pairs of adjoint functors, and they generalize closure operators on partially ordered sets to arbitrary categories. For instance Kleisli minimal algebra and Eilenberg More maximal algebra emerge. The coarse graining between extrema being able to be scaled from a unique law, the main features of ordered lattice are naturally the self-similarity. Originally considered by Lawvere [24] the set of natural numbers $\mathbb{N}$ can be seen as a peculiar object in small category, able to be used for indexation of scaling. Further, the monoidal distinction between $(\mathbb{N}, +)$ and $(\mathbb{N}, \times)$ allows to infer that the usual arithmetic is the study of finite sets in which the categorical aspects have been put away. In category theory, namely the theory of diagrams, the sum (namely the co-product

[2]A small category is defined from the fact that $Obj(C)$ and Hom(C) are sets and a locally small category involves: if $a, b \in C$ then Hom $-$ class[Hom(a, b)]ın sets. An enriched category over a monoidal M-category (with bi-functors and tensorial product $\otimes$) is a category where in particular Hom$(a, b) \in M$ and Hom$(a, b) \otimes$ Hom$(b, c) =$ Hom(a, c).

or the set disjoint sum) of two integers, n and $m \in \mathbb{N}$, must be depicted according to the following diagram of arrows $n \to n + m \leftarrow m$. Applied with arrows by changing the direction the monomorphisms$f \circ g = f \circ h \Rightarrow g = h$ are as injections. Likewise, the multiplication in $\mathbb{N}$ is the cartesian product and the multiplication of two integers n and m must be depicted according to the diagram $n \leftarrow n \times m \to m$. Applied with arrows the epimorphisms $g \circ f = h \circ f \Rightarrow g = h$ involve surjections. In this framework, the construction of multiplicative operator by the iteration of additive operators requires a subsidiary partition implementation: namely *the transposition in the scope of categories of the usual set-exponentiation.* In set theory the matching between addition and multiplication is given by a monotonous exponential function through the set $A^B : A \to \Upsilon(B)$ of all subsets of B. In category theory exponentiation requires a diagrammatic representation based on the following generalization $\xleftarrow[\times]{b^a \times a \to b =_+ f(a) \leftarrow C \times a}$. Therefore the definition of a categorical exponential requires a stage of adjunction. This diagrammatic extension overcomes the monoidal distinction $(\mathbb{N}, +)$ *versus* $(\mathbb{N}, \times)$ involving $\mathbb{N}$ within non-local analysis, namely an integral form.[3] The structure that emerges from the algebraic recursive constraints is a self-similar structure. These aspects were considered by Dana Scott [30] when this mathematician introduced, at the end of 60s, the *theory of domains* [31] namely an algebraic theory of complete and partly ordered lattices. This theory provides a suitable framework for formal computation and lambda calculus [32]. This methodology was reinterpreted and generalized by Plotkin and Smyth [33] in a note disclosed in 1982 and since then was regularly cited especially in Computer Science. Henceforth, Scott's methodology is mainly used as a canonical procedure to study partial order recursive equations in the framework of bi-categories theory [34]. The methods derived from the above approach implement the concepts of $T-$monads and *Kleisli* Algebras.

13.2.3 Monads: Categorical Foundations of Coarse Graining

According to category theory, let us consider a functor T and a $T-$algebra upon an object A. The structure of the $T-$algebra can be reduced to a given number of initial arrows: $1 \to A$ combined with an endomorphism $A \to A$ which preserves the finite product. If $f : TA \to A$ is initial in the category—i.e. provides a general framework for induction and recursion—the $T-$algebra is said to be free algebra. As highlighted by Lawvere [24], the set of natural numbers $\mathbb{N}$ is the archetypal example of an initial $T-$algebra [25,35], and it is the reason for his interest in physics. According to Lambek lemma [36], the initiality implies that f is an isomorphism. The advantage with such an isomorphism is the rise of a successor functor $T(X) = X + 1$ (the recursion is given by a predecessor), that is to say the disjoint sum of a set X with an external element: namely a shift of a unit. The adjunction implies that the $T-$co-algebra is an application characterized by an inverse arrow:$f^{-1} : A \to TA$. The dual Lambek lemma asserts that, to assure a final $T-$co-algebra, f^{-1} must be an isomorphism. This constraint is crucial. *For instance, the method that bases above heuristic approach of Riemann Hypothesis requires a final co-algebra able to assure a self-consistency of* $\mathbb{N}$ *self-similarity up to a boundary fix point given by* $T_+^\infty X = X$. The starting structure being the monoid $(N, +)$ (co-product) and the usual addition being given via $T_+(X) = X + 1$ as initial algebra, both aspects must be combined with a "co-recursion operator"

[3]Generalization of set exponentiation is a requisite to define a Grothendieck topos, which is the structure needed to define any mathematical truth. Additional elements are the existence of limits and the rle of an overall operator for classification.

(namely a partition) so that $(N, \times) : T_{\times}X = A \times X$. Nevertheless a difficulty appears immediately at this stage: a function that maps X upon $A \times X$ does not possess any final co-algebra over $\mathbb{N}$. Indeed, even if it has a fix point $T_{+}^{\infty}X = X$ at infinity, the "co-recursion" function is undefined at zero. Therefore the categorical construction of any approach based on matching dual categories over $\mathbb{N}$ requires sophisticated analysis. *It must be based on a control of the compactness of the topology* in such a way that any endofunctor may have an *initial algebra and a final co-algebra canonically isomorphic.* Only this adjunction can ensure the closure of a categorical coarse graining over a self-similar structure. This construction requiring the validity of the commutativity of diagrams leads naturally to the use of the concept of monad. As recalled above a monad [23,25] is a mathematical autonomous and self-consistent structure which is made up of (i) an endofunctor T defined upon the category of C with $T : C \to C$; (ii) a natural transformation termed unity $e : id \Rightarrow T$ that is to say a function e_X: $1_C \to T$ for any X characterized by appropriate natural characteristics; (iii) a natural transform termed product given by μ_Z: $T^2Z \to TZ$ while $f : X \to TY$, $g : Y \to TZ$, and $Tg : TY \to T^2Z$. Due to the rules of composition of arrows whose result is always the same type of arrow, one can anticipate that any $T-$functor must lead to the emergence of a categorical self-similarity. The natural unity e_X and the natural product μ_X fulfill the following cross conditions: (i) $\mu_X \circ e_{TX} = id$ and $\mu_X \circ Te_{TX} = id$ and (ii) $\mu_X \circ T\mu_X = \mu_X \circ \mu_{TX}$. In addition it can prove that $f : X \to TY$ may then be written as the conjunction of two adjoint functors $G : GY \to TY$ and $F : FX \to X$. Assimilated to the $Y-$branching of a tree (also to a pants) and enriched by associative properties and Kleisli algebra, the monad plays the role of a "stem cell" for the structure where quantitative reference involves the introduction of ordered bi-category namely filters and ideals (spectral analysis).

The usual way to implement a recursive computation is to consider a power endofunctor Υ associated to an exponential operator: $T \cong \Upsilon$. Indeed due to the product relation μ_X : $\Upsilon \circ \Upsilon Z \to \Upsilon Z$ and through the transformation$((x, y), (z)) \to (x, y, z)$; the monad-algebra fits a branching mechanism and scaling in which the product μ merges two consecutive branches. Therefore the monad based upon this power functor Υ involves explicitly the tree structure associated to a coarse graining (from categories approximations and divisions involve universal limit and co-limit). The unit is given by the singleton. The multiplication mimics the set-union operator $\cup$. Accordingly the monad bases the models of recursive operations occurring when a certain object is defined by starting itself. Recursivity is extensively used, for instance, in Computer Sciences, Language Theory, and Fractal Geometry. Let us now specify its relationship with the Kleisli minimal algebra. The reasoning that we would like to develop borrows its main elements to the works carried out by Hasuo, Jacobs, and Sokolova [37]. Given a Υ-monad with morphism Υ upon the objects of the category C, its *Kleisly category* C_Υ is defined upon a $\Upsilon-$algebra: The C_Υ objects are the object of the $\Upsilon-$monad. An arrow $X \to Y$ in C_Υ is mapped over the arrow $X \to \Upsilon Y$ in C. Units and composition of arrows are then matched through the unit e and the multiplication μ of the $C-$monad with Υ. The composition of the arrows in C_Υ is obtained in the following manner: $g \circ f : X \to Y \to Z$ in C_Υ is equivalent to $\mu_Z \Upsilon_{g \circ f:} : X \to \Upsilon Y \to \Upsilon Z$. These operations involve a branching expansion associated to a scaling reduction (for ensuring convergence). Furthermore, a canonical adjunction between sets and C_Υ may be considered at this step: $J : Sets \rightleftharpoons C_\Upsilon$, K^J is such that J transforms$f : X \to \Upsilon Y$ in set onto $e_Y \circ f : X \to Y$ in C_Υ. The mapping $X \to \Upsilon Y$ must then be understood as a scaling process over *sets* (zooming or change of the resolution) because according to the definition of a power functor Υ, a member $f(x) \in Y$ is substituted by its mapping upon a subcategory (then over subset). This operation represents a thickening of the image of $f(x)$. *Therefore, Kleisli categories based on power functor* Υ *are algebraic categories in which, for monadic reasons, the branching phenomenon, and therefore the tree like algebra, is intrinsically a coarse graining over sets. Therefore the algebra runs in the framework of a scaling.*[4]

Order structures and lattices: It is perfectly clear that the definition of monadic extremal algebras (Kleisli algebra is initial algebra as category C_{Υ} while Eilenberg Moore is final algebra as category C^{Υ}) are important when the applications of such concepts of mathematics must be related to coarse graining, because this coarse graining can itself be related to a mathematical lattice characterized by highest and lowest universal limits and therefore to filters and ideals on which computation could be screened. As previously noted, the main interest of any use of the monadic algebras as categories lies in the conditions of fulfillment of matching between initial algebra and final co-algebra. The main tool to preserve the pulling back is the status of the structure of order in play. Due to the use of the power functor Υ, we can prove that $\Upsilon-$monads always ensure this commutation.

At this stage the main issue is to know how the scaling properties, power laws, and associated convolution operators $f(ac) = \int^{b\in B} \varphi(a,b)\psi(bc)$ can emerge from monadic algebras involving $(\mathbb{N},+)$ and $(\mathbb{N},\div)$, namely what happens when division of a functor involves the multiplication of paths between a pair of objects. These facts naturally drive chaotic phenomena whose self-similar features can be disputed. The concept of Kan extension will help us to understand this complex issue.[5]

13.2.4 Kan Extension and Functorial Division

As usual in category theory we must introduce duality based on the distinguo between the target and the empennage of the arrows. There is a major difference between left and right (up and down for a loop). The Working Mathematician, Mac Lane says "The notion of Kan Extensions subsumes all the other fundamental concepts of category theory". The concept of Kan extension is another way for outlooking the functorial divisions required by partitions of functors $(f,\div)$ and their limits. *Actually the concept of extension carries partition, approximation and/or concentration of information within an accurate and relevant mathematical form for any complex structure characterized by a set of internal morphisms. The notions of adjunction, extensions of Kan, and limit & co-limit are then mutually exchanged.* In few words Ran, namely the Kan extension at right, *is the functorial image of the set of all possible morphisms (paths)* previously scattered over an annex category $\mathcal{L}$ and Lan the Kan extension at left *is the set of images of the possible previous*

---

[4]Given a functor F: Set$\rightarrow$ Set, if (and only if) a distributive law exists with λ: $FT \rightarrow TF$, it is possible to lift F into the Kleisli category C_T for obtaining a new functor $F : C_T \rightarrow C_T$. T being, at this stage, restricted to the sole power functor Υ, $C_T = Rel$, which is the category of set binary relations Hom. If F:Set$\rightarrow$ Set is a functor which preserves the pullbacks, the distributive law exists λ: $F\Upsilon \Rightarrow \Upsilon F$ and is given by equation $\lambda_X(u) = [v \in FX|(u;v) \in \mathrm{Rel}_F(\in_X)]$ where $u \in F\Upsilon X$ and $\mathrm{Rel}_F(\in_X \subseteq FX \times F\Upsilon X)$ is the lift of the membership $\in_X$. For an arrow $R : X \rightarrow Y$ in C_{Υ} that is to say a binary relation between X and Y, the functor F : Rel $\rightarrow$ Rel transforms it into a lift $\mathrm{Rel}_F(R)$, $FR = \mathrm{Rel}_F : FX \rightarrow FY$ in C_{Υ}.

[5]The identification between the initial and the final co-algebra in Kleisli category was obtained in the peculiar case in which the monad taken into account is based upon the set power functorΥ . In the framework of this hypothesis the canonical duality is easy to be considered. Nevertheless, it is also interesting to generalize the categorical analysis by taking into account, in accordance to the same constraints, the coincidence between limit and co-limit functors. This approach already mentioned and released in 1981 by Smyth and Plotkin [33] extends the reasoning based on Scott's domain [30]. This approach was considered for making easier the formal recursive calculus required for software developments and computer applications. These developments are considered in the following references [12,38,39]. Due to the fact that zeta function is inherently involved in the construction of final co-algebra, the coincidence between limit and co-limit of chains of a couple of adjoint morphisms also ensures the categorical conditions of the validation of Bagchi Lemma, namely, the self-similarity [19]

scattered morphisms. Within distribution theory such division/integration (sum) is related to convolution. Given C a small category [namely $\mathrm{Hom}(c, c') \in$ set] and a couple of functors $F_{\pm} : C \to \mathcal{L}$ et $f_{\partial} : C \to D$ the resulting diagram let a new functor to F^* emerge, and there is an universal functor in a cone named Kan extension of f_{∂} at left with respect to $F*$ termed $\mathrm{Lan}_{F_{\pm}} f_{\partial}$ which is the adjoint functor of $ef_{\partial} \cong F*$ with a unit natural transform $\exists e : \mathrm{Lan}_{F_{\pm}}.f_{\partial} \dashv ef_{\partial}$ and if $h : \mathcal{L} \to \mathcal{D}$ then: $Lan_{F_{\pm}}.f_{\partial} \Rightarrow h$ where $\Rightarrow$ represents a natural transform over the co-cone. f_{∂} possesses a left adjoint if and only if the right Kan extension of the identity $I_d : C \to C$ along F exists and is preserved by F. Using duality: $\exists \varepsilon : \varepsilon f_{\partial} \dashv \mathrm{Ran}_{F_{\pm}}.f_{\partial}$a universal functor of the cone: $h \Rightarrow \mathrm{Ran}_{F_{\pm}}.f_{\partial}$. In addition, if C is a small category and D is co-complete we can write $\mathrm{Lan}_{F_{\pm}} f_{\partial}(l) = \mathrm{colim} f_{\partial}(a)|_{f:F(c)\to l}$ while $\mathrm{Ran}_{F_{\pm}} f_{\partial} = \lim f_{\partial}$. Due to Yoneda lemma, these expressions involve a *specific role of the convolution operator* based on set through and Hom morphism namely $\mathrm{Lan}_F f_{\partial}(i) = \sum_i \mathrm{Hom}[f(i), -] \otimes F(i)$; in other words for physicists: Lan as a co-limit may be written through a convolution equation $f_{\partial}(l) = \sum_{c \in C} f(l, c) \times F_{\pm}(c)$. All these functors could be composed, and for instance, $G(\mathrm{Lan}_F f_{\partial}) \cong \mathrm{Lan}_F G f_{\partial}$ if G out of the co-domain of f_{∂}. At the end of this short outlook we can expect that the concept of Kan extension applied to space-time leads to enlighten the link experimentally observed between the structure of a physical hyperbolic space of reaction in fractal media, the arrow of time of irreversible process, and a coarse graining in the medium. The complex scale required by the universal Riemann zeta function could then be the parameter able to merge all these relationship by taking into account the self-similarity of the integers which mimics the implementation of sums and partitions within a unique category.

13.2.5 Adjunction, Order Structures: Emergence of a Pair of Scaling Parameters

The concept of order, namely of lattice, is fundamental to introduce the notion of scaling and renormalization operator. An ordered category is characterized by the existence of only one arrow between two objects in the category. Let us analyze this notion using Figure 13.1. Given, P and Q two posets related through two morphisms ρ and λ with $P \overset{\lambda}{\to} Q$ and $Q \overset{\rho}{\to} P$, P and Q form an adjunction if and only if: $\forall p \in P$, $\forall q \in Q$; $p \leq \rho(q) \Leftrightarrow \lambda(p) \leq q$. It can

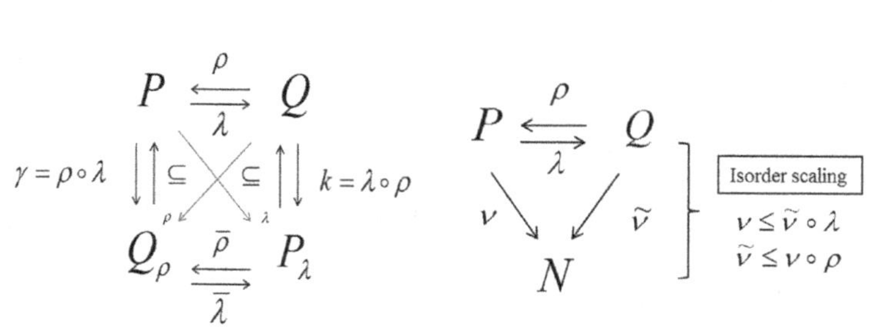

FIGURE 13.1
Schematic representation of the emergence of scaling factors from fonctorial adjunction. Panel (a) schematized the commutativity of the squared structure of adjunction based on inclusion reduced to an order morphism. In Panel (b), the previous diagram is glued on the self-similar structure of the integers $\mathbb{N} + \mathbb{N} = \mathbb{N} = \mathbb{N} \times \mathbb{N}$. The duality of $\mathbb{N}$ joined to the Cauchy completion explains the emergence of a couple of scaling parameters in both $\mathbb{R}$ and $\mathbb{C}$.

be shown that not only a set of subobjects exists but also that this set has an upper bound (closure) and a lower bound (kernel).[6]

This characteristic is due to the fact that the morphism using order (inclusion and division in sets) is also equipped with a co-product (addition). The closure is ensured through the extension of Kan at left, while the existence of the kernel is ensured by the extension of Kan at right. Moreover, the analogy of this structure with the structure of the set of integers being based on the total order of $\mathbb{N} = \mathbb{N} + \mathbb{N}$: P_λ leads that not only a set of subobjects (adjunction) exists but also that this set has an upper bound (closure). In addition a lower bound (kernel) exists based on the partial order of division $\mathbb{N} = \mathbb{N} \times \mathbb{N}$: Q_ρ. Over $\mathbb{N}$ the fixed points on both sides of a Galois connection are called Galois closure. The scale required for the coarse graining is a function defined over a poset P such as if $p < p$' then $\nu(p) \neq \nu(p')$.[7] Indeed *there is* $\nu : P \to N$ *such as* $\nu \circ \rho$ *is a first scale of domain of coarse graining,* but there is also $\nu : Q \to N$ such as $\tilde{\nu} = \nu \circ \lambda \circ \rho$ which represents a second scale referred to as the co-domain. In fact we can take as first scale: $\tilde{\nu} : P \to P \times Q : p \mapsto (p, \lambda(p))$ but also $\tilde{\nu} : Q \to P \times P : q \mapsto (\rho(q), q)$. *This configuration of duality corresponds precisely to the situation of self-similarity which thus naturally emerges from the conjunction of a relation of order associated to an adjunction* [12,13]. Therefore and conversely to the simpler intuition, there is not only one coarse grain parameter (one scale) but a pair of scaling parameters related together through the "collage" of domain and co-domain, namely (i) a family coming from the co-product (sum) and (ii) monadic or Kleisli gluing taking into account the tree structure. Thanks to the cohesive status of the surjection over $\mathbb{N}$, this distinction gives birth to a "differential functor" while scaling. The physical models evoked above point out the non-linear difference between space and time, which is precisely based on the conjunction of both scaling parameters. Through Cauchy completion the first family of scaling can be built over the field $\mathbb{R}$ and the second family over the field $\mathbb{C}$. The last one involves as usual the rle of a phase $[A, \varphi]$, and the polar representation splits the fundamental variables which are able to distinguish what we call "space" and "time"

---

[6]This property is equivalent to the following: ρ and λ are isotone applications (namely preserving the order) with: $p \leq \rho(\lambda(p))$ and $\lambda(\rho(q)) \leq q$. It is then useful to introduce the following concepts: (i) *Principal Ideal* $(p] = \{x \in P : x \leq p\}$ as a set of so-called "small" elements, (ii) *Principal Filter* $[p) = \{x \in P : p \leq x\}$ as a set of so-called "big" elements. We then obtain the following characterizations: $P \overset{\lambda}{\to} Q$ has an adjoint at right if and only if λ is *residue*, that is, if the pullback images of main ideals by λ are principal ideals. Its adjoint at right $Q \overset{\rho}{\to} P$ is then defined by: $\rho(q) = \max\{p \in P : \lambda(p) \leq q\}$. Using duality, $Q \overset{\rho}{\to} P$ has an adjoint at left if and only if ρ is *residual* , namely the pullback images of the principal filters through ρ are principal filters. Its adjoint at left $P \overset{\lambda}{\to} Q$ is therefore : $\lambda(p) = \min\{q \in Q : p \leq \rho(q)\}$. Any *residual* applications between partially ordered sets preserve all existing joints (unions); and any *residue* application between partially ordered sets preserves all existing meets (intersections). Conversely, any application preserving the joints (respectively the meets) between complete lattice is residue (respectively residual). If: $P \overset{\lambda}{\to} Q$ application between complete lattice preserves the joints: $Q \overset{\rho}{\to} P : q \mapsto \vee\{p \in P : \lambda(p) \leq q\}$ is adjoint at right for λ. If $Q \overset{\rho}{\to} P$ application between complete lattice preserves the meets $P \overset{\lambda}{\to} Q : p \mapsto \wedge p \in P : \{\lambda(p) \leq q\}$ is adjoint at left for ρ [44].

[7]For an adjunction (λ, ρ) the fix points of $\lambda \circ \rho$ are called Galois open domain because$\lambda \circ \rho$ is then a kernel operation. Consider a mere following example. Either a subset Y of a complete lattice $L : \lambda_Y : L \to L : x \mapsto \wedge\{y \in Y : x \leq y\}$ is a closure operation; for any isotone opération γ of L we write $\rho(\gamma) = \{x \in L : \gamma(x) \leq x\}$ which is the domain of closure. The pair (λ, ρ) is then a Galois connection between the power set of L, or $\Im L$ (set of all subset L) is the local ordered complete lattice for the internal isotone applications of L. Closed Galois subsets are the closing domains; closed Galois applications are closing operations. For an addition, or a Galois connection, (λ, ρ) between two posets P and Q, we have (i) $\lambda \circ \rho = id_Q$, (ii) $\lambda \circ \rho$ injective, (iii) $\lambda \circ \rho$ surjective, (iiv) ρ injective, and (iv) λ surjective. At this stage, it is useful to recall a major result established by Birkhoff: the Galois connections between the powers of sets are only based on the relationships between the set of subsets. If R is a relation between sets A and B, it gives rise to Galois connection $(R^{\to}, R^{\leftarrow})$ between $\Im A$ and $\Im B$; therefore, $X \subseteq A$ is sent on $R^{\to}(X) = X^R = \{y \in B : xRy \forall x \in X\}$ and $Y \subseteq B$ is sent on $R^{\leftarrow}(Y) = Y_R = \{x \in A : xRy \forall y \in Y\}$.

in physics. This conceptual knot not only questions the physical recursive models but also the universal Riemann function related to them. This duality between scaling domains and co-domains bases the disappearance of the linear relationship between "space" and "time" in recursive processes occurring in complex environments and, in particular, in self-similar geometries (based on polycyclic groups).

The images of the pair of applications lead then to dually isomorphic systems of closure. In return any Galois connection between power sets, i.e. any pair of dual isomorphism Φ between two systems of closure, comes from a unique relation R defined on subsets so that the equation xRy means that y belongs to the image of the closure of $\{x\}$ via the isomorphism Φ. Thus an adjunction has a close link with the closure and the kernel, which via Cauchy completion can map the topological notions of closed and open domains. *In return, we can associate this adjunction scheme to topology.* These relationships can be used as a starting point for the theory of *local* and *frame* [40,41]. The concept of ideal, whose origin comes from algebra and goes back to Gauss with the optimal decomposition of integers n into products of primes $n = \sum_{i\in\wp} p_i^{r_i}$, leads for its part the notion of *localization,* proved to allow the introduction of the differential calculus, for instance in the way followed in the development of the algebraic geometry. *When objects are replaced by morphisms, the above dual concepts are then enriched through the concepts of ultrafilter and monad of cod-ensity. The concept of probability measures and also the introduction of integral calculus can be based on these scaling notions, which require obviously the taking into account of boundary conditions.* The below analysis will focus on the particular case that corresponds to a situation of coarse graining in the vicinity of a fixed point ($T = \lambda \circ \rho$ and $TX = X$). Due to their relationships with differential calculus, above notions introduces tacitly the properties of non-additive integrals [42]. This additional step is, likewise, in accordance with a generalization of differential calculus in hyperbolic environment [43,45–48]. In addition the hypothetical existence of commutativity through mapping could suggest a generalization of the concept of steady state and equilibrium while the differential localizations could then be similar to a distributed set of phases (or singular transitions). The analogy between arrow (mathematical schematic representation) and action (will, wish in engineering and physics) as well as the natural transform ensuring the links between elementary steps and tasks seem ready to be used for a schematizing complex relationship in structures appearing in engineering applications, management, sociology, or data mining [49].[8]

13.3 Physical Application: Fractional Differentiation, α−Exponential, and Arrow of Time

According to Mandelbrot and Tricot approaches of measure in self-similar geometries [57,58] the engineering model of transfer of energy on fractal structure, the TEISI model (TEISI is the acronym of Energy Transfer on Self-Similar Interface in French) initially dedicated to the optimization of Lithium Ion batteries, leads naturally to the use of non-integer differential

[8]In the framework of the functors related to *sets* F : *sets* $\rightarrow$ *sets* is said analytic [37,50–54,56] if it is the left extension of a function $f : Bij(\mathbb{N}) \rightarrow$ *sets* (the species of the functor F) in such a way that if the functor $T \cong\supset$ representation of $F(T) = \int^n T^n f(n)$, as a *coend*, namely here a Taylor type of series. T^n is the product of n copies of T. This representation can be generalized by replacing *sets* with any symmetric monoidally co-complete category. In the same context, due to exponentiation n^{-X}, we have to define the notion of virtual species "$-X''$ that by analogy with analytic functors leads $\forall X, \forall Y$: coin$Z(X) = \int^{n\in \text{setsfinit}} \text{Hom}(-X, n)\times n$ equivalent to Rel$[Z(X), Y] \simeq [-X, Y]$; therefore, the virtual species "$-X''$ is equivalent [55]to the definition of the image of $Z(X)$.

operators [59,60] a typical non-local differential operator. The aim of the TEISI model inter alia is the depiction of dynamics that, though locally exponential, are overall non-linear and able to be rescaled (renormalization) because they are related to the transfer of any type of extensities (electron, ions, momentum, entropy, etc.) across a self-similar (fractal) interface characterized by a non-integer dimension $d_f + 1$ [5,6,61,62].

To sum up, TEISI is based on a convolution between an exponential local dynamics-transferring energy with exchange $\phi(t)$ of extensities (ion, electron, photon, momentum, spin, etc.) which can also be of several nature (first order, diffusive, or α-brownian) and a self-similar interface $L_d(t)$ characterized by a fractal dimension d_f. While the fractal discretization requires a treatment directly in the Fourier space, the geometry appears as "shaping" the transfer function that merges into the entropy production both the local irreversibility related to the randomness of quantum exchanges and the irreversibility involved by the duality of the fractal coarse graining. The *space-time* relationship is—in accordance with Mandelbrot relation relating the Hausdoff content $\eta^d$and the partition variable N namely: $N\eta^d \sim 1$—extended to a model involving the complex field: namely $(i\omega\tau) \times \eta^{d_f}(\omega) \sim 1$ [5,6,61,62] with obviously $1 = e^{i\theta} \times e^{-i\theta}$ and $\theta(\text{mod}\pi) = \pi\nu_\theta$. The dynamics is then expressed through the non-integer differential form $d^\alpha U/dt^\alpha \sim U - U_{\max}$, where U is an external field (physical measure) associated with a transfer function derived from the transfer of the test function ($Z_1 = 1/[1 + i\omega\tau]$ Fourier exponential) as fractional universal type $Z_\alpha = 1/[1 + (i\omega\tau)^\alpha]$. This mathematical form involves the so-called "α−exponential" behavior observed in many Complex Systems [4,6,7,63] fitting power law dynamics, for instance $\varphi(\sigma) + \delta_t \sim (\sigma/\tau)^{1-\alpha}$, and renormalization properties. By taking into account not only the non-integer derivative but also the exponential form of the test function is not neutral. Due to Axcel theorem asserting that the overall behavior of any system must be only determined by only one monodal control [64–66] the universal behavior implies a link between the pair of arithmetic monoids $(\mathbb{N}, +)$ and $(\mathbb{N}, \times)$ involved within any dynamical measure. As proved this constraint leads to self-similar properties of the overall system and vice versa. This feature not only implies the relevance of some degraded form of additivity considered for example through Choquet integrals [42,43] but, moreover, suggests the existence of a meta-structure related to Grothendieck Topos [67] . This structure is itself related to (i) a zeta-hyperbolic variety [45,47] required for covering the dynamics whose base is given by $Z_\alpha(n)$ while (ii) a dual non-causal term named thereafter $\overleftarrow{Z}_{1-\alpha}(n)$ appears to close the Topos, namely to suppress any uncompleteness.

Also used in mechanics and automation for instance for the CRONE (robust control device characterized by non integer order) control system [7,68–70], these models unveil some unexpected links between universal fractional transfer function (mainly for theoretical reasons basing the Cole and Cole transfer function in Fourier space [4,6,59]) and Riemann zeta function [3,6,17,18]. This link is explained through connections using Artin gluing with the theory of fibered bi-categories. These connections are backed on bi-modular arithmetics related to polycyclic semi-groups [71–74] defining the bi-products of categorical projections and co-projection operations. These operations are usually required for building self-similar structures while categorical properties [12–14,33,36] link this connection with zeta Riemann function characteristics [3,6,17,18,84] and furthermore with non-integer differential operators [76–78]. In the above approach the dynamic of exchanges (electrons, ions, momenta etc) determines the coarse graining and in return the complexity, as already suggested by Ilia Prigogine [5,22] and is coarse-grained through irreversible dissipation of any physical processes. These last characteristics imply specific delays between action and reaction, delays introduced by the means of a phase constant $\varphi(d_f)$ angle, which, in the complex plane, plays a role in internal morphism while time is reversed and in implementation of scaling. Fundamentally the phase relates the limits and co-limits of endomorphisms while the Cauchy theorem concerns the residues attached to the polycyclic analytic functions. Obtained from an arc of semicircle as one of the main invariant of the dynamics, this phase appears, in the framework of the TEISI model, as related to fractal dimension, (d_f) namely to the self-similarity of the internal automorphisms of transfer function, otherwise referred to as standard exponential relaxation (semicircle in Fourier space $d_f = 1$). Due to its convergence, this exponential can be used as a test function able to unveil the geometrical complex distribution of the interface of energy exchange.[9] In the framework of the completion of algebraic geometry $\varphi(d_f)$ can be connected with a dynamic commutator termed $[Z_\alpha, \overleftarrow{Z}_{1-\alpha}]$, the non-causal second term reversing the direction of time flow as depicted in Figure 13.2. This commutator emphasizes the role of non-commutative geometries [79–82] ruling $[Z_\alpha, \overleftarrow{Z}_{1-\alpha}] \cong [\overleftarrow{Z}_{1-\alpha}, Z_\alpha e^{i\psi}]$, where $\psi = \pi - 2\varphi$ plays the role of non-commutative factor as suggested by Alain Connes for the transition from the discrete structure of $\mathbb{N}$ to the continuous structure of $\mathbb{R}$. Let us recall that a non-commutative group emerges naturally in the physics of complex media when a determinist process involves a test function convoluted with a fractal geometry (emergence of non-commutativity also considered through Cauchy convergence). Let us observe that in von neumann algebra (quantum algebra) this relation is associated with the emergence of a group related to an automorphism to a parameter of type Tomita-Takesaki [75,79,83]. The fact that the quantum time unit (arrow of time) for A. Connes and C. Rovelli must be considered as a thermal time is only the consequence of the tensor character of the physical problem when energy is assumed to be the key paradigmatic variable, namely when $(i\omega\tau) \times \eta^2(\omega) \sim 1$. The existence of quanta is the result of the lattice structure of the topos defined by this dynamic diagram. The dynamical origin of Zeta function is hidden when obtained through a fibration of symmetrical $Z_\alpha(\eta)$ along the $\theta-$stalk: $\alpha \to s = (1/2) + ik\theta$. In the general case $\alpha \neq 1/2$ the automorphism is certainly always ruled via only one variable, but the parameter must then be written in the complex field. The residual non-commutative group maps the role of the mediator played by the non-integer differential operator based on $\alpha = 1/d_f$, which not only entangles $\mathbb{N} \to \mathbb{R}$, but also according to coarse grain categories $\mathbb{N} \times \mathbb{N} \to \mathbb{R} \times \mathbb{R} \cong \mathbb{C}$, and moreover, $\mathbb{C} \to \mathbb{C}^{op} \times \mathbb{C}$ extension is used to unfold the irreversible processes over the suitable space given by the $\zeta_\alpha(s)-$variety. Furthermore and at the limits, the closure over a squared topos will introduce an "*arrow of time*" which will be neither the usual time of the physicists nor, because the energy loss its physical meaning, the "*thermal time*" of Connes–Rovelli only obtained in the frame of $\zeta_{1/2}(s)-$variety.

In spite of relevance and efficiency of TEISI and CRONE models, some scientists, P.G. de Gennes among them through mentor of one author of this note [5], have been dubious about the validity of the suggested approach. Two main reasons justified this opinion: (i) the unexpected and direct link between the non-integer dimension and the power laws (though confirmed experimentally) and (ii) the absence of involvement of the diffusion transport within the processes analyzed. In the framework of particular electrochemical devices of energy storage in which diffusive control is a kinetic limitation which must be avoided, the debate was artificial. Nevertheless sideways, some more important interrogations remained without answer, and among them (i) the impotence of Noether symmetries to account for TEISI invariants and, in return, the impossibility of such a model to be intrinsically closed. Therefore this context puts aside the invariance

[9]Let us observe that a category in which Hom is also a category is said to be enriched category. Let us also observe that due to the similarity between Cauchy formula on homomorphic functions on $\mathbb{C}$ and Yoneda's lemma [if, indeed, there is a morphism between z et c: $\mathrm{Hom}(c,z) \sim 1/(z-c)$ then$F(c)$: $f(c) \sim \int_\Gamma f(z)dz/(z-c)$], holomorphic analysis contains a surjection of the local upon the global suggesting the design of certain classes of coarse graining possibly having incomplete analytical expression $\mathrm{Hom}(c,z) \sim 1/(z-c)^\alpha$.

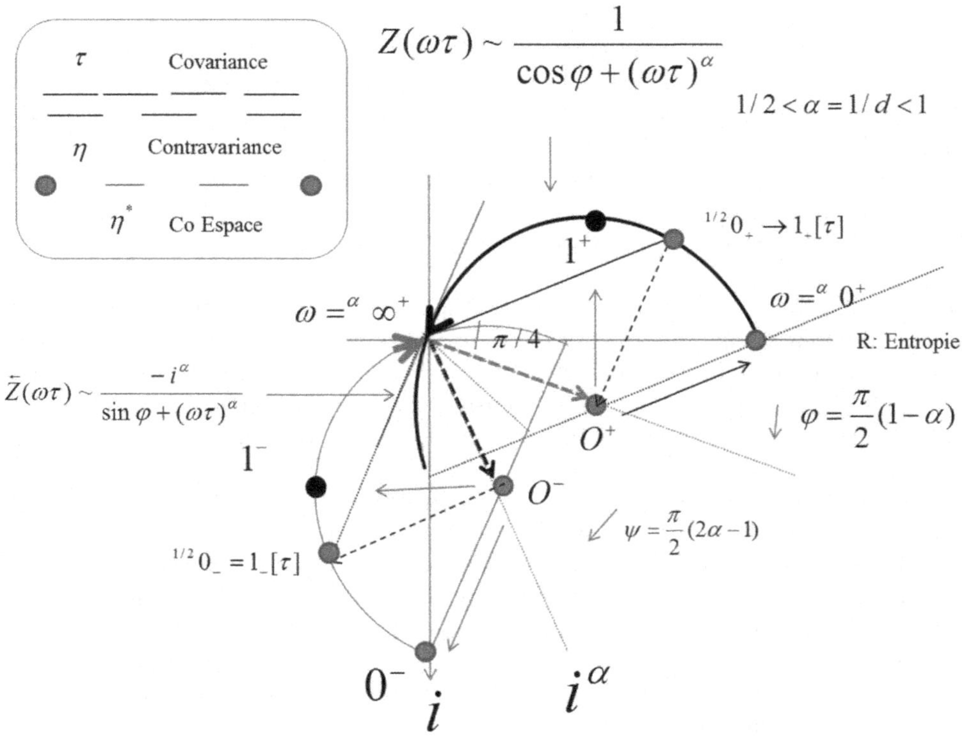

FIGURE 13.2
Tensor structure of fractal dynamics [18,75]. We first consider the complex plane as none oriented in terms of rotation characterized by a modulation locked over $\pi/2$ namely $\theta(mod\pi/2)$. So any physical process oriented will be dual to its symmetrical image, namely $i \rightleftharpoons 1$. The transfer function in the Fourier space, the basis of the TEISI model $\eta \times (i\omega\tau)^{\alpha} \sim 1$ with $1 = e^{i\theta} \times e^{-i\theta}$ is mapped in the arcs of circles. The diagram $\alpha-$exponential (upper arc of circle) has an analytical structure of Cole and Cole type, namely $Z_{\alpha}(\eta) = 1/[1 + (i\omega\tau)^{\alpha}]$ representing the fractional dynamics with $\alpha = 1/d_f$ as Cantor dimension associated with the interface (space $\sqcup$ co-space complete). n is the spatial discrete current point of the impedance diagram. n equals the hyperbolic distance to the singular point of the dynamic arc $u/v = n \in]0, \infty[$ as contravariant variable. Let us observe the internal automorphism with respect to $n = 1$. We see that the spatial unit $1^+$is given at the top of the arc representing the $\alpha-$exponential which is an image of the center of the circle O^+. This center physically points to an entropy and thus a statistical magnitude whose thermodynamical dual variable is a temperature T. At a given temperature, this variable is an energy whose dual variable is time. If we add the self-similarity, namely$\alpha = 1/d_f$ then $n \to m^{\alpha} = |\omega\tau|^{\alpha}$. The co-variant variable associated with the space is a tempo (the time constant τ). The origin of the usual time/temperature physical coupling is the entropy. The representation of the space-time tensor highlights the location of the covariant unit associated with O^+thus $\tau = 1_+$then $u/v = n \in \mathbb{Q}$. So we have $\omega^{\alpha} = u/v$. Due to complex variable involvement the key issue becomes the arithmetic of modules. It will be observed that in physical terms $\tau = 1_+$is strictly arbitrary. It is a choice of the experimenter, and therefore the only variable that counts formally along the coarse grained fractal covering is m, the dynamic variable. According to the definition of elementary fractal geometries $n, m \in \mathbb{Q}$. It is the same with $k, kn, km \in \mathbb{Q}$ because $kn = km^{\alpha} = |\omega k\sigma|^{\alpha}$. Using Hamel's bases considering $\mathbb{R}$ as a $\mathbb{Q}$-vectorial space [85] the extension $\alpha, k \in \mathbb{R}$ can be fixed. The variables $\alpha, k \in \mathbb{R}$ introduce in practice a first implicit relation between the discrete and the continuous set relation.

of the energy and (ii) led to a natural renormalization of the time factor. This result, confirmed by experimental observations, invalidated the fundamental significance of this important factor and confirmed the absence of completeness of this dynamic model. These theoretical difficulties could have been understood and solved using the theory of categories, and especially the theory of Grothendieck's topos, but the authors did not know such a theory. Fortunately, the application of non-additive Choquet measures have already been taken into account [42,43].

Recall finally that in the framework of the functions and the set theory, extensive systems (namely the additive systems) involves a concept of derivation that can always be defined and localized independently of the integration operation. In this context, traditional differential operators verify some strict algebraic conditions, the most important of them being the Leibniz rule, namely$d_t(xy) = xd_t(y) + yd_t(x)$. Conversely this relation is obviously no longer valid for non-extensive systems, in particular the rules of the formal thermodynamics backed on this relation lose most of their relevance, leaving only significant to notion of entropy, which survives in self-similar systems. As we know, this state function, namely the entropy, is in fact exclusively related to the logarithmic function and to some rules of partition (division) that is able to linearize any coarse graining process. Irreversible and non-linear processes as well as some dissymmetry of space induce both non-additive behavior and analogies with the features of ordered categories and bi-categories. Closure principles (over lattices) and commutativity of diagrams (involving extremum algebras, connections, Kan extensions, and universal limits) involve local-global coupling in which construction-partition relationships can then lead specific coarse graining and renormalization opportunities. As we know, entropy is exclusively related to the logarithmic function that here linearizes the coarse graining process. We can therefore conjecture that all fractal dynamics must almost certainly be non-additive and cannot respond to the formalization according to Riemann integrals, which among others should require a strict localization, and therefore the use of functions and not of distributions [27].

13.3.1 From Non-Additive Choquet Integrals to Non-Integer Derivatives

13.3.1.1 Set Derivative and Additive Systems

Let us consider the concept of set-function. A function that associates a real number with every subsets of X is termed $\xi : 2^X \to \mathbb{R}_+$. Due to Cauchy completion theorem, a discretization through the semi-group ($\mathbb{N}+$) can be considered. The set unit is represented by the singleton $\{x\}$ and addition by a set-union $\cup$. The stepping from $n \in \mathbb{N}$ to its successor $n+1$ involves the union covering the set having n for cardinal and the singleton. The notion of cardinal is involved by rooting n on the function ξ that complies with additivity rules of disjoint sets: namely $\xi(A \cup B) = \xi(A) + \xi(B)$ with $A \cap B = \emptyset$. A game involving these rules is said to be *based* if it has an initial categorical kernel: $\xi(\emptyset) = 0$. It is said *normalized*, if it has a final categorical horizon $\xi(1) = 1$. These constraints base the probability theory. An additive set function on the set A must then be written $\xi(A) = \sum_{x \in A} \xi(\{x\})$. Starting from this stage it is easy to think of the concept of derivation ∇ with respect to the set K as the removal of every elements contained in $K \supset A$. The different paths required for fulfilling this withdrawal impose the use of convolutions. Under some restrictions standard in set theory, it is possible to express the derivative using the additive sum: $\nabla\xi(A) = \sum_{L \subseteq K} (-1)^{|K/L|} \xi(A \setminus K) \cup L$. This expression of the set-derivative confirms two important features of this operator: (i) the central role of convolution operators and (ii) the role of parity of the number of paths: namely if its figure is even or odd. The first property relies on the organization of the set of all possible paths. This organization is built on the division of paths into subpaths, while in accordance with the Euler–Poincaré methodology, the parity is used for weighting every path to obtain invariants associated to the hyperbolic polytopes obtained as the limit of the tiling using simplexes occurring from division: $(-1)^{|K/L|}$. At this stage the Mobius transform μ can be introduced as the "inverse" set of the function ξ namely: $\mu^{\xi}(A) = \sum_{B\subseteq A}(-1)^{|A\setminus B|}\xi(B)$. Thus, the set-derivative is given by $\mu^{\xi} : \nabla\xi(A) = \sum_{L\in[K,A\cup K]}\mu^{\xi}(L)$ with $\nabla(\emptyset) = \mu^{\xi}(A)$ [84]. The lack of locality and the sensitivity to initial conditions here appear through a set of boundary constraints. Above analysis is based on the existence of function controlled by the partition on the set of subsets so that a total (over standard arithmetic indexation) and partial order (over ideals) can jointly be considered. Nevertheless, as shown it will be shown concerning stochastic processes, the relevant thought of an order often requires to introduce the very fine Lebesgue point of view. Let us remind his comments about integrals computation: "using the Riemann like integrals, we organize a sum as would implement a merchant without a method, by ranking his coins and notes at random according to the moment in which they arrive in his hands; but we also can implement the ranking like a methodical trader who classes his coins and notes in accordance to their fiduciary values." Let us note that Lebesgue's integral preserves the principle of localization associated to a function. In addition the bank notes are the functor of mutually convertible transactions, so far from the principle of optimal accounting involving "ideal subsets without any remain" namely mapping indexes over primes: n using primes p_i: $n = \prod_{p_i\in\Im}(p_i)^{r_i}$. The image of the decomposition of n on the partially ordered set of primes (Poset) must be considered as a universal arithmetic characterized by non-local properties and very long range correlations. Applied to currencies the involvement of a product of ideals would lead, in a finite realm, to the disappearance of an easy convertibility between bank notes (in general between any objects of the category). The notion of conversion units would disappear. Far from the algebraic Noether ideal realm and in the absence of boundless horizons, the majority of divisions involves, at least implicitly, the existence of residual state of deals involving the revocation of notion of disjoint sum. Any standard set-representation neglects at least the requisite of a meet or of a point of gluing. This constraint involves non-univocal and non-commutative representation of the reality. Any limit is naturally enriched of correlations with respect to the associated field of representation of reality. Standard representation of limits only via a set of addresses without taking into account the morphisms between these addresses (Yoneda lemma) still misses the goal of completeness. Thankfully distributions theory and categorical convolutions open a better approach for any hybrid context controlled through partial orders, notions irreducible to Lesbegue's bank notes. Choquet's capacity appears especially in the circumstances where non-locality and non-additivity lead the revocation of the concept of functions and therefore of the concept of "pointwise value". At this step the conceptual gap between additive and non-additive systems emerges as a pivotal distinguo. While a probability measure characterized by additivity can be based and normalized, the non-additive *Choquet capacity is reduced to a positive monotonous game* without final horizon (we could assert, but wrongly, that "body" cannot be defined while the "skin" cannot define a perfect "shell", and the horizon of being becomes fuzzy and the convergence at boundaries disappears). The singular properties of Choquet measures can only implement the functorial condition of inclusion over sets, namely $\xi(A) \geq \xi(B)$ with $A \subseteq B$ [42,43,84]. These properties only emphasize the monotone character of inclusion chains, namely the existence of only one arrow between a pair of sets. The non-additive sums can be considered by the means of convolution operators using some kind of weighting based on partition of morphisms (arrows) which always stay locked on their objects (as distribution of meets). In category theory the weighting comes from the fact that Hom concerns the arrows themselves, and that Hom is itself a category. The emergence of non-integer differential operators is then due to the existence of successive divisions ensuring a Global over Local monadic connection associated to this enrichment (namely via correlations). By return, the emergence of overall rules is due to the involvement of a hidden memory inside singularities but unveiled in the traces or the remains. As categories theory accounts for, the coupling between "meets and joins" gives birth to a renormalization group (namely a rule over the coarse graining [12,14]). In physical dynamics this kind of groups implies the emergence of power laws such as $f(t) \sim t^{\alpha}$ for both integrals and derivatives, as shown for instance in TEISI and CRONE dynamic models [6,7].

13.3.1.2 Toward Non-Integer Derivatives

Let us remark that up to the discovery of zeta Riemann function involvement in fractional dynamics, the Möbius function contribution was almost[53] ignored even if the geometrical extension of the impedance was seen as an obvious need, and even so without understanding the implication of the cobordism, caused by the combinatorial arithmetic and the algebraic correlations concealed in the geometry of the fractal structures. The disappearance of localization is expressed in the Choquet integral through the measure over subsets of X. The convolution is then, at least implicitly, implemented. This convolution implies $f(x) = \sum_{y \leq x} g(y)$ with $x \in (\mathbb{Q}\div, \leq)$. This formulation leads to the inverse form $g(x) = \sum_{y \leq x} \mu(y,x) \times f(y)$, where the Möbius function $\mu(y,x)$ appears defined through induction with $\mu(y,x) = 1$ if $x = y$, and $\mu(y,x) = -\sum_{y \leq t < x} \mu(t,x)$ if $y < x$; 0 if not. Let us term $B^{+}(F)$ the set of measurable positive and bounded functions F_{μ}. This set equipped with the arithmetic order is none other than a lattice; so $\forall t \in \mathbb{R}, f \in B(F_{\mu})$: $G_{\mu,f}(t) = \mu(\{x \in X : f(x) \geq t\})$ can be defined as a decumulative function. Therefore we can show that the Choquet integral of f with respect to μ is given by $\int f d\mu = \int_0^{\infty} G(t)dt$. The Choquet integral is a transposition of Riemann integral able to control any dynamic process through only one parameter (usually named "time" when the parameter is taken over $\mathbb{R}_{+}$). This formula is understood as an extension of the notion of capacity of F toward $B^{+}(F_{\mu})$ for the set of functions F_{μ}.[10]

The main issue here is the question of universal invariance (therefore the issue of "localization"); indeed $A \in F_{\mu}$ therefore 1_A is measurable while for any capacity μ: $\int 1_A d\mu = \mu(A)$. Nevertheless generally $\int (f+g)dv \neq \int f dv + \int g dv$. Fortunately it should be proved that equality is however obtained if the set function is co-monotone and especially if $f = u \circ h$ and $g = v \circ h$ constraints which characterizes the categorical limits and co-limits. So in the framework of monotonic function there is a simple link between Choquet and Riemann integrals. *This link is obviously related to the coarse graining. For instance let us refer here to a practical issue required by the TEISI model: we can consider the bi-partition of the irreversible dynamics (monotonic) $n = \omega\tau$ in $\Xi_{\alpha}(s)$ while a set derivation is implemented; $\Xi_{\alpha}(s)$ is then shared into a pair of spaces characterized by one parameter but linked together. The measure of each of these sets is given by a convolution as clearly shown by immersion into the Fourier or Laplace space* ($n = i\omega\tau$). Even if the result of a set derivative is scattered, a certain co-limit representation must exist such that the data can give birth to a monotonic distribution. Figure 13.2 shows for instance that this

---

[10]Once f is monotonous and positive the concept will be easily extended; indeed writing that f is measurable on subsets indexed to $i : \{a_1, a_2, a_3, ..a_i...\}$ by ranking the a_i in the increasing order $A_i = \{x \in X : f(x) \geq a_i\}$ with $A_1 = X$, then $\int f d\mu = \sum_{i \in [0,n]} (a_i - a_{i-1})\mu(A_i) = \sum_{i \in [0,n]} a_i(\mu(A_i) - \mu(A_{i+1}))$ with $a_0 = 0$ et $A_{n+1} = \emptyset$. If for simplifying X is a finite set and $F = 2^X$ then f is a vector over the basis $[...f_i...]$ and we can implement a permutation σ_i namely $f_{\sigma(1)} \leq f_{\sigma(2)} \leq ... \leq f_{\sigma(n)}$ in such a way that *in fine* $\int f d\mu = \sum_{i \in [0,n]} f_{\sigma(i)} \mu(\{\sigma(i)\}) = \sum_{i \in [0,n]} f_i \mu(i)$. This integral verifies the right Choquet integral properties.

distribution is $\overleftarrow{Z}_{1-\alpha}(t)$, the non-causal component of the topos which locks the fractional differential dynamics, scopes the set of time constants to the set of all possible time constants in coherence with the scaling constraints. This extension introduce new fractal motives characterized by the same dimension.

From this point of our arguing the whole theory of Choquet integrals can be easily developed. Nevertheless we will only be interested here in aspects related to the fractional derivation, i.e. to a theory of Choquet integrals on the real straight line $\mathbb{R}_+$. By using the power function for testing the fractal distribution of the sites of "reaction" (set of subsets), we get a distorted Lebesgue measure that involves the implementation of a kind of Radon–Nikodym derivatives over $[a, b] \subset \mathbb{R}_+$; namely if the simplest Lebesgue measure λ over $\mathbb{R}_+$ such as $\forall[ab] \subset \mathbb{R}_+$: $\lambda[a, b] = b - a$ by introducing distortion h: $\mathbb{R}_+ \to \mathbb{R}_+$ as a continuous morphism with $h(0) = 0$ we can build a distorted Lebesgue measure through the application $\mu = h \circ \lambda$ which is a Choquet capacity over $\mathbb{R}_+$. Then, given a function f: $\mathbb{R}_+ \to \mathbb{R}_+$ not decreasing continuously differentiable and equipped with a Choquet integral on a subdomain $[0, t]$, if $\mu([\tau, t])$ is a differentiable capacity with respect to τ and if$\forall t > 0; \mu(\{t\}) = 0$ then $\int f_{[0,t]} d\mu = \int_0^t f(\tau) \times (d\mu([\tau, t])/d\tau).d\tau$, where the integral at the right is a distorted Riemann integral. Therefore$\int f_{[0,t]} d\mu = \int_0^t (dh(t - \tau)/d\tau) \times f(\tau) d\tau = \int_0^t \mu'([\tau, t]) \times f(\tau) d\tau$ which is none other than a convolution integral. We can thus use the Laplace transform $L : t \to p$ with $p = 1 + i\omega\tau \in \mathbb{C}$ to express in a simple way above the expression namely $G(t) = \int f_{[0,t]} d\mu_h$; thus, $L(G(t)) = G(p) = pH(p)F(p)$. This formula leads to consider a function $F(t) = L^{-1}(G(p)/pH(p))$ which amounts to get the derivative of Radon–Nikodym $d_{RN}(\nu, \mu) = d\nu/d\mu$ as a measure in the classical sense. It is recalled that for two measures ν and μ over (X, F_μ) once the measure ν is continuous with respect to μ, namely $\nu \ll \mu$ then $\nu(A) = \int_A f d\mu$ with $f = d\nu/d\mu$. Although the Choquet integrals were developed for monotonic functions we can show that under weak restrictions allowing ordered $\sigma-$rearrangement on $f(i)$ and an equi-measurability namely the existence of a constant measure λ, then $f^\sim(\tau) = G_{\lambda f}^{-1}(t - \tau)$ with $f^\sim(\tau)$, where $f^\sim$ is a rearranged distribution to make it a non-decreasing function. If the function can be reversed and decreasing $G_{\lambda f}(x) = \alpha$ then $x = f^\sim(t - \tau)$ namely $G_{\lambda f^\sim}(t - \tau) = \lambda(f^\sim \geq x)$ which is nothing less than a Lebesgue measure. *In the framework of a physical appoach we can consider the measure λ like the test function of an exchange of extensity (test function of the distribution of sites of reaction) able to be convoluted with a function h serving for its part for featuring the structure of the interface of reaction (or vice versa). This is exactly the mathematical principles which based since the 70s the model TEISI (electrons and ions being exchanged) and in mechanics the CRONE models (momentum exchange)*[6,7]. *The expected contribution of α, the cantor dimension plays a role in the expression of no integer power functions while $F(t)$ was later termed $\alpha-$exponential.*

At this step, we can refer to the link between Choquet integrals and fractional integro-differentiation operators as they have been thought from Liouville to Caputo passing by Riemann [59,60,86–89]. These operators require convolution that we interpret as associated with the distorted Lebesgue integrals via Radon–Nykodim derivation, possibly linearized through Fourier (or Laplace) transformations. For example the Riemann–Liouville integral of the functiong over $[0, 1]$ is given by $I_g^\alpha = \dfrac{1}{\Gamma(\alpha)} \int_0^t \dfrac{g(\tau)}{(t - \tau)^{1-\alpha}} d\tau$ with $\alpha > 0$. But due to initial conditions and convergence issues there is no single definition of the fractional derivative termed $d^\alpha f/d^\alpha t$, while $L(t^\alpha) = \Gamma(\alpha+1)/p^{\alpha+1}$. The Riemann–Liouville fractional derivative can be written as $[d^\alpha g/d^\alpha t]_{RL} = \dfrac{1}{\Gamma(n - \alpha)} \dfrac{d^n}{dt^n} \int \dfrac{g(\tau)}{(t - \tau)^{1-\alpha+n}} d\tau$ for $n - 1 < \alpha < n$. So for any constant $a \neq 0 : [d^\alpha a/d^\alpha t]_{RL} = a/\Gamma(1 - \alpha)t^\alpha$. In other words, the derivative of a constant is no longer nil. The Grünwald–Letnikov defines an equivalent operator $[d^\alpha a/d^\alpha t]_{GL}$ via a discrete series; however, even if both definitions give the same result they suffer from lack of standardization of the initial conditions. However, these conditions are crucially required for the computation of Fourier and Laplace transformations. To correct this defect Caputo proposed different definitions, namely $[d^\alpha g/d^\alpha t]_C = \dfrac{1}{\Gamma(n-\alpha)} \int \dfrac{d^n g(\tau) d^n t}{(t-\tau)^{1-\alpha+n}} d\tau$ for $n-1 < \alpha < n$ with $d^n g/d^n t$ for $\alpha = n$. Using this definition, the fractional derivative of a constant becomes nil. It is easy to prove the derivatives of Riemann–Liouville and Caputo applied to a function such as $f^{(k)} = 0$ for $0 \leq k \leq n-1$. Now let us consider the additional practical condition $r_\alpha = t^\alpha/\alpha$ then $L(r_\alpha) = \Gamma(\alpha)/p^{\alpha+1}$. By applying the previous theorems on distorted Lebesgue measures μ_{r_α} then if $\forall g : \mathbb{R}^+ \rightarrow \mathbb{R}^+$ continuous, monotonous, and non-decreasing, it is shown that the Riemann–Liouville integral of g can be computed as a Choquet integral in relation to the distorted measure, i.e. $I_g^\alpha = \dfrac{1}{\Gamma(\alpha)} \int_0^t \dfrac{g(\tau)}{(t-\tau)^{1-\alpha}} d\tau = \dfrac{1}{\Gamma(\alpha)} \int_0^t g(\tau) d\mu_{r_\alpha}(\tau)$. For a differentiable and strictly increasing function such that $m : \mathbb{R}^+ \rightarrow \mathbb{R}^+$ with $m(0) = 0$ and for any positive number $0 \leq \alpha \leq 1$, a generator of order can be built $m^{*\alpha} = L^{-1}[p^{\alpha-1} r_\alpha^\alpha]$ with $M = L(m)$ then $G = L(g)$ and therefore if the convolution ordered generator is a temporal one, i.e. if $m = t$ then $m^{*\alpha} = t^\alpha/\Gamma(\alpha+1) = r_\alpha/\Gamma(\alpha)$. From then the Choquet integral of order α of the g function with respect to μ_m namely $\int g d\mu_{m^{*\alpha}}$ leads $L(\int g d\mu_{m^{*\alpha}}) = (pM)^\alpha G$. Thus the non-integer Choquet integral is the direct extension of any real or fractional number, $m = t$, and it is associated to a Riemann–Liouville integral. Moreover the non-integer derivative with respect to the distorted Lebesgue measure μ_m coincides with the Caputo derivative, which means that the notions of fractional derivation designed from Louville to Caputo are directly related to the concept of non-additive measurement and the associated Choquet integrals. Therefore, these notions are also tightly related to the structures of order and to the Riemann function, and thereby via duality to the Möbius function. Let us now consider the practical applications of these concepts. Now let us focus our attention on the controversial dynamical parametrization through the “time” issues.

13.3.2 Countable versus Real Representation: Hamel Basis, Cauchy Additive Functional, and Extended Autosimilarity

TEISI and CRONE models are mainly based on Hausdorff’s works leading the Mandelbrot equation: $m \times \eta^{d_f} \sim 1$ [57] and hyperbolic coarse graining issues [45–47]. For physical coherence namely the orthogonality between space *versus* time variables, Hausdorff content in Fourier expression ($\eta = 1/n; m = \omega\tau$), Mandelbrot equation is transposed in the following complex form: $(i\omega\tau) \times \eta^{d_f}(\omega) \sim 1$. Converse to the academic approaches , the engineering constraints of experimental devices are considered, namely transfer and not transportation entropy production [5,6,90,91], and therefore the absence of diffusion entropic limitation, these models assert that time and space are linked in the complex plan, through a cantor non-integer dimension $\alpha = 1/d_f$. Figure 13.2 outlines the meaning of the dimension. The models induce obviously a non-linear automorphism given by $(i\omega\tau)^k \times \eta^{kd_f}(\omega) \sim 1^k$ over the ring of the $k-$uples $k \in \mathbb{N}$. *The initial division being unique, defines and scatters the discrete differential structure over many subscales of the coarse graining measure via* η^{d_f}. The ring based on k goes step after step, with a replication of the basic dynamical structure which is locked over *the hybrid Fourier parameter* $i\omega\tau$, *representing the propagation of the signal along the interface* while, ruled over by the factor $k\alpha$, the above equation runs the following cartesian division $i\omega \leftarrow i\omega\tau \rightarrow \tau$. The vectorial distinction (propagation in scale *versus* propagation along the interface) may appear as a paradox because the propagation from a local spatial scale (η) is then linked with macroscopic experimental characteristics in a framework of a modular discrete arithmetic where (τ^α) plays the role of a modulation of space involving α the parameter of fractional metric. If time is set as the referential variable, the model must be built on the relationship defining its "fractal dimension", here Cantor dimension with $\alpha = log(kn)/log(km)$ instead of d_f while the temporal structure driven by the non-linear dynamics through a co-variance equation: $m = \omega\tau$ with $m \in \mathbb{N}$ leads the power law $\alpha = 1/d \Longrightarrow m \rightarrow \eta = m^{\alpha}$. The spatial structure based on η is for its part, driven by the contravariance relation $\eta = u/v$ which fits out the hyperbolic distance attached to the current point of the Cole and Cole universal transfer function. Both variables build a tensorial structure locking together both operators $\oplus \overset{\alpha}{\rightleftharpoons} \otimes$ via the parameter α. The closure of the *global* over the *local* which characterizes the self-similarity is then not only spatial as for fractal geometry but involves the time within a complex form; in other words the dynamic characteristics is determined by a non-linear application $\eta : k \rightarrow \tau$. The metric being itself a category, the only solution to unravel the entanglement is to unlock the knot by using a cohesive fiber (a key) allowing to build above $Z_{\alpha}(\omega\tau)$ through the categorical co-product, the space of representation which appears relevant and which will be later revealed as universal $\alpha \rightarrow s = \alpha + i\theta \leftarrow i\theta$. Taking into account that both complex parameters are nothing else than arithmetic modules the main issue which remains is the relationship between $i\omega = 2i\pi\nu_{\omega}$ the dynamical parameter and $i\theta = 2i\pi\nu_{\theta}$ the complex parameter of the metric leading the universality of the zeta function through the splitting of the scaling factor α. The passage by the fibration of the bases $Z_{\alpha}(n)$ and $\overleftarrow{Z}_{1-\alpha}(\tau)$ makes it possible to understand that the complexity of the cognitive knot, still featured by the Riemann Hypothesis, results not only from the dual self-similarity of the set of $\mathbb{N}$ namely $\mathbb{N}+\mathbb{N} \rightleftharpoons \mathbb{N} \rightleftharpoons \mathbb{N}\times\mathbb{N}$ but also of the analytic status of the zeta function which is ruled by a non-linear mesh between two distinct additive and multiplicative modular coarse graining structures with an involvement of a phase angle $(i)^{\alpha}$ in between. This analytic constraint, which questions the commutation $i \rightleftharpoons -1$ through above self-similarity applied along the complex axis namely $i \rightleftharpoons i \times i$, may easily be expressed by using the categories of infinite vector space Ξ—whose, as shown in Section 2, Riemann zeta function is the exponential measure—namely $\Xi_{\alpha}[\pm i\theta] \overset{?}{\rightleftharpoons} \Xi_{1-\alpha}[(i\omega)^{\alpha}]$. Figures 13.3 and 13.4 schematically give the steps and outlook of the above approach.

The analysis in $\mathbb{N}$ can be extended to the field $\mathbb{R}$ by means of Hamel Basis instead of Ξ joint to the use of Cauchy completion. The belonging $k \in \mathbb{N}$ trivial in the discrete structure of fractal object can nonetheless be extended to the continuous set $k \in \mathbb{R}$, an extension which sets up the major issue of the highly subtle passage from discrete field $\mathbb{N}$ to the continuous field $\mathbb{R}$ (always possible via non-commutative operator [79,80]), here considered via the Yoneda lemma through its equivalence with Cauchy Theorem [84]. Thanks to the axiom of choice used by Georg Hamel [85], it is always possible to equip this $Q-$vectorial space, and therefore $\mathbb{R}$, with a vectorial base. Unlike the subbases $\Xi_{\alpha}(s)$ —already mentioned and whose cardinal is that of the countable set $\aleph_0$, namely the cardinal of $\mathbb{Q}$—the cardinal of Hamel bases is the cardinal of the continuum $\aleph_1 = 2^{\aleph_0}$ which means that Hamel bases are built from a pair of $\mathbb{Q}$ series involving $\mathbb{R}\times\mathbb{R} \cong \mathbb{C}$. *These bases are known for being suitable to the search for all the functions in adequacy with the functional Cauchy additive equation* $f(x+y) = f(x) + f(y)$ $\forall x, y \in \mathbb{R}$, *an equation which explicitly states in additive form the self-similar scaling changes since it is obvious that* $\mathbb{Q}-$*linearity involves* $f(kx) = kf(x)$ $\forall x, k \in \mathbb{Q}$ *or* $\mathbb{R}$ [64–66,92] In other words if $i\omega = 2i\pi\nu_{\omega}$ is the module expressing the automorphism along the time, $i\theta = 2i\pi\nu_{\theta}$ must express the automorphism along the coarse graining, with both being linked not only in $\mathbb{N}$ but also over their completions on $\mathbb{R}$. This question of limit appears fundamental for taking into account the commutativity, and therefore the entropy production in physics. Moreover, the function f is, in its simplest form, a linear morphism for the $Q-$structure, while at the infinite limits a non-linear self-similar morphism is preserved as a renormalization group. It is then associated with

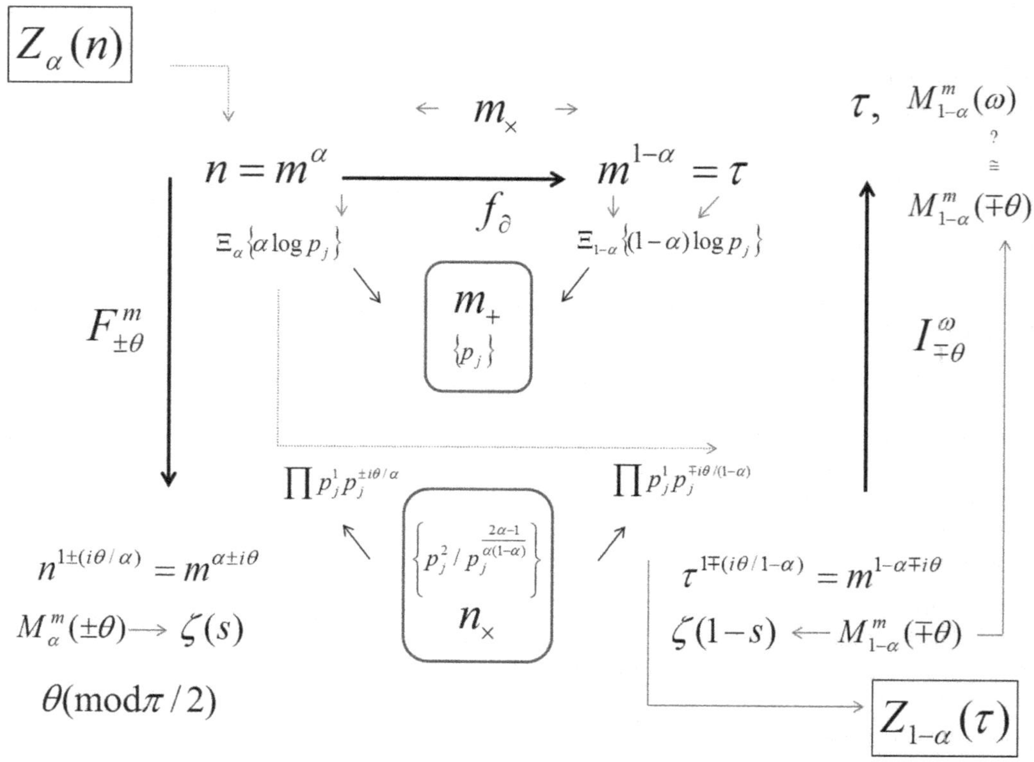

FIGURE 13.3
Details of issues related to Kan Extension: Starting from Z_α we are looking for the "space-time" relations $R(n, m^{-s})$ with $m = \omega\tau$ originally in $\mathbb{N}$. The splitting over primes and exponentiation operator are used to ensure the completeness of Z_α categories. Figure 13.2 rightly related to Figure 13.3 gives a schematic analysis of the dynamic endomorphisms associated with the decomposition of an exponential function into non-symmetric two parts (in accordance with rectangle triangle and Pythagorean theorem expression of $\mathbb{N}+\mathbb{N} = \mathbb{N} = \mathbb{N}\times\mathbb{N}$. Figure 13.3 explains in detail the structure of endomorphisms. The left part shows the category of morphisms of the α-exponential, namely the Z_α dynamics with endomorphism in which m^α the temporal part can be expressed in the infinite vectorial space $\Xi_\alpha\{\alpha\text{Log}(pi)\}$. The relation which concerns us is the distributor (profonctor): $Z_\alpha(n) \nrightarrow m^{-\alpha\mp i\theta}$in the context of dynamical coarse graining. $Z_\alpha$is the basis for fibration $F_{\pm\theta}$ along $\pm i\theta(mod\pi)$ leading the emergence of the manifold of presheaves $M_s^m(\pm\theta)$ The fibration extended the complex plane giving rise to a $\zeta(s)-$variety. Futhermore, using the functor of derivation f_∂ with respect to τ the right part of the diagram is built as the category of presheaves $M_{1-\alpha}^n(\omega)$ with a section at τ and parametrized along ω or in other more relevant respects $i\omega$. In addition the $\zeta(s)-$variety can itself be extended on the right through a functor leading $s \to 1-s$ so that a category termed $Z_{1-\alpha}(\rho)$ associated to $\zeta(1-\alpha)$ can be built as a base of a manifold of presheaves $M_{1-\alpha}^\tau(\mp\theta)$. The open issue is the status of the functor $I_{\mp\theta}^\omega$ between $M_{1-s}^\tau(\mp\theta)$ and $M_{1-\alpha}^n(\omega)$. It is the purpose of the analysis of the Kan extension of f_∂ along $F_{\pm\theta}$ to enlighten this issue. In the above framework Riemann Hypothesis, namely $\alpha = 1/2$ and $\psi = \frac{\pi}{4}(2\alpha-1)$ can be obtained from the general commutator $[\mathbb{N}^{2\alpha}\times\mathbb{N}^{2(1-\alpha)}] = [\mathbb{N}^{2(1-\alpha)}\times\mathbb{N}^{2\alpha}e^{i\psi}] = \mathbb{N} = \mathbb{N}^2$, which under other respects, expresses the mapping between product and co-product within the Pythagorean theorem (Figure 13.2). Nevertheless, this is fundamental if $\alpha \neq 1/2$, and we emphasize the fact that $\mathbb{N} \neq \mathbb{N}^{2\alpha}+\mathbb{N}^{2(1-\alpha)}$. The Hamel basis allows the extension $\mathbb{N} \to \mathbb{R}$.

a distribution of singularities in accordance with the Cauchy additive above equation. We remind that Hamel bases are required when we have to consider the set of real numbers $\mathbb{R}$ as a $Q-$vectorial space. It is the case since the hyperbolic contravariant spatial distance over the Cole and Cole transfer function is given by the ratio $\eta = u/v$ and as well as by its inverse $\eta = v/u$. This inversion involves that $\omega\tau = 1$ also plays a role implying an input of a Grothendieck–Teichmuller deformation [93]. This means that, without worrying about the discrete or continuous nature of the laws of self-similarity, it is always possible to consider the bi-partition of the subbases $\Xi_\alpha(\alpha)$ and $\Xi_{1-\alpha}(\alpha)$ in accordance with the splitting with gluing $\Xi_1(s) \cong \Xi_\alpha(s) \otimes \Xi_{1-\alpha}(s)$. This relation leads a coarse grained involution that binds these spaces and to their categorical complements; while $f : Z_\alpha \to Z_{1-\alpha}$therefore from bi-categorical approach: $f \circ - : \mathrm{Hom}[\zeta_\alpha(s), Z_\alpha] \to hom[\zeta_\alpha(s), Z_{1-\alpha}]$ are related together in the framework of the same Kan extension and schematized in the framework of a square diagram analyzed later.

13.3.3 Kan Extension and Completion of the TEISI and CRONE Models

13.3.3.1 Kan Extension and Completion of the Universal Dynamic Models through the Construction of a Topos

Let us note that Figure 13.4 accounts for the reasoning developed in the current paragraph. Consider the covering functor such as the fibration along θ performed over the basis labeled Cole and Cole transfer function $Z_\alpha(\eta) \sim 1/1 + i(1/\eta)$ with $1/\eta = n = (m)^\alpha$ and $m = \omega\tau$, with base seen as a dynamics category (Figures 13.2 and 13.3). Indeed this dynamics implements an automorphism I_n^α over the successive steps of the dynamics with objects $n, m \in \mathbb{N}$ (or $\mathbb{Q}$ or $\mathbb{R}$ after extension) as the spectral components. $Z_\alpha(\eta)$ is a category. This category can be enriched via the functor $F_{\pm\theta} : Z_\alpha \to M_\alpha^m(\pm\theta)$; the co-domain of $F_{\pm\theta}$ emerges as an universal $\zeta(s)$-manifold heads up along θ.According with the equation $\mathbb{N} = \mathbb{N} \times \mathbb{N}$ this Riemann Zeta function is an exponential measure of $\Xi_\alpha(s)$ [17,18] obtained by changing $\alpha \to s = \alpha + i\theta$ for $m, \omega, \tau \in \mathbb{N}$. Due to the properties of the functor Hom we can assert that $F_{\pm\theta} : I_n^\alpha \overset{/Z_\alpha}{\to} I_m^s$. This relation takes into account the space-time complex scaling coupling regrouped within the dynamics n and m, namely an intrinsic scaling coupling via both variables α, θ and therefore between $\Xi_\alpha(s)$ and $\Xi_1(s/\alpha)$ (Figure 13.3) as if the $\Xi-$basis should be able to be coarse grained in accordance with the intrinsic basis of the logarithms themselves: $\{\log_{f(\theta)}(p_j)\}$. Due to its incompleteness (indeed the fractional derivative depends on the boundary constraints and the energy is no more an invariant), the fractional differential equations leading $Z_\alpha(\eta)$, is epistemically questionable; however, thanks to the geometrical representation of the impedance $Z_\alpha(\eta)_\alpha$ in $\mathbb{C}$, a closure of the models can be foreseen over an exponential function. The closure is then similar to the building of a Grothendieck Topos which locks the number of degrees of freedom at least for the pair of dynamic categories considered and promotes the existence of a cohesive truthfulness without any extra choice. We remind that such a topos requires the presence of an exponential functor achieved here, as suggested by the geometry of $Z_\alpha(\eta)_\alpha$in the complex plane as an arc of semicircle, which, as shown in Figure 13.2, can be amalgamated into an overall category $Z_\alpha(\eta) \oplus \overleftarrow{Z}_{1-\alpha}(\eta)$ by using *the anti-entropic transfer function* $f[Z_\alpha(\eta)] = \overleftarrow{Z}_{1-\alpha}[g(\eta)]$ with f, g applications on $\eta \in \mathbb{Q}$. As a dynamic category $Z_\alpha(\eta)$ can implement a division over the dynamic automorphism, namely I_m^α: $Z_\alpha(m_i) \to Z_\alpha(m_j)$ in particular by using a natural transformation $f_\partial : I_m^\alpha \to I_\tau^{1-\alpha}$ based on the set derivation $\partial n/\partial\tau \sim \tau^{1-\alpha}$. This natural transform f_∂ may be chosen as the main functor to build a Kan extension along the fiber $F_{\pm\theta}$ (Figure 13.4). The localized pointwise function n over

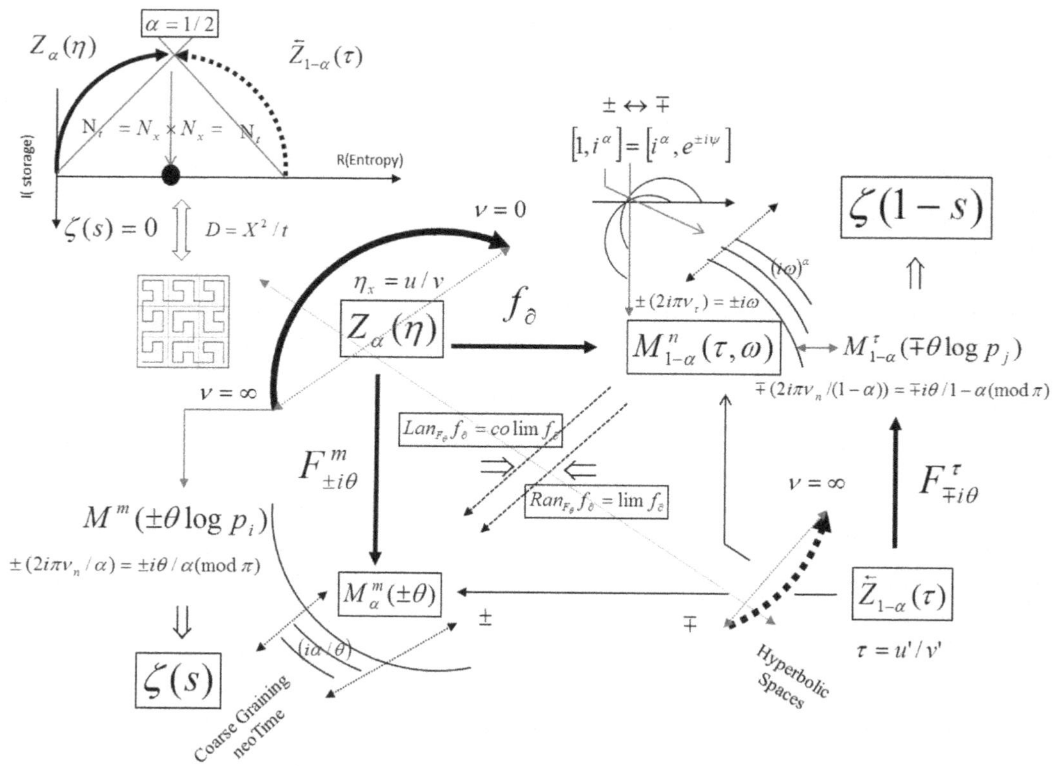

FIGURE 13.4

Schema of the fiber products and co-products: This Topos is built by scattering the components of the causal universal dynamics $Z_\alpha(n)$ along two functors, namely f_∂ and $F(\pm\theta)$. The first one, "derivative" functor, is related to "space" characteristics $m^\alpha = n = (i\omega\tau)^\alpha$ then $|n| = \frac{1}{\eta} = \frac{u}{v}$. and the second one, "integral functor", expends on "time" distribution $m = i\omega\tau$. The proceeding gives birth to two distinct manifolds, the $M^n_{1-\alpha}$ related to f_∂ and parametrized via $(i\omega)^\alpha$ and M^m_α the $\zeta(s)$-manifold related to $F(\pm\theta)$, with both varieties being located above the complex plane. The first manifold appears erected above a non-causal and negentropic virtual dynamics termed $\overleftarrow{Z}_{1-\alpha}(\tau)$ which, notwithstanding the initial scattering along the total order n, can nevertheless be ordered within a monotonic distribution such that $\tau = \frac{1}{T} = \frac{u}{v}$ in accordance with the above notation. Furthermore two distinct natural transforms, Kan extensions, adjoint to f_∂ play a part between both manifolds $M^m_{1-\alpha} \underset{ran}{\overset{lan}{\leftrightarrows}} M^n_\alpha$ and then between $\zeta(1-s)$ and $\zeta(s)$ varieties above the complex plane. The fitting of phase between both manifolds based on variables assigned either to space (via the coarse graining parameterized by using the concept of "neotime") or to time gives birth to a pair of sets scattered in the complex plane, with both being characterized by self-similar complementary metrics and an inverse sign of rotation. In the general case $Lan \neq Ran$, there is a residual group of commutation related to the inversion of the sign of rotation $\pm = \mp$. This group introduces an obstruction to the inversion of the sign of rotation involving in physics an arrow of space-time activated by $\psi = \frac{\pi}{4}(2\alpha - 1)$. The commutativity of this Cartesian , Lan = Ran, introduces the Riemann Hypothesis constraint. This commutativity locks the conditions of validity of the Riemann Hypothesis only realized when $\alpha = 1/2$, namely when, under the disappearance of the dynamical impact of the phase angle $\psi(\alpha) = 0$ (Figure 13.4), while the modulation is given by $\pi/2$ leading to a mere rotation of the axes in accordance with $\pm i^{1/2}$.

Z_α is then scattered (Figure 13.3) via the derivative f_∂ with respect to τ. The category $Z_\alpha(\eta)$ is changed into a manifold $M^{\tau}_{1-\alpha}(\omega)$ heads up along ω. It is characterized by an internal scattered morphism $I^{1-\alpha}_{\tau}$ derived from the dynamics over $Z_\alpha(\eta)$ via a fraction calculus over the morphisms. These last are based on the binary decomposition of $m = \omega\tau$ over the set of prime numbers $\tau \in \mathbb{N}$. The automorphism $I^{1-\alpha}_{m}$ is no more analytic or monotonous but singular. Nevertheless, as for $Z_\alpha(\eta)$ we can consider the projection of the order set of $\{\tau\}$ as leading to the emergence of a "virtual transfer function" $\overleftarrow{Z}_{1-\alpha}(\tau)$ built as a dual of $Z_\alpha(\eta)$ while new kind of dynamics based on the division $\{\tau = u'/v'\}$ emerges via the equation $u'/v' = t^{1-\alpha}$ where t, far from the initial $m-$indexation, matches again a strict order along a morphism I^{1-s}_{ρ}. The completion over the topos is assured through a gluing $Z_\alpha(\eta) \oplus \overleftarrow{Z}_{1-\alpha}(t)$ or disjoint co-product. This amalgamate gives birth as a splitted exponential transfer operator required for building the *Topos* we are looking for. As previously for rising zeta function $Z_\alpha(\eta) \to \zeta(s)$ the projection authorizes the building through a functor $M_\alpha(\pm\theta) \to \zeta(1-s)$ and then $\overleftarrow{Z}_{1-\alpha}(t) \to \zeta(1-s)$ in accordance with the zeta manifold $M^{n}_{1-\alpha}(\mp\theta)$. So that we create a competition between a pair of functors $F_{\mp\theta}$: $\overleftarrow{Z}_{1-\alpha}(t) \to M^{\tau}_{1-\alpha}(\mp\theta)$ and $F_\omega : \overleftarrow{Z}_{1-\alpha}(t) \to M^{n}_{1-\alpha}(\omega)$. The main issue is the relationship between both manifolds. As shown schematically in Figure 13.4 the Kan extension gives the key for this issue.

At this stage the *Topos* we were looking for ensuring the completeness of the dual fractional model of $Z_\alpha(\eta)$ and $\overleftarrow{Z}_{1-\alpha}(t)$ is built on the base of an exponentiation diagram $Z_\alpha(\eta) \oplus \overleftarrow{Z}_{1-\alpha}(t)$. The related pair of manifolds $M^{m}_{\alpha}(\pm\theta)$ and $M^{n}_{1-\alpha}(\omega)$ leads the measures $\zeta(s)$ and $\zeta(1-s)$ over the pair of poset bases $\Xi(s/\alpha)$ and $\Xi(s/1-\alpha)$ characterized by a couple of disorders with respect to $n \in \mathbb{N}$. Nevertheless we can build a new order based on θ and t. Therefore we have at our disposal a pair of classification factors related together through power laws. The main issue we have to deal with is the question of the relationship between these varieties, and notably we have to compare the structures of disorder virtually condensed in a relationship between $M^{\tau}_{1-\alpha}(\mp\theta)$ and $M^{n}_{1-\alpha}(\omega)$. The Kan extensions allow us to answer this issue. $\mathrm{Ran}_{F_{\pm\theta}} f_\partial = \lim f_\partial$ and $\mathrm{Lan}_{F_{\pm\theta}} f_\partial = \mathrm{colimf}_\partial$ must unveil the structure (geometry) of the correlations. In other words, we can assert that there is a pair of universal functor which are adjoint to f_∂ and which authorize different diagrams of commutativity. *Both natural transforms discriminate (Ran) the convolution (sum) of the f_∂-images of the paths determined by $\theta-$involvement and the f_∂-images of the sum (convolution (Lan)) of the paths always determined by $\theta-$involvement.* These transformations confirm that θ as $\alpha-$extension or scaling and ω as $m-$partition of the fractal geometry plays a similar while non-symmetric role. Indeed in very general cases $\mathrm{Ran}_{F_{\pm\theta}} f_\partial \neq \mathrm{Lan}_{F_{\pm\theta}} f_\partial$. The difference involves the role of the Mobius function as a measure of disorder. $F_{\pm\theta}$scatters of the total dynamical order of $n \in Z_\alpha$ leading the image of the sum Lan while to the disorder involved by the functor f_∂ is created by the scattering due to the partition of $m = \omega\tau$, $\tau \in Z_\alpha$ over the set of primes and leading the sum of images Ran. *This difference found the difference between space and irreversible time due to the coarse graining when non-linearity is involved.*

Based on dynamical power laws this analysis which leads to the emergence of $\zeta(s)$ and $\zeta(1-s)$ as discrete universal measures built above the categories $Z_\alpha(\eta)$ and $\overleftarrow{Z}_{1-\alpha}(t)$ confirms that (i) Riemann zeta function is arithmetically self-similar (even by not any paying attention to Bagchi Lemma) and (ii) due to the degree of freedom of α, Riemann Hypothesis must be a locking of the fractal metric. Fortunately the construction of the *Topos* opens the possibility of closing the problem that is not ended yet. This closure based on the equality $\mathrm{Ran}_{F_{\pm\theta}} f_\partial = \mathrm{Lan}_{F_{\pm\theta}} f_\partial$introduces the sole additional constraint of the square diagram (Figure 13.4) able to be related to the non-trivial zeros of the zeta fonction, namely the symmetrization of the *Topos*. Indeed this constraint involves (i) the decreasing of one unit of degree of freedom leading the emergence of an overall additional correlation, (ii) the existence of morphism between integral and derivative manifolds and therefore between the pair of operators ensuring the covering of the complex plane. This situation involves a strict symmetric of the square diagram, namely the phase angle $\varphi_\alpha = \pi/4$, or $\psi = 0$ which leads to $\alpha = 1/2$. This particular formulation of the Riemann Hypothesis reveals that its foundation on the self-similarity of $\mathbb{N}$, namely $\mathbb{N} + \mathbb{N} = \mathbb{N} = \mathbb{N} \times \mathbb{N}$, is also related to the structure of isocel rectangle triangle symmetrie s as a dynamic foundation according to the TEISI model with $\eta^2 \times (i\omega\tau) \sim 1$; in other words, namely the sum of distance is in accordance with the Pythagorean theorem $\mathbb{N} = \mathbb{N} \times \mathbb{N}$ (Space) and the sum over time is additive $\mathbb{N}+\mathbb{N} = \mathbb{N}$(Time) obviously related to the stochastic processes while the geometry resulting of the coarse graining based upon this triangle is nothing else than a 2D Peano fractal structure. In the framework of this reduction, due to Taylors Young relationship between an analytic function and its derivative, this lowering step of the degree of the duality is then necessarily related to an analycity of zeta namely $\{n^{-s}\} \rightleftharpoons \{s^n\}$ and therefore a localization $\zeta(s) = 0$ of the zeta Riemann function over the spectrum of its zeros. This situation is related to a pair of "ping-pong" exchanges concerning $\mathbb{N} \rightleftharpoons \mathbb{N} \times \mathbb{N}$ extended to $\mathbb{R} \rightleftharpoons \mathbb{C}$ between the categories Z_α and $M^{\tau}_{1-\alpha}(\mp\theta)$, namely the space via a coarse grain order ($\mathbb{N}+$) while the categories $M^m_\alpha(\pm\theta)$ and $\overleftarrow{Z}_{1-\alpha}(t)$ namely the time via a partial order of the time constant namely via ($\mathbb{N}\div$). The gap between both natural transforms points out at this step the absence of commutativity and the occurrence of physical irreversibility. Moreover the commutation of the squared topos also relates to "ping-pong" exchanges $\in, \exists \rightleftharpoons \supseteq, \forall$ between the categories $M^m_\alpha(\pm\theta)$ and $M^{\tau}_{1-\alpha}(\mp\theta)$ based on $\in, \exists$, while the categories Z_α and $\overleftarrow{Z}_{1-\alpha}$ are based on $\supseteq, \forall$. All these considerations are clearly explicit in the framework of the infinite vectorial space based on prime numbers, where $\Xi_1(s)$ can be the base of a metrical partition before amalgamation, namely $\Xi_1(s) = \Xi_\alpha(s) \oplus \Xi_{1-\alpha}(s)$. This analysis and the subtleness attached to the time with respect to the space when non-linear link is concerned lead us to term the time involving irreversibility required for physics in a self-similar environment "$d_f - neotime$". When the $d_f = 2$ the "neotime" is nonetheless than the standard time used in physics when the process is controlled through stochastic processes (This last is related to the Pythagorean law extended by using natural number self-similarity $\mathbb{N} + \mathbb{N} = \mathbb{N} = \mathbb{N} \times \mathbb{N}$ to the dynamical growth of Peano curve).

13.4 Conclusions and Outlook

Schematic summary of above approach. The above representation of the links between mathematical tools which shapes our reasoning sketches the graph of the note. We can observe that, ruling the commutativity, adjunction is the key operator when posets are concerned. Starting from this axiom we have to consider the categorical extensions (Integral calculus, Kan extensions, extension of the fields of objects, non-additive & additive sums and their extensions, emergence of filters and ultrafilters, universality of measures through Zeta Riemann, etc.), and conversely the reductions (partition of processes, differential calculus, Keisli algebra, projection over ideals, locals and frame scattering role of the Möbius function, etc.). The cohesive structure of the above diagram leads to not only (i) the automorphism of the local over the global with emergence of fix point if Joyal negative species are considered but also (ii) the emergence of localization of automorphism (lim and co-limit being related together). Figure 13.5 depicts a wonderful mathematical landscape just a little bit montane.

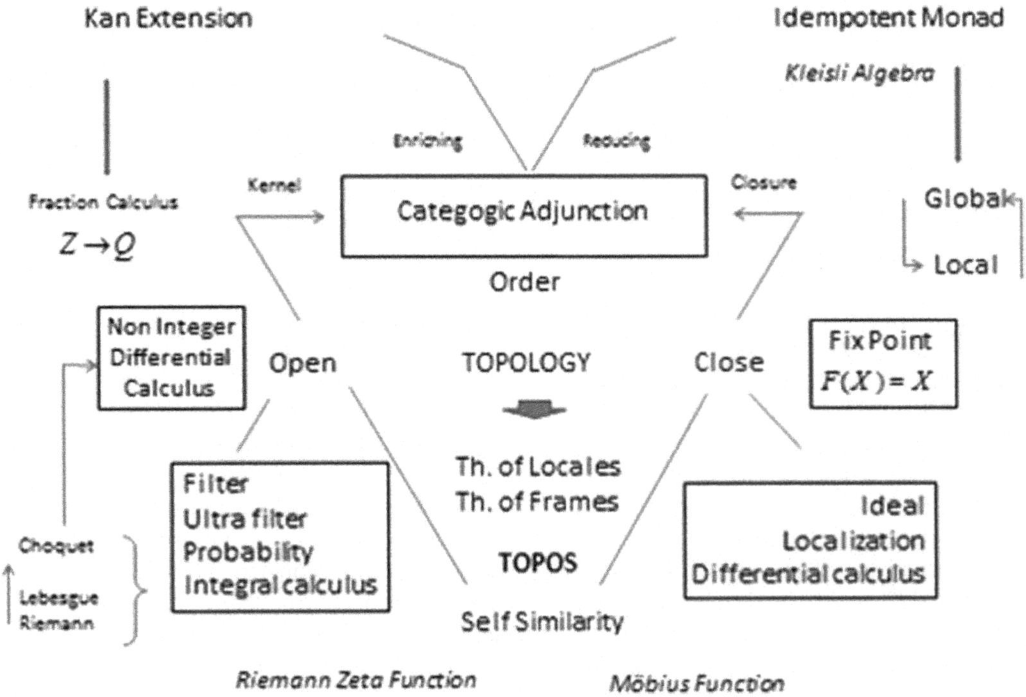

FIGURE 13.5
Schematic synthesis of the categorical approach of dynamical self-similarity.

Categorical approach of zeta function: According to Lawvere [24] Taking categories seriously means as written by Edgar Allan Poe in The Purloined Letter that *Nil sapientiae odiosius acumine nimio* [84]. We can give a synthetical overview of the above analysis. Given the Axzel theorem based on the co-product [65], the additional introduction of a cohesive product changes the initial structure into a semi-ring and therefore to a modular arithmetic. Then, via lattice structures, the fundamental concept of ideal, and especially of prime ideal, as well as the mirror concept of filter emergers (and of ultrafilter from the topology). The categorical translation of the notion of ideal is here termed module or profunctor or distributor, namely the counterpart of a function for standard distributions. There are several ways to realize a coarse graining in category theory: (i) by splitting (division) of arrows (operation measured via a Möbius function); (ii) by the involvement of the metric canonically attached to category if Hom is itself a category [24]; (iii) through the concept of subobject (equivalence class of monomorphisms and existence of a exponential functor as in Topos). In this context, the image of mapping is canonically associated with any type of arrow (exchange between objects and arrows). In category Top, invariance of scaling is obtained through the implementation of Cauchy's completion involving the role of the monad of ultrafilters. The passage implies nevertheless an indexation using the set natural numbers, as the lower monad $\mathbb{N}$, embedded in a bigger category equipped for our specific purpose. This extended approach unifies the role of order ($x \leq y$ and $y \leq z \Rightarrow x \leq z$) the role of a metric that can always be thought or designed ($d(x, y) + d(y, z) \leq d(x, z)$) either through enriched categories or in general via bi-categories, and moreover the role of topological properties of the monadic categories, namely ($\aleph \to y$ and $y \to z \Rightarrow \sum \aleph \to z$). As Jean Bénabou shows for bi-category [94,95], any category is a monadic object in a larger category and in particular into Top categories while the smallest monadic category $\mathbb{N}$, the set of integers, can be used to label any suites. In this framework the minimal representation of the algebra over the endofunctors (namely a collection of arrows), the set of relations $R(TX \to X)$, is the loop a over these arrows (Figure 13.5), while any of them can be categorically splitted and coarse grained by using the pair of monoids $(\mathbb{N},+)$ and $(\mathbb{N},\times)$. This is obtained without harming the categorical coherence by playing on the self-similar archetype of integers $\mathbb{N}+\mathbb{N}=\mathbb{N}=\mathbb{N}\times\mathbb{N}$. At this step we assert $1_X \leq a^{op} \circ a$ where X is the overall unit in the structure. Let us observe that the definitions of a and standard impedance $Z_1(n)$ are very similar. The complex plan is strictly related to an identity loop a. The zeta function $\zeta(s)$ is confirmed as the measure of exponential operator over $\mathbb{N}$ according to the partition $[\mathbb{N}\times\mathbb{N}, X]= [\mathbb{N}, X^{\mathbb{N}}]$ when expressed in the multi-scale space $\Xi_\alpha(s)$. The suite of suites being also a suites, the Cauchy convergence which involves $\mathbb{N} \to \mathbb{R}$ is treated via the sum of Kowalski ultrafilters by using a β functor. One can expand the monadic structure by extending it with a functor preserving the binary co-product. It is then possible to consider the corresponding natural terminal transformation: the monad which, according with the result of Börger [50], is a monad of ultrafilters based on the $\beta-$functor. The previous equation is adjusted in accordance with the new equation $1_{\beta X} \leq a^{op} \circ a$ where the operator β appears as related to the coarse graining or weighting functor. It means that X is compact in the Top category which is the embedding category (so that any ultrafilter converges). The convergence of Cauchy consists in considering all the weighted co-limits. Invariance by co-product then leads to the discrete matrix construction of enriched categories ($Matr$) while the presence of monadic functors (Mod) naturally use these new objects. By combination distributors emerge over enriched categories [94–97]: $Dist(\nu) = ModMatr(\nu)$. This representation is equivalent in terms of collage to the Street families of projections and co-projections [97], in one-to-one correspondence with the polycyclic semi-groups (here equivalent to a renormalization group associated with self-similar isomorphisms) at least reduced to $\mathbb{N}=\mathbb{N}\times\mathbb{N}$. As schematized in the rectangle triangle used to build a Peano curve (Figure 13.4) this relation leads the partition between space and time [98,99]. The transitivity of monadic product is given numerically through the categorical formula Hom : $2 \to [0,\infty] \to [0,\infty]\times[0,\infty] \cong \mathbb{C}$, which gives the coarse graining in accordance with Hausdorff–Gromov scaling formula: $\mathbb{N} \to \mathbb{R}$ and $\mathbb{N}=\mathbb{N}\times\mathbb{N} \to \mathbb{C}$. As Leisner shows [12,13], idempotence is universal because $\mathbb{N}+\mathbb{N}=\mathbb{N}=\mathbb{N}\times\mathbb{N}$. The monads put into play are systematically Kock–Zöberlein monad [100]. $\mathbb{N}$ is always a bi-product, especially in the Rel category of sets and fans; this means in particular that there is a family of projections P_i and a family of co-projections E_i generators of polycyclic semi-groups able to give birth, in accordance with the equations: $P \underset{Ei}{\overset{Pi}{\rightleftharpoons}} A_i$ with $P_i.E_j = 1_{Ai}$ if $i=j$ and $\perp$ if not, with $\vee P_i.E_i = 1_P$, to self-similar structures. It is exactly what is done when we address the question about the duality of bases $Z_\alpha(n) \overset{?(\alpha)}{\rightleftharpoons} Z_{1-\alpha}(n)$. Based on $\mathbb{N}+\mathbb{N}=\mathbb{N}=\mathbb{N}\times\mathbb{N}$ the bi-product achieves the formal identity between product and co-product. In terms of co-product, this structure realizes the equivalence between limit and co-limit [33], domain and co-domain, functors possibly associated with fibered structures, or between existential operator $\exists$ and universal $\forall$. Figure 13.4 introduces a schematic representation of all these results. When $\alpha = 1/2$ value required for a quantization over $\mathbb{N}$ the functional Voronin–Bagchi inequality is obtained. It is fundamentally a fixed-point equation based on the Cauchy completion over $\mathbb{N}$. The structure is a compact K in the critical band and disjoint copies of the basis are expressed along the imaginary axis. The distribution of non-trivial zeros of self-similar zeta function expresses, through the Voronin–Bagchi's functional inequality, the transitivity of the product in the monad of ultrafilters (Figure 13.6). In other words, the veracity of Riemann conjecture is based upon Cauchy convergence. Idempotence, therefore the self-similarity (or fractality), is present at all stages of the monadic approach. The physical intuition behind the non-integer derivation operators is here at the service of a most fundamental characteristic of any category immersed in a

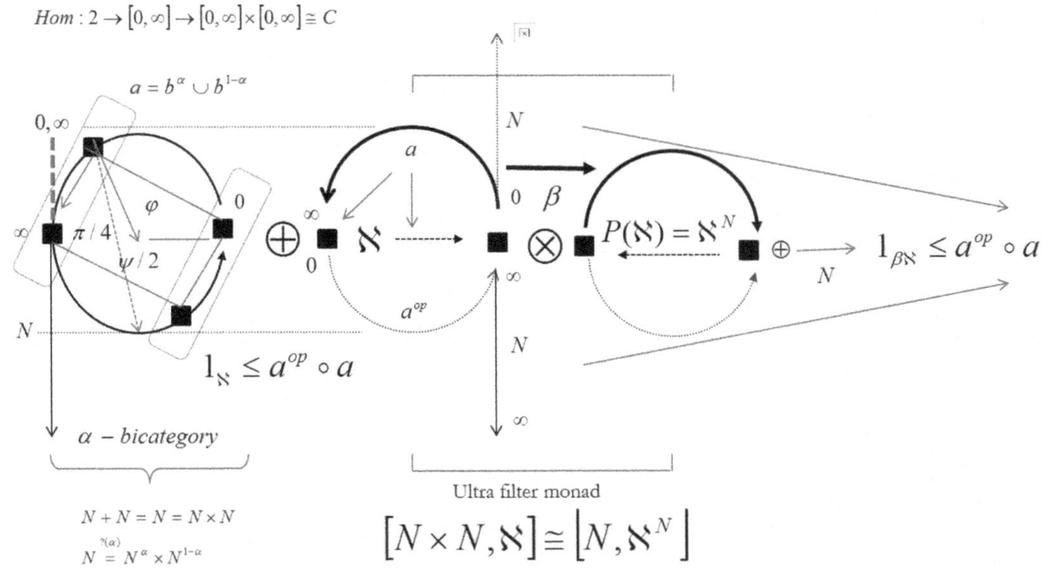

FIGURE 13.6
Schematic synthesis and representation of the equivalence between addition and multiplication schematized using half a circle (the fourier transform of a semi-line) and pointing out the completion involved by the presence of the ultrafilter. Riemann hypothesis appears as the warranty of the cohesive constraint between continuous and discrete structure without any remant attached to a space whose Cantor-Hausdorff dimension would be distinct of 1/2, namely without any external.

larger category and therefore admitting the monadic ascents (in accordance to Bénabou's works), namely a duplication by co-product. This remark explains the universality of self-similar structures and immanent recursive behaviors. In all cases we are looking for an invariant coarse graining model. The choice of the $\mathbb{N}$ for labeling all objects of suitable categories. Indeed the natural numbers is the smallest category satisfying the recurrence principle (successor or sum and recursor or partition). All intuitive constructions admit such kind of formal transcription as concepts and operations in category theory.

What about the arrow of time? The concept of neotime. The analysis developed above questions the legitimacy of the notion of time usually used, especially when dynamical processes are convoluted with self-similar geometries. It is proved in this note that Riemann conditions, which express the self-similarity of the distributions of integers and prime numbers, subject to ergodicity (namely the existence of dual averages and their formal equivalence), lead to analogies with stochastic processes. In this framework, the self-similarity of zeta function is also related with Cauchy convergence. However these characteristics also mimic the Pythagorean properties of the isocel rectangle triangle. Like any rectangle triangle, this one is obviously writable in a circle in the complex plane. It can therefore be associated with amalgamated sums of universal Cole and Cole transfer functions and therefore via a simple coarse grained construction, to Peano geometries $d_f = 2$. The categorical analysis outlined above leads to a new approach of static and dynamic properties of quantum mechanics based on $\zeta(s)$ function. The quantum states correspond to zeros. Both dual zeta functions $\zeta(s)$ and $\zeta(1-s)$ are characterized by dynamical automorphisms constrained by $\mathbb{N}$ self-similarity. The quantum transitions that are attached to this self-similarity respond to von Neumann algebras. As shown by Tomita Takesaki theorem, the self-similarity of internal automorphisms is controlled by a single real parameter. From a categorical perspective this result is paradoxical. Indeed the previous analysis suggests not only one but a pair of scaling parameters (Figure 13.6) or a complex one. The partition of this parameter enriched with the symmetries depicted in the isocel rectangle triangle synthesizes the dual self-similarity of integers: $\mathbb{N} + \mathbb{N} = \mathbb{N} = \mathbb{N} \times \mathbb{N}$.

Nevertheless, the arguments must be revisited when $1/2 < \alpha < 1$. In this case the time stated that the usual clocks of experimentalist t correspond no longer to the internal proper time or "neotime" t_p of the process embedded in fractal media. Indeed this one is the upshot of the underlying existence of the self-similar geometry. Nevertheless the pair of time parameter builds a ring leading on a modular arithmetic whose main feature is the dynamic phase related to the parameter α. This one introduced a scaling law: $t_p \sim (1/\theta)^\alpha$ which has to fit the external clock time resonances. In a physical context which forces a discrimination between dimension and co-dimension of the fractal geometry [58] and unlike the Riemannian hypothesis which confuses both zeta-bases with $\zeta(s) = 0$, the use of square bi-categories (Figure 13.4) brings out the existence of an obstruction to commutativity, namely in physics, the emergence of a hysteretic entropic residue onto the scaling boundaries (co-homology group at the edge [101]). This result justifies the experimental observation that the "neotime" t_p is related with the entropy production, namely the velocity of the process. The proper time is no longer seen as a numerical quantity but as a natural transform, and more precisely as the co-limit of a coarse grained contribution whose limit is none other than the space parameter, moreover, linearly related to the usual time of mechanics. Due to the duality of zeta bases in the framework of Pythagorean representation, the energy recovers its status of universal reference. The obstruction observed when the model would be closed through the coarse graining, namely the obstruction to a complete matching between limit and co-limit, can, in physical terms, be understood as an inversion of the rotation of the complex plane therefore as a reversing of the signature of the internal "neotime". The very definition of categorical limit and of co-limit as well, generate, given the physical meaning attributed here, an irreducible discretization of space-time. Such discretization has already been considered [102]. Now the coarse graining being the expression of the link between categorial dual concepts (domain/co-domain; lim/co-limit; Lan/Ran;$(\mathbb{N}, +)/(\mathbb{N}, \times)$; etc.) the existence of a residual at the edge as a group of co-homology raises the question of the immanence or not of the very notion of time.

In all cases, keeping in presence of the time of stochastic and deterministic mechanics, therefore the even residual meaning of the concept of energy in fractal geometry, means a pivotal role of energy at scaling boundaries. This role is expressed through the linear link between $\psi(\alpha)$ or $\varphi(\alpha)$ the phase angles attached to the self-similarity of $\zeta(s)$ and to the base of the associated dynamics. These angles are themselves attached to the functional automorphisms characterizing the zeta functions. We can thus affirm that it has no causal process without adding to it a dual non-causal contribution. It must be seen as the sum of all the possibilities given to action which chooses only one. The duality*lim* versus *colim* noted above partly disappears under the constraints of the Riemann condition without taking off the automorphism hidden in the $\zeta(s, 1 - s)$ functional relation. This is the case, for instance, in quantum mechanics when the quantum states can be associated with the non-trivial zeros of the zeta function, allowing the transitions to live freely on $\zeta-$variety (whose archetype is the set of integers) as an extension of the complex plane. The canonical reductions of the pairs of states through transitions are nevertheless controlled by the properties of automorphism intrinsic to the set of real numbers $\mathbb{N}$, i.e. a universal equivalence between succession (time aspects) and partition (spatial aspects). The trace of this equivalence class must categorically subsist at scaling boundary in the discrete dynamical space-time (as arrow) even though this distinction disappears of static terms (set of pure states). We assume herein that this categorical trace is carried in physics by the Heisenberg uncertainty principle. But it is expressed in the framework of the above approach by a metrical fluctuation of the space-time structure around the Peano value 2. Since the notion of entropy retains its physical meaning, the fluctuations are naturally associated with the existence of temperature. Therefore the time scale associated to such a fluctuation exists that Connes and Rovelli have called a thermal time [75,83]. Given that the above analysis opens wide applications in complex systems (economy, sociology, finance, etc. [49]), it remains for the mathematicians to consolidate the previous approach and to establish its cognitive limits.

Acknowledgments

We would like to acknowledge Materials Design Inc & SARL (Dr. E. Wimmer), Kazan Federal University, University of Quebec and the Franco Quebecois Institute in Paris (Dr. S. Raynal) for the support of these studies and L. Nivaven (ISMANS) and A El Kaabouchi & Alexandre Wang (ESIEA) for their scientific help.

References

[1] Ivić A. (2003) *The Riemann Zeta Function: Theory and Applications*, Dover, New York.

[2] Edwards H.M. (1974) *Riemann Zeta Function*, Academic Press, London.

[3] Le Mehaute A., El Kaabouchi A. and Nivanen A. (2010) Riemann conjecture and fractional derivative, *Comput. Math. Appl.*, 59(5), 1610–1620.

[4] Jonscher A.K. (1996) *Universal Relaxation Law*, Chelsea Dielectric Press, London.

[5] Le Méhauté A., de Guibert A., Delaye A. and Fillipi M. (1982) *Modèle de transfert d'énergie sur interfaces fractales*, C.R. Acad.Sci., Paris 294(II), 865–872.

[6] Le Méhauté A. (1990) *Les géométries fractales*, Hermes, Paris and (1991) *Fractals Geometry and Applications*, Penton Press, Londre.

[7] Oustaloup A. (2016) *Diversity and Non Integer Differentiation for System Dynamics*, ESTE Wiley, London.

[8] Wolf M. (2015) Will a physicist prove the Riemann Hypothesis? [arxiv:1410-1214 [math Ph]].

[9] Keating J.P. and Snaith N.C. (2000) Ramdom matrix theory, *Comm. Math. Phys.* 214, 57–89.

[10] Lapidus M.L. and van Frankenhuy M. (2000) *Fractal Geometry and Number Theory: Complex Dimensions of Fractal Strings and Zeros of Zeta Function*, Birhauser, Boston, MA.

[11] Steunding J. (2002) On the Universality of Riemann Zeta function, PhD Thesis, J.W. Goethe Universitat.

[12] Leinster T. (2001) A general theory of self similarity, *Adv. Math.* 226, 2935–3017.

[13] Leinster T. (2014) *Basic Category Theory*, Cambridge Studies in Advanced Mathematics, Cambridge.

[14] Hines P. (1999) A categorical theory of self similarity, *Theor. Appl. Categories* 6, 33–46.

[15] Badiou A. (2014) *Mathematics of Transcendental*, Bloomsbury Academics, London.

[16] Luna-Torres J. (2015) Sheaves in elementary mathematics. The case of positive integer numbers, [arxiv150602551 [math CT]].

[17] Riot P. and Le Méhauté A. (2017) Autosimilarité, fonction zta et conjecture de Riemann, *Revue d'Electricité et d'Electronique REE 01/ Fevrier* 17, 118–125.

[18] Le Méhauté A. and Riot P. (2017) Arrows of times, non-integer operators, self similar structures, Zeta functions, and Riemann hypothesis: A synthetic categorical approach, *J. Appl. Non Linear Dyn.* 6(2), 261–299.

[19] Bagchi G. (1982) Join universality theorem for Dirichlet L Functions, *Math. Z.* 181, 319–334.

[20] Voronin S. (1975) Theorem of universality of the Riemann Zeta function, *Izv. Akad. Nauk. SSSR* 39, 475–486; Reprint *Maths URSS Izv* 9, 443–455.

[21] Smolin L. (2013) *Time Reborn: From the Crisis in Physics to the Future of the Universe*, Houghton Mifflin Harcourt, Boston, MA.

[22] Prigogine I. et Glansdorff P. (1971) *Structure, Stabilité et Fluctuations*, Masson, Paris.

[23] Mac Lane S. (1971) *Categories for Working Mathematicians*, Springer-Verlag, New York.

[24] Lawvere F.W. and Schanuel S. (1997) *First Introduction to Categories*, Cambridge University Press, Cambridge (2nd Ed. 2009).

[25] Borceux F. (1994) *Handbook of Categorical Algebra*, Cambridge University Press, Cambridge.

[26] Chatelet G. (2016) *L'enchantement du virtuel*, C. Alunni and C. Paoletti (Eds.), Edition Rue d'Ulm, Paris.

[27] Schwartz L. (1951) *Theorie des Distributions*, Hermann, Paris.

[28] Sabbagh K. (2002) *Dr. Riemann's Zeroes*, Atlantic Books, London, pp. 208–221.

[29] Jedzejewski F. (2011) *Ontologie des Categories*, l'Harmattan, Paris.

[30] Scott D.S. (1975) *Data Types as Lattices in Proceeding of International Summer Institute and Logic Colloquium Lectures Notesin Mathematics*, Springer Verlag, New York, 499, pp. 579–651.

[31] Gierz G., Hofmann K.H., Keimel K., Lawson J.D., Mislove M. and Scott D.S. (2003) *Continuous Lattice and Domains*, Encyclopedia of Mathematics and Its Applications 93, Cambridge University Press, Cambridge.

[32] Abramsky S. and Jung A. (1994) Domain theory, in *Handbook of Logic in Computer Sciences,* S. Abramsky, D.M. Gabbay and T.S.E. Maibaum (Eds.), Clarendon Press, Oxford, pp. 1–168.

[33] Smyth M.B. and Plotkin G. (1982) The category-theoretic solution of recursive domain equations, *SIAM J. Comp.* 11/4, 761–783.

[34] Cattani G.L., Fiore M., Plotkin G. and Winstel G. (1998) A theory of recursive domains with applications to concurrency, in *13th Conference, 214-225 completed by G.L. Cattani, M. Fiore and G. Winstel, dedicated to G. Plotkin, The bicategory-theoretic solution of recursive domains equations*, GDP Festschrift ENTCS 172 (2007).

[35] Freyd P. (1991) *Algebraically Complete Categories* , Como Conference on Category Theory, LNM 1488 Springer-Verlag, New York.

[36] Lambek J. (1968) A fixed point theorem for complete categories, *Mathe. Z.* 103/2, 151–161.

[37] Hasuo I., Jacobs B. and Sokolova A. (2007) Generic trace semantics via coinduction, *Methods Comput. Sci.* 3, 246–260.

[38] May P.M.H. (1997) The algebra of self-similarity and its applications, PhD Thesis Bangor.

[39] Markl M., Shnider S. and Stashe J. (2002) Operad in Algebra, topology and physics, *Mathe. Surv. Monogr. Am. Math. Soc.* 96, 14.

[40] Zariski O. (1943) Foundations of a general theory of birational correspondences, *Trans. Am. Maths AMS* 53(3), 490–542.

[41] Lam T.Y. (2001) *First Course in Non Commutative Rings*, Graduated Texts in Maths, Springer Verlag, New York.

[42] Choquet G. (1953) Theory of capacities, *Ann. Inst. Fourier* 5, 131–295.

[43] Kojadinovic, I. (2006) Contribution l'interpretation des mesures non additives de Choquet, *HDR Laboratoire d'Informatique*, Université de Nantes 11/21/2006.

[44] Lindenhovius A.J.(2014) Grothendieck topology on posets, [arxiv14054408 [math CT]].

[45] Ghys E. and de la Harpe P. (1990) *Sur les groupes hyperboliques d'apres Mikhael Gromov*, Progress in Mathematics, Birkhauser Boston, MA, 83, pp. 1–25 (web: perso.univ-rennes1.fr/serge.cantat/Documents/ disque-poincare.pdf).

[46] Gromov M. (1988) Hyperbolic groups, in Essays in *Group Theory*, Mathematical Sciences Research Institute Springer-Verlag Publications, New York, 8, pp. 75–263.

[47] Gromov M.(2000) Spaces and questions, *GAFA* 2000; *Tel. Aviv.* 1999; *Geom. Funct. Anal.* Special Volume, Part I, 118–161.

[48] Belaiche A. (1996) The tangent space in seb riemannian geometry, in *Progress in Maths*, A. Belaiche and J.S.Risler (Eds.), Singular Trajectories and Their Control Theory, Birkhauser Verlag, Berlin, pp. 233–269.

[49] Le Méhauté A., (2016) A trail between Riemann hypothesis and the founts of currency, *Hyperion Int. J. Econophys. & New Economies*, 9(1), pp. 19–77.

[50] Borger I. (1987) Coproduct and ulfiralters, *J. Pure Appl. Algebra* 46, 35–47.

[51] Joyal A. (1986) Foncteurs analytiques et especes de structure, in *Combinatoire Enumerative*, Springer Lecture Notes in Maths 1234, pp. 126–159.

[52] Trnkova V. (1969) Some properties of set functors, *Comment Math. Univ. Carolinae* 10, 323–352.

[53] Joyal A. (1981) Une theorie combinatoire des series formelles, *Adv. Math.* 10, 42(1), 1–82.

[54] Joyal A., Street R. and Verity D. (1996) Traced monoidal categories, *Math. Proc. Camb. Phil. Soc.* 119, 447–468.

[55] Stay M. (2015) Compact closed bicategories, [Arxiv 1301.1053 [math CT]].

[56] Garcia A.M. and Milne S.C. (1981) A Rogers-Ramanujan bijection, *J. Com. Theor.(A)* 31, 289–339.

[57] Mandelbrot B. (1983) *Fractal, Form Chance and Dimension* (first edition Flammarion Paris 1975 in french), Freeman, San Franscisco.

[58] Tricot C. (2008) *Géométrie et mesure fractale*, Ellipses, Paris.

[59] Oldham K.B. and Spanier J.S. (1974) *The Fractional Calculus*, Academic Press, New York.

[60] Podlubny Y. (1999) *Fractional Differential Equations*, Academic Press, London.

[61] Le Méhauté A. and Crépy, G. (1983) Introduction to transfer and motion in fractal media, the geometry of kinetics, *Solid State Ionics* 9&10, 17–23.

[62] Le Méhauté A. (1984) Transfer process in fractal media, *J. Stat. Phys.* 36(5&6), 665–671.

[63] Le Méhauté A., Nivanen L. and Nigmatullin R. (1998), *Flèche du Temps et Géométrie Fractale*, Hermes, Paris.

[64] Aczel J. (1949) Sur les operations definies sur les nombres reels, *Bull. Soc. Maths. France* 76, 59–64.

[65] Aczel J. (2004) The associated equations revisited, in *Bayesian Inference and Maximum Entropy Method in Science and Engineering* G. Erickson and Y. Zhai (Eds.), American Institute of Physics, College Park, MD, pp. 195–203.

[66] Craigen R. and Pales Z.(1989) The associativity equation revisited, *Aequationes Math.* 37, 306–312.

[67] Ari M., Grothendieck A. and Verdier J.L. (2005) La théorie des topos et cohomologie étale des schémas, http://www.cmls.polytechnique.fr/perso/laszlo/sga4/SGA4-1/sga41.pdf.

[68] Le Méhauté A., Machado B., Sabatier J. and Trigeassou J.-C.(2005) *Fractional Dérivation and Applications, Proceedings of FDA Bordeaux Meeting*, July 2005, E Book Lisbonne.

[69] Oustaloup A. et Mathieu B. (1999) *La Commande CRONE*, Hermes, Paris.

[70] Oustaloup A., Cois O. et le Lay L. (2005) *Représentation et identification par modèles non entiers*, Hermes, Lavoisier, Paris.

[71] Wolf J.A. (1968) Growth of finitely generated solvable groups and curvature of Riemannian manifolds, *J. Diff. Geom.* 2, 421–446.

[72] Milnor J. (1968) Growth of finitely generated solvable groups, *J. Diff. Geom.* 2, 1–7.

[73] Wehrfritz B.A.F. (1973) *Three Lectures on Polycyclic Groups*, Queen Mary College, London.

[74] Segal, D. (1983) *Polycyclic Groups*, Cambridge University Press, Cambridge.

[75] Riot P. and Le Méhauté, A. (2017) From the arrow of time in Badiali's quantum approach to the dynamic meaning of Riemann hypothesis, *Condens. Matter Phys.* 20(3), 1–13.

[76] Guariglia E. and Silvestov, S. (2017) A fractional equation for Riemann zeta function fractional derivative, *INCPAA 2016 World Congress.* doi: 10.1063/1.4972655.

[77] Catani C., Guaraglia E. and Wang D. (2017) On critical strip of the Riemann zeta fractional dérivative, *Fundam. Inf.* 151, 459–472.

[78] Fernandez, A. (2018) The Lerch zeta function fractional derivative, [ArXiv:1804.0793649 [Math CV] 21 April].

[79] Connes A. (2017) *La géométrie et le quantique*, Cours du collège de France Janvier-Février, France.

[80] Connes A. (1990) *Geometrie Non Commutative*, Dunod Interédition << coll. U >>, Paris (http://www.alainconnes.org/docs/lecollege.pdf).

[81] Connes A. (2015) An essay on the Riemann hypothesis, [ArXiv 1509.05576v1, Sept].

[82] Petitot, J. (2015) Non commutative geometry and physics, [arXiv: 1505-001132, [Maths-Ph]] February.

[83] Connes, A. and Rovelli, C. (1994) Von neumann algebra automorphisms and time-thermodynamics relation in general covariant quantum theories, *Class. Quantum Grav.* 11, 2899–2918.

[84] Riot, P. (2019) Scientific, *Blog/Zetainnovation.fr.*

[85] Hamel G. (1905) Eine basis aller Zalhen und die unstetigen der funktionalalgleichung, $f(x+y) = f(x) + f(y)$, *Math. Ann.* 60(3), 459–462.

[86] Samko S., Kilbas A., Marichev O. (1993) *Fractional Integrals and Derivatives: Theory and Applications*, Gordon & Breach Science Publishers, Philadelphia.

[87] Miller K. and Ross B. (1993) *An Introduction to the Fractional Calculus and Fractional Differential Equations*, John Wiley & Sons Inc., New York.

[88] Kokilashvili V., Meskhi A., Rafeiro H., Samko, S. (2016) *Integral Operators in Non-Standard Function Spaces, Volume 1*; *Variable Exponent Lebesgue and Amalgam Spaces, Volume 2*, Birkhuser, Basel.

[89] Caputo M. (1967) Linear model of dissipation whose Z is almost frequency independent, *Geophys. J. R. Astron. Soc.* 13, 529–539.

[90] Le Méhauté A. and Dugast A. (1983) Introduction to the scaling properies in electrochemistry, *J. Power Sources* 9, 359–364.

[91] Fruchter L., Crepy G. and Le Méhauté A.(1986) Batteries Identified fractal objects, *J. Power Sources* 18(1), 51–62.

[92] Bergeron F., Labelle G. and Leroux, P. (1998) *Combinatorial Species and Tree-Like Structures*, Cambridge University Press, Cambridge.

[93] Lochak P., Leila S. and Scheiderer, C. (1997) A cohomological interpretation of the Grothendieck-Teichmüller group, *Inventiones Math.* 127(3), 571–600.

[94] Benabou J., (1968) Structures algèbriques dans les catégories, *Cahier de Topologie et Geometrie Differentielle* 10, 1–118.

[95] Benabou J. (2000) Distributors at work, https://www2.mathematik.tu-darmstadt.de/~streicher/FIBR/DiWo.pdf.

[96] Carboni A., Kasangian S. and Walters R. (1987) An axiomatics for bicategories of modules, *J. Pure Appl. Algebra*, 45, 127–141.

[97] Street R. (2004) Cauchy characterization of enriched categories, *Theor. Appl. Categories* 4, 1–16

[98] Rovelli C. (2018) *L'ordre du temps*, Flammarion, Paris and (2014) *Et si le temps n'existait pas?* [<< *Che cos'è il tempo? Che cos'è lo spazio?* >>], Dunod, Paris.

[99] Knuth K. (2015) The deeper roles of mathematics in physics laws, Arxiv.1504.06686 [Math CT].

[100] Hofmann D. and Sousa L. (2017) Aspects of algebraic algebra, http://sweet.ua.pt/dirk/artigos/2017/algebraic_revised.pdf.

[101] Malsiniotis G. (2000) La theorie de l'homotopie de Grothendieck, https://webusers.imj-prg.fr/~georges.maltsiniotis/ps/prstnew.pdf.

[102] Nottale, L. (1993) *Fractal Space-Time and Micro-Physics*, Editions World Scientific, Singapore.

14

New Contour Surfaces to the (2+1)-Dimensional Boussinesq Dynamical Equation

Haci Mehmet Baskonus
Harran University

Carlo Cattani
University of Tuscia

Armando Ciancio
University of Messina

CONTENTS

14.1 Introduction

In the last decade, many scientists have focused on the mathematical and physical models of real-world problems for a long time. Therefore, some important properties of these evaluation equations have been submitted to the literature to understand the physical phenomena of nonlinear sciences. Special forms of traveling wave solutions may depend only on single independent variables such as soliton, complex, and hyperbolic solutions. *Guy Metivier* has presented some important mathematical and physical models arising in nonlinear optics [1]. Numerous applications in many different branches of physics and applied sciences show many mathematical models [2]. One of such models is the Boussinesq dynamical equation which has of a lot of forms of the Boussinesq equation and is more complicated from the mathematical and numerical point of view [3]. Van Nghi Vu et al. have extended the Boussinesq equations in porous field [4]. They have developed a mathematical model as an *Extended Boussinesq Model* for seeking the propagation of waves in porous media. Important properties of Boussinesq such as the existence and uniqueness have been obtained by Veli B. Shakhmurov [5]. Abdul-Majid Wazwaz has first submitted the logarithmic Boussinesq equation recently, which is another new aspect of Boussinesq equation, and the logarithmic regularized Boussinesq equation in solitary waves [6]. They have investigated some new solutions for Gaussian solitary wave solutions. The Boussinesq equation is used to describe the motion of water with small amplitude and long wave [7]. Another type of Boussinesq equation is the Wick-type stochastic modified Boussinesq equation considered by S. Saha Ray and S. Singh [8]. They have applied the inverse Hermite transform to the solution. Moreover, many new powerful methods have been proposed and presented to the literature for a better understanding of such models [9–38].

The paper is organized as follows. The main steps of the method are mentioned in Section 14.2. In Section 14.3, we apply the present method to the (2+1)-dimensional Boussinesq equation for constructing complex, hyperbolic, and solitary wave solutions, defined as

$$u_{tt} - u_{xx} - \beta \left(u^2\right)_{xx} - u_{yy} - u_{xxxx} = 0, \tag{14.1}$$

where β is a real constant and nonzero [39]. Equation (14.1) is used to describe the propagation of gravity waves on the water surface [39]. Finally, we present a comprehensive conclusion of results after presenting counter surfaces of solutions under suitable values of parameters.

14.2 General Properties of MEFM

Modified Exponential Function Method (MEFM) is summarized as follows [40–42];

$$P\left(u, u_t, u_x, u_{xx}, \ldots\right) = 0, \tag{14.2}$$

where $u = u(x,t)$ is an unknown function and P is a polynomial in $u(x,t)$.

Step 1: Combining the independent variables x and t by a dependent variable ξ

$$u(x,t) = U(\xi), \quad \xi = k(x - ct), \tag{14.3}$$

$$\frac{\partial u}{\partial x} = kU'(\xi), \frac{\partial u}{\partial t} = -cU'(\xi), \\ \vdots \tag{14.4}$$

where k is the wave number and c is the velocity of the wave. Taking the traveling wave transformation Equation (14.3), it can be converted into Nonlinear Ordinary Differential equation (NODE) as follows:

$$\text{NODE}\left(U, U', U'', U''', \ldots\right) = 0, \tag{14.5}$$

where NODE is a polynomial of $U = U(\xi)$.

Step 2: We suppose the solution form of Equation (14.5) in the following form:

$$U = \frac{\sum\limits_{i=0}^{N} A_i \left[\exp\left(-\Omega(\xi)\right)\right]^i}{\sum\limits_{j=0}^{M} B_j \left[\exp\left(-\Omega(\xi)\right)\right]^j} = \frac{A_0 + A_1 \exp\left(-\Omega(\xi)\right) + \cdots + A_N \exp\left(-N\Omega(\xi)\right)}{B_0 + B_1 \exp\left(-\Omega(\xi)\right) + \cdots + B_M \exp\left(-M\Omega(\xi)\right)}, \tag{14.6}$$

in which A_i $(0 \le i \le N)$ and B_j $(0 \le j \le M)$ are real constants with $A_N \ne 0, B_M \ne 0$. Here, $\Omega = \Omega(\xi)$ satisfies the following differential:

$$\Omega' = \mu \exp(\Omega) + \exp(-\Omega) + \lambda. \tag{14.7}$$

Equation (14.7) shows the following results under several conditionals defined as

Family-1: If $\mu \neq 0, \lambda^2 - 4\mu > 0$,

$$\Omega(\xi) = \ln\left(\frac{-\sqrt{\lambda^2 - 4\mu}}{2\mu}\tanh\left(\frac{\sqrt{\lambda^2 - 4\mu}}{2}(\xi + c_1)\right) - \frac{\lambda}{2\mu}\right), \tag{14.8}$$

Family-2: When $\mu \neq 0, \lambda^2 - 4\mu < 0$,

$$\Omega(\xi) = \ln\left(\frac{\sqrt{-\lambda^2 + 4\mu}}{2\mu}\tan\left(\frac{\sqrt{-\lambda^2 + 4\mu}}{2}(\xi + c_1)\right) - \frac{\lambda}{2\mu}\right), \tag{14.9}$$

Family-3: When $\mu = 0, \lambda \neq 0$, and $\lambda^2 - 4\mu > 0$,

$$\Omega(\xi) = -\ln\left(\frac{\lambda}{\exp(\lambda(\xi + c_1)) - 1}\right), \tag{14.10}$$

Family-4: When $\mu \neq 0, \lambda \neq 0$, and $\lambda^2 - 4\mu = 0$,

$$\Omega(\xi) = \ln\left(-\frac{2\lambda(\xi + c_1) + 4}{\lambda^2(\xi + c_1)}\right), \tag{14.11}$$

Family-5: When $\mu = 0, \lambda = 0$, and $\lambda^2 - 4\mu = 0$,

$$\Omega(\xi) = \ln(\xi + c_1), \tag{14.12}$$

being $A_0, A_1, A_2, \ldots, A_N, B_0, B_1, B_2, \ldots, B_M, c_1, \lambda, \mu$ are real constants and nonzero.

Step 3: Setting Equations (14.3 and 14.4) into the Equation (14.2), afterwards, we can find the polynomial of $\exp(-\Omega(\xi))$. Considering all the coefficients of same power of $\exp(-\Omega(\xi))$ to zero gives a system. By solving this system via various computational programs, we can obtain the values of parameters. This process gives many solutions to the model considered.

14.3 Implementation of the Method

In this subsection, we have applied MEFM to find new analytical solutions to Equation (14.1).

Example 1: Let's consider the traveling wave transformation as follows:

$$u(x, y, t) = U(\xi), \quad \xi = x + y + kt, \tag{14.13}$$

where k is the wave velocity. Substituting Equation (14.13) into Equation (14.1), we obtain following NODE:

$$2\beta(U')^2 + 2\beta UU'' + (2 - k^2)U'' + U^{(4)} = 0. \tag{14.14}$$

For simplicity, we consider in Equation (14.14) as

$$2\beta(UU')' + (2 - k^2)U'' + U^{(4)} = 0. \tag{14.15}$$

Integrating this equation and considering the integral constants to zero, this equation can be rewritten as follows:

$$2\beta UU' + (2 - k^2)U' + U''' = 0. \tag{14.16}$$

By integrating Equation (14.16) once with respect to ξ, and taking the integral constants to zero again, we can find the following nonlinear equation:

$$\beta U^2 + \left(2 - k^2\right) U + U^{''} = 0. \tag{14.17}$$

When we rearrange to Equation (14.6), for balance principle between $U^{''}$ and U^2, we obtain the following relationship between M and N:

$$N = M + 2. \tag{14.18}$$

Considering different integer values of M and N, we obtain many, different and new, analytical solutions for Equation (14.1) as follows:

Case-1

If we choose $M = 1$ and $N = 3$, we can write as follows:

$$U = \frac{A_0 + A_1 \exp(-\Omega) + A_2 \exp(-2\Omega) + A_3 \exp(-3\Omega)}{B_0 + B_1 \exp(-\Omega)} = \frac{\Upsilon}{\Psi}, \tag{14.19}$$

and

$$\begin{aligned} U' &= \frac{\Upsilon'\Psi - \Psi'\Upsilon}{\Psi^2}, \\ U'' &= \frac{\Upsilon''\Psi^3 - \Psi^2\Upsilon'\Psi' - \left(\Psi''\Upsilon + \Psi'\Upsilon'\right)\Psi^2 + 2\left(\Psi'\right)^2\Upsilon\Psi}{\Psi^4}, \\ &\vdots \end{aligned} \tag{14.20}$$

where $A_3 \neq 0, B_1 \neq 0$. When we use Equations (14.19 and 14.20) in Equation (14.17), we get a system of equations including coefficients of polynomials the $\exp(-\Omega(\xi))$. Solving this system yields many coefficients.

Case-1.1: We can choose the following suitable coefficients

$$\begin{aligned} A_1 &= -\frac{6\lambda B_0}{\beta} - \frac{A_0(\beta A_2 + 6B_0)}{6\lambda B_0}; A_3 = \frac{\beta A_2 + 6B_0}{\lambda B_0}, B_1 = \frac{\beta A_2 + 6B_0}{6\lambda}, \mu = -\frac{\beta A_0}{6B_0}, \\ k &= -\frac{i\sqrt{-2\beta A_0 - 3\left(2 + \lambda^2\right) B_0}}{\sqrt{3B_0}}. \end{aligned} \tag{14.21}$$

If we put Equation (14.21) along with Equation (14.8) in Equation (14.19), we find the new complex hyperbolic function solution to Equation (14.1) under the terms of *Family-1* which are $\mu \neq 0, \lambda^2 - 4\mu > 0$, as follows:

$$u_1(x,y,t) = -\frac{3A_0\left(2\beta A_0 + 3B_0\lambda^2\right)\operatorname{sech}^2\left(c_1\kappa + \kappa x + \kappa y - i\kappa\varpi t\right)}{\left(3\lambda B_0 + B_0\sqrt{9\lambda^2 + \frac{6\beta A_0}{3B_0}}\tanh\left(c_1\kappa + \kappa x + \kappa y - i\kappa\varpi t\right)\right)^2}, \tag{14.22}$$

where $\kappa = \frac{1}{2}\sqrt{\lambda^2 + \frac{2\beta A_0}{3B_0}}, \varpi = \frac{\sqrt{-2\beta A_0 - 3(2+\lambda^2)B_0}}{\sqrt{3B_0}}$. For the validity, under the terms of *Family-1*, $\lambda^2 + \dfrac{2\beta A_0}{3B_0} > 0$ (Figure 14.1).

Case-1.2: We can consider another coefficient, being suitable one, as follows:

$$\begin{aligned} A_0 &= \frac{-6\mu B_0}{\beta}, A_1 = \frac{\beta\mu A_2 + 6\left(-\lambda^2 + \mu\right)}{\lambda\beta}, A_3 = \frac{\beta A_2 + 6B_0}{\beta\lambda}, \\ B_1 &= -\frac{\beta A_2 + 6B_0}{6\lambda}, k = -\sqrt{2 + \lambda^2 - 4\mu}, \end{aligned} \tag{14.23}$$

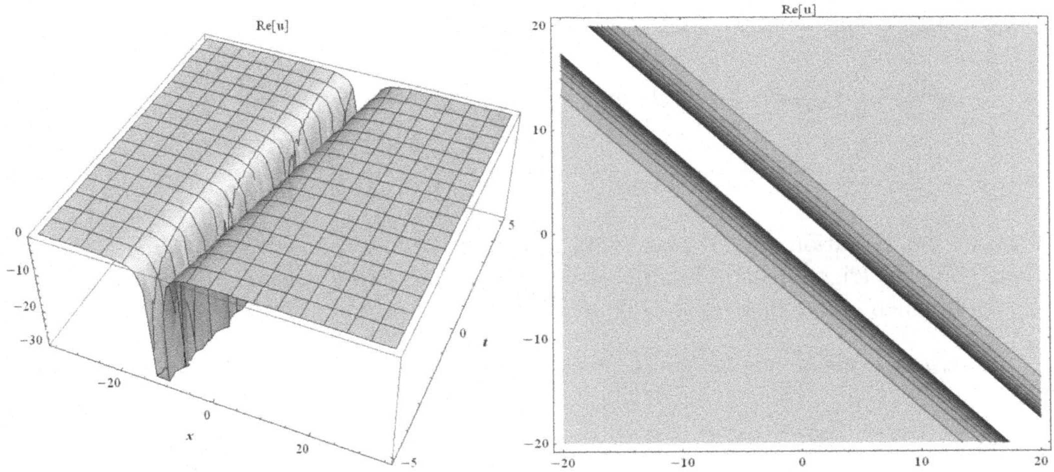

FIGURE 14.1
The 3D and contour surfaces of the real part of Equation (14.22) considering for the values of $\lambda = 0.1, \beta = 0.2, A_0 = -0.3, B_0 = -0.4, c_1 = -0.5, y = 3, -35 < x < 35, -5 < t < 5$, and $t = 0.85, -20 < y < 20$ for contour simulation.

If we put Equation (14.23) along with Equation (14.8) in Equation (14.19), we obtain a new inverse trigonometric and hyperbolic function solution to Equation (14.1) under the condition of *Family-1*, which is $\mu \neq 0, \lambda^2 - 4\mu > 0$, as follows:

$$u_2(x,y,t) = \frac{24\mu\kappa^2}{\beta}\Big(\lambda\cosh(c_1\kappa + \kappa x + \kappa y - \kappa\varpi t) + \sqrt{\lambda^2 - 4\mu}\sinh(c_1\kappa + \kappa x + \kappa y - \kappa\varpi t)\Big)^{-2}, \quad (14.24)$$

where $\varpi = \sqrt{2 + \lambda^2 - 4\mu}$, $\kappa = \frac{\sqrt{\lambda^2-4\mu}}{2}$ (Figure 14.2).

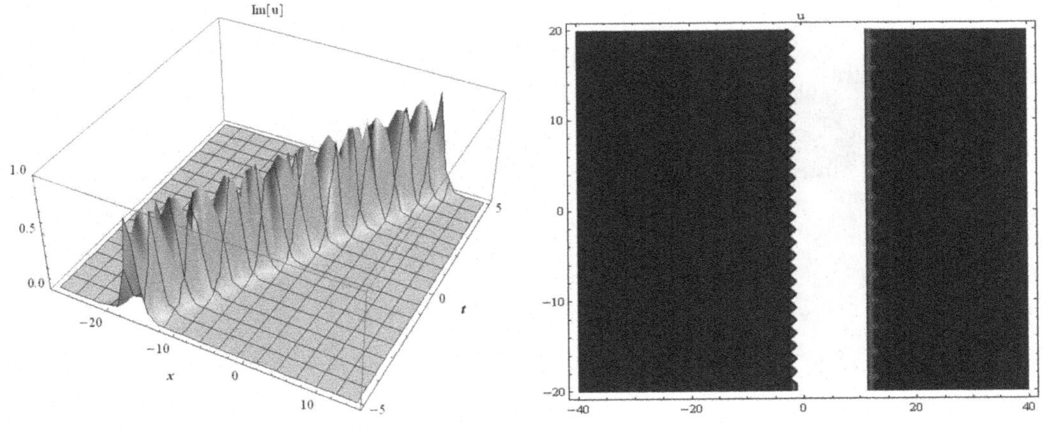

FIGURE 14.2
The 3D and contour surfaces of the real part of Equation (14.24) considering for the values of $\lambda = 2, \mu = 0.1, c_1 = -0.5, \beta = 6, y = 3, -28 < x < 15, -5 < t < 5$, and $t = 4, -40 < x < 40, -20 < y < 20$ for contour simulation.

Case-1.3: It can be taken with another coefficient, which is suitable one, as follows:

$$A_0 = \frac{5}{6}\lambda\left(-2A_1 + 5\lambda A_2\right), A_3 = \frac{2\left(A_1 - \lambda A_2\right)}{3\lambda^2}, B_1 = \frac{2\left(A_1 - \lambda A_2\right)B_0}{\lambda\left(-2A_1 + 5\lambda A_2\right)},$$
$$\mu = \frac{5\lambda^2}{2}, \beta = \frac{18\lambda B_0}{2A_1 - 5\lambda A_2}, k = \frac{i\sqrt{-2A_2 + 9A_2\lambda^2}\sqrt{-A_1 + \lambda A_2}}{\sqrt{A_2\left(-A_1 + \lambda A_2\right)}}, \tag{14.25}$$

When we put Equation (14.25) along with Equation (14.8) into Equation (14.19), we obtain a new complex trigonometric function solution to Equation (14.1) under the condition of *Family-1*, which is $\mu \neq 0, \lambda^2 - 4\mu > 0$, as follows:

$$u_3\left(x,y,t\right) = \psi\left(-5 + 4\cos\left(3\lambda\left(E + x + y + it\varepsilon\right)\right) + 3\sin\left(3\lambda\left(E + x + y + it\varepsilon\right)\right)\right)^{-1}, \tag{14.26}$$

in which $\psi = 15\lambda\left(2A_1 - 5\lambda A_2\right)/2B_0, \varepsilon = \left(\sqrt{A_2}\sqrt{-2 + 9\lambda^2}\sqrt{-A_1 + \lambda A_2}\right)/\left(\sqrt{A_2}\sqrt{-A_1 + \lambda A_2}\right)$. This solution is valid only under the condition of *Family-1*. Therefore, we have to choose as follows:

$$\lambda^2 - 4\mu = \lambda^2 - 4\left(\frac{5\lambda^2}{2}\right) = \lambda^2 - 10\lambda^2 = -9\lambda^2 > 0,$$

in here, λ has to be chosen as $\lambda = \pm i$. When we consider the suitable values of parameters in Equation (14.26), we can plot 2D, 3D, and contour surfaces as follows (Figures 14.3–14.5).

Case-1.4: It can be considered with another coefficient, as follows:

$$A_0 = -\frac{6B_0}{\beta^2 A_3}\left(\beta A_1 + 6\lambda B_0\right), A_2 = \lambda A_3 - \frac{6B_0}{\beta}, B_1 = \frac{\beta A_3}{6},$$
$$\mu = \frac{\beta A_1 + 6\lambda B_0}{\beta A_3}, k = \frac{\sqrt{-4\beta A_1 + \beta\left(2 + \lambda^2\right)A_3 - 24\lambda B_0}}{\sqrt{\beta}\sqrt{A_3}}, \tag{14.27}$$

Putting Equation (14.25) along with Equation (14.8) into Equation (14.19), we obtain another new hyperbolic function solution to Equation (14.1) under the condition of *Family-1*, which is $\mu \neq 0, \lambda^2 - 4\mu > 0$, as follows:

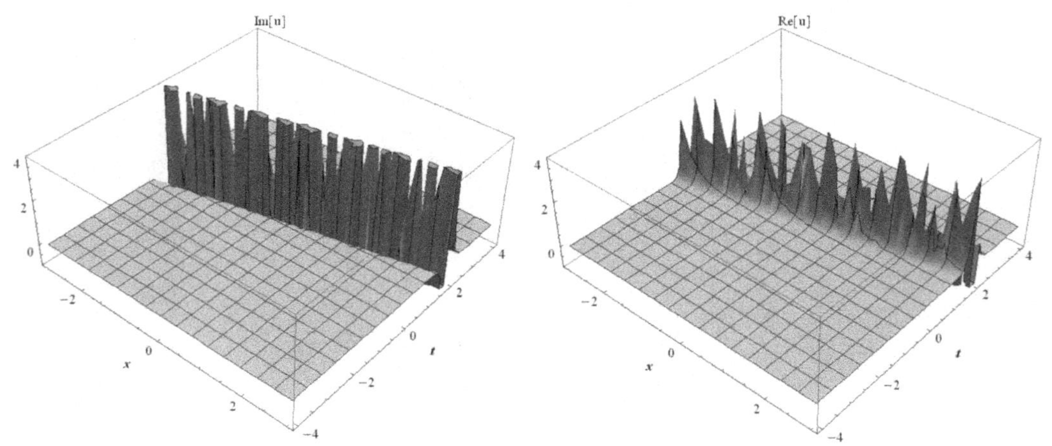

FIGURE 14.3
The 3D surfaces of new complex trigonometric function solution Equation (14.26) considering for the values of $\lambda = i, \beta = 0.2, B_0 = 0.6, A_1 = 0.3, A_2 = 0.4, c_1 = 0.5, y = 3, -3 < x < 3, -4 < t < 4$.

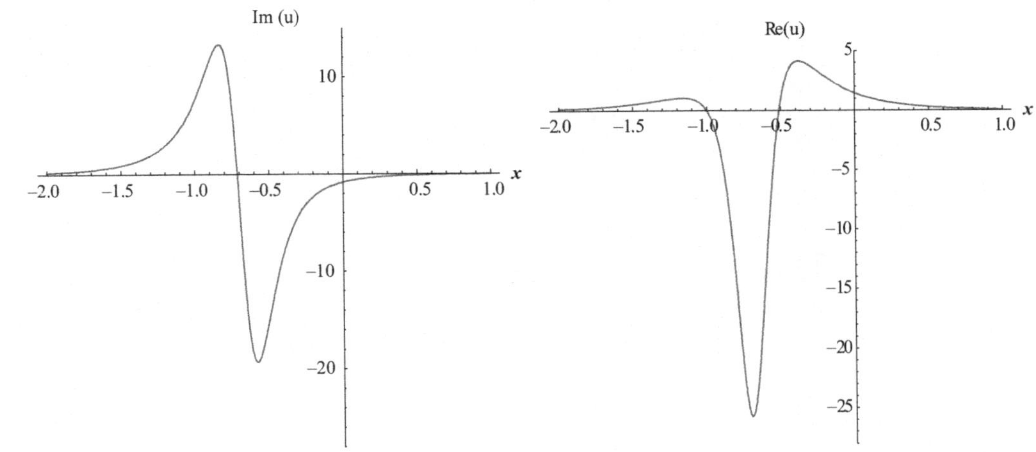

FIGURE 14.4
The 2D surfaces of new complex trigonometric function solution Equation (14.26) considering for the values of $\lambda = i, \beta = 0.2, B_0 = 0.6, A_1 = 0.3, A_2 = 0.4, c_1 = 0.5, y = 3,$ $t = 0.85, -3 < x < 3$.

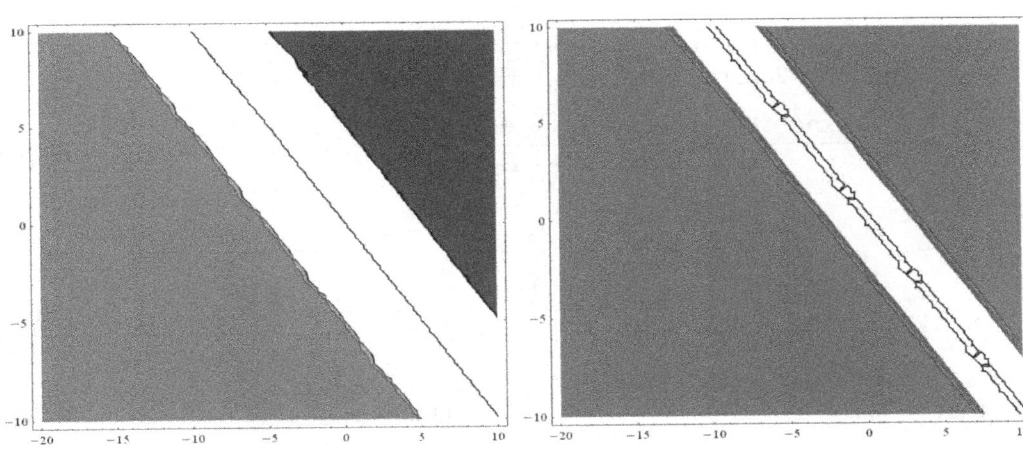

FIGURE 14.5
Contour surfaces of Equation (14.26) considering for the values of $\lambda = i, \beta = 0.2, B_0 = 0.6,$ $A_1 = 0.3, A_2 = 0.4, c_1 = 0.5, t = 0.2, -20 < x < 10, -10 < x < 10$.

$$u_4(x, y, t) = \frac{6 \sec h^2 (c_1\omega + \omega x + \omega y + t\omega k) \left(-4\beta A_1 + \lambda \left(\beta \lambda A_3 - 24 B_0\right)\right) \left(\beta A_1 + 6\lambda B_0\right)}{\beta^3 A_3^2 \left(\lambda + 2\omega \tanh (c_1\omega + \omega x + \omega y + t\omega k)\right)^2}, \tag{14.28}$$

in which $\omega = \sqrt{(\lambda^2 \beta A_3 - 4\beta A_1 - 24\lambda B_0) / (4\beta A_3)}$. This solution is only valid under the condition of *Family-1*. Hence, we have to choose as

$$\lambda^2 - 4\mu = \lambda^2 - 4\left(\frac{\beta A_1 + 6\lambda B_0}{\beta A_3}\right) = \frac{\lambda^2 \beta A_3 - 4\beta A_1 - 24\lambda B_0}{\beta A_3} > 0.$$

When we consider the suitable values of parameters in Equation (14.28), we can plot 3D and contour surfaces as follows (Figure 14.6).

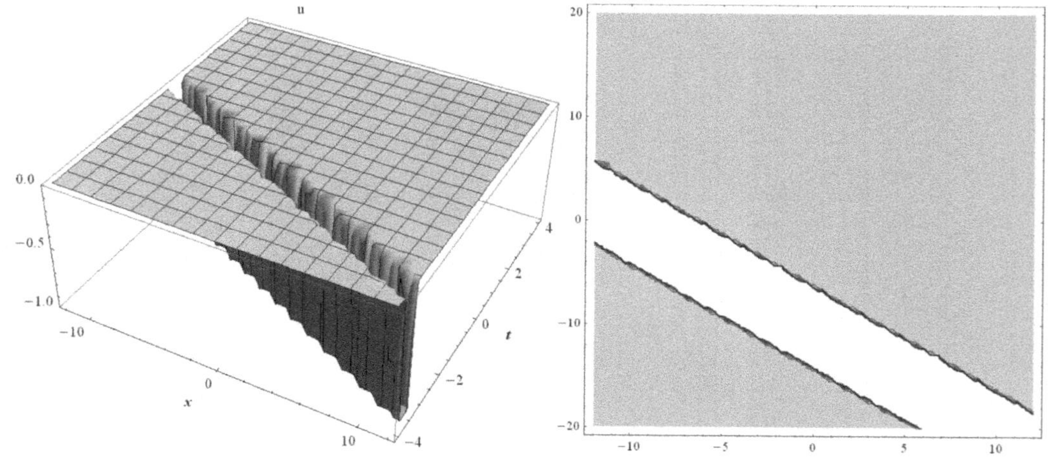

FIGURE 14.6
The 3D and contour surfaces of the real part of Equation (14.28) considering for the values of $\lambda = 0.1, \beta = 0.2, A_0 = -0.3, B_0 = -0.4, c_1 = -0.5, y = 3, -35 < x < 35, -5 < t < 5$, and $t = 0.85, -20 < y < 20$ for contour simulation.

Case-2

If we choose $M = 2$ and $N = 4$, we can write as follows:

$$U = \frac{A_0 + A_1 \exp(-\Omega) + A_2 \exp(-2\Omega) + A_3 \exp(-3\Omega) + A_4 \exp(-4\Omega)}{B_0 + B_1 \exp(-\Omega) + B_2 \exp(-2\Omega)} = \frac{\Upsilon}{\Psi}, \tag{14.29}$$

and

$$\begin{aligned} U' &= \frac{\Upsilon'\Psi - \Psi'\Upsilon}{\Psi^2}, \\ U'' &= \frac{\Upsilon''\Psi^3 - \Psi^2\Upsilon'\Psi' - (\Psi''\Upsilon + \Psi'\Upsilon')\Psi^2 + 2(\Psi')^2\Upsilon\Psi}{\Psi^4}, \\ &\vdots \end{aligned} \tag{14.30}$$

where $A_4 \neq 0, B_2 \neq 0$. When we use Equations (14.29 and 14.30) in Equation (14.17), we get algebraic equations ofexp $(-\Omega(\xi))$. Solving this system by using various computer programs yields many results.

Case-2.1: We can choose the following suitable coefficients:

$$\begin{aligned} A_0 &= \frac{-\left(A_3^3 - 48A_1A_4^2\right)\left(A_3^3 + 16A_1A_4^2\right)}{1024A_3^2A_4^3}, \quad A_2 = \frac{A_3^2}{4A_4} + \frac{2A_1A_4}{A_3}, \\ k &= \frac{i\sqrt{3A_3^3 - 16A_4^2(3A_1 + A_3)}}{2A_4\sqrt{2A_3}}, \quad B_0 = \frac{-A_3B_1}{16A_4} + \frac{3A_1A_4B_1}{A_3^2}, B_2 = \frac{2A_4B_1}{A_3}; \\ \beta &= -\frac{12B_1}{A_3}, \mu = \frac{3A_1}{2A_3} - \frac{A_3^2}{32A_4^2}, \quad \lambda = \frac{A_3}{2A_4}. \end{aligned} \tag{14.31}$$

If we take into account Equation (14.31) along with Equation (14.8) in Equation (14.29), we find the new complex hyperbolic solution to Equation (14.1) under the terms of *Family-1*, which are $\mu \neq 0, \lambda^2 - 4\mu > 0$, as follows:

$$u_5(x,y,t) = \frac{\kappa\varpi \tanh(f(x,y,t)) - 3\varepsilon + \varepsilon \tanh^2(f(x,y,t)) + 7A_3^3 + 3A_3^3 \tanh^2(f(x,y,t))}{32A_3B_1\left[2A_3A_4 + \omega A_4^2 \tanh^2(f(x,y,t))\right]^2}, \tag{14.32}$$

in here, $\varpi = A_3^3 - 16A_1A_4^2, \omega = \sqrt{\frac{-96A_1}{A_3} + \frac{6A_3^2}{A_4^2}}, f(x,y,t) = \frac{1}{4}\sqrt{\frac{-24A_1}{A_3} + \frac{3A_3^2}{2A_4^2}}(c_1 + x + y + it\sigma)$, $\kappa = 4A_3^2A_4\omega, \sigma = \sqrt{3A_3^3 - 16A_4^2(3A_1 + A_3)}/\left(2A_4\sqrt{2A_3}\right), \varepsilon = 48A_1A_4^2$, For the validity, under the terms of *Family-1*, $\frac{3A_3^2}{4A_4^2} - \frac{6A_1}{A_3} > 0$. With the suitable values of parameters in Equation (14.28), we can draw 3D and contour surfaces as follows (Figures 14.7–14.9).

Case-2.2: Taking another coefficient, being a suitable one, as follows:

$$
\begin{aligned}
A_0 &= -\frac{A_3^4 - 32A_1A_3A_4^2}{256A_4^3}, A_2 = \frac{A_3^2}{4A_4} + \frac{2A_1A_4}{A_3}, k = -\frac{\sqrt{A_3^3 + 4(-4A_1 + A_3)A_4^2}}{\sqrt{2}\sqrt{A_3}A_4}, \\
B_0 &= \frac{A_3B_1}{8A_4}, B_2 = \frac{2A_4B_1}{A_3}, \beta = -\frac{12B_1}{A_3}, \mu = \frac{2A_1}{A_3} - \frac{A_3^2}{16A_4^2}, \lambda = \frac{A_3}{2A_4},
\end{aligned}
\tag{14.33}
$$

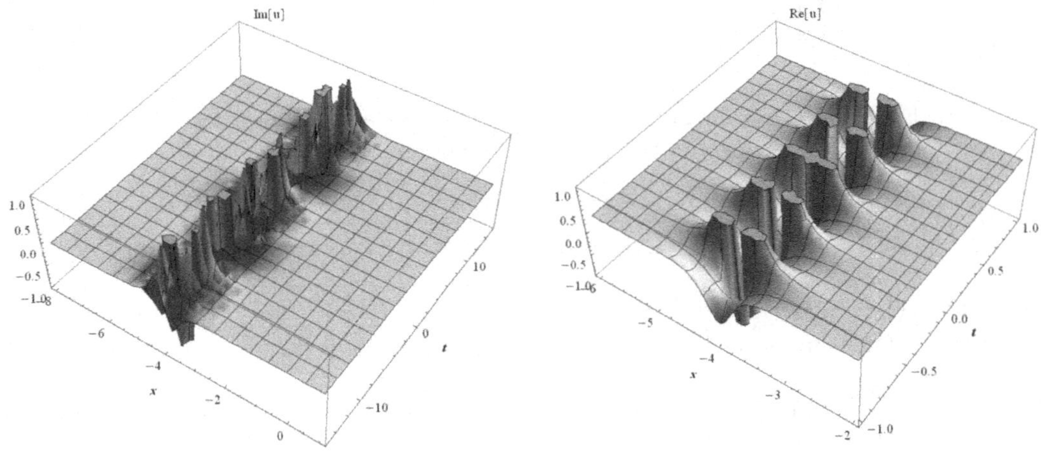

FIGURE 14.7
The 3D surfaces of a new solution for Equation (14.32) considering for the values of $A_1 = -3, A_3 = 1, A_4 = 2, c_1 = 3, B_1 = 4, y = 1$.

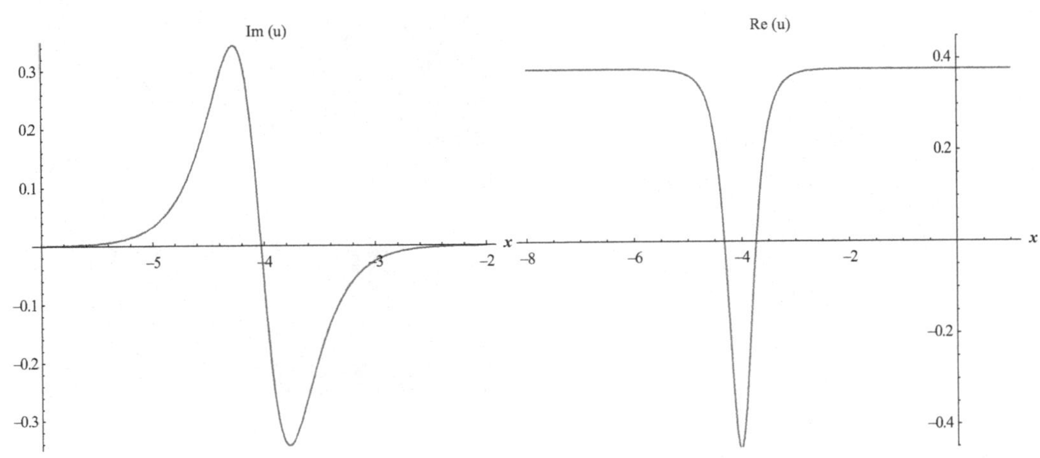

FIGURE 14.8
The 2D surfaces of a new solution for Equation (14.32) considering for the values of $A_1 = -3, A_3 = 1, A_4 = 2, c_1 = 3, B_1 = 4, y = 1, t = 0.85$.

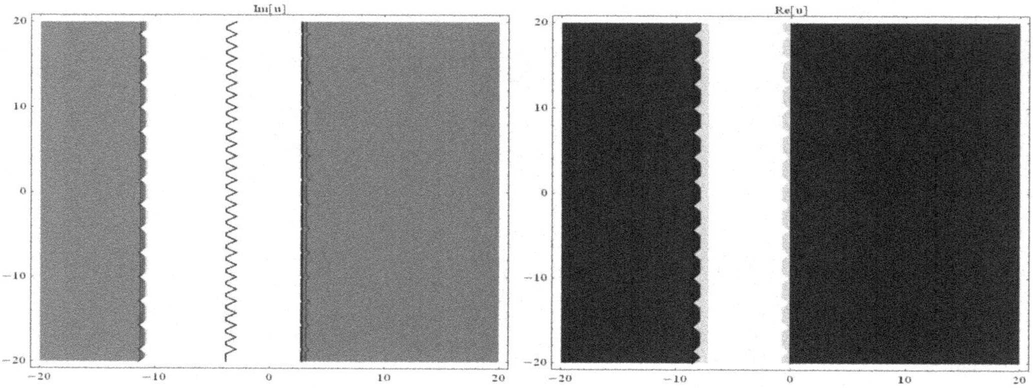

FIGURE 14.9
Contour surfaces of Equation (14.32) considering for the values of $A_1 = -3, A_3 = 1, A_4 = 2, c_1 = 3, B_1 = 4, t = 3$.

along with Equation (14.8) in Equation (14.29) produces another new hyperbolic solution to Equation (14.1) under the rules of *Family-1*, which is $\mu \neq 0, \lambda^2 - 4\mu > 0$, as follows:

$$u_6(x,y,t) = \frac{\left(A_3^3 - 32A_1A_4^2\right)\left(A_3^3 - 16A_1A_4^2\right)\sec h^2\left(\phi\left(c_1 + x + y - t\zeta\right)\right)}{16A_3B_1\left(A_3A_4 + \nu A_4^2 \tanh\left(\phi\left(c_1 + x + y - t\zeta\right)\right)\right)^2}, \tag{14.34}$$

where $\phi = \frac{1}{2}\sqrt{\frac{-8A_1}{A_3} + \frac{A_3^2}{2A_4^2}}, \zeta = \frac{\sqrt{A_3^3 + 4(-4A_1 + A_3)A_4^2}}{\sqrt{2}\sqrt{A_3}A_4}, \nu = \sqrt{-\frac{32A_1}{A_3} + \frac{2A_3^2}{2A_4^2}}$ (Figure 14.10).

Case-2.3: One can be taken with another coefficient as follows:

$$A_0 = \frac{A_1^4 A_4}{A_3^2}, A_2 = \frac{A_3^2}{4A_4} + \frac{2A_1A_4}{A_3}, k = -i\frac{\sqrt{3A_3^3 - 8\left(6A_1 + A_3\right)A_4^2}}{2\sqrt{A_3}A_4}, \tag{14.35}$$

$$B_0 = \frac{2A_1A_4B_1}{A_3^2}, B_2 = \frac{2A_4B_1}{A_3}, \beta = -\frac{12B_1}{A_3}, \mu = \frac{3A_1}{A_3} - \frac{A_3^2}{8A_4^2}, \lambda = \frac{A_3}{2A_4},$$

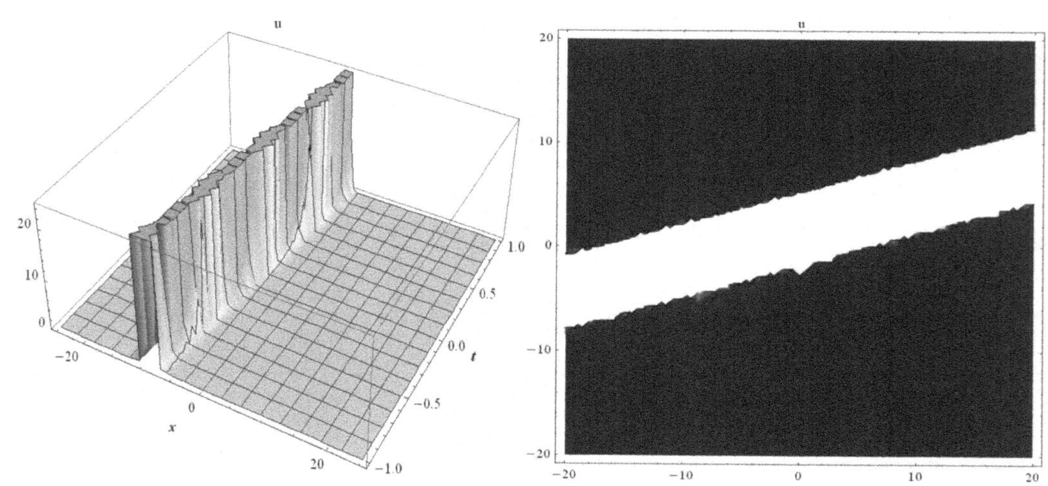

FIGURE 14.10
The 3D and contour surfaces of Equation (14.34) considering for the values of $A_1 = -3$, $A_3 = 6, A_4 = 2, c_1 = 0.3, B_1 = 0.4, y = 5, -25 < x < 25, -1 < t < 1$, and $y = 5, -20 < x < 20, -20 < t < 20$ for contour simulation.

Putting the Equation (14.35) along with Equation (14.8) into Equation (14.29) gives rise to the new complex hyperbolic solution to Equation (14.1) under the condition of *Family-1*, which is $\mu \neq 0, \lambda^2 - 4\mu > 0$, as follows:

$$u_7(x,y,t) = \frac{\varpi\left[2A_3^3 + A_3^2 \varsigma A_4 \tanh(f(x,y,t))\right] + 12A_1A_4^2\left[-3 + \tanh^2(f(x,y,t))\right]}{8A_3B_1\left[A_3A_4 + A_4^2\varsigma \tanh(f(x,y,t))\right]^2}, \tag{14.36}$$

in here, $f(x,y,t) = \frac{1}{4}\sqrt{\frac{-48A_1}{A_3} + \frac{3A_3^2}{A_4^2}}(c_1 + x + y - it\sigma), \sigma = \sqrt{3A_3^3 - 8A_4^2(6A_1 + A_3)}/(2A_4\sqrt{A_3}), \varsigma = \sqrt{\frac{-48A_1}{A_3} + \frac{3A_3^2}{A_4^2}}, \varpi = A_3^3 - 16A_1A_4^2$. The solution for Equation (14.36) is only validity, under the terms of *Family-1*,

$$\lambda^2 - 4\mu = \left(\frac{A_3}{2A_4}\right)^2 - \frac{12A_1}{A_3} + \frac{A_3^2}{2A_4^2} = \frac{3A_3^2}{4A_4^2} - \frac{12A_1}{A_3} > 0,$$

With the suitable values of parameters in Equation (14.36), we can see 2D, 3D, and contour surfaces as follows (Figures 14.11–14.13).

14.4 Conclusions

In this study, we have investigated complex, trigonometric, hyperbolic, and solitary wave solutions of an important model of the (2+1)-dimensional Boussinesq dynamical equation by successfully applying the algorithm of MEFM. These complex, trigonometric, and hyperbolic solutions obtained in this paper help us to understand the physical meaning of complex physical models and have vital importance in applied sciences. It has been checked whether these solutions have satisfied the (2+1)-dimensional Boussinesq dynamical equation with the help of some computational program (Maple, Mathematica, MATLAB®, and so on).

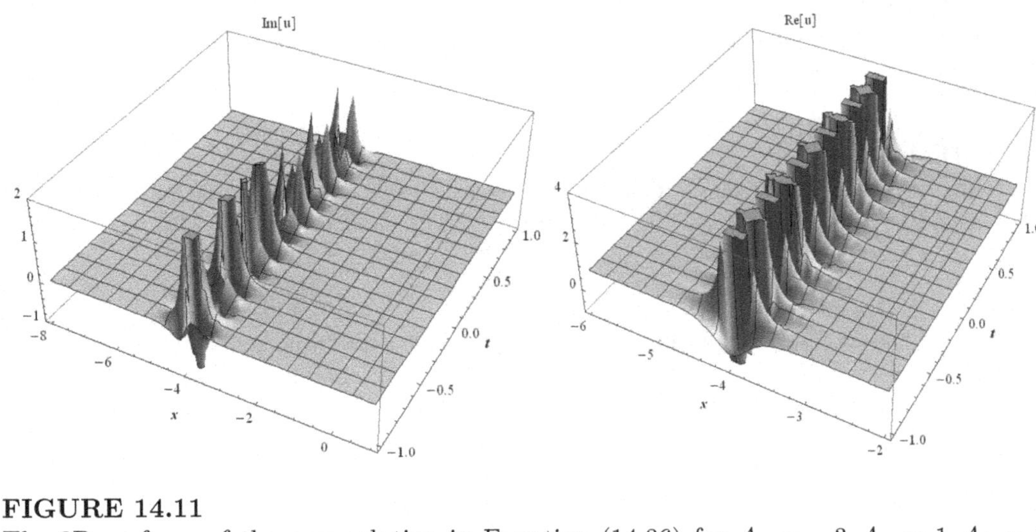

FIGURE 14.11
The 3D surfaces of the new solution in Equation (14.36) for $A_1 = -3, A_3 = 1, A_4 = 2, c_1 = 3, B_1 = 4, y = 1$.

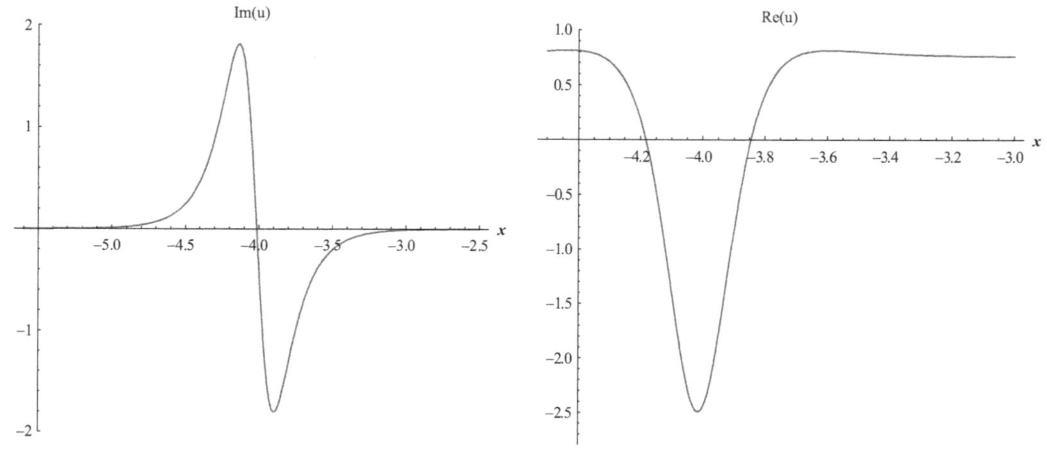

FIGURE 14.12
The 2D graph of the new solution in Equation (14.36) for $A_1 = -3, A_3 = 1, A_4 = 2, c_1 = 3, B_1 = 4, y = 1, t = 0.5$.

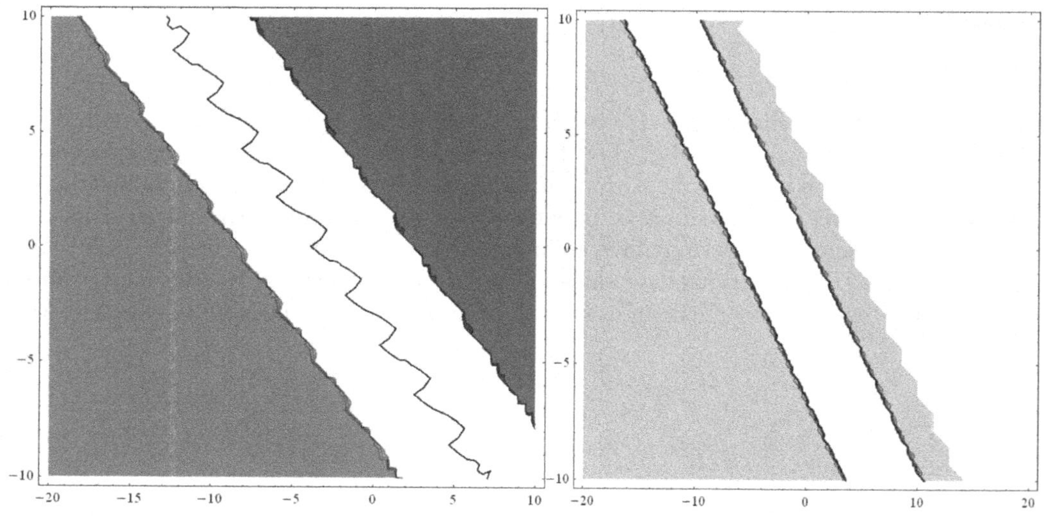

FIGURE 14.13
Contour surfaces of the new complex hyperbolic solution in Equation (14.36) for $A_1 = -3, A_3 = 1, A_4 = 2, c_1 = 3, B_1 = 4, t = 1, -20 < x < 20, -10 < y < 10$.

These solutions such as complex, trigonometric, and hyperbolic solutions are new when we compare them with the existing results in literature. The comparison has been supported by graphical plots along with contour surfaces which were first submitted. The main advantage of this method is that it gives rise to many coefficients and different physical meanings of the model considered. After all results and to the best of our knowledge, the application of MEFM to the (2+1)-dimensional Boussinesq dynamical equation has not been submitted beforehand.

Conflict of Interest

The authors declare that they have no conflict of interest.

References

[1] G. Metivier, *The Mathematics of Nonlinear Optics*, Chapter 3, 1–145, 2009, http://cachescan.bcub.ro/2008_05_28/E-book/580770-169-313.pdf.

[2] M. A. Jafarizadeh, A. R. Esfandyari, M. Moslehi-Fard, Exact solutions of Boussinesq equation, *Third International Conference on Geometry, Integrability and Quantization*, June pp. 304–314, Sofia, 2001.

[3] D. Dutykh, F. Dias, Dissipative Boussinesq equations, arXiv:physics/0703192,2007.

[4] V.N. Vu, C. Lee, T.-H. Jung, Extended Boussinesq equations for waves in porous media, *Coastal Engineering*, 139 (2018) 85–97.

[5] V.B. Shakhmurov, Nonlocal problems for Boussinesq equations, *Nonlinear Analysis*, 142 (2016) 134–151.

[6] A.-M. Wazwaz, Gaussian solitary waves for the logarithmic Boussinesq equation and the logarithmic regularized Boussinesq equation, *Ocean Engineering*, 94 (2015) 111–115.

[7] M.T. Darvishi, M. Najafi, A.M. Wazwaz, Soliton solutions for Boussinesq-like equations with spatio-temporal dispersion, *Ocean Engineering*, 130 (2017) 228–240.

[8] S. Saha Ray, S. Singh, New exact solutions for the Wick-type stochastic modified Boussinesq equation for describing wave propagation in nonlinear dispersive systems, *Chinese Journal of Physics*, 2017. doi: 10.1016/j.cjph.2017.04.017.

[9] W. Gao, H.F. Ismael, A.M. Husien, H. Bulut, H.M. Baskonus, Optical Soliton solutions of the nonlinear Schrödinger and resonant nonlinear Schrödinger equation with parabolic law, *Applied Science*, 10(1) (2020) 219, 1–20.

[10] J. Singh, D. Kumar, D. Baleanu, R. Sushila, On the local fractional wave equation in fractal strings, *Mathematical Methods in the Applied Sciences*, 42(5) (2019) 1588–1595.

[11] N. Raza, U. Afzal, A.R. Butt, H. Rezazadeh, Optical solitons in nematic liquid crystals with Kerr and parabolic law nonlinearities, *Optical and Quantum Electronics*, 51(4)(2019) 107.

[12] J. L. García Guirao, H.M. Baskonus, A. Kumar, M.S. Rawat, G. Yel, Complex soliton solutions to the (3+1)-dimensional B-type Kadomtsev-Petviashvili-Boussinesq equation, *Symmetry*, 12(1) (2020)17.

[13] H. Rezazadeh, D. Kumar, T.A. Sulaiman, H. Bulut, New complex hyperbolic and trigonometric solutions for the generalized conformable fractional Gardner equation, *Modern Physics Letters B*, 33(17) (2019) 1950196.

[14] Z. Zhao, Böcklund transformations, rational solutions and soliton-cnoidal wave solutions of the modified Kadomtsev-Petviashvili equation, *Applied Mathematics Letters*, 89 (2019) 103–110.

[15] W. Gao, G. Yel, H.M. Baskonus, C. Cattani, Complex solitons in the conformable (2+1)-dimensional Ablowitz-Kaup-Newell-Segur equation, *Aims Mathematics*, 5(1) (2020) 507–521.

[16] Z. Zhao, B. Han, Lump solutions of a (3+1)-dimensional B-type KP equation and its dimensionally reduced equations, *Analysis and Mathematical Physics*, 9(1) (2019) 119–130.

[17] Z. Velioğlu, Supalgebras and Quotient Algebras of Parafree Lie Algebras, *International Journal Pure and Applied Mathematics*, 83(3) (2013) 507–514.

[18] W. Gao, H.F. Ismael, S.A. Mohammed, H.M. Baskonus, H. Bulut, Complex and real optical soliton properties of the paraxial nonlinear Schrödinger equation in Kerr media with M-fractional, *Frontiers in Physics* (2019). doi: 10.3389/fphy.2019.00197.

[19] Z. Zhao, L. He, Y. Gao, Rogue wave and multiple lump solutions of the (2+1)-dimensional Benjamin-Ono equation in fluid mechanics, *Complexity*, 8249635 (2019) 1–18.

[20] Z. Velioğlu, Soluble product of Para free Lie algebras and its residual properties, *Applied Mathematics and Nonlinear Sciences*, 4(1) (2019) 1–5.

[21] W. Gao, P. Veeresha, D.G. Prakasha, H.M. Baskonus, G. Yel, A powerful approach for fractional Drinfeld–Sokolov–Wilson equation with Mittag-Leffler law, *Alexandria Engineering Journal*, 58, 1301–1311, 2019.

[22] Z. Zhao, L. He, Multiple lump solutions of the (3+1)-dimensional potential Yu-Toda-Sasa-Fukuyama equation, *Applied Mathematics Letters*, 95 (2019) 114–121.

[23] M. Guedda, Z. Hammouch, On similarity and pseudo-similarity solutions of Falkner–Skan boundary layers, *Fluid Dynamics Research* 38(4) (2006) 211–223.

[24] W. Gao, H. Rezazadeh, Z. Pinar, H.M. Baskonus, S. Sarwar, G. Yel, Novel Explicit solutions for the nonlinear Zoomeron equation by using newly extended direct algebraic technique, *Optical and Quantum Electronics*, 52(52) (2020) 1–13.

[25] H. Gültekin Çitil, On the exact solutions and the approximate solutions by Adomian Decomposition Method of the Second-Order Linear Fuzzy initial value problems using the generalized differentiability, *Scholars Journal of Physics, Mathematics and Statistics*, 4(4) (2017) 145–161.

[26] W. Gao, H.F. Ismael, H. Bulut, H.M. Baskonus, Instability modulation for the (2+1)-dimension paraxial wave equation and its new optical soliton solutions in Kerr media, *Physica Scripta* (2019). doi: 10.1088/1402-4896/ab4a50.

[27] Z. Zhao, B. Han, Residual symmetry, Böcklund transformation and CRE solvability of a (2+1)-dimensional nonlinear system, *Nonlinear Dynamics*, 94(1) (2018) 461–474.

[28] W. Gao, M. Partohaghighi, H.M. Baskonus, S. Ghavi, Regarding the Group preserving scheme and method of line to the Numerical Simulations of Klein-Gordon Model, *Results in Physics*, 15(102555) (2019) 1–7.

[29] H. Gültekin Çitil, On the exact and the approximate solutions of second-order Fuzzy initial value problems with constant coefficients, *Malaya Journal of Matematik*, 6(1) (2018) 61–68.

[30] W. Gao, B. Ghanbari, H.M. Baskonus, New numerical simulations for some real world problems with Atangana-Baleanu fractional derivative, *Chaos, Solitons and Fractals*, 128 (2019) 34–43.

[31] B. Sharma, S. Kumar, C. Cattani, D. Baleanu, Nonlinear dynamics of Cattaneo-Christov heat ux model for third-grade power-law fluid, *Journal of Computational and Nonlinear Dynamics* (2019) CND-19-1131.

[32] H.M. Baskonus, J.F. Gómez-Aguilar, New singular soliton solutions to the longitudinal wave equation in a Magneto Electro-elastic circular rod with local M-derivative, *Modern Physics Letters B*, 33(21) (2019) 1–16.

[33] A. Ciancio, A. Quartarone, A hybrid model for tumor-immune competition, *UPB Scientific Bulletin, Series A*, 75(4) (2013) 125–136.

[34] H.M. Baskonus, C. Cattani, A. Ciancio, Periodic complex and Kink-type solitons for the nonlinear model in microtubules, *Journal Applied Sciences*, 21 (2019) 34–45.

[35] A. Ciancio, Analysis of time series with wavelets, *International Journal of Wavelets, Multire Solution and Information Processing (IJWMIP)*, 5(2) (2007) 241–256.

[36] H.M. Baskonus, Complex surfaces to the fractional (2+1)-dimensional Boussinesq dynamical model with local M-derivative, *European Physical Journal Plus*, 134(322) (2019) 1–10.

[37] E.İ. Eskitaşçıoğlu, M.B. Aktaş, H.M. Baskonus, New complex and hyperbolic forms for Ablowitz-Kaup-Newell-Segur Wave Equation with fourth order, *Applied Mathematics and Nonlinear Sciences*, 4(1) (2019) 105–112.

[38] H.M. Baskonus, Complex Soliton solutions to the Gilson-Pickering Model, *Axioms*, 8(1) (2019) 18.

[39] A. Ali, A.R. Seadawy, D. Lu, New solitary wave solutions of some nonlinear models and their applications, *Advances in Difference Equations*, 232 (2018) 2018.

[40] H. M. Baskonus, H. Bulut, Analytical studies on the (1+1)-dimensional nonlinear dispersive modified Benjamin-Bona-Mahony equation defined by seismic sea waves, *Waves in Random and Complex Media* (2015) 1–13. **doi:**10.1080/17455030.2015.1062577.

[41] M.G. Hafez, M.N. Alam, M.A. Akbar, Travelling wave solutions for some important coupled nonlinear physical models via the coupled Higgs equation and the Maccari system, *Journal of King Saud University - Science*, 27 (2015) 105–112.

[42] S. Uddin, N. Alam, S.M.S. Hossain, H. Samiu, M.A. Akbar, Some new exact traveling wave solutions to the (3+1)-dimensional Zakharov-Kuznetsov equation and the Burgers equations via Exp-ϕ-Expansion Method, *Frontiers of Mathematics and Its Applications*, 1(1) (2014) 1–8.

15

Statistical Approach of Mixed Convective Flow of Third-Grade Fluid towards an Exponentially Stretching Surface with Convective Boundary Condition

Anum Shafiq
Preston University

Zakia Hammouch
Université Moulay Ismaïl

Tabassum Naz Sindhu
Quaid I Azam University

Dumitru Baleanu
Cankara University

CONTENTS

15.1 Introduction

Nowadays, non-Newtonian fluids are well suited in processing industries such as foods, plastics, pharmaceuticals, and dyestuffs. Most biological fluids are also non-Newtonian [1–6]. These fluids occur in diverse natural and synthetic forms and have many important applications in the oil industry [7–11]. On the other hand the studies of non-Newtonian fluids over a stretching sheet are further important with reference to several applications. Recently, some investigators discussed such flows through convective conditions at the sheet [12–14].

The objective of this work is to address the magnetohydrodynamics mixed convection flow past an exponential stretching surface with convective boundary condition through mathematical as well as statistical approaches. An incompressible non-Newtonian fluid named as third-grade fluid is considered. Constitutive relations of thermodynamically compatible third-grade fluid are used for problem development. The governing non-linear analysis is computed by homotopy analysis method (HAM) [14]. The correlation coefficients for the skin friction and Nusselt number are computed and used to calculate the probable error.

15.2 Mathematical Formulation

Here the characteristic of heat transfer in two-dimensional boundary layer flow of an incompressible third-grade fluid over an exponential stretching surface in the presence of a convective boundary condition is considered. Also, mixed convective phenomena is taken into account. It is assumed that the velocity of the stretching surface is $U_w(x) = U_0 \exp(\frac{x}{l})$. We denote U_0 the reference velocity, $T_f(x)$ the surface temperature, and T_∞ the ambient uniform temperature. It should be pointed out that $T_f(x) > T_\infty$ corresponds to a heated surface case (assisting flow) and $T_f(x) < T_\infty$ corresponds to a cooled surface situation (opposing flow). The fluid is electrically conducting in the presence of a magnetic field with strength B_0 applied in the transverse direction to the flow, i.e. along the $y-$axis. There is no electric field, and the induced magnetic field is neglected under the assumption of small magnetic Reynolds number. The governing two-dimensional boundary layer flow equations are given by

$$\frac{\partial u}{\partial x} + \frac{\partial v}{\partial y} = 0, \tag{15.1}$$

$$\begin{aligned} u\frac{\partial u}{\partial x} + v\frac{\partial u}{\partial y} &= \nu\frac{\partial^2 u}{\partial y^2} + \frac{\alpha_1^*}{\rho}\left[u\frac{\partial^3 u}{\partial x \partial y^2} + \frac{\partial u}{\partial x}\frac{\partial^2 u}{\partial y^2} + 3\frac{\partial u}{\partial y}\frac{\partial^2 v}{\partial y^2} + v\frac{\partial^3 u}{\partial y^3}\right] \\ &+ 2\frac{\alpha_2^*}{\rho}\frac{\partial u}{\partial y}\frac{\partial^2 v}{\partial y^2} + 6\frac{\beta_3}{\rho}\left(\frac{\partial u}{\partial y}\right)^2\frac{\partial^2 u}{\partial y^2} + \frac{g\,\beta_T}{\rho}(T - T_\infty) - \frac{\sigma\, B_0^2}{\rho}u, \end{aligned} \tag{15.2}$$

$$u\frac{\partial T}{\partial x} + v\frac{\partial T}{\partial y} = \frac{k}{\rho c_p}\frac{\partial^2 T}{\partial y^2} + \frac{\sigma\, B_0^2}{\rho c_p}u^2, \tag{15.3}$$

with the boundary conditions

$$\begin{aligned} &u(x,0) = U_w(x) = U_0 \exp\left(\frac{x}{l}\right), \; v(x,0) = 0, \; -k\frac{\partial T(x,0)}{\partial y} = h\left(T_f - T\right), \\ &u \to 0, \quad T \to T_\infty \quad \text{as} \quad y \to \infty. \end{aligned} \tag{15.4}$$

In the above expressions u and v are the velocity components in the $x-$ and $y-$ directions, respectively, $(\alpha_1^*, \alpha_2^*, \beta_3)$ are the fluid parameters, ρ is the fluid density, σ is the electric charge density, K is the thermal conductivity, and c_p is the specific heat. The dimensionless transformation is defined as follows:

$$\eta = \sqrt{\frac{U_0}{2\nu l}} \exp\left(\frac{x}{2l}\right) y, \; \psi = \sqrt{2\nu l U_0} f(\eta) \exp\left(\frac{x}{2l}\right)$$
$$u(x,y) = U_0 \exp\left(\frac{x}{l}\right) f'(\eta), \; v(x,y) = -\sqrt{\frac{\nu U_0}{2l}} \exp\left(\frac{x}{2l}\right) \left[f(\eta) + \eta f'(\eta)\right],$$
$$\theta = \frac{T - T_\infty}{T_f - T_\infty}. \tag{15.5}$$

Using these transformation, Equation (15.1) is identically satisfied and the other governing transformed equations can be written as follows:

$$f''' - 2(f')^2 + ff'' + \alpha_1 \left\{3f'f''' - ff^{(iv)} - 2\eta f'' f''' - 9(f'')^2\right\}$$
$$- \alpha_2 \left\{3(f'')^2 + \eta f'' f'''\right\} + 3\beta \operatorname{Re}(f'')^2 f''' + \lambda\theta - 2M^2 f' = 0, \tag{15.6}$$

$$f'(0) = 1, \; f(0) = 0, \; f'(\infty) = 0, \tag{15.7}$$

$$\theta'' + \Pr f\,\theta' + 2M^2 \Pr Ec\,(f')^2 = 0, \tag{15.8}$$

$$\theta'(0) = -\gamma\left[1 - \theta(0)\right], \; \theta(\infty) = 0, \tag{15.9}$$

where Pr denotes the Prandtl number, Ec the Eckert number, M the magnetic parameter, and λ the mixed convection parameter. These quantities can be put into the following forms:

$$\alpha_1 = \frac{U_0 \alpha_1^* \exp\left(\frac{x}{l}\right)}{\mu l}, \quad \alpha_2 = \frac{U_0 \alpha_2^* \exp\left(\frac{x}{l}\right)}{\mu l}, \quad \beta = \frac{U_0^2 \beta_3 \exp\left(\frac{2x}{l}\right)}{\mu l^2},$$
$$\Pr = \frac{\mu_0 c_p}{K}, \; Ec = \frac{U_w^2}{c_p (T_w - T_\infty)}, \; \lambda = \frac{Gr}{\operatorname{Re}_x^2}, \; \operatorname{Re}_x = \frac{U_w l}{\nu}, \; M^2 = \frac{\sigma B_0^2}{\rho U_w}. \tag{15.10}$$

Skin friction coefficient C_f and the local Nusselt number Nu_x can be defined by

$$C_f = \frac{\tau_{xy}}{\rho (cx)^2}, \qquad Nu_x = \frac{x q_w}{K(T - T_\infty)}, \tag{15.11}$$

where the wall shear stress (τ_{xy}) and the wall heat flux (q_w) are

$$\tau_{xy} = \left[\mu_0 \frac{\partial u}{\partial y} + \frac{\alpha_1^*}{\rho}\left(2\frac{\partial u}{\partial x}\frac{\partial u}{\partial y} + v\frac{\partial^2 u}{\partial y^2} + u\frac{\partial^2 u}{\partial x \partial y}\right) + 2\frac{\beta_3}{\rho}\left(\frac{\partial u}{\partial y}\right)^3\right]_{y=0},$$
$$q_w = -K\left(\frac{\partial T}{\partial y}\right)_{y=0}. \tag{15.12}$$

After simplification we have

$$\left(\operatorname{Re}_x\right)^{1/2} C_f = \sqrt{2}\left[f'' + \alpha_1\left(\frac{7}{2}f'f'' - \frac{1}{2}ff'''\right) + \beta f''^3\right]_{\eta=0}, \quad \left(\operatorname{Re}_x\right)^{-1/2} Nu_x = -\sqrt{\frac{X}{2}}\theta'(0), \tag{15.13}$$

where $X = \frac{x}{l}$, the Reynolds number $\operatorname{Re}_x = \frac{U_w\, x}{\nu}$, and ν is the kinematic viscosity.

15.3 Homotopy Analytic Solutions

The initial guesses $f_0(\eta)$ and $\theta_0(\eta)$ and the auxiliary linear operators L_1 and L_2 are chosen in the following forms:

$$f_0(\eta) = 1 - \exp(-\eta), \qquad \theta_0(\eta) = \frac{\gamma}{1+\gamma}\exp(-\eta), \tag{15.14}$$

$$L_1[f(\eta)] = \frac{d^3 f}{d\eta^3} - \frac{df}{d\eta}, \qquad L_2[\theta(\eta)] = \frac{d^2\theta}{d\eta^2} - \theta. \tag{15.15}$$

The operators satisfy the following properties:

$$L_1[C_1 + C_2\exp(\eta) + C_3\exp(-\eta)] = 0, \tag{15.16}$$

$$L_2[C_4\exp(\eta) + C_5\exp(-\eta)] = 0, \tag{15.17}$$

where $C_i\ (i = 1-5)$ are the constants determined from the boundary conditions.

15.3.1 Zeroth-Order Deformation Problem

The zeroth-order deformation problems are given by

$$(1-q)L_1\left[\hat{f}(\eta,q) - f_0(\eta)\right] = q h_f \mathcal{N}_1\left[\hat{f}(\eta,q)\right], \tag{15.18}$$

$$(1-q)L_2\left[\hat{\theta}(\eta,q) - \theta_0(\eta)\right] = q h_\theta \mathcal{N}_2\left[\hat{\theta}(\eta,q), \hat{f}(\eta,q)\right], \tag{15.19}$$

$$\hat{f}(0,q) = 0, \qquad \left.\frac{\partial \hat{f}(\eta,q)}{\partial\eta}\right|_{\eta=0} = 1, \qquad \left.\frac{\partial \hat{f}(\eta,q)}{\partial\eta}\right|_{\eta\to\infty} = 0, \tag{15.20}$$

$$\left.\hat{\theta}(\eta,q)\right|_{\eta=0} = -\gamma[1-\theta(0,q)], \qquad \left.\hat{\theta}(\eta,q)\right|_{\eta\to\infty} = 0. \tag{15.21}$$

In the above expressions $q \in [0,1]$ denotes an embedding parameter and $h_f \neq 0$, $h_\theta \neq 0$ are the auxiliary parameters, respectively. The non-linear operators $\mathcal{N}_1$ and $\mathcal{N}_2$ are

$$\begin{aligned}\mathcal{N}_1[f(\eta,q)] = {} & \frac{\partial^3 \hat{f}}{\partial\eta^3} - 2\left(\frac{\partial \hat{f}}{\partial\eta}\right)^2 + f\frac{\partial^2 \hat{f}}{\partial\eta^2} + \alpha_1\left[3\frac{\partial \hat{f}}{\partial\eta}\frac{\partial^3 \hat{f}}{\partial\eta^3} - f\frac{\partial^4 \hat{f}}{\partial\eta^4} - 2\eta\frac{\partial^2 \hat{f}}{\partial\eta^2}\frac{\partial^3 \hat{f}}{\partial\eta^3}\right.\\ & \left. - 9\left(\frac{\partial^2 \hat{f}}{\partial\eta^2}\right)^2\right] - \alpha_2\left[3\left(\frac{\partial^2 \hat{f}}{\partial\eta^2}\right)^2 + \eta\frac{\partial^2 \hat{f}}{\partial\eta^2}\frac{\partial^3 \hat{f}}{\partial\eta^3}\right] + 3\beta\,\mathrm{Re}\left(\frac{\partial^2 \hat{f}}{\partial\eta^2}\right)^2\frac{\partial^3 \hat{f}}{\partial\eta^3}\\ & + \lambda\theta - 2M^2 f',\end{aligned} \tag{15.22}$$

$$\mathcal{N}_2[\theta(\eta,q), f(\eta,q)] = \frac{\partial^2\hat{\theta}}{\partial\eta^2} + \Pr\ f\,\frac{\partial\hat{\theta}}{\partial\eta} + 2\,M^2 \Pr\ Ec\left(\frac{\partial \hat{f}}{\partial\eta}\right)^2. \tag{15.23}$$

For $q = 0$ and $q = 1$ one obtains

$$\hat{f}(\eta,0) = f_0(\eta),\ \hat{\theta}(\eta,0) = \theta_0(\eta) \text{ and } \hat{f}(\eta,1) = f(\eta),\ \hat{\theta}(\eta,1) = \theta(\eta). \tag{15.24}$$

In view of Taylor series, one can express $\hat{f}(\eta,q)$ and $\hat{\theta}(\eta,q)$ in the following forms:

$$\hat{f}(\eta,q) = f_0(\eta) + \sum_{m=1}^{\infty} f_m(\eta)q^m, \qquad \hat{f}_m(\eta) = \frac{1}{m!}\left.\frac{\partial^m f(\eta,q)}{\partial q^m}\right|_{q=0}, \tag{15.25}$$

$$\hat{\theta}(\eta,q)=\theta_0(\eta)+\sum_{m=1}^{\infty}\theta_m(\eta)\,q^m,\quad \hat{\theta}_m(\eta)=\frac{1}{m!}\frac{\partial^m\theta(\eta,q)}{\partial q^m}\bigg|_{q=0}. \tag{15.26}$$

Where the convergence depends upon h_f and h_θ. The convergence of the series (25) and (26) is strictly based upon h_f and h_θ. The values of h_f and h_θ are chosen in such a way that the series (25) and (26) behave convergent at $q=1$. Hence

$$\hat{f}(\eta)=f_0(\eta)+\sum_{m=1}^{\infty}f_m(\eta), \tag{15.27}$$

$$\hat{\theta}(\eta)=\theta_0(\eta)+\sum_{m=1}^{\infty}\theta_m(\eta). \tag{15.28}$$

15.3.2 mth-Order Deformation Problems

The mth-order deformation problems are given by

$$L_1\left[\hat{f}_m(\eta)-\chi_m\hat{f}_{m-1}(\eta)\right]=h_f\mathcal{R}_m^f(\eta), \tag{15.29}$$

$$\hat{f}_m(0,q)=0,\quad \frac{\partial\hat{f}_m(\eta,q)}{\partial\eta}\bigg|_{\eta=0}=0,\quad \frac{\partial\hat{f}_m(\eta,q)}{\partial\eta}\bigg|_{\eta\to\infty}=0, \tag{15.30}$$

$$L_2\left[\hat{\theta}_m(\eta)-\chi_m\hat{\theta}_{m-1}(\eta)\right]=\hbar_\theta\mathcal{R}_m^\theta(\eta), \tag{15.31}$$

$$\hat{\theta}_m(\eta,q)\Big|_{\eta=0}-\gamma\ \theta_m(\eta,q)|_{\eta=0}=0,\quad \hat{\theta}_m(\eta,q)\Big|_{\eta\to\infty}=0, \tag{15.32}$$

$$\begin{aligned}\mathcal{R}_m^f(\eta)&=f'''_{m-1}(\eta)-2\sum_{k=0}^{m-1}f'_{m-1-k}f'_k+\sum_{k=0}^{m-1}f_{m-1-k}f''_k+\alpha_1\sum_{k=0}^{m-1}\Big[3f'_{m-1-k}f'''_m\\&-f_{m-1-k}f_k^{(iv)}-2\eta\ f''_{m-1-k}\ f'''_k-9f''_{m-1-k}\ f''_k\Big]+\alpha_2\sum_{k=0}^{m-1}\Big[3f''_{m-1-k}f''_k\\&+\eta f''_{m-1-k}f'''_k\Big]+3\beta\,\mathrm{Re}\sum_{k=0}^{m-1}f''_{m-1-k}\sum_{l=0}^{k}f''_{k-l}f'''_l+\lambda\theta_{m-1}-2M^2f'_{m-1},\end{aligned} \tag{15.33}$$

$$\mathcal{R}_m^\theta(\eta)=\theta''_{m-1}(\eta)+\Pr\sum_{k=0}^{m-1}f_{m-1-k}\theta'_k+2M^2\Pr\ Ec\sum_{k=0}^{m-1}f'_{m-1-k}f'_k, \tag{15.34}$$

where

$$\chi_m=\begin{cases}0, & m\leqslant 1\\ 1, & m>1\end{cases}, \tag{15.35}$$

The general solutions (f_m,θ_m) of the mth-order Equations (15.29) and (15.31) in terms of special solutions (f_m^*,θ_m^*) are

$$f_m(\eta)=f_m^*+C_1+C_2\exp(\eta)+C_3\exp(-\eta), \tag{15.36}$$

$$\theta_m(\eta)=\theta_m^*+C_4\exp(\eta)+C_5\exp(-\eta), \tag{15.37}$$

where the arbitrary constants C_i $(i = 1 - 5)$ through boundary conditions (15.30) and (15.32) have the values

$$C_3 = C_4 = 0, \qquad C_2 = \left.\frac{\partial f_m^*(\eta)}{\partial \eta}\right|_{\eta=0}, \qquad C_1 = -C_2 - f_m^*(0),$$
$$C_5 = \frac{1}{1+\gamma}\left[\theta_m^{*\prime}(\eta)\big|_{\eta=0} - \gamma\,\theta_m^*(\eta)\big|_{\eta=0}\right]. \tag{15.38}$$

15.4 Convergence Analysis

The involved auxiliary parameters $\hbar_f$ and $\hbar_\theta$ in the series solutions are used to control the convergence. The $\hbar$-curves in Figure 15.1 are plotted for 16th-order of approximations just to determine the admissible values of the auxiliary parameters $\hbar_f$ and $\hbar_\theta$. From Figure 15.1 we observed that the admissible ranges of $\hbar_f$ and $\hbar_\theta$ are $-0.8 \leq \hbar_f < -0.2$ and $-0.9 \leq \hbar_\theta < -0.1$. Table 15.1 shows the convergence of series solutions numerically. It is noted that 7th order of approximation is enough for velocity and 12th order of approximation is enough for temperature.

15.5 Results and Discussion

The main objective of this section is to analyze the influence of various physical parameters on the velocity and temperature profiles. The graphical results are displayed in Figures 15.2–15.6. The numerical values of skin friction coefficient $\mathrm{Re}_x^{1/2}\,C_f$ and the local Nusselt number $\mathrm{Re}_x^{-1/2}\,Nu_x$ are presented and examined for various values of second-grade parameter $(\alpha_1,\ \alpha_2)$, third-grade parameter β, Reynolds number Re, Hartman number M, mixed convection parameter λ, Biot number γ, Prandtl number Pr, and Eckert number Ec in Table 15.2.

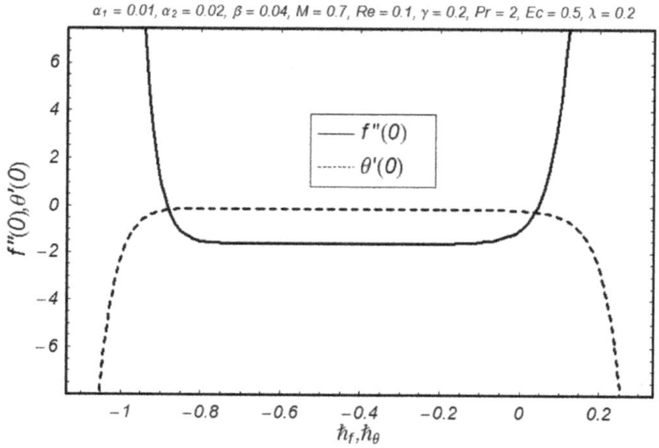

FIGURE 15.1
h-curve for the functions $f(\eta)$ and $\theta(\eta)$.

TABLE 15.1
Convergence of HAM Solutions for Different Orders of Approximation When $\alpha_1 = 0.01$, $\alpha_2 = 0.02$, $\beta = 0.04$, $\mathrm{Re} = 0.1$, $M = 0.7$, $\lambda = 0.2$, $\gamma = 0.2$, $Ec = 0.5$ and $\Pr = 2$

Order of Approximation	$-f''(0)$	$-\theta'(0)$
1	1.419	0.1371
2	1.534	0.1245
5	1.613	0.1099
7	1.616	0.1072
12	1.616	0.1057
18	1.616	0.1057
26	1.616	0.1057
30	1.616	0.1057

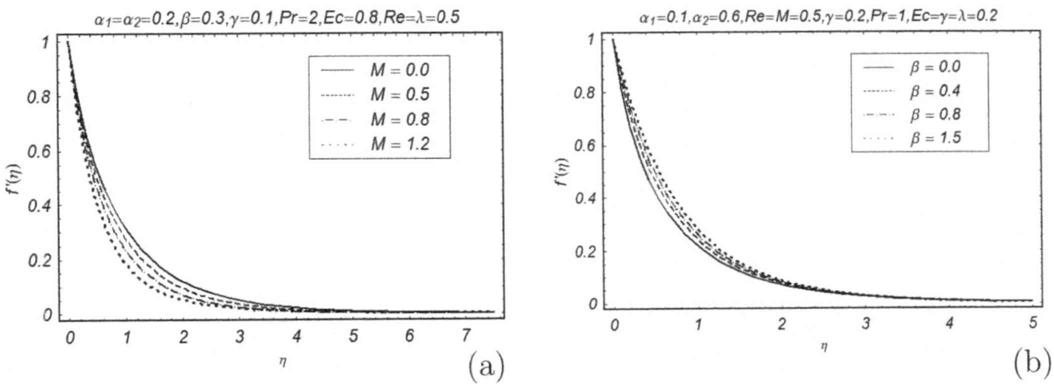

FIGURE 15.2
(a) Variation of velocity component $f'(\eta)$ vs. η for magnetic parameter M, (b) Variation of velocity component $f'(\eta)$ vs.** η for fluid parameter β.

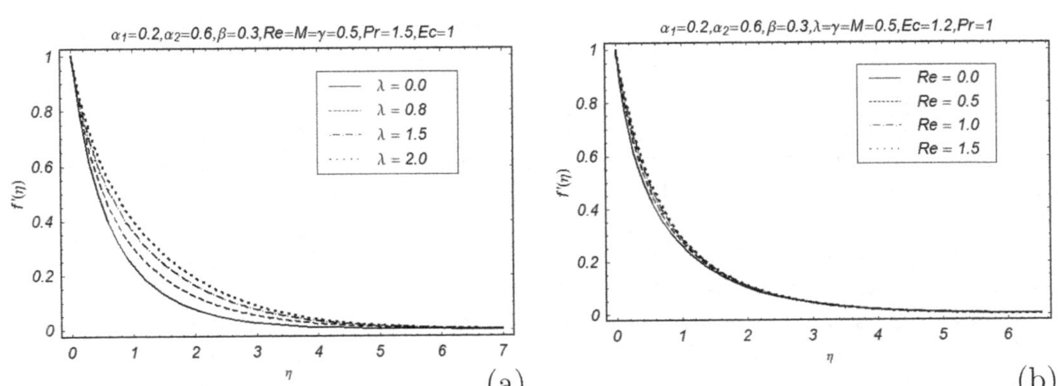

FIGURE 15.3
(a) Variation of velocity component $f'(\eta)$ vs η for mixed convection parameter λ, (b) Variation of velocity component $f'(\eta)$ vs η for Reynold number Re.

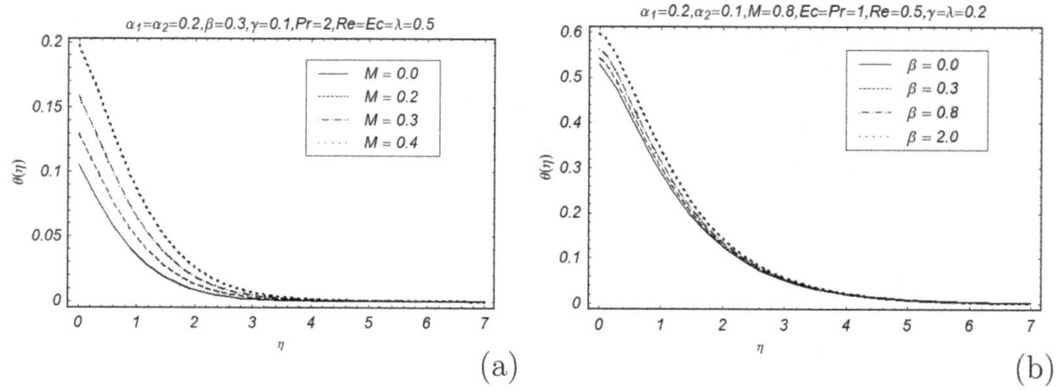

FIGURE 15.4
(a) Variation of temperature component $\theta(\eta)$ vs. η for magnetic parameter M, (b) Variation of temperature component $\theta(\eta)$ vs. η for fluid parameter β.

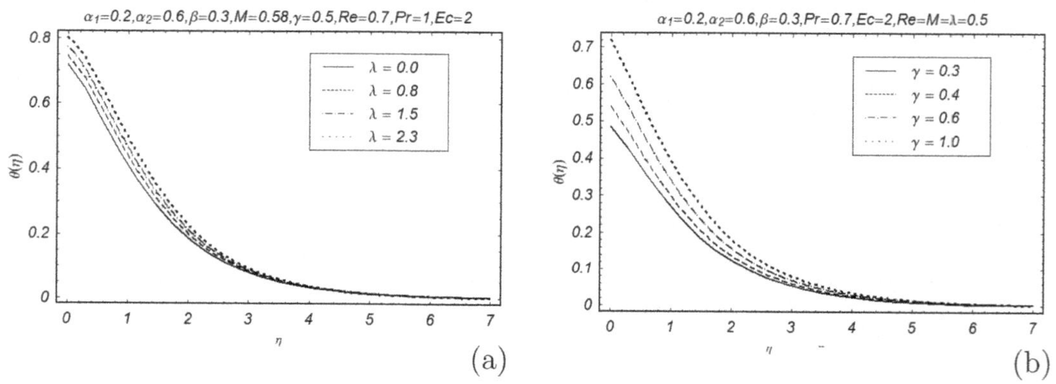

FIGURE 15.5
(a)Variation of temperature component $\theta(\eta)$ vs. η for mixed convection parameter λ, (b) Variation of temperature component $\theta(\eta)$ vs. η for Biot number γ.

15.5.1 Analysis of Velocity Profile

In Figure 15.2a the influence of the different values of Hartman number M is displayed by keeping the other parameters fixed. It is noted that the velocity profile is a decreasing function of Hartman number M. Besides, the boundary layer thickness also decreased. Due to the increase in magnetic force, the Lorentz force increases and provides the resistance to the fluid flow. Therefore the velocity profile decreases.

Figure 15.2b shows the variation of velocity u versus η when $\beta = 0.0$, $\beta = 0.4$, $\beta = 0.8$, and $\beta = 1.5$. It is noted that the velocity increases for larger third-grade parameter β. All profiles asymptotically tend to the horizontal axis. The longitudinal velocity is maximum at the wall $\eta = 0$ for all cases. The velocity near the wall rapidly increases, whereas it vanishes away from the wall.

Figure 15.3a illustrates the profile of velocity $f'(\eta)$ versus η for different values of λ. It is observed that with an increase in λ the velocity profile increases along with the thickness of the boundary layer. The maximum velocity is obtained very near to the sheet.

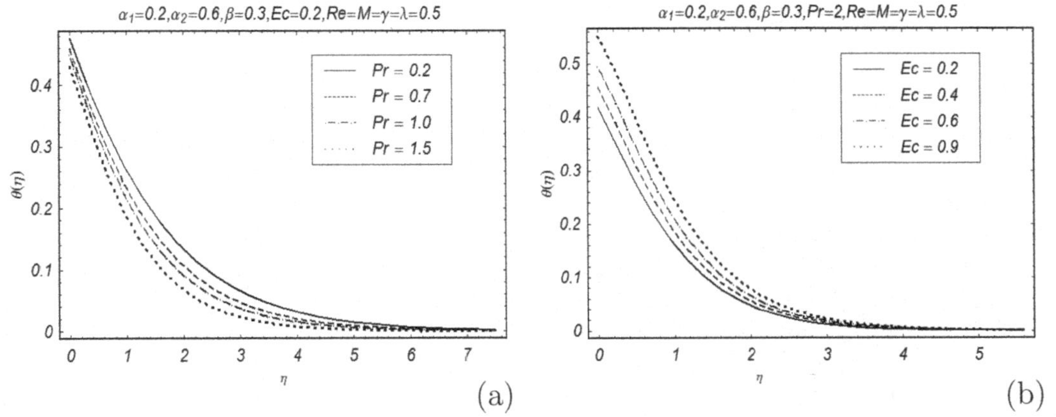

FIGURE 15.6
(a) Variation of temperature component $\theta(\eta)$ vs. η for Prandtl number Pr, (b) Variation of temperature component $\theta(\eta)$ vs. η for Eckert number Ec.

TABLE 15.2
Numerical Values of Skin Friction Coefficient and Nusselt Number for Different Values of Physical Parameters

α_1	α_2	β	M	λ	γ	Pr	Ec	$-\mathrm{Re}_x^{1/2} C_x$	$\mathrm{Re}_x^{-1/2} Nu_x$
0.0	0.1	0.1	0.7	0.2	0.2	2	0.5	0.10612	0.10638
0.1								0.10691	0.10708
0.2								0.10738	0.10749
0.1	0.0	0.1	0.7	0.2	0.2	2	0.5	0.10620	0.10576
	0.1							0.10691	0.10643
	0.2							0.10759	0.10707
0.1	0.1	0.0	0.7	0.2	0.2	2	0.5	0.10708	0.10708
		0.1						0.10691	0.10691
		0.2						0.10674	0.10675
0.1	0.1	0.1	0.5	0.2	0.2	2	0.5	0.13238	0.13236
			0.6					0.12042	0.12034
			0.7					0.10691	0.10674
0.1	0.1	0.1	0.7	0.2	0.2	2	0.5	0.10691	0.10674
				0.3				0.10662	0.10645
				0.4				0.10631	0.10620
0.1	0.1	0.1	0.7	0.2	0.2	2	0.5	0.10691	0.10674
					0.3			0.14559	0.14539
					0.4			0.17777	0.17756
0.1	0.1	0.1	0.7	0.2	0.2	2.0	0.5	0.10691	0.10674
						2.4		0.10737	0.10717
						2.7		0.10740	0.10756
0.1	0.1	0.1	0.7	0.2	0.2	2	0.3	0.12800	0.12794
							0.4	0.11750	0.11736
							0.5	0.10691	0.10674

Figure 15.3b plots the velocity profile for various values of Reynolds number Re. It is found that the velocity profile increases with an increase in Reynold number Re. Further the boundary layer thickness also increases.

15.5.2 Analysis of Temperature Profile

The variation of magnetic parameter M on the temperature profile is demonstrated in Figure 15.4a. It is analyzed that the temperature profile increases with an increase in magnetic parameter M. The magnetic parameter M leads to the enhancement of the thermal boundary layer thickness. Figure 15.4b shows the effect of third-grade parameter β on the temperature profile. It is depicted that the temperature profile increases while the thermal boundary layer thickness decreases when β is increased. Figure 15.5a depicts the graph of the variations in temperature $\theta(\eta)$ versus η. Here different values of mixed convection parameter $\lambda = 0, 0.8, 1.5, 2.3$ are employed. The temperature profile is increased for a larger mixed convection parameter. Figure 15.5b is plotted for the variations of Biot number $\gamma = 0.3, 0.4, 0.6, 1$ in temperature $\theta(\eta)$ via η. Both the temperature profile and thermal boundary layer thickness increased when γ increases. In fact we have a uniform Biot number which is the ratio of convection to conduction.

The influence of Prandtl number Pr on the temperature profile is given in Figure 15.6a. Here we noticed that the temperature profile and thermal boundary layer thickness decreased for larger values o f Pr.

Figure 15.6b describes the influence of Eckert number Ec on temperature field $\theta(\eta)$. Upon increasing Ec, the temperature field and the thermal boundary layer are enhanced.

15.5.3 Skin Friction and Nusselt Number

Table 15.2 shows the values of skin friction coefficient and local Nusselt number. We noted that by increasing the values of γ and Pr, the values of skin-friction coefficient and the local Nusselt number increase. However by increasing the values of β, M, λ, and Ec, both the skin friction coefficient and the local Nusselt number decreased. Increase in α_1 and α_2 causes an increase in both the skin friction coefficient and the local Nusselt number.

15.6 Statistical Paradigm

In this section, we extend our study to the various consequences of pertinent parameters on the considered problem. In order of priority to understand the relationship between various physical parameters and the skin friction coefficient and also for the Nusselt number. We reported the values of skin friction coefficient and heat transfer rate in Table 15.2. The values of correlation coefficients are calculated and cataloged in Table 15.3. It is obvious

TABLE 15.3
Numerical Values of Correlation Coefficient for Skin Friction Coefficient and Nusselt Number

r	$\mathbf{Re}_x^{1/2} C_x$	$\mathbf{Re}_x^{-1/2} Nu_x$
α_1	−0.9894203	0.9888143
α_2	−0.9999224	0.9999126
β	1	0.999847
M	−0.9993833	−0.9993651
λ	0.9998149	−0.9990868
γ	−0.9986005	0.9986076
Pr	−0.926172	0.998535
Ec	0.999997	−0.9999994

that the value of correlation coefficient is bounded between -1 and 1. In addition the absolute value is bounded between 0 and 1. The correlation coefficient not only explores the relation between two variates but also reveals the inverse and direct correspondence between them. The correlation coefficient has the following interpretations:

- If $r = 1$, then there is a perfect positive linear relationship between the two variables.
- If $r = -1$, then there is a perfect negative linear relationship between the two variables.
- If $0.7 \leq r \leq 1$, then there is a strong positive linear relationship between the two variables.
- If $-1 \leq r \leq -0.7$, then there is a strong negative linear relationship between the two variables.
- If $r = 0$, then there is no linear relationship between the two variables.

From Table 15.3, it is noted that there is strongly positive, strongly negative, and perfect positive correlation between the physical attributes and skin friction coefficient as well as the local Nusselt number.

15.7 Probable Error

The probable error is the value which is added or subtracted from the correlation coefficient to get the upper limit and the lower limit, respectively, within which the value of correlation coefficient expectedly lies. The probable error (P.E.) of the correlation coefficient can be obtained by applying the following formula:

$$P.E\,(r) = 0.6745\frac{(1 - r^2)}{\sqrt{n}},$$

where r denotes the correlation coefficient and n denotes the number of observations. The correlation coefficient is not significant if the value of r is less than $P.E.$, and this reveals that there is no correlation between the variables. The correlation is said to be certain when the value of r is 6 times more than the $P.E.$ and insignificant when r is less than $P.E.\,(r)$. This shows that the value of r is significant. Thus the $P.E.$ is calculated to check the reliability of the value of the correlation coefficient. Table 15.4 shows the values of $P.E.$ for the rate of skin friction and local Nusselt number.

15.7.1 Statistical Proclamation

The values of $\frac{r}{P.E.\,(r)}$ are dispensed in Table 15.5. From this table it is obvious that when few values are satisfied the relation $\frac{r}{P.E.\,(r)} > 6$. On the other hand some values of r are less than $P.E.\,(r)$, which indicates that the correlation coefficient is statistically insignificant. In case of a perfect correlation that is $r = 1$, we get the perfect significant correlation. As a consequence we come to the conclusion that some correlation coefficients are remarkable, and the parameters are greatly interconnected to the physical attributes.

TABLE 15.4
Numerical Values of *P.E.* for Skin Friction Coefficient and Nusselt Number

P.E. (r)	$\mathrm{Re}_x^{1/2} C_x$	$\mathrm{Re}_x^{-1/2} Nu_x$
α_1	0.0008196352	0.0008663224
α_2	6.045686×10^{-6}	6.806504×10^{-6}
β	0	1.191624×10^{-5}
M	4.801421×10^{-5}	4.943143×10^{-5}
λ	1.441773×10^{-5}	7.109498×10^{-5}
γ	0.0001089201	0.0001083747
Pr	0.005537802	0.0001140184
Ec	2.363901×10^{-7}	4.621126×10^{-8}

TABLE 15.5
Numerical Values of $\dfrac{r}{P.E.(r)}$ for Skin Friction Coefficient and Nusselt Number

$\frac{r}{P.E.(r)}$	$\mathrm{Re}_x^{1/2} C_x$	$\mathrm{Re}_x^{-1/2} Nu_x$
α_1	−1207.147	1141.393
α_2	−1.65394.3655	146905.4598
β	∞	83906.24895
M	−20814.32784	−20217.11692
λ	69346.20776	−14052.84593
γ	−9168.193015	9214.397825
Pr	−167.2454161	8757.665429
Ec	4230282.909	−21639734.56

15.8 Concluding Remarks

Here we investigated the magnetohydrodynamic mixed convective flow of third-grade fluid towards an exponentially stretching sheet with convective condition. The key points of the considered study are mentioned below.

- The velocity field $f'(\eta)$ decreases with an increase in M, and the temperature profile $\theta(\eta)$ increases when M increases.
- An increase in Pr corresponds to a decrease in the temperature distribution.
- Effects of fluid parameter β on the velocity and temperature fields are quite similar.
- Numerical values of skin friction coefficient and the local Nusselt number increases with an increase in γ and Pr.
- Both the skin friction coefficient and the local Nusselt number decrease for larger values of Ec, β, M, and λ.
- There is perfect positive correlation between the β and the skin friction coefficient.
- Correlation coefficient is statistically insignificant between skin friction coefficient and physical parameters $(\alpha_1, \alpha_2, M, \gamma, \mathrm{Pr})$

References

[1] A. Choudhary, D. Kumar, J.Singh (2016), A fractional model of fluid flow through porous media with mean capillary pressure, *Journal of the Association of Arab Universities for Basic and Applied Sciences* 21, 59–63.

[2] D. Kumar, J. Singh, D. Baleanu (2017), A new fractional model for convective straight fins with temperature-dependent thermal conductivity, *Thermal Science.* doi: 10.2298/TSCI170129096K.

[3] J. Singh, Y. S. Shishodia (2013), An efficient analytical approach for MHD viscous flow over a stretching sheet via homotopy perturbation sumudu transform method, *Ain Shams Engineering Journal* 4, 549–555.

[4] H. M. Srivastava, D. Kumar, J. Singh (2017), An efficient analytical technique for fractional model of vibration equation, *Applied Mathematical Modelling* 45, 192–204.

[5] J. Singh, D. Kumar, R. Swroop, S. Kumar (2017), An efficient computational approach for time-fractional Rosenau-Hyman equation, *Neural Computing and Applications.* doi: 10.1007/s00521-017-2909-8.

[6] J. Choi, D. Kumar, J. Singh, R. Swroop (2017), Analytical techniques for system of time fractional nonlinear differential equations, *Journal of the Korean Mathematical Society* 54(4), 1209–1229.

[7] M. Guedda, Z. Hammouch (2006), On similarity and pseudo-similarity solutions of FalknerSkan boundary layers. *Fluid Dynamics Research*, 38(4), 211–223.

[8] M. Guedda, Z. Hammouch (2008), Similarity flow solutions of a Non-Newtonian Power-law fluid. *International Journal of Nonlinear Science*, 6(3), 255–264.

[9] T. Mekkaoui, Z. Hammouch (2012), Approximate analytical solutions to the Bagley-Torvik equation by the fractional iteration method. *Annals of the University of Craiova-Mathematics and Computer Science Series*, 39(2), 251–256.

[10] Z. Hammouch, T. Mekkaoui (2012), A Laplace-variational iteration method for solving the homogeneous Smoluchowski coagulation equation. *Applied Mathematical Sciences*, 6(18), 879–886.

[11] R. Ul Haq, Z. Hammouch, et W. A. Khan (2014), Water-based squeezing flow in the presence of carbon nanotubes between two parallel disks, *Thermal Science*, 00, 148–148.

[12] A. M. Rashad, A. J. Chamkha and M. Modather (2013), Mixed convection boundary-layer flow past a horizontal circular cylinder embedded in a porous medium filled with a nanofluid under convective boundary condition. *Computers and Fluids*, 86(5), 380–388.

[13] T. Hayat, A. Shafiq, M. Nawaz, A. Alsaedi (2012), MHD axisymmetric flow of third grade fluid between porous disks with heat transfer. *Applied Mathematics and Mechanics (English Edition)*, 33(6), 749–764.

[14] A. Shafiq, M. Nawaz, T. Hayat and A. Alsaedi (2013), Magnetohydrodynamic axisymmetric flow of a third-grade fluid between two porous disks. *Brazilian Journal of Chemical Engineering*, 30(3), 599–609.

16

Solvability of the Boundary-Value Problem for a Third-Order Linear Loaded Differential Equation with the Caputo Fractional Derivative

Praveen Agarwal
International Center for Basic and Applied Sciences
Anand International College of Engineering
Harish-Chandra Research Institute (HRI)
Netaji Subhas University of Technology

Umida Baltaeva
Khorezm Mamun Academy
Urgench State University

Jessada Tariboon
King Mongkut's University of Technology North Bangkok

CONTENTS

16.1 Introduction and Formulation of the Problem

It is well known that many mathematical models of applied problems require investigations of mixed-type equations. The interest for these equations arises intensively due to both theoretical and practical uses of their applications. At present, the range of problems under consideration for partial differential equations as well as for mixed-type equations has significantly expanded. Along with studying the basic boundary-value problems for such equations, since the 70s, much attention has been paid to the formulation and study of non-local boundary-value problems. This is explained by the fact that many practically important problems related to the dynamics of soil moisture [1,2], with the diffusion of particles in a turbulent plasma with the cooling of an inhomogeneous curved rod [3–5], modeling of the laser radiation process [6,7], and problems of population dynamics [8,9], are reduced to the problems with non-local boundary conditions.

On the other hand, another class of researchers confirming the actuality of loaded differential equations is a phenomena in complex evolutionary systems with memory essentially depending on the prehistory of this system [10] and these phenomena are described by the following loaded integro-differential equations of elliptic type [8]

$$\Delta_x u(x,t) + \int_0^t \sum_{j=1}^n k_j(t,\tau)\frac{\partial^2 u(x,\tau)}{\partial x_j^2}d\tau = 0,$$

parabolic type

$$\frac{\partial u(x,t)}{\partial t} = \Delta_x u(x,t) + \int_0^t k_0(t,\tau)\frac{\partial u(x,\tau)}{\partial \tau}d\tau,$$

where Δ_x- Laplace operators at $x = (x_1, x_2, x_3)$, $k_0(t,\tau), k_j(x,\tau)$ are given real-valued functions, $j = 1, 2, \ldots, n$.

The following loaded partial differential equation

$$\frac{\partial^2 u}{\partial t^2} - c_0^2\frac{\partial^2 u}{\partial x^2} = mc_0^2\frac{\partial^2}{\partial x^2}\int_{-\infty}^t \exp\left(-\frac{t-\eta}{\tau_0}\right)\frac{\partial u(x,\eta)}{\partial \eta}d\eta,$$

is found in the theory of wave propagation in dispersive environments [11], where $u(x,t)$ is the density, c_0 is the speed, m is the mass, and τ_0 is the time of relaxation.

The following loaded equation of the first order [4]

$$\frac{\partial \varphi}{\partial t} + y\frac{\partial \varphi}{\partial x} = (1-\rho)\varphi\int_\alpha^\beta (y-\eta)\varphi(x,t,\eta)d\eta - \frac{1}{\tau_0}\left[\varphi - G(y)\int_\alpha^\beta \varphi(x,t,\eta)d\eta\right],$$

is the equation modeling the movement of vehicles on the highway, where $\varphi(x,t,y)$ is the density of the car in the points $x \in [a,b] \subset R$ having a speed, m is the mass, τ_0 is the time of relaxation, and p and G are the given values; moreover, the function $G = G(y)$ satisfies the condition $\int_\alpha^\beta G(y)dy = 1$.

We can cite by many examples as which emphasized the development of the theory of loaded equations in the last several decades. As we know [8,12], the mathematical models of non-local physical-biological fractal processes are represented by loaded integro-differential equations or loaded differential equations, especially with Riemann–Liouville and Caputo operators [13,14,16].

The main goal of the paper is to formulate and investigate the unique solvability of boundary-value problems for a third-order loaded partial differential equation of mixed type, with the Caputo operators.

Integro-differential equations with the operators Caputo and Riemann–Liouville have been recommended and used by many authors. We note that boundary-value problems for mixed-type equations involving these operators were investigated in literature [17–21]. The boundary-value problems for the second-order loaded equations with an integro-differential operator were studied for the hyperbolic and parabolic equations in the works of V.N. Borisov, A.M. Nakhushev [22], V.M. Kaziev [23], and I.G. Mamedov [24], for the equations of parabolic-hyperbolic and elliptic-hyperbolic types in the works of A.V. Dzarakhohov, V.A. Eleev [25], B.S. Kishin, O.Kh. Abdullaev [26–28]. The construction of the theory of unique solvability of problems for a loaded third-order differential equations of parabolic-hyperbolic and elliptic-hyperbolic operators with real and variable fractional order coefficients are required both for the internal completeness of the fractional integro-differentiation theory and numerous applications.

Problems for such equations in some cases are possible to investigate, reducing to the new non-local boundary-value problems for the classical partial differential equations [8,12]. In this work, we reduce the boundary-value problems for the loaded integro-differential equation to non-local problems for the classical partial differential equation. Taking into account the unique solvability of non-local problems, we prove the solvability of the local boundary-value problems for the loaded equation.

We consider the following linear loaded [27] integro-differential equation

$$\frac{\partial}{\partial x}\left(\tilde{L}u\right) = cD_{0x}^{\alpha}\, u\,(x,0), \tag{16.1}$$

where

$$\tilde{L}u \equiv \begin{cases} \widetilde{L_1}u \equiv u_{xx} + a_1u_x + b_1u_y + c_1u, & y > 0, \\ \widetilde{L_2}u \equiv u_{xx} - u_{yy} + a_2u_x + b_2u_y + c_2u, & y < 0, \end{cases}$$

$cD_{0x}^{\alpha}f(x)$ is a fractional differential operator of Caputo type,

$$cD_{0x}^{\alpha}f(x) = \frac{1}{\Gamma(1-\alpha)}\int_0^x (x-t)^{-\alpha}f'(t)dt, \quad 0 < \alpha < 1,$$

$a_k,\ b_k,\ c_k(k = 1,2)$ are the given real parameters.

Let $\Omega_1 \subset R^2$ be a bounded domain by the segments AB, BB_0, AA_0, A_0B_0 of straight lines $y = 0,\ x = 1,\ x = 0,\ and\, y = h$, respectively, when $y > 0$. Ω_2 is a characteristic triangle bounded by the segment AB, the axis OX, and two characteristics

$$AC : x + y = 0, \quad BC : x - y = 1$$

of Equation (16.1) for $y < 0$.

Let us introduce the following notation:

$$I = \{(x,y) : 0 < x < 1,\ y = 0\}, \quad \Omega = \Omega_1 \cup \Omega_2 \cup I.$$

Problem AT**.** Find a regular solution $u(x,y)$ of the Equation (16.1) in Ω, which is continuous in $\overline{\Omega}$, has continuous derivatives $u_x(u_y)$, up to $AA_0 \cup AB \cup AC\,(AB \cup AC)$, and satisfies the boundary-value conditions

$$u(0,y) = \varphi_1(y),\ \ u(1,y) = \varphi_2(y),\ \ u_x(0,y) = \varphi_3(y),\ \ 0 \le y \le h, \tag{16.2}$$

$$u(x,-x) = \psi_1(x), \quad \left.\frac{\partial u(x,y)}{\partial n}\right|_{y=-x} = \psi_2(x), \quad 0 \le x \le \frac{1}{2}, \tag{16.3}$$

and the gluing condition

$$u_y\,(x,+0) = u_y\,(x,-0)\,, \quad (x,0) \in I,$$

where $\varphi_1(y),\ \varphi_2(y),\ \varphi_3(y),\ \psi_1(x)$, and $\psi_2(x)$ are the given functions.

16.2 Representation of Solution of the Equation

One of the most important aspect of the investigation of the boundary-value problem is shown by the following theorem. The following theorem holds.

Theorem 16.1. *Any regular solution of Equation (16.1) (when $y \neq 0$) is represented in the form*

$$u(x,y) = z(x,y) + w(x), \tag{16.4}$$

where $z(x,y)$ is a solution to the equation

$$0 = \frac{\partial}{\partial x}(\widetilde{L}z)\frac{\partial}{\partial x}\begin{cases} \widetilde{L}_1 z \equiv z_{xx} + a_1 z_x + b_1 z_y + c_1 z, & y > 0, \\ \widetilde{L}_2 z \equiv z_{xx} - z_{yy} + a_2 z_x + b_2 z_y + c_2 z, & y < 0, \end{cases} \tag{16.5}$$

where $w(x)$ is a solution of the following integro-differential equation

$$w'''(x) + a_i w''(x) + c_i w'(x) - cD_{0x}^{\alpha} w(x) = cD_{0x}^{\alpha} z(x,0). \tag{16.6}$$

Proof. Let $z(x,y)$ be a regular solution of Equation (16.5) and $w(x)$ be a solution of Equation (16.6). Then, we must prove that the function represented by formula (16.4) will be the solution for Equation (16.1). Thus, in (16.1), $u(x,y)$ instead of $z(x,y) + w(x)$ takes into account (16.5) and (16.6), and it is easy to verify that the relation

$$\frac{\partial}{\partial x}(u_{xx} + a_1 u_x + b_1 u_y + c_1 u) - cD_{0x}^{\alpha} u(x,0) = \frac{\partial}{\partial x}(z_{xx} + a_1 z_x + b_1 z_y + c_1 z)$$

$$+ w'''(x) + a_1 w''(x) + c_1 w'(x) - cD_{0x}^{\alpha} w(x) - cD_{0x}^{\alpha} z(x,0) = 0,$$

that is, represented by formula (16.4), is a solution for Equation (16.1) for $y > 0$.

Vice versa, let $u(x,y)$ be a regular solution of Equation (16.1), and $w(x)$ be a certain solution of the following integro-differential equation

$$w'''(x) + a_1 w''(x) + c_1 w'(x) = cD_{0x}^{\alpha} u(x,0). \tag{16.7}$$

Let us prove the validity of the relation (16.4).

Obviously the function

$$u(x,y) = z(x,y) + \left(-\frac{2}{a_1}\right)\int_0^x \left(\left(x - t + \frac{2}{a_1}\right)e^{-\frac{a_1}{2}(x-t)} - \frac{2}{a_1}\right) cD_{0t}^{\alpha} u(t,0)dt,$$

is a solution of Equation (16.1), as for equation with the right side at $a_1^2 = 4c_1 \neq 0$, where $z(x,y)$ is a solution of Equation (16.5), and the function

$$u(x,y) = -\frac{2}{a_1}\int_0^x \left(\left(x - t + \frac{2}{a_1}\right)e^{-\frac{a_1}{2}(x-t)} - \frac{2}{a_1}\right) cD_{0t}^{\alpha} u(t,0)dt,$$

is a partial solution to Equation (16.1). Hence, (16.1) highlights the validity of the representation (16.4), i.e. $u(x,y) = z(x,y) + w(x)$.

It follows from the latter representation that $u(x,0) = z(x,0) + w(x)$. Thus, Equation (16.7) provides

$$w'''(x) + a_1 w''(x) + c_1 w'(x) - cD_{0x}^{\alpha} w(x) - cD_{0x}^{\alpha} z(x,0) = 0,$$

and the function $z(x,y) = u(x,y) - w(x)$ satisfies Equation (16.5) for $y > 0$ in cases $a_1^2 > 4c_1$ and $a_1^2 < 4c_1$, respectively.

Analogously it was proved in the case for $y < 0$. Theorem 16.1 is proved. □

Remark 16.1. *Taking into account, the function $z(x) = ae^{\sqrt{\lambda}x} + be^{-\sqrt{\lambda}x} + c$ satisfies Equation (16.5) for the investigation of the problem AT, and we can assume without loss of generality that*

$$w(0) = w'(0) = w''(0) = 0. \tag{16.8}$$

Denoting $w'(x) = y(x)$ in Equation (16.6) we have

$$y''(x) + a_i y'(x) + c_i y(x) - \frac{1}{\Gamma(1-\alpha)} \int_0^x (x-t)^{-\alpha} y(t)dt = cD_{0x}^{\alpha} z(x,0), \tag{16.9}$$

or

$$y''(x) + a_i y'(x) + c_i y(x) - \frac{1}{\Gamma(1-\alpha_i)} \int_0^x (x-t)^{-\alpha} y(t)dt$$
$$= \frac{1}{\Gamma(1-\alpha)} \int_0^x (x-t)^{-\alpha} z'(t,0)dt.$$

The characteristic equation, corresponding to the homogeneous equation, has the form

$$k^2 + a_i k + c_i = 0. \tag{16.10}$$

Let us introduce the notation $\triangle = a_i^2 - 4c_i$:

1. If $\triangle = 0$, then Equation (16.10) has two real roots; moreover, they are equal. Thus, at $a_i^2 = 4c_i$, Equation (16.9) is equivalent to the following form:

$$y(x) - \frac{1}{\Gamma(1-\alpha)} \int_0^x y(t)dt \int_t^x (x-s)(s-t)^{-\alpha} e^{-\frac{a_i}{2}(x-s)} ds$$
$$= l_1(0)e^{-\frac{a_i}{2}x} + l_2(0)xe^{-\frac{a_i}{2}x} + \int_0^x (x-t)\, e^{-\frac{a_i}{2}(x-t)} cD_{0x}^{\alpha} z(t,0)dt, \tag{16.11}$$

   where $l_1(0)$ and $l_2(0)$ are unknown constants to be defined.

2. If $\triangle > 0$, then Equation (16.10) has two different real roots, which can be written in the form $k_{1,2} = (-a_i \pm \sqrt{\triangle})/2$, and Equation (16.9) can be written as follows:

$$y(x) - \frac{2}{(k_1-k_2)\Gamma(1-\alpha)} \int_0^x y(t)dt \int_t^x (s-t)^{-\alpha} e^{-\frac{a_i}{2}(x-s)} sh\frac{\sqrt{\triangle}}{2}(x-s)ds$$
$$= l_1(0)e^{k_1 x} + l_2(0)e^{k_2 x} + \frac{2}{(k_1-k_2)} \int_0^x e^{-\frac{a_i}{2}(x-t)} sh\frac{\sqrt{\triangle}}{2}(x-t) cD_{0x}^{\alpha} z(t,0)dt, \tag{16.12}$$

   where $l_1(0)$ and $l_2(0)$ are unknown constants to be defined.

3. If $\triangle < 0$, then it is known that Equation (16.10) has two complex conjugate roots, and Equation (16.9) is equivalent to

$$y(x) - \frac{2}{\sqrt{\triangle}\Gamma(1-\alpha)} \int_0^x y(t)dt \int_t^x (s-t)^{-\alpha} e^{-\frac{a_i}{2}(x-s)} \sin\frac{\sqrt{\triangle}}{2}(x-s)ds$$
$$= l_1(0)e^{-\frac{a_i}{2}x} \cos\frac{\sqrt{\triangle}x}{2} + l_2(0)xe^{-\frac{a_i}{2}x} \sin\frac{\sqrt{\triangle}x}{2}$$
$$+ \frac{2}{\sqrt{\triangle}} \int_0^x e^{-\frac{a_i}{2}(x-t)} \sin\frac{\sqrt{\triangle}}{2}(x-t) cD_{0x}^{\alpha} z(t,0)dt, \tag{16.13}$$

   where $l_1(0)$ and $l_2(0)$ are an unknown constants to be defined.

In Equation (16.11) at $s = t + (x-t)v$, with respect to $y'(0) = y(0) = 0$, we obtain the Volterra integral equation relative to $y(x)$:

$$y(x) - \int_0^x K_1(x,t)y(t)dt = \int_0^x (x-t)\, e^{-\frac{a_i}{2}(x-t)} cD_{0x}^{\alpha} z(t,0)dt, \tag{16.14}$$

where

$$K_1(x,t) = \frac{(x-t)^{2-\alpha}}{\Gamma(1-\alpha)} \int_0^1 v^{-\alpha} (1-v)\, e^{-\frac{a_i}{2}(x-t)(1-v)} ds.$$

Solving the next integral Equation (16.14) we get

$$y(x) = \int_0^x \left(K_1(x,t) + \int_t^x R_1(x,s) K_1(s,t) ds \right) z'(t,0) dt, \tag{16.15}$$

where $R_1(x,s)$ is the resolvent [24] of the kernel $K_1(x,t)$.

Hence, by virtue of (16.8), (16.15), and the designation $w'(x) = y(x)$, after some transformation, we have

$$w(x) = \int_0^x K(x,t) z'(t,0) dt, \tag{16.16}$$

where

$$K(x,t) = \int_t^x \left\{ K_1(s,t) + \int_t^z R_1(s,z) K_1(z,t)\, dz \right\} ds.$$

Similarly we can receive in the cases 2) $a_i^2 > 4c_i$, and 3) $a_i^2 < 4c_i$.

16.3 The Main Results

Theorem 16.2. *If $b_1 < 0$, $c_1 \leq 0$, in Ω_1, and $\varphi_1(0) = \psi_1(0)$,*

$$\varphi_j(y) \in C^1[0,h], (j=1,2),\ \varphi_3(y) \in C[0,h] \cap C^1(0,h), \tag{16.17}$$

$$\psi_1(x) \in C^1[0,1/2] \cap C^3(0,1/2),\ \psi_2(x) \in C[0,1/2] \cap C^2(0,1/2), \tag{16.18}$$

then there exists a unique solution to the problem AT in the domain Ω.

Proof. First, by virtue of the representation (16.4) and in view of (16.8), Equation (16.1) and boundary-value conditions (16.2) and (16.3) are reduced to the form (16.4),

$$z(0,y) = \varphi_1(y),\ z(1,y) = \varphi_2(y) - w(1),\ z_x(0,y) = \varphi_3(y),\ \ 0 \leq y \leq 1, \tag{16.19}$$

$$z(x,-x) = \psi_1(x) - \int_0^x K(x,t) z'(t,0) dt,\ \ 0 \leq x \leq \frac{1}{2}, \tag{16.20}$$

$$\frac{\partial z(x,-x)}{\partial n} = \psi_2(x) - \frac{1}{\sqrt{2}} \int_0^x K'(x,t) z'(t,0) dt,\ 0 \leq x \leq \frac{1}{2}, \tag{16.21}$$

where $w(1)$ is expressed in terms of (16.16).

Since problem AT was reduced to the equivalent non-local problem for a third-order equation of mixed type (16.5), we may conclude that Problem AT has a unique solution, as a direct result of the unique solvability of the non-local problem.

Bearing in mind [28], after integration of Equation (16.5), with respect to x, we have the following form:

$$z_{xx} + a_1 z_x + b_1 z_y + c_1 z = \omega_1(y),\ \ y > 0, \tag{16.22}$$

$$z_{xx} - z_{yy} + a_2 z_x + b_2 z_y + c_2 z = \omega_2(y),\ y < 0, \tag{16.23}$$

where $\omega_1(y)$, $\omega_2(y)$ are arbitrary continuous functions.

First, we find the main functional relations on I deduced from the domains Ω_1 and Ω_2. We introduce the following notations:

$$z(x,y)|_{y=-0} = \tau(x), \qquad (x,0) \in \overline{I}, \tag{16.24}$$

$$\left.\frac{\partial z(x,y)}{\partial y}\right|_{y=-0} = \nu(x), \quad (x,0) \in I, \tag{16.25}$$

where $\tau(x)$ and $\nu(x)$ are unknown functions.

The solution of the Cauchy problem for Equation (16.23) with the conditions (16.24) and (16.25) can be represented in the form

$$z(x,y) = \frac{1}{2}\left[\tau(x-y)\exp\left(-\frac{a_2-b_2}{2}y\right) + \tau(x+y)\exp\left(-\frac{a_2+b_2}{2}y\right)\right]$$

$$-\frac{1}{2}\exp\left(\frac{b_2}{2}y\right)\int_{x-y}^{x+y}\left(\frac{b_2}{2}J_0\left[\sqrt{\lambda_2\left((x-\xi)^2-y^2\right)}\right] - \lambda_2 y\,\bar{J}_1\left[\sqrt{\lambda_2\left((x-\xi)^2-y^2\right)}\right]\right)$$

$$*\exp\left(-\frac{a_2}{2}(x-\xi)\right)\tau(\xi)\,d\xi \tag{16.26}$$

$$+\frac{1}{2}\exp\left(\frac{b_2}{2}y\right)\int_{x-y}^{x+y} J_0\left[\sqrt{\lambda_2\left((x-\xi)^2-y^2\right)}\right]\exp\left(-\frac{a_2}{2}(x-\xi)\right)\nu(\xi)\,d\xi$$

$$+\frac{1}{2}\int_0^y \omega_2(\eta)\,d\eta\int_{x-y+\eta}^{x+y-\eta} J_0\left[\sqrt{\lambda_2\left((x-\xi)^2-(y-\eta)^2\right)}\right]d\xi,\ \ \lambda_2 = \frac{1}{4}\left(4c_2^2 - a_2^2 - b_2^2\right),$$

where $J_0[z]$, $J_1[z]$ is Bessel function [26], and $\bar{J}_1[z] = J_1[z]/z$.

Then using condition (16.21) from (16.26), taking account of $\tau'(0)+\nu(0) = \sqrt{2}\psi_2(0)$, the property of Bessel function was established in [28] and, by replacing the argument x for $-y$ in the relation, we get the following functional relation, transferred from the domain Ω_2 to I:

$$\omega_2(y) + \lambda_2\int_0^y L(\eta,y)\omega_2(\eta)\,d\eta = \Phi'(y), \tag{16.27}$$

$$L(\eta,y) = \int_{\eta-2y}^{-\eta}\left[\bar{J}_1\left[z(-y,y;\xi,\eta)\right] - \frac{1}{2}J_2\left[z(-y,y;\xi,\eta)\right]\right]d\xi, \tag{16.28}$$

$$\Phi(y) = \sqrt{2}\psi_2(-y) - \omega'(-y) + \frac{a_2-b_2}{4}\tau(-2y)\exp\left(-\frac{a_2-b_2}{2}y\right) - \frac{1}{2}\exp\left(\frac{a_2+b_2}{2}y\right)$$

$$*\left\{\int_{-2y}^{0}\left[\frac{a_2-b_2}{2}p_0(\xi,y) - \lambda_2(2y+\xi)p_1(\xi,y) + \lambda_2\bar{J}_1\left[z(-y,y;\xi)\right]\right]\exp\left(\frac{a_2}{2}\xi\right)\tau(\xi)\,d\,\xi\right.$$

$$\left.+\int_{-2y}^{0}\left[-\frac{a_2-b_2}{2}J_0\left[z(-y,y;\xi)\right] + \lambda_2(2y+\xi)\,\bar{J}_1\left[z(-y,y;\xi)\right]\right]\exp\left(\frac{a_2}{2}\xi\right)\nu(\xi)\,d\xi\right\}$$

$$-\left[\nu(0) + \frac{2\lambda_2 y + a_2 - b_2}{4}\tau(0) + \tau'(0)\right]\exp\left(\frac{a_2+b_2}{2}y\right),$$

$$z(x,y;\xi,\eta) = \sqrt{\lambda_2\left((x-\xi)^2-(y-\eta)^2\right)},\ \ z(x,y;\xi,0) = z(x,y;\xi),$$

$$p_i(\xi,y) = \frac{b_2}{2}\bar{J}_i\left[z(-y,y;\xi)\right] - \lambda_2 y\,\bar{J}_{i+1}\left[z(-y,y;\xi)\right].$$

Hence, if we reckon the right-hand side known, with regard to the properties of special functions [28] and the theory of integral equations [26], the solution of Equation (16.27) can be written in the form

$$\omega_2(y) = \Phi'(y) + \int_0^y R_2(y,t)\Phi'(t)dt, \tag{16.29}$$

where $R_2(y,t)$ is the resolvent of the kernel $L(\eta, y)$. Further, from (16.26), bearing in mind (16.20), $\tau(0) = \psi_1(0)$, we deduce the next functional relation, transferred from the Ω_2 to I:

$$\begin{aligned}
&\tau(x) + \int_0^x p_0\left(\frac{x}{2}, \xi\right) \exp\left(-\frac{a_2}{2}(x-\xi)\right) \tau(\xi)\, d\xi \\
&- \int_0^x J_0\left[z\left(\frac{x}{2}, -\frac{x}{2}; \xi\right)\right] \exp\left(-\frac{a_2}{2}(x-\xi)\right) \nu(\xi)\, d\xi \\
= \; & -\varphi_1(0) \exp\left(-\frac{a_2}{2}x\right) + 2\exp\left(-\frac{a_2 - b_2}{4}x\right) \\
&\times \left\{\psi_1\left(\frac{x}{2}\right) - \omega\left(\frac{x}{2}\right) - \frac{1}{2}\int_0^{-\frac{x}{2}} \omega_2(\eta)\, d\eta \int_{x+\eta}^{-\eta} J_0\left[z(x, -x; \xi, \eta)\right] d\xi\right\}.
\end{aligned} \tag{16.30}$$

The equality (16.30) with regard to (16.29) is a second functional relation between $\tau(x)$ and $\nu(x)$, transferred from the domain Ω_2 to I.

Now that, passing through the limit $y \to +0$ in (16.22) considering (16.24), (16.25) and the conditions of the problem (1) and (4), we obtain a second functional relation between the functions $\tau(x)$ and $\nu(x)$ transferred from Ω_1 to I:

$$\tau''(x) + a_1\tau'(x) + b_1\nu(x) + c_1\tau(x) = \omega_1(0), \tag{16.31}$$

where $\omega_1(0)$ is an unknown constant, which will be found later on.

Excepting function $\nu(x)$ from the main functional relations (16.30) and (16.31), with regard to (16.29), we get

$$\tau(0) = \varphi_1(0), \quad \tau'(0) = \varphi_3(0), \quad \tau(x) = \varphi_1(0) + \int_0^x \tau'(t)dt, \tag{16.32}$$

and after some transformations, we get the Volterra integral equation with shift regarding the unknown function $\tau'(y)$:

$$\tau'(x) - \int_0^x L_1(x,t)\tau'(t)dt - \int_0^{\frac{x}{2}} L_2(x,t)\tau'(t)dt = \Phi_2(x), \tag{16.33}$$

where

$$\Phi_2(x) = \omega_1(0)\int_0^x J_0[\sqrt{\lambda_2\xi(\xi - x)}]\exp\left(-\frac{a_2}{2}(x-\xi)\right)d\xi + \Phi_1(x), \tag{16.34}$$

$L_1(x,t)$, $L_2(x,t), and\, \Phi_1(x)$ are continuously differentiable functions.

Since (16.28), (16.29), (16.30), and (16.34), taking account of (16.17), (16.18) enclose that

$$|L_i(x,t)| \le \text{const}, (i = 1, 2), \quad |\Phi_2(x)| \le \text{const}. \tag{16.35}$$

Hence, we use the method of successive approximations for Equation (16.33). Let us consider the following sequence

$$\tau_0'(x) = \Phi_2(x),$$

$$\tau_1'(x) = \Phi_2(x) + \int_0^x L_1(x,t)\Phi_2(t)dt + \int_0^{\frac{x}{2}} L_2(x,t)\Phi_2(t)dt,$$

$$\tau_2'(x) = \Phi_2(x) + \int_0^x L_1(x,t)\tau_1'(t)dt + \int_0^{\frac{x}{2}} L_2(x,t)\tau_1'(t)dt,$$

....

$$\tau_n'(x) = \Phi_2(x) + \int_0^x L_1(x,t)\tau_{n-1}'(t)dt + \int_0^{\frac{x}{2}} L_2(x,t)\tau_{n-1}'(t)dt. \tag{16.36}$$

In order to prove the existence of the solution of Equation (16.33) it is enough to prove the uniform convergence of $\left\{\tau_n'(x)\right\}_{n=1}^{\infty}$. It is enough to prove the convergence of the following series [26]

$$\tau_0'(x) + \sum_{i=1}^{n}\left[\tau_n'(x) - \tau_{n-1}'(x)\right]. \tag{16.37}$$

By denoting $\sup\{\max|L_1(x,t)|, \max|L_2(x,t)|\} = M$, $\max\limits_{x\in[0,1]}|\Phi_2(x)| = m$, we have the estimates

$$\left|\tau_0'(x)\right| \le m,$$

$$\left|\tau_1'(x) - \tau_0'(x)\right| \le \left|\int_0^x L_1(x,t)\tau_0'(t)dt + \int_0^{\frac{x}{2}} L_2(x,t)\tau_0'(t)dt\right| \le \frac{3x}{2}mM,$$

$$\left|\tau_2'(x) - \tau_1'(x)\right| \le \left|\int_0^x L_1(x,t)\left[\tau_1'(t) - \tau_0'(t)\right]dt + \int_0^{\frac{x}{2}} L_2(x,t)\left[\tau_1'(t) - \tau_0'(t)\right]dt\right|$$
$$\le \frac{(2^1+1)(2^2+1)}{2!\,2^3}x^2 mM^2,$$

....

$$\left|\tau_n'(x) - \tau_{n-1}'(x)\right| \le \left|\int_0^x L_1(x,t)\left[\tau_{n-1}'(t) - \tau_{n-2}'(t)\right]dt\right.$$
$$\left. + \int_0^{\frac{x}{2}} L_2(x,t)\left[\tau_{n-1}'(t) - \tau_{n-2}'(t)\right]dt\right| \le \frac{\prod_{i=1}^{n}(2^i+1)x^n}{n!2^{\frac{n(n+1)}{2}}}\cdot mM^n.$$

Consequently, based on the recent estimates, the set (16.37) absolutely and uniformly converges, and the function $\tau'(x) = \lim\limits_{n\to\infty}\tau_n'(x)$ is a solution of Equation (16.33) in I. For proving the uniqueness of the solution of Equation (16.33) in class $C[0,1]$ it is enough to prove that the corresponding homogeneous equation has only a trivial solution in this class.

Further, to write the solution of the integral Equation (16.33) in the explicit form, presuppose

$$\alpha(x) = \Phi_2(x) + \int_0^{\frac{x}{2}} L_2(x,t)\tau'(t)dt, \tag{16.38}$$

and Equation (16.33) has a form

$$\tau'(x) - \int_0^x L_1(x,t)\tau'(t)dt = \alpha(x). \tag{16.39}$$

Considering the right side as known and by virtue of the properties of special functions, based on the theory of integral equations [26], we can say that (16.39) has a unique solution, which is represented as

$$\tau'(x) = \alpha(x) + \int_0^x S_1(x,t)\alpha(t)dt, \tag{16.40}$$

where $S_1(x,t)$ is the resolvent of the kernel $L_1(x,t)$.

Equality (16.40) by considering (16.38) takes into account the formula Dirichlet that has the form

$$\tau'(x) - \int_0^{\frac{x}{2}} L_2^*(x,t)\tau'(t)dt = \Phi_2^*(x), \tag{16.41}$$

where

$$L_2^*(x,t) = L_2(x,t) + \int_{2t}^{x} R_1(x,s)L_2(s,t)ds, \tag{16.42}$$

$$\Phi_2^*(x) = \Phi_2(x) + \int_0^x R_1(x,t)\Phi_2(t)dt. \tag{16.43}$$

By virtue of (16.17), (16.18), (16.35), (16.42), and (16.43) conclude that

$$|L_2^*(x,t)| \le \text{const}, \qquad |\Phi_2^*(x)| \le \text{const}. \tag{16.44}$$

From here, Equation (16.41) has a unique solution, and it can be represented in the form

$$\tau(x) = \Phi_2^*(x) + \int_0^x \Phi_2^*(t)\,dt + \int_0^{\frac{x}{2}} \Phi_2^*(t)\,dt \int_{2t}^{x} S_2(z,t)dz, \; 0 \le x \le 1, \tag{16.45}$$

where $S_2(x,t)$ is the resolvent of the kernel $L_2^*(x,t)$.

Thus, using relation $\tau(1) = \varphi_2(0) - w(1)$, we uniquely define $\omega_1(0)$.

After the determination $\tau(x)$ we uniquely define the functions $\nu(x)$ and $w(x)$, respectively, from the relations (16.31) and (16.16).

After finding the functions $\nu(x)$ and $w_2(y)$, the solution of problem AT in the domain Ω_2 will be found as a solution of the Cauchy (16.26) or Darboux [28] problem, respectively, for Equation (16.23).

It is obvious that the problem for Equation (16.1) in Ω_1 is an undetermined real function which is equivalent to the following problem AT_1 for Equation (16.22) with the boundary conditions (16.2) and $u(x,+0) = \tau(x)$.

Thus, the solution of the problem in Ω_2 with regard to (16.16) and (16.26) is uniquely determined by the formula (16.4), and in the domain Ω_1 we arrive at the problem Γ_{11} [27] for a non-loaded equation of the third order, which, taking into account (16.17) and $b_1 < 0$, $c_1 \le 0$, is uniquely solvable.

Thus, the solution of the problem AT in the domain Ω_2 in view of (16.16) and (16.26) is determined uniquely according to the formula (16.4), and in the domain Ω_1, we arrive to the problem for a non-loaded equation of the third order Γ_{11} [27].

Therefore, we can conclude that there exists a regular solution of problem AT. □

Similarly we can prove solvability of this analogous problem Tricomi.

Problem AT^*. Find a regular solution $u(x,y)$ Equation (16.1) in Ω, which has continuous derivatives $u_x(u_y)$, up to $AA_0 \cup AB \cup BC\,(AB \cup BC)$, and satisfies the boundary-value conditions (16.2),

$$u(x,x-1) = \psi_3(x), \quad \left.\frac{\partial u(x,y)}{\partial n}\right|_{y=x-1} = \psi_4(x), \;\; \frac{1}{2} \le x \le 1, \tag{16.46}$$

and the gluing condition

$$u_y(x,+0) = u_y(x,-0), \;\; (x,0) \in I,$$

where $\varphi_1(y)$, $\varphi_2(y)$, $\varphi_3(y)$, $\psi_3(x)$ and $\psi_4(x)$ are given real-valued functions, moreover $\varphi_2(0) = \psi_3(1)$.

Theorem 16.3. *If* $b_1 < 0$, $c_1 \leq 0$, *and condition is satisfied (16.8),* $\psi_3'(1) = -\sqrt{2}\psi_4(1) + 2\varphi_2'(0)$, $\varphi_j(y) \in C^1[0,h], (j = 1,2)$, $\varphi_3(y) \in C[0,h] \cap C^1(0,h)$, $\psi_3(x) \in C^1[1/2,1] \cap C^3(1/2,1)$, $\psi_4(x) \in C[1/2,1] \cap C^2(1/2,1)$, *then there exists a unique solution to the problem* AT^* *in the domain* Ω.

Theorem 16.3 is proved in the same way as Theorem 16.2. That is, by virtue of Theorem 16.1, Equation (16.1) and the boundary-value conditions (16.2), (16.40), in view of (16.8) are reduced to the form (16.5), (16.19) and

$$z(x, x-1) = \psi_3(x) - w(x), \quad \left.\frac{\partial z(x,y)}{\partial n}\right|_{y=x-1} = \psi_4(x) - \frac{1}{\sqrt{2}}w'(x), \ 0 \leq x \leq \frac{1}{2}, \tag{16.47}$$

where $w(x)$ is expressed in terms of (16.16).

Since problem AT^* was reduced to the equivalent non-local problem, Equation (16.5) can be represented in the form (16.22) and (16.23). Thus, using the solution of the Cauchy problem (16.26), taking into account (16.46),

$$\tau(1) = \psi_3(1), \ \tau(x) = \psi_3(1) - \int_x^1 \tau'(t)dt, \ \psi_3'(1) = -\sqrt{2}\psi_4(1) + 2\varphi_2'(0),$$

we obtain the first functional relation, transferred from the Ω_2 to I. On the other side, the second functional relation (16.31) between the functions $\tau(x)$ and $\nu(x)$ transferred from the Ω_1 to I is correct. Further, conducting an analogous reasoning to that of $\tau(x)$, from the main functional relations with regard to the conditions of the problem, we can find the $\nu(x)$ and $w_2(y)$. Subsequent investigations are performed by analogy with problem AT.

16.4 Conclusion

In this paper, we investigated the solvability of boundary-value problems with the continuous gluing conditions for the linear loaded integro-differential equation. The boundary-value problems for loaded integro-differential equations connect to the non-local boundary value problems, wherein boundary-value problems are reduced to the integral equation of Volterra type with the shift, and using the successive approximations method was proven an existence of unique solutions of equations.

Problems AT and AT^* can be studied with generalized gluing conditions.

Remark 16.2. *Problems* AT *and* AT^* *are possible to examine with the general discontinuous gluing conditions. In this case the problem* AT *(resp. Problem* AT^**) changes in the following way: the function* $u(x,y)$ *continuous in each closed domains* Ω_1 *and* Ω_2 *and conditions (16.2) and (16.3)(resp.(16.46)) remains invariable. Indeed, the conditions are fulfilled:*

$$\begin{aligned} u(x,+0) &= \alpha_1(x)\,u(x,-0) + \gamma_1(x), && 0 < x < 1, \\ u_y(x,+0) &= \beta_1(x)\,u_y(x,-0) + \alpha_2(x)\,u_y(x,-0) + \gamma_2(x), && 0 < x < 1, \end{aligned}$$

where α_1, $\gamma_1 \in C^3$, α_2, β_1, $\gamma_2 \in C^2$ are given functions, and also $\alpha_1\beta_1 \neq 0$ for $0 < x < 1$.

Problem AT and AT^* in this case exist as unique solutions as well.

Acknowledgments

The Praveen Agarwal would like to thanks the worthy referees and editor for their valuable suggestions for our Chapter in this Book. This work was supported by under the first author research grant supported by SERB Project Number: TAR/2018/000001, DST(project DST/INT/DAAD/P-21/2019) and DST (project INT/RUS/RFBR/308).

References

[1] Serbina LI. *Nonlocal Mathematical Models of Movement in Watertransit Systems*, Nauka Publishers, Moscow (2007).

[2] Ashyralyev A, Ozdemir Y. On nonlocal boundary value problems for hyperbolic-parabolic equations. In: Tas K, Tenreiro Machado JA, Baleanu D (eds) *Mathematical Methods in Engineering*, Springer, Dordrecht (2007): 131–140.

[3] Repin OA, Kumykova, SK. Nonlocal problem for an equation of mixed type in a domain whose elliptic part is a half-strip, *Differential Equations* 50, (2014): 805. doi: 10.1134/S0012266114060081.

[4] Nakhushev AM. *Loaded Equations and Their Applications*, Nauka Publishers, Moscow (2012) [in Russian].

[5] Agarwal P, Berdyshev A, Karimov E. Solvability of a non-local problem with integral transmitting condition for mixed type equation with caputo fractional derivative, *Results in Mathematics*, 71(3–4), (2017): 1235–1257.

[6] Daniel F. Gennaro Infante and Feliz Manuel Minhos, nonlocal boundary value problems, *Boundary Value Problems*, 2012, (2012): 23. doi: 10.1186/1687-2770-2012-23.

[7] Dzhenaliev MT. *A Remark on the Theory of Linear Boundary Value Problems for Loaded Differential Equations*, Institute of Theoretical and Applied Mathematics, Almaty, (1995) 270 p.

[8] Nakhushev AM. *Equations of Mathematical Biology*, Vishaya shkola, Moscow, (1995) 302 p.

[9] Maity D, Tucsnak M, Zuazua E. Controllability and positivity constraints in population dynamics with age structuring and diffusion, (2018) 1–23. *HAL Id: hal*-01764865, https://hal.archives-ouvertes.fr/hal-01764865v2.

[10] Vladimirov VS. Mathematical problems of one-speed particle transport theory, *Proceedings of the Steklov Institute of Mathematics, AN SSSR*, 61, (1961):3–158 (Russian).

[11] Vinogradova MB, Rudenko OV, Sukhorukov AP. *The Wave Theory*, Nauka Publishers, Moscow (1990).

[12] Baltaeva UI. The loaded parabolic-hyperbolic equation and its relation to non-local problems, *Nanosystems Physics, Chemistry, Mathematics*, 8(4), (2017): 413–419.

[13] Miller KS, Ross B. *An Introduction to the Fractional Calculus and Fractional Differential Equations*, Wiley & Sons, New York (1993).

[14] Podlubny I. Geometric and physical interpretation of fractional integration and fractional differentiation. *Dedicated to the 60th Anniversary of Prof. Francesco Mainardi. Fractional Calculus and Applied Analysis* 5(4), (2002): 367–386.

[15] Kilbas AA, Srivastava HM, Trujillo JJ. *Theory and Applications of Fractional Differential Equations*, Elsevier, Amsterdam (2006).

[16] Lazarevic MP, Rapaic MR, Sekara TB. Introduction to fractional calculus with brief historical background, In *Advanced Topics on Applications of Fractional Calculus on Control Problems.* WSEAS Press Editors, Valeri Mladenov, Nikos Mastorakis, (2014), 202 pp.

[17] Kilbas AA, Repin OA. An analog of the Tricomi problem for a mixed type equation with a fractional derivative, *Fractional Calculus and Applied Analysis*, 13(1), (2010): 69–84.

[18] Berdyshev AS, Cabada A, Karimov ET. On a non-local boundary problem for a parabolic-hyperbolic equation involving a Riemann-Liouville fractional differential operator, *Nonlinear Analysis*, 75, (2012): 3268–3273.

[19] Karimov ET, Al-Salti N, Kerbal, S. Optional supplementary materials inverse source non-local problem for mixed type equation with Caputo fractional differential operator (2016), *arXiv:1603.09476v1 [math.AP] 31 March.*

[20] Agarwal P, Karimov E, Mamchuev M, Ruzhansky M. On boundary value problems for a partial differential equation with Caputo and Bessel operators, In *Novel Methods in Harmonic Analysis*, Vol. 2, pp. 707–719, Editors: Isaac Pesenson, Quoc Thong Le Gia, Azita Mayeli, Hrushikesh Mhaskar, Ding-Xuan Zhou. Applied and Numerical Harmonic Analysis, Birkhauser, Basel (2017).

[21] Baltaeva U, Torres PJ. Analog of the Gellerstedt problem for a loaded equation of the third order, *Mathematical Methods in the Applied Sciences*, 42, (2019): 3865–3876.

[22] Nakhushev AM, Borisov VN. Boundary value problems for loaded parabolic equations and their applications to the prediction of ground water level, *Differential Equations*, 13(1), (1977): 105–110.

[23] Kaziev VM. Darboux's problem for a loaded second-order integrodifferential equation, *Differential Equations*, 14(1), (1978): 181–184.

[24] Mamedov IG. Generalization of multipoint boundary-value problems of Bitsadze-Samarski and Samarski-Ionkin type for fourth order loaded hyperbolic integro-differential equations and their operator generalization, *Proceedings of IMM of NAS of Azerbaijan*, 56(1), (2005): 77–84.

[25] Dzarakhohov AV, Eleev VA. On a nonlocal boundary value problem for a third-order loaded equation, *Vladikavkazskii Matematicheskii Zhurnal*, 6(3), (2004): 36–46.

[26] Sadarangani KB, Abdullaev OKh. Non-local problems with integral gluing condition for loaded mixed type equations involving the Caputo fractional derivative, *Electronic Journal of Differential Equations*, 2016(164), (2016): 1–10.

[27] Islomov B, Baltaeva UI. Boundary value problems for a third-order loaded parabolic-hyperbolic equation with variable coefficients. *Electronic Journal of Differential Equations*, 2015(221), (2015): 1–10.

[28] Baltayeva U, Agarwal P. Boundary-value problems for the third order loaded equation with noncharacteristic type-change boundaries, *Mathematical Methods in the Applied Sciences* 41, (2018): 3307–3315.

17

Chaotic Systems and Synchronization Involving Fractional Conformable Operators of the Riemann–Liouville Type

J. E. Solís-Pérez

Tecnológico Nacional de México/CENIDET

J. F. Gómez-Aguilar, R. F. Escobar-Jiménez, and J. Reyes-Reyes

CONACyT-Tecnológico Nacional de México/CENIDET

CONTENTS

17.1 Introduction

Chaos synchronization is an interesting topic to address for the scientific community because of chaos being present in a wide variety of phenomena of different nature. Examples of synchronization include communication among fireflies [1,2], financial time series [3], biosocial interactions [4], brain cells [5,6], circadian rhythms [7], electric grids [8,9], robotic swarms [10,11], encryption schemes [12,13], mobile communication [14,15], among others. In literature, different approaches have been proposed and studied to investigate synchronization of two or more chaotic oscillators such as back-stepping [16,17], sliding mode control method [18,19], active control method [20,21], linear control method [22], adaptive control method [23], master–slave synchronization [24,25], among others.

Introduction of fractional calculus (FC) in nonlinear models and chaotic systems has rendered a new dimension to the existing problems. Fractional derivatives, which are a generalization of classical derivatives, have been extensively used in chaotic systems and chaos synchronization, and examples and applications can also be found in [26]–[31]. In these examples, fractional derivatives have been effectively applied to the applications of chaos synchronization problem.

Fundamental definitions of fractional derivatives include Coimbra, Riesz, Riemann–Liouville (RL), Hadamard, Weyl, Grünwald–Letnikov, Marchaud, Liouville–Caputo, and so on [32]. An interesting idea that extends the ordinary limit definitions of the derivatives of a function called conformable derivative was proposed in [33]. Considering the conformable derivative, the authors in [34] proposed new fractional operators in RL-Caputo and Hadamard sense, these operators were obtained iterating Riemann improper integrals.

In this chapter, we will focus on master–slave synchronization topology. In this case, one system acts as the "master", and consequently, the other system acts as the "slave". The aim is to synchronize the complete response of the slave (described by a state observer) using the master (described by a dynamic system in state-space representation) by driving the slave with a signal provided by the master system.

17.2 Mathematical Preliminaries

The RL operator with order ($\alpha > 0$) is defined as follows [32]:

$$ {}_{a}^{RL}\mathscr{D}_{t}^{\alpha} f(t) = \frac{1}{\Gamma(n-\alpha)} \frac{d^n}{dt^n} \int_a^t f(\theta)(t-\theta)^{n-\alpha-1} d\theta, \qquad n-1 < \alpha < n. \tag{17.1} $$

Let $f : [a, \infty) \longrightarrow \mathbb{R}$ be a function of αth order, the conformable derivative with order ($\alpha > 0$) of $f(t)$ is given by [33]

$$ {}_{a}\mathscr{D}_{t}^{\alpha} f(t) = \lim_{\epsilon \to 0} \frac{f\left(t + \epsilon t^{1-\alpha}\right) - f(t)}{\epsilon}. \tag{17.2} $$

The conformable integral of a function $f(t)$ is defined as

$$ {}_{a}I_{t}^{\alpha} f(t) = \int_a^t \frac{f(x)}{(x-a)^{1-\alpha}} dx, \quad x \geq a, \quad 0 < \alpha \leq 1, \tag{17.3} $$

and the above integral is the Riemann improper integral [35].

Iterating n-times the integral (17.3) and changing the order of the integral, we have

$$ {}_{a}I_{t}^{n\ \alpha} f(t) = \frac{1}{\Gamma(n)} \int_a^t \left(\frac{(t-a)^{\alpha} - (x-a)^{\alpha}}{\alpha}\right)^{n-1} \frac{f(x)}{(x-a)^{1-\alpha}}\, dx. \tag{17.4} $$

Replacing the integer n, for $\beta \in \mathbb{C}$, with $\mathrm{Re}(\beta) > 0$, we define the following fractional conformable integral [34]

$$ {}_{a}^{\beta}I_{t}^{\alpha} f(t) = \frac{1}{\Gamma(\beta)} \int_a^t \left(\frac{(t-a)^{\alpha} - (x-a)^{\alpha}}{\alpha}\right)^{\beta-1} \frac{f(x)}{(x-a)^{1-\alpha}}\, dx. \tag{17.5} $$

Let $\text{Re}(\beta) \geq 0$, $n = [\text{Re}(\beta)] + 1$, $f \in C^n_{\alpha,a}\Big([a,b]\Big)$, $\Big(f \in C^n_{\alpha,b}\big([a,b]\big)\Big)$. The left fractional conformable derivative in the RL sense is given by [34]

$$
\begin{aligned}
{}^{RL}{}^{\beta}_{a}\mathscr{D}^{\alpha}_{t} f(t) &= \frac{{}^{n}_{a}\mathscr{D}^{\alpha}_{t}}{\Gamma(n-\beta)} \int_a^t \left(\frac{(t-a)^{\alpha} - (x-a)^{\alpha}}{\alpha}\right)^{n-\beta-1} \frac{f(x)}{(x-a)^{1-\alpha}}\, dx, \\
&= {}^{RL}{}^{n}_{a}\mathscr{D}^{\alpha}_{t}\Big({}^{RL}{}^{n-\beta}_{a}I_t f(t)\Big).
\end{aligned} \tag{17.6}
$$

17.3 Design of the Slave System

For consideration, the master system can be represented as an observable nonlinear system (17.7)

$$
\begin{aligned}
\dot{\varsigma}(t) &= A\varsigma(t) + \varphi\left(\varsigma(t)\right), \\
y(t) &= C\varsigma(t),
\end{aligned} \tag{17.7}
$$

where $\varsigma(t)$ is the n states vector given by ς_i with $i = 1, 2, \ldots, n$. The output signal $y(t)$ represents the measured state or the driven signal if we talk about synchronization. The matrices A and C are given by

$$
A = \begin{bmatrix} 0 & 1 & \cdots & 0 & 0 \\ 0 & 0 & 1 & \cdots & 0 \\ \vdots & & & \ddots & \vdots \\ 0 & & \cdots & & 1 \\ 0 & & \cdots & & 0 \end{bmatrix}, \tag{17.8}
$$

$$
C = \begin{bmatrix} 1 & 0 & \cdots & 0 \end{bmatrix}, \tag{17.9}
$$

where the Lipschitz nonlinearities vector $\varphi(\varsigma)$ is given by

$$
\varphi(\varsigma(t)) = \begin{bmatrix} \varsigma_1(t) \\ \varsigma_1(t), \varsigma_2(t) \\ \varsigma_1(t), \varsigma_2(t), \varsigma_3(t) \\ \vdots \\ \varsigma_1(t), \varsigma_2(t), \ldots, \varsigma_n(t) \end{bmatrix}. \tag{17.10}
$$

In this proposal, we consider the fractional operator (17.6) in the RL sense to represent the system given in Eq. (17.7), i.e.

$$
\begin{aligned}
{}^{RL}{}^{\beta}_{a}\mathscr{D}^{\alpha}_{t} f(t)\varsigma(t) &= A\varsigma(t) + \varphi\left(\varsigma(t)\right), \\
y(t) &= C\varsigma(t).
\end{aligned} \tag{17.11}
$$

A fractional conformable slave system can be designed if the system (17.7) is observable, then the state observer is given by

$$
{}^{RL}{}^{\beta}_{a}\mathscr{D}^{\alpha}_{t} f(t)\hat{\varsigma}(t) = A\hat{\varsigma}(t) + \varphi\left(\varsigma(t)\right) + K\left(y(t) - C\hat{\varsigma}(t)\right), \tag{17.12}
$$

where $\hat{}$ means estimation, and K is a correction term given by

$$
K = S^{-1}C^{T}, \tag{17.13}
$$

where S is the solution of the algebraic Lyapunov equation

$$-\lambda S - A^T S - SA + C^T C = 0, \tag{17.14}$$

where λ is a parameter to ensure the convergence of the observer (17.12).

17.4 Numerical Method for Fractional Conformable Derivative in the RL Sense

We can get the second derivative of a function $f(t)$ from the first derivative of a continuous function

$$\frac{d}{dt}f(t) \equiv \dot{f}(t) = \lim_{h\to 0} \frac{f(t) - f(t-h)}{h}, \tag{17.15}$$

i.e.

$$\frac{d^2}{dt^2}f(t) \equiv \ddot{f}(t) = \lim_{h\to 0} \frac{f(t) - 2f(t-h) + f(t-2h)}{h^2}. \tag{17.16}$$

From Eqs. (17.15) and (17.16) we can obtain the third derivative of the function $f(t)$

$$\frac{d^3}{dt^3} \equiv \dddot{f}(t) = \lim_{h\to 0} \frac{f(t) - 3f(t-h) + 3f(t-2h) - f(t-3h)}{h^3}. \tag{17.17}$$

If we consider $n \in N$, $j > n$, then we can generalize the n-th derivative from the function $f(t)$ in the following way

$$\frac{d^n}{dt^n}f(t) \equiv f^{(n)}(t) = \lim_{h\to 0} \frac{1}{h^n} \sum_{j=0}^{n} (-1)^j \binom{n}{j} f(t - jh), \tag{17.18}$$

where the binomial coefficients are defined as

$$\binom{n}{j} = \frac{n(n-1)(n-2)\cdots(n-j+1)}{j!} = \frac{n!}{j!(n-j)!}. \tag{17.19}$$

The fractional-order derivative of order $\beta \in \mathbb{R}$ is expressed according to Eqs. (17.15–17.18), i.e.

$${}^{GL}\mathscr{D}_t^\beta f(t) = \lim_{h\to 0} \frac{1}{h^\beta} \sum_{j=0}^{\infty} (-1)^j \binom{\beta}{j} f(t - jh), \tag{17.20}$$

where ${}^{GL}\mathscr{D}_t^\beta f(t)$ is the operator of Grnwald-Letnikov (GL) type. Using the Euler's Gamma function, the binomial coefficients are defined as

$$\binom{\beta}{j} = \frac{\beta!}{j!\,(\beta-j)!} = \frac{\Gamma(\beta+1)}{\Gamma(j+1)\Gamma(\beta-j+1)}, \tag{17.21}$$

with $\binom{\beta}{0} = 1$.

We can express the limit of Eq. (17.20) if we consider $n = \frac{t}{h}$, 0 and t as the bounds of operation for the following expression

$${}_0^{GL}\mathscr{D}_t^\beta f(t) = \lim_{h\to 0} \frac{1}{h^\beta} \sum_{j=0}^{n} (-1)^j \binom{\beta}{j} f(t - jh). \tag{17.22}$$

The numerical solution of fractional-order derivatives is obtained for the relation (17.23) derived from GL definition (17.22)

$$_{(k-L_m/h)}{}^{GL}\mathscr{D}_{t_k}^{\beta} f(t) \approx h^{-\beta} \sum_{j=0}^{k} (-1)^j \binom{\beta}{j} f(t_{k-j}), \tag{17.23}$$

where L_m is the "memory length", $t_k = kh$, h time step, and $(-1)^j \binom{\beta}{j}$ are binomial coefficients $c_j^{(\beta)}$ $(j = 0, 1, \dots)$ which can be calculated for the following expression

$$c_0^{(\beta)} = 1, \quad c_j^{(\beta)} = \left(1 - \frac{1+\beta}{j}\right) c_{j-1}^{(\beta)}. \tag{17.24}$$

From the solution expressed in (17.23), the numerical algorithm to obtain a solution to Eq. (17.6) is expressed as follows

$$y(t_k) = \left({}^{RL}{}_a^{\beta}\mathscr{D}_t^{\alpha} f(y(t_k), t_k)\right) h^{\beta} - \sum_{j=1}^{k} c_j^{(\beta)} y(t_{k-j}). \tag{17.25}$$

At this point, to the best of the authors' knowledge, there is no known or published numerical methods for solving Eq. (17.25).

17.5 Examples

17.5.1 Moore–Spiegel System

The Moore–Spiegel system is one of the general classes of dynamical systems proposed in [36]. In this work, we propose a new approach involving Eq. (17.6), whose representation is given by the following fractional conformable equations in the RL sense

$$\begin{aligned} {}^{RL}{}_a^{\beta}\mathscr{D}_t^{\alpha} x_1 &= \frac{1}{t^{1-\alpha}} x_2, \\ {}^{RL}{}_a^{\beta}\mathscr{D}_t^{\alpha} x_2 &= \frac{1}{t^{1-\alpha}} x_3, \\ {}^{RL}{}_a^{\beta}\mathscr{D}_t^{\alpha} x_3 &= \frac{1}{t^{1-\alpha}} \left[-x_3 - \left(T - R + Rx_1^2\right) x_2 - Tx_1\right], \end{aligned} \tag{17.26}$$

where R and T are constants, and $\alpha \in (0, 1]$.

The numerical solution to the previous equations to generate new chaotic behavior derives from the scheme (17.25), i.e.

$$x_1(t_k) = \left(\frac{1}{t_{k-1}{}^{(1-\alpha)}} x_2(t_{k-1})\right) h^{\beta} - \sum_{j=1}^{k} c_j^{(\beta)} x_1(t_{k-j}), \tag{17.27a}$$

$$x_2(t_k) = \left(\frac{1}{t_{k-1}{}^{(1-\alpha)}} x_3(t_{k-1})\right) h^{\beta} - \sum_{j=1}^{k} c_j^{(\beta)} x_2(t_{k-j}), \tag{17.27b}$$

$$\begin{aligned} x_3(t_k) = \Bigg(\frac{1}{t_{k-1}{}^{(1-\alpha)}} & \left[-x_3(t_{k-1}) - \left(T - R + Rx_1(t_{k-1})^2\right) x_2(t_{k-1})\right. \\ & \left. - Tx_1(t_{k-1})\right]\Bigg) h^{\beta} - \sum_{j=1}^{k} c_j^{(\beta)} x_3(t_{k-j}), \end{aligned} \tag{17.27c}$$

with the following parameters $T = 20$, $R = 100$, step size $h = 1 \times 10^{-3}$, and initial conditions $x_1(0) = -1$, $x_2(0) = 1$, $x_3(0) = 0$. The orders of fractional conformable derivative were fixed at $\alpha = 0.8$, $\beta = 0.9$ and the time simulation was $t = 300$ [s].

Figure 17.1 shows the behavior of the system given by Eq. (17.27) where we can observe the generation of "new chaotic behavior" involving two different orders. The behavior of fractional conformable Moore–Spiegel system is different according to the conformable and

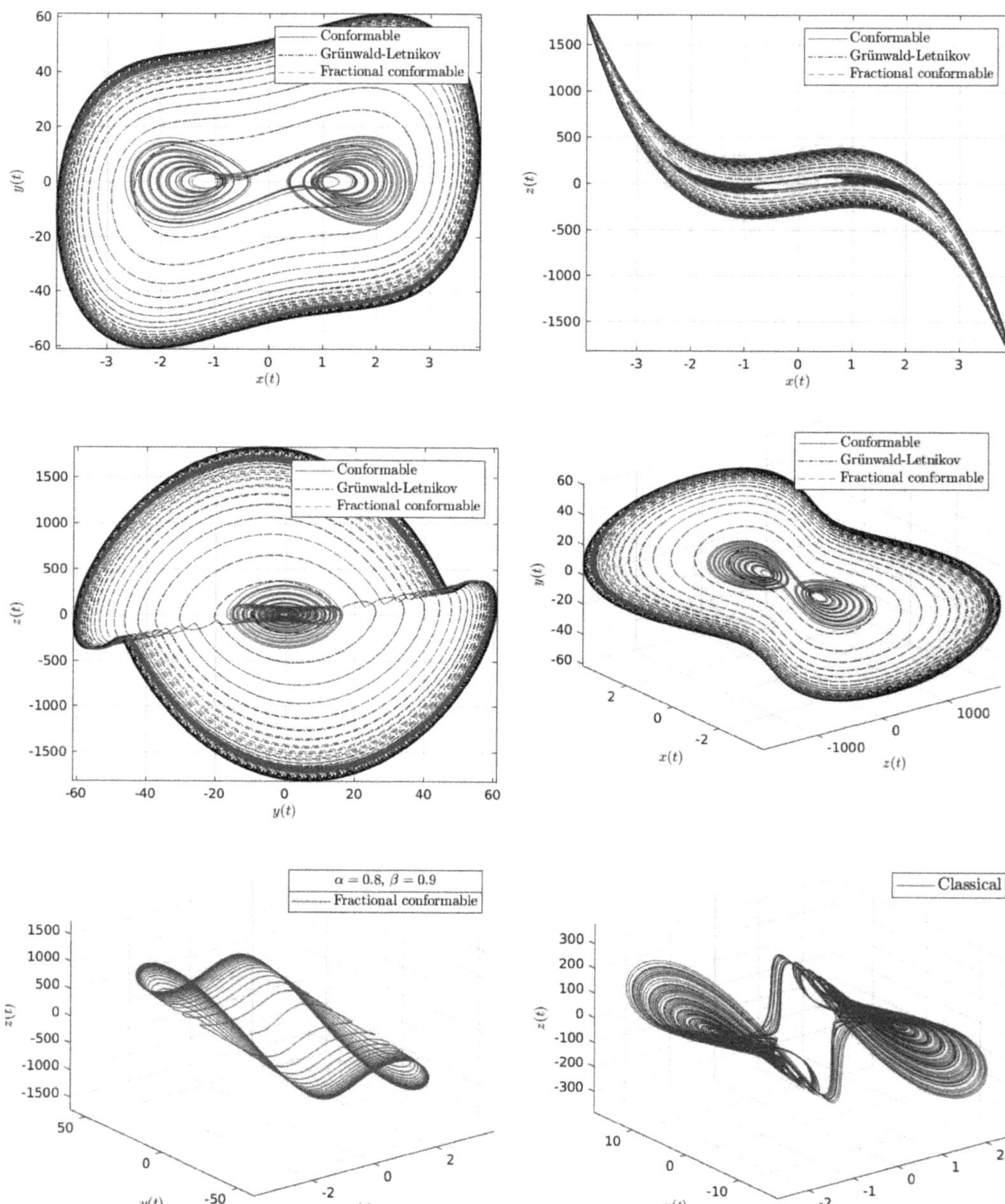

FIGURE 17.1

— Conformable Moore–Spiegel system with $\alpha = 0.8$, ···· Fractional Moore–Spiegel system in GL sense with $\beta = 0.9$, - - - Fractional conformable Moore–Spiegel system in Riemann–Liouville sense with $\alpha = 0.8$ and $\beta = 0.9$.

fractional formulation even when the parameters and initial conditions are the same. This new behavior will be used for the synchronization under the master–slave scheme using the formulation given by Eq. (17.12).

17.5.2 Arneodo's System

The Arneodo system [37] in the fractional conformable sense is represented by the following set of equations

$$\begin{aligned}
{}^{RL}{}^{\beta}_{a}\mathscr{D}^{\alpha}_{t} x_1 &= \frac{1}{t^{1-\alpha}}\, x_2, \\
{}^{RL}{}^{\beta}_{a}\mathscr{D}^{\alpha}_{t} x_2 &= \frac{1}{t^{1-\alpha}}\, x_3, \\
{}^{RL}{}^{\beta}_{a}\mathscr{D}^{\alpha}_{t} x_3 &= \frac{1}{t^{1-\alpha}}\Big[-\beta_1 x_1 - \beta_2 x_2 - \beta_3 x_3 + \beta_4 x_1^3\Big],
\end{aligned} \tag{17.28}$$

where β_1, β_2, β_3, and β_4 are constant parameters. If $\beta_1\beta_4 > 0$, then the Arneodo system has three fixed points. When we choose $\beta_1 = -5.5$, $\beta_2 = 3.5$, $\beta_3 = 1$, and $\beta_4 = -1$, the system given by Eq. (17.28) has a chaotic attractor; however, we find that a new chaotic behavior exists in Arneodo's system if we involve the orders α and β.

In this example, we consider that the master system is represented by Eq. (17.28), and we assume that the available state is $y(t) = x_1(t)$, hence, making a change of coordinates

$$\begin{aligned}
\varsigma_1(t) &= x_1(t), \\
\varsigma_2(t) &= x_2(t), \\
\varsigma_3(t) &= x_3(t).
\end{aligned} \tag{17.29}$$

In this way, the fractional conformable Arneodo's system in canonical form is given by the following equations

$$\begin{aligned}
t^{1-\alpha}\, {}^{RL}{}^{\beta}_{a}\mathscr{D}^{\alpha}_{t} \begin{bmatrix} \hat{\varsigma}_1(t) \\ \hat{\varsigma}_2(t) \\ \hat{\varsigma}_3(t) \end{bmatrix} &= \begin{bmatrix} 0 & 1 & 0 \\ 0 & 0 & 1 \\ 0 & 0 & 0 \end{bmatrix} \begin{bmatrix} \hat{\varsigma}_1(t) \\ \hat{\varsigma}_2(t) \\ \hat{\varsigma}_3(t) \end{bmatrix} + \begin{bmatrix} 0 \\ 0 \\ -\beta_1\hat{\varsigma}_1(t) - \beta_2\hat{\varsigma}_2(t) - \beta_3\hat{\varsigma}_3(t) + \beta_4\hat{\varsigma}_1(t)^3 \end{bmatrix} \\
&\quad + K \begin{bmatrix} y(t) - \hat{\varsigma}_1(t) \\ y(t) - \hat{\varsigma}_1(t) \\ y(t) - \hat{\varsigma}_1(t) \end{bmatrix},
\end{aligned} \tag{17.30}$$

where $t^{1-\alpha}$ is given by the following property [33]

$${}_{a}\mathscr{D}^{\alpha}_{t} f(t) = t^{n+1-\alpha}\frac{d^{n+1}}{dt^{n+1}} f(t), \qquad \alpha \in [n, n+1], \tag{17.31}$$

and $K = \begin{bmatrix} k_1 & k_2 & k_3 \end{bmatrix}^T$ is calculated by Eq. (17.13); also, system (17.30) represents the slave system.

The numerical solution of the systems (17.28) and (17.30) can be obtained if we consider the numerical algorithm (17.25) obtaining

$$x_1(t_k) = \left(\frac{1}{{t_{k-1}}^{(1-\alpha)}}\, x_2(t_{k-1})\right) h^{\beta} - \sum_{j=1}^{k} c_j^{(\beta)} x_1(t_{k-j}), \tag{17.32a}$$

$$x_2(t_k) = \left(\frac{1}{{t_{k-1}}^{(1-\alpha)}}\, x_3(t_{k-1})\right) h^{\beta} - \sum_{j=1}^{k} c_j^{(\beta)} x_2(t_{k-j}), \tag{17.32b}$$

$$x_3(t_k) = \left(\frac{1}{t_{k-1}{}^{(1-\alpha)}}\left[-\beta_1 x_1(t_{k-1}) - \beta_2 x_2(t_{k-1}) - \beta_3 x_3(t_{k-1}) + \beta_4 x_1(t_{k-1})^3\right]\right)h^\beta - \sum_{j=1}^{k} c_j^{(\beta)} x_3(t_{k-j}), \tag{17.32c}$$

for the master system and

$$\hat{\varsigma}_1(t_k) = \left(\frac{1}{t_{k-1}{}^{(1-\alpha)}}\left[\hat{\varsigma}_2(t_{k-1}) + k_1\left(y(t_{k-1}) - \hat{\varsigma}_1(t_{k-1})\right)\right]\right)h^\beta - \sum_{j=1}^{k} c_j^{(\beta)} \hat{\varsigma}_1(t_{k-j}), \tag{17.33a}$$

$$\hat{\varsigma}_2(t_k) = \left(\frac{1}{t_{k-1}{}^{(1-\alpha)}}\left[\hat{\varsigma}_3(t_{k-1}) + k_2\left(y(t_{k-1}) - \hat{\varsigma}_1(t_{k-1})\right)\right]\right)h^\beta - \sum_{j=1}^{k} c_j^{(\beta)} \hat{\varsigma}_2(t_{k-j}), \tag{17.33b}$$

$$\hat{\varsigma}_3(t_k) = \left(\frac{1}{t_{k-1}{}^{(1-\alpha)}}\Big[-\beta_1\hat{\varsigma}_1(t_{k-1}) - \beta_2\hat{\varsigma}_2(t_{k-1}) - \beta_3\hat{\varsigma}_3(t_{k-1}) + \beta_4\hat{\varsigma}_1(t_{k-1})^3 + k_3\left(y(t_{k-1}) - \hat{\varsigma}_1(t_{k-1})\right)\Big]\right)h^\beta - \sum_{j=1}^{k} c_j^{(\beta)} \hat{\varsigma}_3(t_{k-j}), \tag{17.33c}$$

for the case of the slave system.

To synchronize the fractional conformable system given by Eq. (17.30) we consider the following initial conditions: $x_1(0) = -0.2$, $x_2(0) = 0.5$, $x_3(0) = 0.2$ and $\varsigma_1(0) = 0$, $\varsigma_2(0) = 0$, $\varsigma_3(0) = 0$ for the master and slave systems, respectively. The step size was $h = 1 \times 10^{-2}$ and the time simulation $t = 500$ $[s]$. The gain of the observer obtained by optimizing Eq. (17.14) is $K = \begin{bmatrix} 7.4061 & 18.2834 & 15.0454 \end{bmatrix}^T$.

Figure 17.2 (top left) shows the dynamics of the master and slave systems using the fractional conformable derivative of order $\alpha = 0.90$ and $\beta = 0.89$, where we can appreciate the available state in Figure 17.2 (top right), whereas Figure 17.2 (bottom left and right) show the synchronized states.

In order to show the effect of the fractional conformable derivative, in Figure 17.3 (top left) a new behavior is shown considering the same parameters and initial conditions and varying only the orders α and β of the master system (17.28). In Figures 17.3 (top right and bottom) it is possible to appreciate that the synchronization is not affected; on the contrary, the presence of the orders in Eq. (17.6) provides two additional degrees of freedom.

17.5.3 Van der Pol Oscillator (VPO)

The Van der Pol oscillator (VPO) model was proposed by Van der Pol in 1920 with the aim to study oscillations in vacuum tube circuits. VPO is an oscillator with nonlinear damping represented by a second-order differential equation [38]

$$\ddot{y}(t) + \mu\left(y(t)^2 - 1\right)\dot{y}(t) + y(t) = 0, \tag{17.34}$$

where μ is the control parameter. To get an interesting behavior we consider the operator (17.6) to represent Eq. (17.34) in state-space representation, i.e.

$$\begin{aligned} {}^{RL\ \beta}_{\ \ a}\mathscr{D}_t^\alpha x_1 &= \frac{1}{t^{1-\alpha}}\, x_2, \\ {}^{RL\ \beta}_{\ \ a}\mathscr{D}_t^\alpha x_2 &= \frac{1}{t^{1-\alpha}}\left[\mu\left(1 - x_1^2\right)x_2 - x_1\right], \end{aligned} \tag{17.35}$$

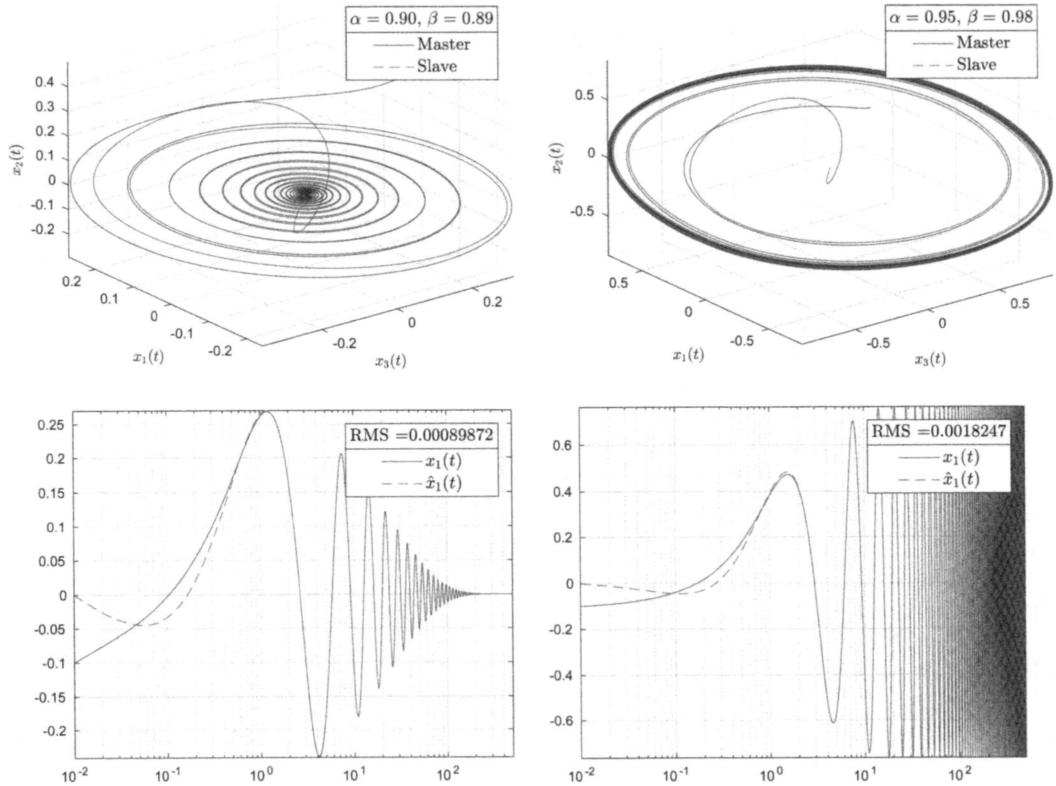

FIGURE 17.2
Arneodo's system with $\alpha = 0.90$ and $\beta = 0.89$. (top left) Portrait phase of the master and slave systems, (top right) measured and estimated state, and (bottom left and right) synchronization of the states in the master and slave systems, respectively.

with an equilibrium point in the origin. For synchronization purpose, we derived a numerical solution of the fractional conformable VPO in RL sense, obtained by using the relation (17.25), which has the following form

$$x_1(t_k) = \left(\frac{1}{{t_{k-1}}^{(1-\alpha)}}\, x_2(t_{k-1})\right) h^\beta - \sum_{j=1}^{k} c_j^{(\beta)} x_1(t_{k-j}), \tag{17.36a}$$

$$x_2(t_k) = \left(\frac{1}{{t_{k-1}}^{(1-\alpha)}} \left[\mu\left(1 - x_1(t_{k-1})^2\right) x_2(t_{k-1}) - x_1(t_{k-1})\right]\right) h^\beta - \sum_{j=1}^{k} c_j^{(\beta)} x_2(t_{k-j}), \tag{17.36b}$$

with $\mu = 3.5$, initial conditions $x_1(0) = 0.5$, $x_2(0) = 0$, $h = 1 \times 10^{-2}$ is the simulation time, and time simulation fixed at $t = 500$ $[s]$.

The slave system derives from the following change of coordinates

$$\begin{aligned} \varsigma_1(t) &= x_1(t), \\ \varsigma_2(t) &= x_2(t), \end{aligned} \tag{17.37}$$

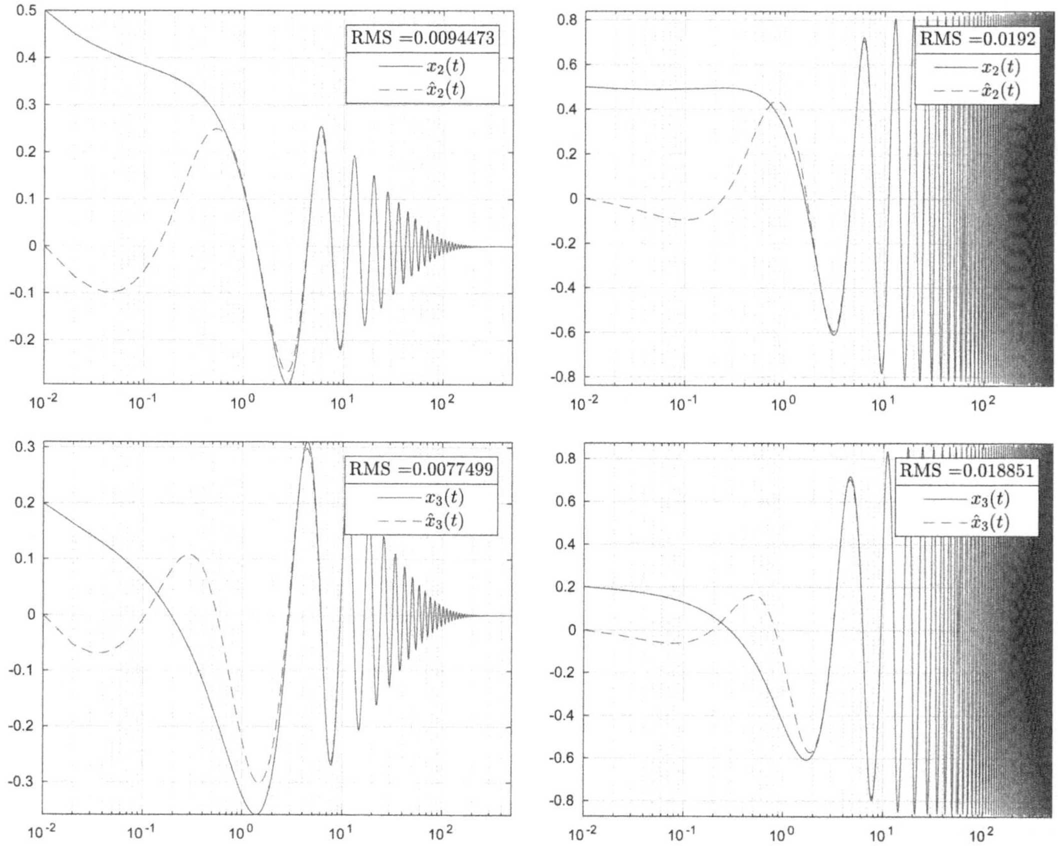

FIGURE 17.3
Arneodo's system with $\alpha = 0.95$ and $\beta = 0.98$. (top left) Portrait phase of the master and slave systems, (top right) measured and estimated state, and (bottom left and right) synchronization of the states in the master and slave systems, respectively.

Consequently, we have the following system

$$t^{1-\alpha}\ {}^{RL}\,{}_{a}^{\beta}\mathscr{D}_t^{\alpha}\begin{bmatrix}\hat{\varsigma}_1(t)\\ \hat{\varsigma}_2(t)\end{bmatrix} = \begin{bmatrix}0 & 1\\ 0 & 0\end{bmatrix}\begin{bmatrix}\hat{\varsigma}_1(t)\\ \hat{\varsigma}_2(t)\end{bmatrix} + \begin{bmatrix}0\\ \mu\left(1-\hat{\varsigma}_1(t)^2\right)\hat{\varsigma}_2(t)-\hat{\varsigma}_1(t)\end{bmatrix} + K\begin{bmatrix}y(t)-\hat{\varsigma}_1(t)\\ y(t)-\hat{\varsigma}_1(t)\end{bmatrix}, \tag{17.38}$$

whose numerical solution is given by

$$\hat{\varsigma}_1\left(t_k\right) = \left(\frac{1}{{t_{k-1}}^{(1-\alpha)}}\left[\hat{\varsigma}_2(t_{k-1}) + k_1\left(y(t_{k-1})-\hat{\varsigma}_1(t_{k-1})\right)\right]\right)h^{\beta} - \sum_{j=1}^{k} c_j^{(\beta)}\hat{\varsigma}_1\left(t_{k-j}\right), \tag{17.39a}$$

$$\hat{\varsigma}_2\left(t_k\right) = \left(\frac{1}{{t_{k-1}}^{(1-\alpha)}}\left[\mu\left(1-\hat{\varsigma}_1(t_{k-1})^2\right)\hat{\varsigma}_2(t_{k-1}) - \hat{\varsigma}_1(t_{k-1}) + k_2\left(y(t_{k-1})-\hat{\varsigma}_1(t_{k-1})\right)\right]\right)h^{\beta} - \sum_{j=1}^{k} c_j^{(\beta)}\hat{\varsigma}_2\left(t_{k-j}\right). \tag{17.39b}$$

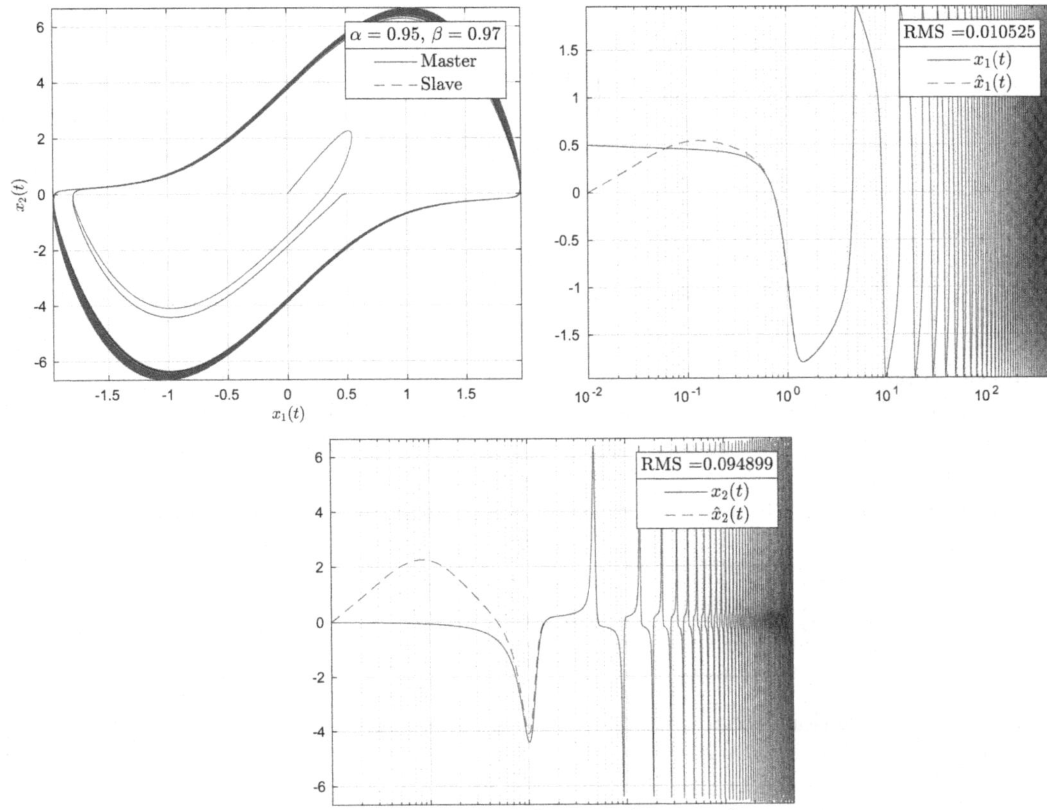

FIGURE 17.4
VPO with $\alpha = 0.95$ and $\beta = 0.97$. (top left) Portrait phase of the master and slave systems, (top right) measured and estimated state, and (bottom) synchronization of the state in master and slave systems.

For this test, the observer gain was $K = \begin{bmatrix} 20 & 100 \end{bmatrix}^T$ and the initial conditions of the slave system are $\varsigma_1(0) = 0$, $\varsigma_2(0) = 0$. Figure 17.4 (top left) shows the synchronization between the master (17.35) and the slave (17.38) systems, whereas Figures 17.4 (top right and bottom) show the "new chaotic behavior" as a result of applying Eq. (17.6).

17.5.4 Chua's Circuit Sine Function Approach

In this example, the nonlinear characteristic function of Chua's circuit is replaced by a sine function; however, we generalize the system proposed in [39] to generate n-fractional conformable scrolls attractor. Hence, the fractional conformable modified Chua's circuit can be expressed as follows

$$
\begin{aligned}
{}^{RL\ \beta}_{\ \ 0}\mathscr{D}^{\alpha}_{t} x_1 &= \frac{1}{t^{1-\alpha}} \Big[\gamma \left(x_2 - f(x_1) \right) \Big], \\
{}^{RL\ \beta}_{\ \ 0}\mathscr{D}^{\alpha}_{t} x_2 &= \frac{1}{t^{1-\alpha}} \Big[x_1 - x_2 + x_3 \Big], \\
{}^{RL\ \beta}_{\ \ 0}\mathscr{D}^{\alpha}_{t} x_3 &= \frac{1}{t^{1-\alpha}} \Big[-\delta x_2 \Big],
\end{aligned}
\tag{17.40}
$$

where

$$f(x_1) = \begin{cases} \frac{b\pi}{2a}(x_1 - 2ac), \\ -b\sin\left(\frac{\pi x_1}{2a} + d\right), \\ \frac{b\pi}{2a}(x_1 + 2ac), \end{cases} \tag{17.41}$$

and

$$d = \begin{cases} \pi, & \text{for } c \text{ is even}, \\ 0, & \text{for } c \text{ is odd}, \end{cases} \tag{17.42}$$

moreover, γ, δ, a, b, c, and d are the real parameters.

The system (17.40) can generate $(c+1)$-scroll chaotic attractors in classical formulation; nevertheless, we found that by involving the orders α and β does not satisfy the previous relationship.

Supposing that $x_1(t)$ is the available state, then, we propose a change of coordinates

$$\begin{aligned} \varsigma_1(t) &= x_1(t), \\ \varsigma_2(t) &= \gamma x_2(t), \\ \varsigma_3(t) &= x_3(t). \end{aligned} \tag{17.43}$$

In this way, the fractional conformable system in canonical form is given by the following equations

$$\begin{aligned} t^{1-\alpha}\, {}^{RL\ \beta}_{\ 0}\mathscr{D}^{\alpha}_{t} \begin{bmatrix} \hat{\varsigma}_1(t) \\ \hat{\varsigma}_2(t) \\ \hat{\varsigma}_3(t) \end{bmatrix} &= \begin{bmatrix} 0 & 1 & 0 \\ 0 & 0 & 1 \\ 0 & 0 & 0 \end{bmatrix} \begin{bmatrix} \hat{\varsigma}_1(t) \\ \hat{\varsigma}_2(t) \\ \hat{\varsigma}_3(t) \end{bmatrix} + \begin{bmatrix} -\gamma f(\hat{\varsigma}_1(t)) \\ \hat{\varsigma}_1(t) - \frac{1}{\gamma}\hat{\varsigma}_2(t) \\ -\frac{\delta}{\gamma}\hat{\varsigma}_2(t) \end{bmatrix} \\ &\quad + K \begin{bmatrix} y(t) - \hat{\varsigma}_1(t) \\ y(t) - \hat{\varsigma}_1(t) \\ y(t) - \hat{\varsigma}_1(t) \end{bmatrix}. \end{aligned} \tag{17.44}$$

Applying the proposed scheme in Eq. (17.25), the numerical solution to Eq. (17.44) is given as follows

$$\begin{aligned} \hat{\varsigma}_1(t_k) = {} & \left(\frac{1}{{t_{k-1}}^{(1-\alpha)}} \left[\hat{\varsigma}_2(t_{k-1}) - \gamma f(\hat{\varsigma}_1(t_{k-1})) + k_1 (y(t_{k-1}) - \hat{\varsigma}_1(t_{k-1}))\right]\right) h^{\beta} \\ & - \sum_{j=1}^{k} c_j^{(\beta)} \hat{\varsigma}_1(t_{k-j}), \end{aligned} \tag{17.45a}$$

$$\begin{aligned} \hat{\varsigma}_2(t_k) = {} & \left(\frac{1}{{t_{k-1}}^{(1-\alpha)}} \left[\hat{\varsigma}_3(t_{k-1}) + \hat{\varsigma}_1(t_{k-1}) - \frac{1}{\gamma}\hat{\varsigma}_2(t_{k-1}) + k_2 (y(t_{k-1}) - \hat{\varsigma}_1(t_{k-1}))\right]\right) h^{\beta} \\ & - \sum_{j=1}^{k} c_j^{(\beta)} \hat{\varsigma}_2(t_{k-j}), \end{aligned} \tag{17.45b}$$

$$\begin{aligned} \hat{\varsigma}_3(t_k) = {} & \left(\frac{1}{{t_{k-1}}^{(1-\alpha)}} \left[-\frac{\delta}{\gamma}\hat{\varsigma}_2(t_{k-1}) + k_3 (y(t_{k-1}) - \hat{\varsigma}_1(t_{k-1}))\right]\right) h^{\beta} \\ & - \sum_{j=1}^{k} c_j^{(\beta)} \hat{\varsigma}_3(t_{k-j}), \end{aligned} \tag{17.45c}$$

where the function $f(\varsigma_1(t))$ is given by

$$f\left(\varsigma_1(t_{k-1})\right) = \begin{cases} \frac{b\pi}{2a}\left(\varsigma_1(t_{k-1}) - 2ac\right), & \text{if } \varsigma_1(t_{k-1}) \geq 2ac, \\ -b\sin\left(\frac{\pi\varsigma_1(t_{k-1})}{2a} + d\right), & \text{if } -2ac < \varsigma_1(t_{k-1}) < 2ac, \\ \frac{b\pi}{2a}\left(\varsigma_1(t_{k-1}) + 2ac\right), & \text{if } \varsigma_1(t_{k-1}) \leq -2ac. \end{cases} \tag{17.46}$$

The solution for the fractional conformable master system (17.40) can be also obtained using the numerical algorithm (17.25), and the result is shown below

$$x_1(t_k) = \left(\frac{1}{t_{k-1}{}^{(1-\alpha)}}\left[\gamma\left(x_2(t_{k-1}) - f\left(x_1(t_{k-1})\right)\right)\right]\right)h^\beta - \sum_{j=1}^{k} c_j^{(\beta)} x_1(t_{k-j}), \tag{17.47a}$$

$$x_2(t_k) = \left(\frac{1}{t_{k-1}{}^{(1-\alpha)}}\left[x_1(t_{k-1}) - x_2(t_{k-1}) + x_3(t_{k-1})\right]\right)h^\beta - \sum_{j=1}^{k} c_j^{(\beta)} x_2(t_{k-j}), \tag{17.47b}$$

$$x_3(t_k) = \left(\frac{1}{t_{k-1}{}^{(1-\alpha)}}\delta x_2(t_{k-1})\right)h^\beta - \sum_{j=1}^{k} c_j^{(\beta)} x_3(t_{k-j}), \tag{17.47c}$$

with $\gamma = 10.814$, $\delta = 14$, $a = 1.3$, $b = 0.11$, $c = 3$, initial conditions $x_1(0) = 0.1$, $x_2(0) = 0$, $x_3(0) = 0$ and $\varsigma_1(0) = 0$, $\varsigma_2(0) = 0$, $\varsigma_2(0) = 0$ for the master and slave systems, respectively, total simulation time $t = 500$ $[s]$, computational time step $h = 1 \times 10^{-2}$ and gain were $K = \begin{bmatrix}3.3261 & 3.6876 & 1.3628\end{bmatrix}^T$.

In this example we perform two tests: one with $c = 3$ to generate a 4-scroll attractor and $c = 8$ for a 8-scroll system under the same initial conditions. Figures 17.5 and 17.6 (top left) show the synchronization of the master (17.40) and slave systems (17.45). Figures 17.5 and 17.6 show the synchronization of the states in both systems.

This system, in particular, is sensitive to the change of initial conditions or parameters; therefore, the relation $n = (c + 1)$ is affected by the orders α and β; consequently, we have a different behavior not only in the states but also in the phase portrait.

17.6 Conclusions

In this work, we propose a numerical algorithm based on GL derivative and Euler method to get a numerical solution to fractional conformable derivative in the RL sense. We observe that it is possible to generate a new chaotic behavior in the master system by varying only the orders α and β; nevertheless, this orders offer us two degrees of freedom for the synchronization of the slave systems proposed in this work. The optimization of the correction term ensures the estimation error, and as a consequence, the gain obtained from this process ensures the synchronization between the master and slave systems. From the numerical simulations, the performance of the proposed methodology to generate new fractional conformable dynamics seems to be satisfactory and may have future applications in different fields such as chaotic maps and communications to encode and decode information.

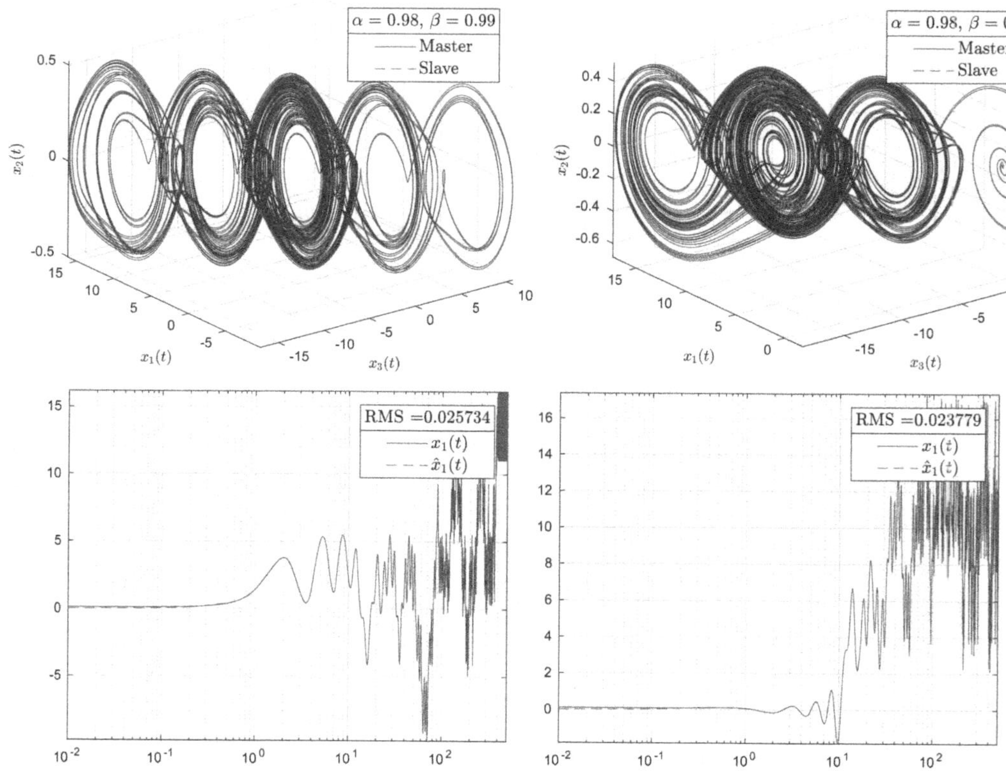

FIGURE 17.5
Chua's circuit sine function with $\alpha, \beta(c + 1)$-scroll attractors. The fractional orders considered are $\alpha = 0.98$ and $\beta = 0.99$. (top left) Portrait phase of the master and slave systems, (top right) measured and estimated state, and (bottom left and right) synchronization of the states in master and slave systems, respectively.

Acknowledgments

Jess Emmanuel Solís Pérez acknowledges the support provided by CONACyT through the assignment doctoral fellowship. José Francisco Gómez Aguilar acknowledges the support provided by CONACyT: ctedras CONACyT para jóvenes investigadores 2014. José Francisco Gómez Aguilar, Ricardo Fabricio Escobar Jiménez, and Juan Reyes Reyes acknowledges the support provided by SNI-CONACyT.

Conflicts of Interest

The authors declare no conflict of interest.

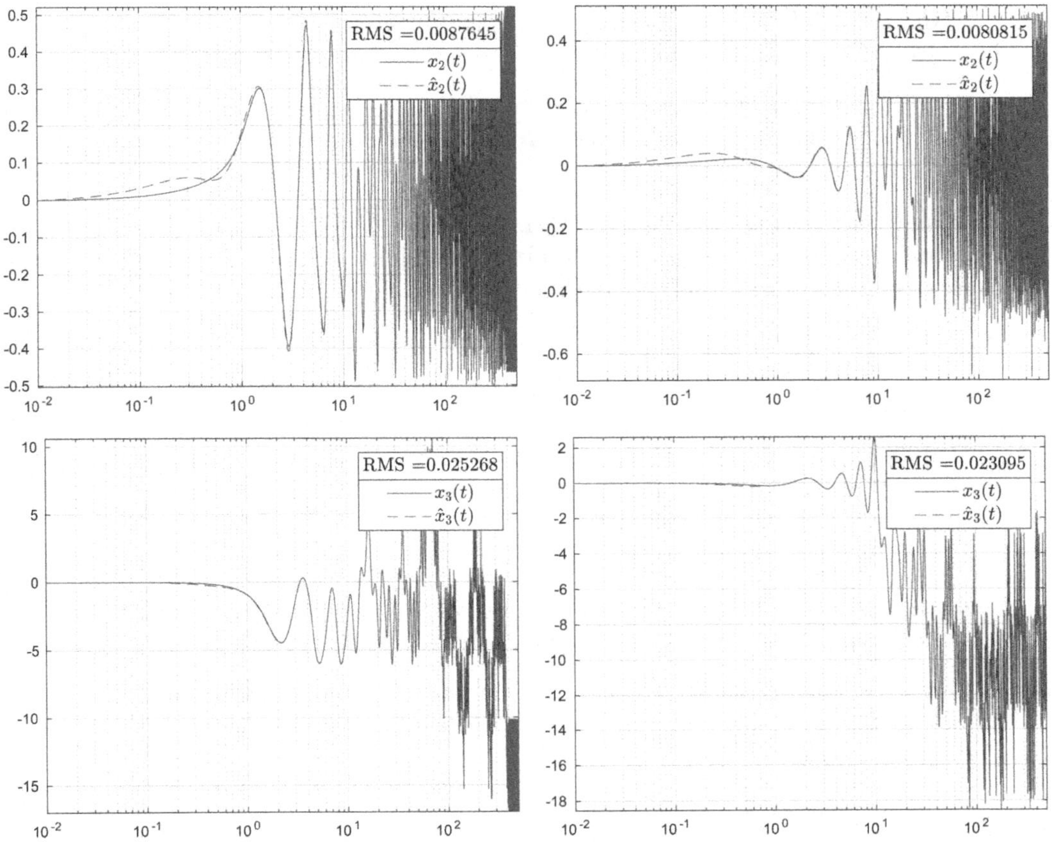

FIGURE 17.6
Chua's circuit sine function with $\alpha, \beta(c+1)$-scroll attractors. The fractional orders considered are $\alpha = 0.98$ and $\beta = 0.99$. (top left) Portrait phase of the master and slave systems, (top right) measured and estimated state, and (bottom left and right) synchronization of the states in master and slave systems, respectively.

References

[1] G. Brandner, U. Schilcher, C. Bettstetter. Firefly synchronization with phase rate equalization and its experimental analysis in wireless systems. *Computer Networks*, 97, (2016), 74–87.

[2] G.M. Ramírez-Ávila, J. Kurths, J.L. Deneubourg. Fireflies: A paradigm in synchronization. In *Chaotic, Fractional, and Complex Dynamics: New Insights and Perspectives*. Springer, Cham. Edelman M., Macau E., Sanjuan M. (eds), (2018), 35–64.

[3] S. Radhakrishnan, A. Duvvuru, S. Sultornsanee, S. Kamarthi. Phase synchronization based minimum spanning trees for analysis of financial time series with nonlinear correlations. *Physica A: Statistical Mechanics and its Applications*, 444, (2016), 259–270.

[4] S.J. Guastello, A.F. Peressini. Development of a synchronization coefficient for biosocial interactions in groups and teams. *Small Group Research*, 48(1), (2017), 3–33.

[5] S. Vaidyanathan. Hybrid chaos synchronization of 3-cells cellular neural network attractors via adaptive control method. *International Journal of PharmTech Research*, 8(8), (2015), 61–73.

[6] Z. Huang, J. Zhang, J. Wu, P. Qin, X. Wu, Z. Wang, Z. Yang. Decoupled temporal variability and signal synchronization of spontaneous brain activity in loss of consciousness: an fMRI study in anesthesia. *Neuroimage*, 124, (2016), 693–703.

[7] A. Ramkisoensing, J.H. Meijer. Synchronization of biological clock neurons by light and peripheral feedback systems promotes circadian rhythms and health. *Frontiers in Neurology*, 6, (2015), 1–16.

[8] L. Hadjidemetriou, E. Kyriakides, F. Blaabjerg. A robust synchronization to enhance the power quality of renewable energy systems. *IEEE Transactions on Industrial Electronics*, 62(8), (2015), 4858–4868.

[9] B. Wen, D. Dong, D. Boroyevich, R. Burgos, P. Mattavelli, Z. Shen. Impedance-based analysis of grid-synchronization stability for three-phase paralleled converters. *IEEE Transactions on Power Electronics*, 31(1), (2016), 26–38.

[10] H. Gutiérrez, A. Morales, H. Nijmeijer. Synchronization control for a swarm of unicycle robots: Analysis of different controller topologies. *Asian Journal of Control*, 19(5), (2017), 1822–1833.

[11] R.M. de Mendonca, N. Nedjah, L. de Macedo Mourelle. Efficient distributed algorithm of dynamic task assignment for swarm robotics. *Neurocomputing*, 172, (2016), 345–355.

[12] H. Liu, H. Wan, K.T. Chi, J. Lü. An encryption scheme based on synchronization of two-layered complex dynamical networks. *IEEE Transactions on Circuits and Systems I: Regular Papers*, 63(11), (2016), 2010–2021.

[13] W.H. Chen, S. Luo, W.X. Zheng. Impulsive synchronization of reaction-diffusion neural networks with mixed delays and its application to image encryption. *IEEE Transactions on Neural Networks and Learning Systems*, 27(12), (2016), 2696–2710.

[14] W. Sun, E.G. Ström, F. Brännström, M.R. Gholami. Random broadcast based distributed consensus clock synchronization for mobile networks. *IEEE Transactions on Wireless Communications*, 14(6), (2015), 3378–3389.

[15] N. Fujiwara, J. Kurths, A. Díaz-Guilera. Synchronization of mobile chaotic oscillator networks. *Chaos: An Interdisciplinary Journal of Nonlinear Science*, 26(9), (2016), 1–9.

[16] H. Jian-Bing, H.Y.Z. Ling-Dong. A novel stablility theorem for fractional systems and its applying in synchronizing fractional chaotic system based on back-stepping approach. *Acta Physica Sinica*, 4, (2009), 1–18.

[17] B.A. Idowu, U.E. Vincent, A.N. Njah. Control and synchronization of chaos in nonlinear gyros via backstepping design. *International Journal of Nonlinear Science*, 5(1), (2008), 11–19.

[18] S. Vaidyanathan, S. Sampath. Anti-synchronization of four-wing chaotic systems via sliding mode control. *International Journal of Automation and Computing*, 9(3), (2012), 274–279.

[19] Y. Chi-Ching. Synchronization of second-order chaotic systems via adaptive terminal sliding mode control with input nonlinearity. *Journal of the Franklin Institute*, 349(6), (2012), 2019–2032.

[20] Y. Lei, W. Xu, H. Zheng. Synchronization of two chaotic nonlinear gyros using active control. *Physics Letters A*, 343(1–3), (2005), 153–158.

[21] S. Bhalekar, V. Daftardar-Gejji. Synchronization of different fractional order chaotic systems using active control. *Communications in Nonlinear Science and Numerical Simulation*, 15(11), (2010), 3536–3546.

[22] Z. Wang, X. Shi. Synchronization of a four-dimensional energy resource system via linear control. *Communications in Nonlinear Science and Numerical Simulation*, 16(1), (2011), 463–474.

[23] M.T. Yassen. Adaptive control and synchronization of a modified Chua's circuit system. *Applied Mathematics and Computation*, 135(1), (2003), 113–128.

[24] M.E. Yalcin, J.A. Suykens, J. Vandewalle. Master-slave synchronization of Luré systems with time-delay. *International Journal of Bifurcation and Chaos*, 11(06), (2001), 1707–1722.

[25] S. Mobayen, F. Tchier. Composite nonlinear feedback control technique for master/slave synchronization of nonlinear systems. *Nonlinear Dynamics*, 87(3), (2017), 1731–1747.

[26] J. Sun, W. Deng, G. Cui, Y. Wang. Real combination synchronization of three fractional-order complex-variable chaotic systems. *Optik-International Journal for Light and Electron Optics*, 127(23), (2016), 11460–11468.

[27] A.K. Singh, V.K. Yadav, S. Das. Synchronization between fractional order complex chaotic systems with uncertainty. *Optik-International Journal for Light and Electron Optics*, 133, (2017), 98–107.

[28] H. Su, R. Luo, Y. Zeng. The exponential synchronization of a class of fractional-order chaotic systems with discontinuous input. *Optik-International Journal for Light and Electron Optics*, 131, (2017), 850–861.

[29] A. Ouannas, Z. Odibat, T. Hayat. Fractional analysis of co-existence of some types of chaos synchronization. *Chaos, Solitons & Fractals*, 105, (2017), 215–223.

[30] G.M. Mahmoud, M.E. Ahmed, T.M. Abed-Elhameed. On fractional-order hyperchaotic complex systems and their generalized function projective combination synchronization. *Optik-International Journal for Light and Electron Optics*, 130, (2017), 398–406.

[31] S.T. Kingni, V.T. Pham, S. Jafari, P. Woafo. A chaotic system with an infinite number of equilibrium points located on a line and on a hyperbola and its fractional-order form. *Chaos, Solitons & Fractals*, 99, (2017), 209–218.

[32] I. Podlubny. *Fractional Differential Equations: An Introduction to Fractional Derivatives, Fractional Differential Equations, to Methods of Their Solution and Some of Their Applications*. Academic Press, San Diego, CA, (1999).

[33] R. Khalil, M. Al Horani, A. Yousef, M. Sababheh. A new definition of fractional derivative. *Journal of Computational and Applied Mathematics*, 264, (2014), 65–70.

[34] F. Jarad, E. Ugurlu, T. Abdeljawad, D. Baleanu. On a new class of fractional operators. *Advances in Difference Equations*, 2017(1), (2017), 1–16.

[35] A. Atangana, D. Baleanu A. Alsaedi. New properties of conformable derivative. *Open Mathematics*, 13(1), (2015), 1–10.

[36] D.W. Moore, E.A. Spiegel. A thermally excited non-linear oscillator. *The Astrophysical Journal*, 143, (1966), 1–15.

[37] A. Arneodo, P. Coullet, E. Spiegel, C. Tresser. Asymptotic chaos. *Physica D*, 14(3), 1985, 327–347.

[38] M.S. Tavazoei, M. Haeri, M. Attari, S. Bolouki, M. Siami. More details on analysis of fractional-order Van der Pol oscillator. *Journal of Vibration and Control*, 15(6), (2009), 803–819.

[39] W.K. Tang, G.Q. Zhong, G. Chen, K.F. Man. Generation of n-scroll attractors via sine function. *IEEE Transactions on Circuits and Systems I: Fundamental Theory and Applications*, 48(11), (2001), 1369–1372.

Index